Lecture Notes in Computer Science 533

E. Börger H. Kleine Büning
M. M. Richter W. Schönfeld (Eds.)

Computer Science Logic

4th Workshop, CSL '90
Heidelberg, Germany, October 1-5, 1990
Proceedings

Springer-Verlag
Berlin Heidelberg New York
London Paris Tokyo
Hong Kong Barcelona
Budapest

Series Editors

Gerhard Goos
GMD Forschungsstelle
Universität Karlsruhe
Vincenz-Priessnitz-Straße 1
W-7500 Karlsruhe, FRG

Juris Hartmanis
Department of Computer Science
Cornell University
Upson Hall
Ithaca, NY 14853, USA

Volume Editors

Egon Börger
Dipartimento di Informatica, Università di Pisa
Corso Italia, 40, I-56100 Pisa, Italia

Hans Kleine Büning
Fachbereich 11, Praktische Informatik, Universität-GH-Duisburg
Postfach 10 16 29, W-4100 Duisburg 1, FRG

Michael M. Richter
Fachbereich Informatik, Universität Kaiserslautern
Postfach 30 49, W-6750 Kaiserslautern, FRG

Wolfgang Schönfeld
IBM Deutschland GmbH, Wissenschaftliches Zentrum
Wilckensstraße 1a, W-6900 Heidelberg, FRG

CR Subject Classification (1991): D.3, F, G.2, H.2, I.1, I.2.2-8

ISBN 3-540-54487-9 Springer-Verlag Berlin Heidelberg New York
ISBN 0-387-54487-9 Springer-Verlag New York Berlin Heidelberg

Printed in Germany

Typesetting: Camera ready by author
Printing and binding: Druckhaus Beltz, Hemsbach/Bergstr.
2145/3140-543210 - Printed on acid-free paper

Preface

The workshop CSL'90 (Computer Science Logic) was held at the Max-Planck-Haus in Heidelberg, Germany, from October 1-5, 1990. It was the fourth in a series of workshops, following CSL'89 at the University of Kaiserslautern, CSL'88 at the University of Duisburg, and CSL'87 at the University of Karlsruhe. Thirty-five talks were presented at the workshop, consisting of invited presentations and those selected from eighty-nine submissions.

As was the case for CSL'87 (see Lecture Notes in Computer Science, Vol. 329), CSL'88 (see Lecture Notes in Computer Science, Vol. 385), and CSL'89 (see Lecture Notes in Computer Science, Vol. 440), we collected the original contributions after their presentation at the workshop and began a review procedure which resulted in the selection of the papers in this volume. They appear here in final form.

We would like to thank the referees without whose help we would not have been able to accomplish the difficult task of selecting among the many valuable contributions.

We gratefully acknowledge the financial sponsorship and organizational support of IBM Germany.

July, 1991 E. Börger, H. Kleine Büning, M. M. Richter, W. Schönfeld

Table of Contents

Monadic Second Order Logic, Tree Automata and Forbidden Minors

Stefan Arnborg* Andrzej Proskurowski† Detlef Seese‡

Abstract

N.Robertson and P.D.Seymour proved that each minor closed class K of graphs is characterized by finitely many minimal forbidden minors. If these minors are given then they can be used to find an efficient membership test for such classes (see [Rob Sey 86b]).From these minors one can get a monadic second order description of the class K. Main result of the article is that from a monadic second order description of K the minimal forbidden minors can be constructed, when K contains only graphs of universally bounded tree width. The result is applied to the class of partial 2-pathes.

1 Introduction

A graph H is a minor of graph G if H results from a subgraph of G by contracting of some edges. N.Robertson and P.D.Seymour developed a theory to describe the structure of graphs which avoid a given graph as a minor (see [Rob Sey 86] [Rob Sey 86a] [Rob Sey 86b]). One of their main results is the proof of the Wagner conjecture, that each set K of graphs, which has the property, that no graph of K is isomorphic to a minor of another graph of K, is finite. Another result (see [Rob Sey 86b]) is that they proved for each fixed graph H the existence of a polynominal time algorithm with the help of which it is possible to decide for an arbitrary graph G whether H is isomorphic to a minor of G. Combining both results they obtained the famous result, that each property P of graphs which is closed under minor taking (i.e. if a graph G has property P then also all its minors have property P) can be tested in polynominal time. Unfortunately the proof of this result is purely existential, i.e. it is only shown that there exists a polynominal time algorithm, but the algorithm itself is not known (see [Fel Lan 87]). To make this method constructive it is necessary to find for a given P the minimal forbidden minors for the given class of graphs with property P. A pioneering work was done here in the articles of [Fel Lan 87][Fel Lan 89][Fel Lan 89a][Fel 89][Fel 89a] [Abr Fel 89].

Another source of our research were the investigations of minimal forbidden minors for the set of partial 3-trees and graphs of path width < 3 (see [Arn Pr 86] and [Kin 89]).

*address:NADA,KTH, S-100 44 Stockholm, Sweden

†address: Department of Computer and Information Science, University of Oregon, Eugene, Oregon 97403, USA

‡address: Karl-Weierstrass Institute for Mathematics Mohrenstr. 39, PF 1304, O-1086 Berlin, Germany.
The research to this article was started in summer 1989 while the first and the third author were visiting the University of Oregon in Eugene. The third author gratefully announces support from the Universities of Toronto, Idaho (Moscow), the Washington State University (Pullmann) and the University of Oregon (Eugene).

For a fixed value of the integer parameter k, partial k-trees are exactly the subgraphs of those chordal graphs that have at most $k + 1$ completely interconnected vertices. Thus, partial 1-trees are the acyclic graphs (forests), and partial 2-trees are the series-parallel graphs (graphs with no K_4 minors or homeomorphs).

Partial k-trees have been in the focus of attention in recent years because of their interesting algorithmic properties. Namely, for a large number of inherently difficult (on general graphs) discrete optimization problems, partial k-trees admit a linear time solution algorithm when the value of k is fixed and any partial k-tree given with its k-tree embedding (see [Arn Pr 89] [Arn La Se 88] [Bod 87][Bod 88] [Cou 88] [Schef 89] [Sees 85] [Sees 86] [Wim 88]). Recently in [Bod Klo 90] was given for each fixed k an $O(n\ log^2n)$ algorithm to decide whether the tree (path-) width of graph is $\leq k$ and to construct a corresponding decomposition in the positive case. In case $k \leq 3$ linear time algorithms can be found in [Mat Tho 88] (see also [Arn Pr 86]).

In [Cou 88] it is proved that for graphs of tree width $\leq k$ (this class of graphs is equivalent to the class of partial k-trees) properties which can be formalized in monadic second order logic can be decided in linear time, when the graphs are given together with a corresponding tree decomposition. The proof of this result was simplified in [Arn La Se 89], where it was also extended to a larger class of graph properties, the EMS-properties. This proof based on two key techniques: firstly a uniform method, the linear time interpretability, to reduce graph properties described in special languages from classes of graphs of tree width $\leq k$ to other similar properties on trees and secondly the reduction of tree properties, which are formalized in this languages, to recognizability problems for tree automata.

Exploiting these techniques also in this paper we give a simple proof that for each minor closed class K of graphs of bounded tree width it holds that K is monadic second order definable if and only if the number of minimal forbidden minors is finite. Moreover from a monadic second order description of K the minimal forbidden minors can be constructed. This result is an interesting supplement of the results on t-finite state and t-cutset regular graph families from [Abr Fel 89] and on finite congruences and reduction systems from [Arn Lag 90]. Moreover its simple proof demonstrates again (see also [Arn La Se 89]) the power and usefulness of the interpretability method and other ideas developed to prove the decidability of theories in mathematical logic. It shows again how results for graphs of bounded tree width can be reduced via interpretability to similar results for trees. But trees are a well investigated class of structures. Hence one can shed light into the theory of graphs of bounded tree width just by looking to trees and using interpretability. We use the advantage of this method and can so give a natural explantation of many surprising results from the last years in this area.

The paper ends with a short investigation of how to come to an explicite monadic second order description for the class of partial 2-pathes, which does not use minimal forbidden minors, which are not known in case of partial 2-pathes. As a consequence one can effectively construct an $O(n^2)$-algorithm to decide whether a graph is a partial 2-path. Moreover for each EMS-property P of graphs (see [Arn La Se 89]) one can effectively construct an $O(n^2)$-algorithm to solve P for all partial 2-pathes, when the EMS-description of the problem is given. The proof exploits a fine structure investigation of partial k-trees started in [Pros 84].

2 Definition and terminology

We will use standard graph theory terminology, as found, for instance, in [Bon Mur 76]. Some basic concepts of special importance for the paper are defined in the following.

All graphs in this paper will be finite, if not stated otherwise. For a graph G the set of its vertice will be denoted by $V(G)$, the set of its edges by $E(G)$. A graph H is a *subgraph* of a graph G if $V(H) \subseteq V(G)$ and $E(H) \subseteq E(G)$. H is a subgraph *induced* by set $W \subseteq V(G)$, if $V(H) = W$ and $E(H)$ is the set of all those edges from $E(G)$ which join vertices from W only; in this case H is sometimes also denoted as $G|W$. The size $|G|$ of a graph G is $|V(G)| + |E(G)|$.

If G is a graph and e is an edge of G with endvertices a and b, then a new graph H results by identifying a and b in such a way, that the resulting vertex is incident to all those edges (other than e) which were originally incident to a or b in G. In case that we regard simple graphs we have to substitute multiple edges resulting in this way by a single representative. This operation is called *contracting the edge e*.

If a graph H can be obtained from G by a succession of such edge contractions, then G is contractible to H. H is called a *minor of* G if there is a subgraph of G which is contractible to H. H is a *proper minor* of G if H is a minor of G and H is not isomorphic to G. It is easy to see that:

Theorem 2.1 *H is isomorphic to a minor of G if there is a function f such that:*

For each $a \in V(H) f(a)$ is a connected subgraph of G.

$f(a) \cap f(b) = \emptyset$ for all vertices $a \neq b$ from $V(H)$.

For each edge $e \in E(H)$ with ends a and b there are vertices $a' \in f(a)$ and $b' \in f(b)$ such that a' and b' are adjacent in G

H is a *1-step minor* of a graph G if H results from G by omission of one vertex or omission of one edge or by one edge contraction. Let F be a class of graphs which is closed under isomorphism. A graph H is said to be a *minimal forbidden minor for F* if H is not in F but each proper minor of H is in F. A class I of graphs is called a *lower ideal* [Rob Sey 89] if with each graph $G \in I$ also all graphs H which are isomorphic to a minor of G are in I. The notion of a minor for graphs was introduced by N.Robertson and P.D.Seymour, and in a famous series of papers on graph minors (see [Rob Sey 86][Rob Sey 86a] [Rob Sey 86b]) they investigated the structure of graphs which have no minor isomorphic to an arbitrary given fixed graph. The series culminated in a proof of Wagner's conjecture, which we give here in a formulation with lower ideals.

Theorem 2.2 *[Rob Sey 89]): Each lower ideal I has up to isomorphism a finite number of minimal forbidden minors.*

A *walk* is a sequence of vertices such that every two consecutive vertices are adjacent. If all the vertices are different, we have a *path* A walk forms a cycle if only its first and last vertices are identical. A set of k vertices, every two of which are adjacent, is called a *k-clique*. A (minimal) subset of vertices of a graph such that their removal disconnects the graph is a (minimal) *separator*. A *k-tree* is either a complete graph with k or $k+1$ vertices (denoted as K_k or K_{k+1} respectively) or a connected graph such that every minimal separator is a k-clique. Equivalently, the complete graph on k vertices, K_k, is a k-tree, and any k-tree with more than k, say n, vertices can be constructed from a k-tree with $n-1$ vertices by adding a new vertex adjacent to all vertices of a k-clique of that graph. A vertex of degree k in a k-tree is a *k-leaf* of that k-tree. A *partial k-tree* is any subgraph of a k-tree.

A full ***k*-path** is a k-tree that has exactly 2 k-leaves. We shall denote a 1-path also as path. A *partial k-path* is a subgraph of a (full) k-path. In analogy to the pathes there is an equivalent definition. A graph G is a k-path if there is an allternating sequence $(S_0, E_1, S_1, E_2, \dots, S_n)$ of distinct k- and $k+1$-complete subgraphs starting and ending with a K_k and such that E_i contains exactly S_i and $S_{i-1} (1 \leq i \leq n)$. In this case n is denoted as the length of the k-path.

A *caterpillar* is a tree that can be partitioned into two subgraphs: the *body*, which is a path, and *hairs*, which are sets of vertices adjacent to vertices of the body. In [Pros 84] this notion is generalized to k-trees by defining a *k-caterpillar* as a k-tree that can be likewise partitioned into the *body*, which is a k-path and *hairs*, each of which is adjacent to all k vertices of a minimal seperator of the body. Our notion of k-caterpillar is originally in [Pros 84] denoted as *interior k-caterpillar* (k-intercat).

In [Rob Sey 86] a *tree decomposition* of a graph G is defined to be a pair (T, S), where T is a tree and S is a family $(X_t : t \in V(T)$ of sets, such that

i $\cup(X_t : t \in V(T)) = V(G)$

ii Every edge of G has its ends in some $X_t\, (t \in V(T))$.

iii For $t, t', t'' \in V(T)$, if t' lies on the path of T from t to t'' then $X_t \cap X_{t''} \subseteq X_{t'}$.

The *width* of the tree decomposition is $\max(|X_t| - 1 : t \in V(T))$. The *tree width* of G is the minimum $w \geq 0$ such that G has a tree decomposition of width $\leq w$. In the same way in [Rob Sey 83] *path decomposition* and the *path width* of a graph are defined, by demanding T to be a path instead of a tree.

The notion of tree width and partial k-trees are closely related.

Lemma 2.3 *A graph G is a partial k-tree if and only if its tree width is $\leq k$.*

A proof of this fact can be found in [Schef 89]. For path width a similar result holds.

Lemma 2.4 *A graph G has path width $\leq k$ if and only if G is a partial k-caterpillar.*

The proof is a simple variant of the proof from [Schef 89]. Hence we shall give some hints only. In case that G is a subgraph of a k-caterpillar H then the needed path decomposition of H results from corresponding representation of H as an alternating sequence of k- and $(k+1)$-complete graphs. To prove the other direction one starts with a path decomposition of G of width k and transforms it to a path decomposition (T, X) which fulfills moreover the following two additional conditions:

$$(iv)\ |X_t| = k+1, \quad \text{for all} \quad t \in V(T).$$
$$(v)\ |X_s \cup X_t| = k \quad \text{for each edge} \quad (s,t) \in E(T).$$

Then one defines the graph $H := (V(G), E')$, where

$$E' := \{(u,v) : \text{there is a } t \in V(T) \text{ with } \{u,v\} \subseteq X_t\}.$$

It is easy to show that this graph H is a k-caterpillar and that G is a subgraph of it.

Path decompositions and tree decompositions which have the additional properties (*iv*) and (*v*) will be denoted as *proper decompositions.*

3 Second order logic and lower ideals

Monadic second order logic results from first order logic by adding to a regarded elementary language L a sequence of quantifiable set variables and allowing new atomic formulas $x \in Y$, where x is an individual variable and Y is a set variable. The intended interpretation here is that $\in$ is the membership relation and the set variables range over all subsets of the domain of a structure for L. In the following we shall assume, that $x, y, z, x_1, x_2, x_3, \ldots$ are the individual variables and $X, Y, Z, X_1, X_2, X_3, \ldots$ are set variables of the regarded languages. To fix the signature for graphs we shall regard graphs as relational structures (A, V, E, R_{in}), where A is a nonemty set (set of vertices and edges), V, E are unary predicates and R_{in} is a binary predicate, with their intended meaning that:

V designates the set of vertices,
E designates the set of edges and
R_{in} holds if vertex a is incident with edge b.

Monadic second order logic has a large expressive power. For instance the property that a subgraph of G induced by a set Z is connected can be expressed by a formula $\mathrm{Con}(Z)$, which is constructed in the following way:

$$\begin{aligned}
\mathrm{Part}(X,Y,Z) &\equiv (Z = X \cup Y) \wedge (X \cap Y = \emptyset) \wedge (X \neq Z) \wedge (Y \neq Z) \\
\mathrm{Adj}(X,Y) &\equiv \exists x \exists y \exists z (V(x) \wedge V(y) \wedge E(z) \wedge y \in Y \wedge x \in X \\
&\quad \wedge R_{in}(x,z) \wedge R_{in}(y,z)) \\
\mathrm{Con}(Z) &\equiv \forall X \forall Y \mathrm{Part}(X,Y,Z) \rightarrow \mathrm{Adj}(X,Y),
\end{aligned}$$

where $Z = X \cup Y$ and $X \cap Y = \emptyset$ are obviously expressible in monadic second order logic.

A property P for structures of a class K is a *monadic second order property* if there is a monadic second order formula φ of the language corresponding to K, such that an arbitrary structure $G \in K$ has property P if and only if $G \models \varphi$. Hence the connectivity of a graph is a monadic second order property for the class of graphs.

Using this it is easy to prove the following lemma.

Lemma 3.1 *For each graph H there is a monadic second order formula φ_H such that for each graph G it holds:*

H is isomorphic to a minor of $G \Leftrightarrow G \models \varphi_H$

Proof: Assume that $a_1, \ldots, a_l$ is an enumeration of all vertices of $V(H)$. By (2.1) it is sufficient to find a formula which expresses:

'There exists sets $A_1, \ldots, A_l$ of vertices of G such $G|A_i$ is connected for each $i (1 \leq i \leq l)$ and such that for each edge $e \in E(H)$ with endvertices a_i and a_j there are vertices $b_i \in A_i$ and $b_j \in A_j$ such that b_i and b_j are adjacent in G'

But this is obviously expressed by the following formula:

$$\exists X_1 \ldots \exists X_l \bigwedge_{1\leq i\leq 1} \mathrm{Con}(X_i) \wedge \bigwedge_{\substack{i,j: a_i \text{ and } a_j \\ \text{adjacent in } H}} Adj(X_i, X_j)$$

The following lemma is a consequence of this result.

Lemma 3.2 *Each lower ideal I is definable by a monadic second order formula.*

Proof: By (2.2) there are up to isomorphism finitely many minimal forbidden minors $H_1, \ldots, H_l$ for I. Then I is defined by the following formula:

$$\neg\varphi_{H_1} \wedge \ldots \wedge \neg\varphi_{H_l}$$

where φ_{H_l} is defined as in (3.1).

When the minimal forbidden minors for I are given then the formula can be constructed effectively. But the problem with N.Robertson and P.D.Seymours result (2.2) is that there is no effective way to find the minimal forbidden minors for an arbitrarily given lower ideal I (see [Fel Lan 87][Fel Lan 89a]). Even for the lower ideal of graphs of tree width bounded by k the minimal forbidden minors are not known when $k > 3$ (see [Arn Cor Pr 86]). Hence the following result is of interest.

Theorem 3.3 *Assume that $m \geq 0$ is an integer and let J be a lower ideal of graphs of tree-width $\leq m$. Then from a monadic second order definition of J one can effectively construct the minimal forbidden minors of J.*

The basic idea of the proof is first to find a formula which holds in a graph G if and only if G is isomorphic to a minimal forbidden minor for J. The next step is to reduce the problem to trees via the method of model interpretability. But for trees the transformed problem can be solved via tree automata. The details of the proof are given in the following three sections.

4 Interpretability

The basic concept of interpretability which we shall use here was introduced in [Rab 64] as a useful tool to transfer decidability and undecidability results from one theory to another. We shall use here a variant of it which was introduced in [Arn La Se 88]. In the present section we shall give the necessary definitions and results which we need for our application. For the missing details we refere the reader to [Arn La Se 89].

Let K_1 and K_2 be two classes of relational structures and let L_1 and L_2 be corresponding monadic second order languages. Assume that there are L_2-formulas $\alpha(x)$ (intended to define the domain of the new structures) and $\beta_{R_i}(x_1, \ldots, x_{l_i})$ (for each relational symbol R_i of L_1 of arity l_i, where $R_1, \ldots, R_k$ are assumed to be all relational symbols of L_1), which have only the indicated variables as free variables. For each structure G of K_2 and each L_2 formula $\varphi(x_1, \ldots, x_s)$ define $G(\varphi)$ to be the s-ary relation which is defined by φ in G. For simplicity denote the sequence $(\varphi, \beta_{R_1}, \ldots, \beta_{R_k})$ of L_2 formulas as interpretation I and define $I(G)$ for each structure $G \in K$ by

$$I(G) := (G(\alpha), G(\beta_{R_1}), \ldots, G(\beta_{R_k})).$$

Here for a formula $\delta(x_1,\ldots,x_s)$ and a corresponding structure G, $G(\delta)$ is defined to be $\{(a_1,\ldots,a_s) : a_1,\ldots,a_s$ are elements of the domain of G and $G \models \delta[a_1,\ldots,a_s]\}$.

K_1 is said to be *linear (polynomial) time interpretable into* K_2 with respect to L_1, L_2 if there is an I as above and an algorithm A which constructs for each $G \in K_1$ a structure $A(G) \in K_2$ in time linear (polynomial) in $|G|$ such that: G is isomorphic to $I(A(G))$.

Each interpretation gives a canonical way to transform L_1 formulas into L_2 formulas:

$$\begin{aligned}
(x \in X)^I &:= x \in X \\
(x = y)^I &:= x = y \\
(R(x_1,\ldots,x_l))^I &:= \beta_R(x_1,\ldots,x_l) \text{ for each l-ary relational symbol } R \text{ from } L_1 \\
(\neg\varphi)^I &:= \neg(\varphi^I) \\
(\varphi \wedge \gamma)^I &:= \varphi^I \wedge \gamma^I \\
(\exists x\varphi)^I &:= \exists x(\alpha(x) \wedge \varphi^I) \\
(\exists X\varphi)^I &:= \exists X(\varphi^I)
\end{aligned}$$

If P is a monadic second order property for K_1, defined by the formula φ, then a property $I(P)$ for the structures of K_2 is defined by the formula φ^I.

The usefulness of linear time interpretability is based on the following result.

Theorem 4.1 *[Arn La Se 89]: Assume that K_1, K_2, L_1, L_2 are as above and that K_1 is linear (polynomial) time interpretable into K_2 with respect to L_1, L_2, where the corresponding interpretation is denoted by I and that P is a monadic second order property over K_1. If $I(P)$ can be solved over K_2 in linear (polynomial) time, then P can also be solved in linear (polynomial) time over K_1.*

The main step in the proof is the following standard lemma for interpretability (see [Rab 64] or [Arn La Se 89].

Lemma 4.2 *Assume that K_1, K_2, L_1, L_2 and I are as above. Moreover let φ be a monadic second order sentence over K_1 and let G be a structure from K_2. Then*

$$G \models \varphi^I \textit{ if and only if } I(G) \models \varphi.$$

To prove theorem 3.3 we need as a key step a result with the help of which we are able to reduce graphs of bounded tree width to very simple trees, the binary trees. We need some notations first.

For a finite set $\sum \neq \emptyset$ let $\sum^*$ be the set of all finite words (finite sequences) of elements from $\sum$. The empty word is denoted by Λ. Assume in the following that $\Delta := \{0,1\}$. Let $A \subseteq \Delta^*$ be closed under the initial segment relation, i.e.

$$\text{if } uv \in A, \text{ then also } u \in A$$

With each such A we connect in a canonical way a binary tree:

$$\begin{aligned}
(A, S_0, S_1), \text{ with } S_0 &:= \{(u, u0) : u0 \in A\} \\
\text{and } S_1 &:= \{(u, u1) : u1 \in A\}.
\end{aligned}$$

Such structures we shall denote as binary trees. The empty word is the root of such a tree. Structures, which are isomorphic to a binary tree will be denoted also as binary trees. The initial segment relation is used for such trees in a similar way, i.e. u is an initial segment of v if u is in the path from the root of the tree to v.

Theorem 4.3 *[Arn La Se 89]: For each natural number m every class K_n of graphs whose tree width is bounded by m is polynomial (linear) time interpretable into the class of binary trees. If the graphs from K_n are given together with a tree decomposition then the corresponding transformations can be performed in linear time.*

The original result proved in [Arn La Se 89] is slightly different since it uses a notion of interpretability, which allowed the interpretation of the equality $=$ by an equivalence relation definable by a monadic second order formula. For our more restrictive notion of interpretability almost the same proof as in [Arn La Se 89] can be used. One has only to keep in mind that one can select a unique element from the equivalence class by choosing the minimal element under the initial segment relation, which is definable in monadic second order logic in binary trees (see [Rab 69]). Moreover also the use of additional unary predicates in the original proof can be easily avoided.

Let I and A be this interpretation and the corresponding algorithm transforming a graph G from K into the needed binary tree $A(G)$. This A depends on a given tree decomposition (T, X) of G. Then it is easy to see that $A(G)$ can be choosen in such a way that it has the additional property, that (T, X) is definable by monadic second order formulas in $A(G)$. To be more precise there are formulas $\gamma(x), \delta(x, y)$ and $\varphi(x, y, z)$ such that the following holds:

$(A(G)(\gamma), A(G)(\delta)) = T$ and $A(G)(\varrho)$ is the relation
$\{(a, b, t) : t \in V(T),\ a \in X_t \text{ and } b \in X_t\}$

From this it is easy to deduce the following fact.

Lemma 4.4 *There is a formula θ such that for each $G \in K_m$ it holds:*

$A(G) \models \theta$ *if and only if the tree decomposition choosen for G in the interpretation is proper.*

This technique of interpretability will enable us to reduce our problem from graphs of bounded tree width to binary trees. For binary trees the related problem will be solved then via tree automata, which are handled in the next section.

5 Tree automata

For a finite set $\sum \neq \emptyset$, a $\sum$-tree is a structure (A, S_0, S_1, τ) such that (A, S_0, S_1) is a binary tree and $\tau : A \to \sum$ is a function.

The following concept of tree automata was developed in [Don 66] and [Tha Wrig 68].

A tree automaton M is a quintuple $(S, \sum, \delta, s, F)$, where

S is a finite set of states
$\sum$ is a finite set, the alphabet, $\sum \cap S = \emptyset$

δ is a transition function $S \times S \times \sum \to S$
s is the initial state, $s \in S$
F is the set of accepting states, $F \subseteq S$.

For a given $\sum$ such an automaton will be denoted also as $\sum$-automaton.

A run of M on a binary tree $T = (A, S_0, S_1, \tau)$ is a function $r : A \to S$ defined by

$$r(a) := \delta(s, s, \tau(a)) \text{ if } a \text{ is a leaf}$$

$$r(b) := \delta(s_0, s_1, \tau(a)) \text{ if } \begin{cases} s_0 &= r(b0) \text{ if } b0 \in A \\ s_1 &= r(b1) \text{ if } b1 \in A \\ s_0 &= s \text{ if } b0 \notin A \\ s_1 &= s_1 \text{ if } b1 \notin A \end{cases}$$

M accepts T if there is a run r of M on T with $r(\Lambda) \in F$.

Now let L be a monadic second order language corresponding to binary trees. Such a language has a high expressive power. For instance the prefix relation in a binary tree can be defined by the following formula:

$$\varphi(x, y) \Leftrightarrow \forall X(y \in X \wedge \forall z \forall w(z \in X \wedge (S_0(w, z) \vee S_1(w, z))) \to w \in X) \to x \in X)$$

Assume now that (A, S_0, S_1) is a binary tree and that $B_1, \ldots, B_m$ are subset of A. Let $X_{B_i} : A \to \{0, 1\}$ be the characteristic function of each B_i :

$$\chi_{B_i}(a) := \begin{cases} 0 & \text{if} \quad a \notin B_i \\ 1 & \text{if} \quad a \in B_i \end{cases}$$

Define Δ^m to be the set of all words from Δ^* which have length m. Then define $\tau_{B_1, \ldots, B_m}$: $A \to \Delta^m$ by: $\tau_{B_1, \ldots B_n}(a) := \chi_{B_1}(a) \ldots \chi_{B_m}(a)$.

The key result showing the connection between monadic second order formulas and tree automata is the following theorem.

Theorem 5.1 *[Don 66][Tha Wrig 68]: For each formula $\varphi(X_1, \ldots, X_m)$ of L there can be constructed an automaton M_φ such that for each binary tree (A, S_0, S_1) and all sets $B_1 \subseteq A, \ldots, B_m \subseteq A$*

$(A, S_0, S_1) \models \varphi[B_1, \ldots, B_n]$ if and only if M_φ accepts $(A, S_0, S_1, \tau_{B_1, \ldots, B_m})$.

Remark: Obviously each automaton works in linear time on the corresponding trees, i.e. for a given automaton one can decide in a time linear in the number of the vertices of the tree whether it accepts the tree or not.

Corollary 5.2 *(implicitely in [Tha Wrig 68]): Let m be a fixed natural number and let K be a class of binary trees with m additional sets (or unary predicates). Then for each fixed monadic second order property P over K there is an algorithm deciding $P(G)$ for $G \in K$ in time linear in $|V(G)|$.*

This result is the basis to get efficient algorithms for graphs whose structure is closely related to trees, e.g. the graphs of bounded tree width (see [Arn La Se 89]).

Let $T = (B, S_0, S_1)$ be a binary tree and let t and t' be elements of B assuming that t is an initial segment of t'. Then define:

$$\begin{aligned} B(t) &:= \{s : s \in B \text{ and } t \text{ is an initial segment of } s\}, \\ B(t,t') &:= \{s : s \in B \text{ and } t \text{ is an initial segment of } s \\ &\quad\ \text{and } t' \text{ is not an initial segment of } s\}, \\ B(-t) &:= B - B(t), \\ B(-t,+t') &:= B(-t) \cup B(t'). \end{aligned}$$

Moreover let $T(t), T(t,t')$ and $T(-t)$ be the structures induced in T by $B(t), B(t,t')$ and $B(-t)$ respectively. The structure $T(-t,+t')$ results from the structure induced by $B(-t,+t')$ in T

by adding (t'', t') to S_0 if $t''0 = t$ and
by adding (t'', t') to S_1 if $t''1 = 0$.

Here t'' will be denoted as predecessor of t. t is then the new root of $T(t)$. $T(-t,+t')$ can be viewed as the result of the following operation. Take T, cut $T(t)$ away and substitute for it $T(t')$, connecting t' via S_0 or S_1 with the corresponding predecessor of t. In a similar way define $T(-t',+t)$ by cutting $T(t')$ away, substituting it by an isomorphic copy of $T(t)$ which has to be disjoint from T and connect the root of this copy with the predecessor of t'.

In case that T is a $\sum$-tree the structures $T(t)$, $T(t,t')$, $T(-t)$, $T(-t,+t')$ and $T(-t',+t)$ are defined in a similar way.

The *depth* of the binary tree T is the length of the largest word in B.

Lemma 5.3 *([Don 66][Tha Wrig 68]): Assume that M is a $\sum$-automaton which accepts a binary tree T and that r is a corresponding run. Moreover assume that t and t' are elements of the domain of T such that t is an initial segment of t' and $r(t) = r(t')$. Then M accepts also $T(-t,+t')$ and $T(-t'+t)$.*

The lemma follows immedialety from the above definitions and it will be usefull in the next section together with the following corollary.

Corollary 5.4 *([Don 66] [Tha Wrig 68]): Let M be the $\sum$-automaton $(S, \sum, \delta, s, F)$. Then there is a $\sum$-tree accepted by M if and only if M accepts a $\sum$-tree of depth $< |S|$. Moreover M accept a $\sum$-tree of depth $\geq |S|$ if and only if M accepts infinitely many pairwise non-isomorphic $\sum$-trees.*

6 The proof of theorem 3.3

Building on the ideas presented in the previous sections we can now present the main part of the proof of theorem 3.3.
So assume that J is defined by a monadic second order formula φ. First we observe that there is a formula ψ which is true in a graph G if and only if G is isomorphic to a minimal forbidden minor for J. ψ is defined as follows

$$\neg\varphi \wedge \text{'for each 1-step minor it holds } \varphi\text{,'}$$

where to be a 1-step minor is obviously expressible in monadic second order logic.

Lemma 6.1 *Let K be a class of graphs of tree width, path width $< m$ then each minimal forbidden minor for K has tree width, path width $< m+1$ respectively.*

Proof: Assume that the tree width of each graph of K is $< m$ and let H be a minimal forbidden minor for K. Let a be a vertex of H and let H' be the graph resulting from H by omitting a. Since H' is in K it has a tree decomposition (T, X) of width $< m$. Define X' as family

$$(X'_t : t \in V(T)) \text{ with } X'_t := X_t \cup \{a\}.$$

Then (T, X') is a tree decomposition of H' of width $< m+1$. The same simple idea works also for path width.

Now assume that each graph of J has tree width $< m$. Then we know by (6.1) that the tree width of all minimal forbidden minors for J is $< m+1$ By (4.3) there is a polynominal time interpretation of the class of all graphs of tree width $< m+1$ into the class of binary trees. Let A be the corresponding algorithm as in (4.3). By the definition of an interpretation A is a 1-1 mapping of the class of graphs of tree width $< m+1$ into the class of binary trees.

We show first that under this assumptions one can prove that up to isomorphism there are only finitely many minimal forbidden minors for J.

Hence assume that there is an infinitite number of pairwise non-isomorphic minimal forbidden minors for J and denote this set by $F(J)$. Hence by (4.2) there are infinitely many pairwise non-isomorphic binary trees $A(G)$ with $G \in F(J)$ which
fulfill: $A(G) \models \psi^I$

But in the definition of $A(G)$ we started with a proper tree decomposition (T, X) of G. Hence the formula θ from (4.4) is also true in $A(G)$.

By (5.1) there is a tree automaton $M = (S, \{\emptyset\}, \delta, s, F)$ such that for each binary tree T it holds: M accepts T if and only if $T \models \gamma^I \wedge \theta$.

Hence by our assumption and (16) there is a minimal forbidden minor G for J and a binary tree $T = A(G)$ of depth $> |S|$, which is accepted by M. Let r be an accepting run of M on T. Hence one can find vertices $t(0)$ and $t(1)$ of T such that $t(0)$ is a an initial segment of $t(1)$ and $r(t(0)) = r(t(1))$.

Following the definition of I and A it is not difficult to see that $G, T, t(0)$ and $t(1)$ can be choosen in such a way that $t(0)$ and $t(1)$ do not correspond via the interpretation to vertices of G and that each of $T(-t(0)), T(t(0), t(1))$ and $T(t(1))$ contains an element which corresponds to a vertex of G and there is no vertex in the other two which corresponds to the same vertex of G. To see this one has to remember that our definition of A started with a proper tree descomposition and then one has to distinguish the case when the corresponding tree of the tree descomposition contains a path of length $\geq |S|$ from the case when it contains a vertex of degree $\geq |S|$. One of these cases appears, since there are infinitely many pairwise non-isomorphic minimal forbidden minors for J of tree width $< m+1$ and the regarded tree descompositions are all proper.
Now define $T1$ to be the binary tree $T(-t(1), +t(0))$. The isomorphic copy of $T(t(0))$ which is used here contains a vertex $t(2)$ which corresponds under the corresponding isomorphism to $t(1)$. The root of $T(t(0))$ is defined to be $t(1)$. Now we define by induction on n a binary tree Tn and a sequence $t(0), t(1), t(2), \ldots, t(n), t(n+1)$ of vertices of Tn.
Assume that $Tn-1$ and $t(0), t(1), \ldots, t(n)$ are already defined. Then define Tn to be the binary tree $Tn-1(-t(n), +t(n-1))$ and define the root of the isomorphic copy of

$Tn-1(t(n-1))$ to be $t(n)$ and let $t(n+1)$ be the vertex corresponding to $t(n)$ under the corresponding isomorphism.

This Tn can be constructed for each natural number $n > 0$ and by (5.4) all these Tn are accepted by M. Now let Gn be the graph $I(Tn)$. Then regard the subgraphs of Gn induced by the vertices of Gn corresponding to $Tn(t(i), t(j))$ and denote it by $Gn(t(i), t(j))$. But $V(Gn(t(i), t(j)))$ separates the set of vertices of G corresponding to $Tn(-t(0))$ from the set of vertices of G corresponding to $Tn(t(n+1))$.
By our definition all $Gs(t(i), t(i+1))$ (for $s > 0$ and $i \leq s$) are isomorphic, since the corresponding $Tn(t(i), t(j))$ are. Using this it is not difficult to see that there is an n and an $n' < n$ that Gn' is a minor of Gn. To see this choose for each i $(0 < i < n)$ *a* minimal subset of $V(Gn(t(i), t(i+1)))$ which is a seperator of Gn seperating $Gn(t(i-1), t(i))$ from $Gn(t(i+1), t(i+2))$. This can be done in such a way that the vertices $v1, \ldots, vl$ of the minimal seperator inside $Gn(t(i), t(i+1))$ are mapped by the choosen isomorphism between $Gn(t(i), t(i+1))$ and $Gn(t(i+1), t(i+2))$ to the corresponding vertices $v'1, \ldots, v'l$ of the minimal seperator $Gn(t(i+1)), t(i+2))$. Let some l vertex disjoint paths between the vertices of these two seperators connect vj with $v'\tau(j)$ for a certain permutation τ of l element. Now let $s > 0$ be such that τ^s is the identity. By contracting the edge of these s pathes one can show that Gn has a minor isomorphic to $Gn-s$. This contradicts the assumption that the formula ψ holds in Gn and hence G is a minimal forbidden minor of J.

It remains to show that the minimal forbidden minors can be constructed if the monadic second order definition of J is given. But this follows from the above proof, since one has in the first step simply to construct the above automaton M, which is a standard construction. Then one checks for each binary tree of depth $< |S|$ whether it is accepted by M or not. For each such accepted tree T one constructs $I(T)$. All such graphs $I(T)$ are minimal forbidden minors of J and moreover each minimal forbidden minor of J is isomorphic to one of the $I(T)$. This ends the proof of (3.3).

Recently [Arn Lag 90] showed that monadic second order descriptions of lower ideals are equivalent to definitions by finite decidable congruences or graph reduction systems and gave a decidable congruence for the family of partial k-trees. Hence for the class of partial k-trees itself the minimal forbidden minors can be found via (3.3).

It is interesting to remark that the above proof of (3.3) is also a proof of Theorem 2.2 in case that the lower ideal I is monadic second order definable and contains only graphs of universally bounded tree width. Moreover this proof gives an upper bound on the number of the minimal forbidden minors for the lower ideal. This bound is the number of all possible binary trees (up to isomorphism) of depth $< |S|$, where S is the set of states of the corresponding automaton. Unfortunately this bound is not very practical, since the size of S grows exponenttially with the quantifier depth of the formula.

7 Partial 2-paths

To apply this result one has to find monadic second order definitions of interesting classes of graphs. It will be demonstrated here how one can come to such descriptions in case of partial 2-paths. The idea is to find a structural characterization of this class which can be described by a monadic second order formula.

At first we need the following results from [Pros 84].

Theorem 7.1 *([Pros 84]): In a k-tree Q, the vertices of minimal subgraphs separating two nonadjacent vertices u and v induce a k-path.*

Theorem 7.2 *([Pros 84]): Given a k-tree Q ($k \geq 2$) and two nonadjacent vertices u and v. The union S of k-complete graphs separating u and v is a $(k-1)$-caterpillar.*

Lemma 7.3 : *Let P' and P'' be the vertex-disjoint paths of a full 2-path H. Contraction of some edges of P' and P'' and deletion of their remaining edges gives a minor of H that is a caterpillar.*

Proof: By (7.2) the union of edges induced by the minimal separators of a full 2-path is a caterpillar. Since this union includes all edges of H exept the edges of P' and P'', deletion of the latter leaves a caterpillar. Contraction of an edge e of, say P', gives a full 2-path H' with the two vertex-disjoint paths $Q' = P' - \{e\}, Q'' = P''$, unless e is adjacent to a 2-leaf s of H and a vertex of degree > 3. In the latter case, the vertex adjacent to s through the other edge, f, will become the new 2-leaf and $Q' = P' - \{e\} \cup \{f\}, Q'' = P'' - \{f\}$. The lemma holds by (7.2).

Let $G = (V, E)$ be a connected graph and V_1, V_2 be a partition of V into two sets, such that V_i induces a collection of paths, P_i $(1 \leq i \leq 2)$. Define a bipartite graph $B(G; V_1, V_2) = (N, E')$ as follows: $N = P_1 \cup P_2$

$$E' = \{(p, q) : \text{ there are } u \in p \in P_1 \text{ and } v \in q \in P_2 \text{ with } (u, v) \in E\}.$$

Let a be a vertex of a component of P_i of degree 1. Then a *natural order* of this component is a linear ordering $\leq$ defined by: $b \leq c$ if and only if each subpath of the component containing a and c contains also b.

Theorem 7.4 : *G is a partial 2-path if and only if its vertices can be partitioned into two sets V_1 and V_2 such that the following conditions are fulfilled:*

(global condition)

$B(G; V_1.V_2)$ is a caterpillar. V_i induces a collection of paths, $P_i (1 \leq i \leq 2)$ and for each component of P_i there is a natural order $\leq$ of its vertices.

(local condition)

- *(a) for all vertices $u \leq v \in p \in P_1$ and all vertices $a, b \in q \in P_2$: $(u, a), (v, b) \in E$ implies $a \leq b$*
- *(b) for all $u \leq v \in p \in P_1$, $a \in q_1 \in P_2$, $b \in q_2 \in P_2, c \in q \in P_2$ with (u, a), $(v, b) \in E$ and $q_1 \neq q_2$ we have that if there is $w \leq p_1 \in P_1$ with $(w, c) \in E$ then $a < c$ implies that $u \leq w \leq v$ and $c < b$ implies that $u \leq w \leq v$.*

Here $<$ is the natural order from the global condition.

Proof:
(necessity) If G is a partial 2-path, consider its full 2-path embedding, H; the two vertex-disjoint paths of H (augmented by the 2-leaves, each put into the set to which it is adjacent) define a partition of V into V_1 and V_2. The global condition follows by lemma (7.3), since the edges of the vertex-disjoint path of H missing in G define the conditions

of (7.3). Since H is planar, both (a) and (b) of the local condition postulating a planar embedding of G hold.

(sufficiency) We will show by double induction on the number n of nodes in N and the number m of cross-edges in G. If $n = 1$ then G is trivially a partial 2-path. Assume that the hypothesis holds for all graphs such that $|N| < n$ and consider G such that $G(B; V_1, V_2)$ has n nodes and exactly one edge to a path $q \in P_1$. Assume also that the global and local conditions hold. The graph G' represented by $B(G; V_1, V_2)$ is embeddable in a full 2-path H'. Consider vertices of a path $p \in P_1$ incident with that edge $((w, c) \in E$ where $w \in p$, $c \in q)$. If there are vertices $u, v \in p$ such that $u < w < v$, and $a \in q_1 \in P_2, b \in q \in P_2$ with (u, a), $(v, b) \in E$ then $q_1 \neq q_2$, since this would violate (a) of the local condition. Thus, G can be embedded in a full 2-path Hderived from H' by 'fitting' q between q_1 and q_2 on P_2. If such u and v are not present, q can be easily made the last (or the first) part of a path in H.

Assume now that the hypothesis is true for graphs with fewer than m edges represented by caterpillars with n nodes.
Consider G obtained from G' (embeddable in H') by adding an edge
$(w, c) \in E$ $(w \in p \in P_1, c \in q \in P_2)$ that preserves the global and the local condition. Consider the largest v, $v < w$ and the smallest $u, w < u$ such that there are $a \in q_1 \in P_2$, $b \in q_2 \in P_2$, (u, a) $(v, b) \in E$. If both exist, then either $a \leq c \leq b$ $(q_1 = q = q_2)$ and u, a, v, b in H' can be triangulated to include (w, c) or $q_1 \neq q_2$ and $a \leq c$ or $c \leq b$, with a simular triangulation of the corresponding quadrangle. If, without loss of generality, v does not exist, then either $a \leq c$ and there is no vertex t, $a \leq t \leq c$ incident to a cross-edge of G and u, a, w, c can be triangulated, or $q_1 \neq q$ and there is no such $t \leq c$ and G can be embedded in a full 2-path, completing the proof.

Now it is obvious that the global as well as the local condition of theorem (7.4) can be formalized in monadic second order logic. But this gives together with theorem (3.3) the existence of an algorithm to construct the minimal forbidden minors for the class of partial 2-pathes.

As a consequence one gets the following results (see also [Arn Cou Pr Se 90])

Theorem 7.5 : *There is an $O(n^2)$-algorithm to decide whether a graph is a partial 2-path. For each (extended) monadic second order property P (EMS-property) one can effectively construct an $O(n^2)$-algorithm to solve P for all partial 2-paths.*

It is possible to extend this method here to find a simple monadic second order description of the class of graphs of path width ≤ 2. This monadic second order description is shorter than those which would result via (3.2) from the 110 minimal forbidden minors found for this class of graphs in [Kin 89].

For $n < 4$ it is easy to find monadic second order descriptions of the class of graphs of tree width $\leq n$ using the minimal forbidden minors found in [Arn Cor Pr 86]. It is to hope that using the methods of the last section monadic second order descriptions of the class of graphs of tree width $< k$, for $k > 4$, can be found, which are more efficient than those which would result from the minimal forbidden minors for this class.

References

[Abr Fel 89] M.R. Fellows and K. Abrahamson 1989, *Cutset-Regularity*

Beats Well-Quasi-Ordering for Bounded Tree-width (Extended Abstract), preprint Nov. 1989.

[Arn Cor Pr 86] **S. Arnborg, A. Proskurowski and D.G. Corneil** 1986, *Forbidden minor characterization of partial 3-trees*, Discrete Math., to appear.

[Arn Cou Pr Se 90] **S.Arnborg, B.Courcelle, A. Proskurowski and D. Seese** 1990, *An algebraic theory of graph reduction*, preprint January 17, 1990.

[Arn Lag 90] **S. Arnborg and J. Lagergren** 1990, *Finding minimal forbidden minors using a finite congruence*, preprint November 13, 1990.

[Arn La Se 88] **S. Arnborg, J. Lagergren and D. Seese** 1988, *Problems Easy for Tree-descomosable graphs* (extended abstract). Proc. 15th ICALP, Springer Verlag, Lect. Notes in Comp. Sc. 317 38–51.

[Arn La Se 89] **S. Arnborg, J. Lagergren and D. Seese** 1989, *Problems Easy for Tree-descomposable graphs* to appear in J. of Algorithm.

[Arn Pr 86] **S. Arnborg and A. Proskurowski** 1986, *Characterization and Recognition of Partial 3-trees*, SIAM J.Alg. and Discr. Methods 7, 305–314.

[Arn Pr 89] **S. Arnborg and A. Proskurowski** 1989, *Linear Time Algorithm for NP-hard Problems on Graphs Embedded in k-trees* Discr. Appl. Math. 23, 11–24.

[Arn Pr See 89] **S. Arnborg, A. Proskurowski, and D. Seese** 1989, *Logical description of graphs of path-width 2 and their minimal forbidden minors* (Draft), preliminary version, preprint July 25, 1989.

[Bod 87] **H.L. Bodlaender** 1887, *Dynamic Programming on Graphs with Bounded Tree-width*, MIT/LCS/TR-394, MIT.

[Bod 88] **H.L. Bodlaender** 1988, *Improved self-reduction algorithms for graphs with bounded tree-width*, Technical Report RUU-CS-88-29, September 1988, University of Utrecht.

[Bod Klo 90] **H.L. Bodlaender, T. Kloks** *Better Algorithms for the Pathwidth and Treewidth of Graphs* (extended abstract), preprint 1990.

[Bon Mur 76] **J.A. Bondy and U.S.R. Murty** 1976, *Graph Theory with Applications*, North Holland.

[Cou 88] **B. Courcelle** 1988, *The monadic second order logic of graphs III: Tree-width, forbidden minors, and complexity issues*, Report $I-$ 8852, Bordeaux-1 University.

[Don 66] **J.E. Doner** 1966, *Decidability of the Weak Second Order Theory of two Successors*, Abstract 65T-468, Notices Amer. Math. Soc. 12, 819,ibid., 513.

[Fel 89] **M. Fellows** 1989, *Nonconstructive Proofs of Polynominal-Time Complexity: Algorithms for Computing Obstructing Sets*, draft, preprint April 12, 1989.

[Fel 89a] **M. Fellows** 1989, *Applications of an Analogue of the Myhill-Nerode Theorem, In Obstruction Set Computation*, preprint April 28, 1989.

[Fel Lan 87] **M. Fellows and M. Langston** 1987, *Nonconstructive Advances in Polynominal Time Complexity*, Info. Proc. Letters 26, 157–162.

[Fel Lan 89] **M. Fellows and M. Langston** 1989, *An Analogue of the Myhill-Nerode Theorem and Its Use in Computing Finite-Basis Characterizations* (Extended Abstract), to appear, FOCS 89.

[Fel Lan 89a] **M. Fellows and M.Langston** 1989, *Exploiting RS Posets: Con-*

structive Algorithms from Nonconstructive Tools, preprint revised February 1989.

[Kin 89] N.Kinnersley 1989, *Obstruction set isolation for layout permutation problems*, Ph.D. thesis, Washington State University.

[Mat Tho 88] J. Matoušek, R. Thomas *Algorithms finding tree - decompositions of graphs*, preprint 1988.

[Pros 84] A. Proskurowski 1984, *Separating subgraphs in k-trees : cables and caterpillar*, Discrete Mathematics 49, 275-285.

[Rab 64] M.O. Rabin 1964, *A simple Method of Undecidablility proofs and some applications*, in Log. Meth. Phil. Sci. Proc. Jerusalem, 58–68.

[Rab 69] M.O. Rabin 1969, *Desidability of second order and automata on infinite trees*, Trans. Am. Math. Soc. 141, 1–35.

[Rob Sey 83] N. Robertson and P.D. Seymour 1983, *Graph Minors I. Excluding a forest*, J. Combin. Theory Ser.B. 35, 39–61.

[Rob Sey 86] N. Robertson and P.D. Seymour 1986 *Graph Minors II. Algorithmic Aspects of Tree Width* Journal of Algorithms 7, 309–322.

[Rob Sey 86a] N. Robertson and P.D. Seymour 1986, *Graph Minors V. Excluding a planar graph* J. Combinatorial Theory, Ser. B, 41, 92–114.

[Rob Sey 86b] N. Robertson and P.D. Seymour 1986, *Graph Minors XIII. The Disjoint Path Problem* Preprint.

[Rob Sey 88] N. Robertson and P.D. Seymour 1988, *Graph Minors XV. Wagners conjecture* Preprint.

[Rob Sey 89] N. Robertson and P.D. Seymour 1989, *Personal Communication*, Toronto, Eugene.

[Schef 86] P. Scheffler 1986, *Dynamic programming algorithms for tree-descomposition problems*, Karl-Weierstrass-Institut für Mathematik, Preprint P-Math-28/86, Berlin.

[Schef 89] P. Scheffler 1989 *Die Baumwerte von Graphen als ein Maß für die Kompliziertheit algorithmischer Probleme*, Dissertation (A), AdW d. DDR, Berlin 1989.

[Sees 85] D. Seese 1985, *Tree-partite graphs and the complexity of algorithms* (extended abstract), in FCT'85, ed. L. Budach, LNCS 199, Springer, Berlin, 412–421.

[Sees 86] D. Seese 1986, *Tree-partite graphs and the complexity of algorithms*,preprint P-Math 08/86, Karl-Weierstrass-Institute für Mathematik.

[Tha Wrig 68] J.W. Thatcher and J.B. Wright 1968, *Generalized Finite Automata Theory with an Application to a Decision Problem in Second-Order Logic*, Mathematical Systems Theory 2, 57-81.

[Wim 88] T.V. Wimer 1988, *Ph D Thesis URI-030*, Clemson.

On the reduction theory for average case complexity

Andreas Blass[1] and Yuri Gurevich[2]

Abstract. This is an attempt to simplify and justify the notions of deterministic and randomized reductions, an attempt to derive these notions from (more or less) first principles.

1. Introduction

Let us review the notion of a decision problem. First of all, one has a set – the set of *instances*, the *universe* of the decision problem. For simplicity, we stick to the Turing machine model and suppose that the universe is always a set of strings in some finite alphabet. Of course, objects of interest are not necessarily strings. They may be graphs for example. But they should be coded as strings.

Second, the instances of a given decision problem split into positive and negative; this can be formalized by means of a *characteristic function* from the universe to $\{0,1\}$. But this is not all. In order to discuss e.g. polynomial time algorithms, we need that instances have sizes. Ordinarily, the size of a string is its length [GJ], but this isn't always convenient. For example, if objects of interest are graphs then one may prefer to define the size of the encoding string as the number of vertices in the graph rather than the length of the encoding string.

Definition. A *size function* on a set U is a function from U to natural numbers. A size function $x \mapsto |x|$ is *conventional* if the set $\{x : |x| \leq n\}$ is finite for each n.

Typically, the size of an instance is polynomially related to the length, but unconventional size functions turn out to be useful as well. In some situations, it may be convenient to allow non-integer sizes; for simplicity, we shall stick here to integer sizes. The notation $|x|$ will be used to denote the size of an element x.

---

[1]Partially supported by NSF grant DMR 88-01988. Address: Mathematics Department, University of Michigan, Ann Arbor, MI 48109-1003, USA; andreas_blass@ub.cc.umich.edu

[2]Partially supported by NSF grant CCR 89-04728. Address: Electrical Engineering and Computer Science Department, University of Michigan, Ann Arbor, MI 48109-2122, USA; gurevich@eecs.umich.edu

In the rest of this paper, a decision problem is a set of strings (not necessarily the set of all strings) in some finite alphabet together with a size function and a characteristic function.

In the average case approach, pioneered by Leonid Levin, a decision (resp. search) problem is considered together with a probability distribution on instances. Such a pair is called a *randomized*, or *distributional*, *decision* (resp. *search*) *problem*. For simplicity, we restrict attention to randomized decision problems. In order to obtain completeness results, Levin defined reductions among randomized decision problems. His notion of reduction seems *ad hoc*. In the first part of this paper (Sections 2–4), we attempt to derive a simpler version of it from (more or less) first principles.

Even though that original notion of reduction, deterministic in nature, was sufficient to establish the completeness of a number of natural problems [Le, Gu1, Gu2], it turned out to be too restrictive. Many randomized decision problems of interest are *flat* in the following technical sense: There exists $\varepsilon > 0$ such that the probability of any instance of sufficiently large size n is bounded from above by $2^{-n^{\varepsilon}}$. However, no randomized decision problem with a flat domain is complete unless deterministic exponential time equals nondeterministic exponential time [Gu1]. To overcome this difficulty, Levin suggested more general randomizing reduction. Versions of randomizing reduction were defined and successfully used in [VL] and [BCGL]. A simple version of randomizing reduction was defined in greater detail and proved transitive in [Gu1]. That version was too simple however, and in the second part of the paper (Section 5–6) we attempt to justify a more general version of randomizing reductions.

Remark. The randomizing reductions of Section 5 can be further generalized. Our goal is not the most general notion of reduction, but rather a simpler notion of reduction sufficient for most applications.

2. The notion of a domain

Definition [Gu2]. A *domain* D is a nonempty set U_D (the *universe* of D) of strings in some finite alphabet with a size function $|x|$ (or $|x|_D$) and a probability distribution $\mathbf{P}_D$ such that there are only finitely many elements of positive probability of any given size.

Since the elements of a domain are strings, there is a well-defined notion of computability for functions on domains. The requirement that there are only finitely many elements of positive probability of any given size will be relevant in Section 5 where we deal with unconventional size functions.

Remark. One could work at a more abstract level, where the universes of domains have some prescribed (or assumed) notions of computability and computation time subject to suitable axioms. For example, the size of an output is bounded by the computation time, composing two computable functions yields a computable function, and the computation time is additive for composition. We shall stick here to the simpler definition of domains given above which allows us to use the standard Turing machine model. The particular axioms mentioned above are assumed to be true in that model.

It is natural to say that a function T from a set with a size function to nonnegative reals is *polynomially bounded* if $T(x)$ is bounded by a polynomial of $|x|$.

Definition. A function T from a domain D to the set $\bar{\mathcal{R}}^+$ of nonnegative reals augmented with ∞ is AP (or *polynomial on average* or *polynomially bounded on average*) with respect to D if, for some $\varepsilon > 0$, $\sum_x (Tx)^\varepsilon |x|^{-1} \mathbf{P}_D(x) < \infty$. We will say that T is *linear on average* if the witness ε can be chosen to be 1.

Here x ranges over elements of positive size; to avoid the nuisance of dealing with elements of size 0, we restrict attention to domains without elements of size 0 in the rest of the paper. One can, without loss of generality, restrict x to range over elements satisfying $(Tx)^\varepsilon > x$. This is because the omitted terms sum to at most $\sum_x \mathbf{P}_D(x) = 1$. The probability $\mathbf{P}_D(x)$ of an element x is of course the probability $\mathbf{P}_D[\{x\}]$ of the set (or *event*) $\{x\}$.

The notion of polynomiality on average is due to Levin [Le]; it is motivated and discussed in [Jo, Gu1, BCGL, Gu2]. We have generalized this definition slightly by allowing ∞ as a possible value of the function T in question. In the case when T is the computation time of some algorithm, the infinite value allows us to consider algorithms that may diverge at some inputs: If the algorithm diverges at a point x then $T(x) = \infty$. We suppose that $\infty \cdot 0 = 0$. If the algorithm diverges at some set of probability zero, the computation time still may be polynomial on average. Because we required that there are only finitely many elements of positive probability of any given size n, we have that, for each n, $\sum_{x,|x|=n} (Tx)^\varepsilon |x|^{-1} \mathbf{P}_D(x) < \infty$ provided that the probability of the event $T(x) = \infty$ is zero. This property is desirable and consistent with the spirit of the asymptotic approach.

Many complexity experts prefer to deal with instances of a fixed size. Can the definition of AP be reformulated in such terms? For a wide range of domains, the answer is yes; see [Gu1] in this connection.

Definition. A partial function on a domain D is AP-*time* (or AP-*time computable*) if it is computable in time polynomial on average with respect to D.

The following lemma justifies the use of more convenient size functions:

Lemma 2.1. Suppose that D_1, D_2 are two domains with the same universe U and the same probability distribution. Let S_i be the size function of D_i and T be a function from U to $\bar{\mathcal{R}}^+$.

1. Suppose that S_1 is bounded by a polynomial of S_2 and T is polynomial on average with respect to D_1. Then T is polynomial on average with respect to D_2.

2. If S_1, S_2 are bounded each by a polynomial of the other then T is polynomial on average with respect to D_1 if and only if it is polynomial on average with respect to D_2.

Proof. It suffices to prove (1). Fix some $k \geq 1$ such that $S_1(x) \leq (S_2x)^k$. We suppose that ε witnesses that T is polynomial on average with respect to D_1 and prove that ε/k witnesses that T is polynomial on average with respect to D_2. Ignoring points x such that $(Tx)^{\varepsilon/k} < S_2(x)$, we have

$$\sum_x (Tx)^{\varepsilon/k} \cdot (S_2x)^{-1} \cdot \mathbf{P}(x) \leq \sum_x (Tx)^{\varepsilon} \cdot (S_2x)^{-k} \cdot \mathbf{P}(x) \leq$$
$$\sum_x (Tx)^{\varepsilon} \cdot (S_1x)^{-1} \cdot \mathbf{P}(x) < \infty. \text{ QED}$$

The notion of uniform probability distribution plays an important role in the theory of finite probability spaces. On a set of nonzero finite cardinality m, the uniform probability distribution assigns the probability $1/m$ to each element. In order to generalize this notion to infinite probability spaces, one needs to fix a default probability distribution on positive integers; somewhat arbitrarily, we choose the default probability of n to be proportional to $n^{-1} \cdot (\log n)^{-2}$ for $n > 1$. Some comments related to this issue can be found in [Gu1] and [Gu2].

Definition. A domain D is *uniform* if elements of the same size have the same probability and $\mathbf{P}_D[\{x : |x| = n\}]$ is proportional to $n^{-1} \cdot (\log n)^{-2}$ for $n > 1$.

Definition. The direct product $A \times B$ of domains A and B is the domain of pairs (a, b), where $a \in A$ and $b \in B$, such that $|(a, b)| = |a| + |b|$ and $\mathbf{P}(a, b) = \mathbf{P}_A(a) \times \mathbf{P}_B(b)$.

Given a probability distribution μ on some set U and a subset $V \subseteq U$ of positive probability, we define the *restriction* $\mu|V$ of μ to V to be the conditional probability $\nu(x) = \mu(x \mid x \in V)$. In other words, $\nu(x) = 0$ for every $x \in U - V$, and $\nu(x) = \mu(x)/\mu[V]$ for every $x \in V$.

Definition. A domain B is a *subdomain* of a domain A if $U_B \subseteq U_A$, and the size function of B is the restriction of the size function of A, and $\mathbf{P}_B$ is the restriction of $\mathbf{P}_A$ to U_B. Further, A^+ is the subdomain of A comprising the elements of positive probability.

3. Domain reductions

Consider a function f from a set A of strings in some finite alphabet to a set B of strings in some finite alphabet, and let T range over functions from B to $\bar{\mathcal{R}}^+$. If f is computable and T is bounded by a computable function then the composition $T \circ f$ is bounded by a computable function. Suppose that the sets A and B come equipped with size functions. If f is polynomial time computable and T is polynomially bounded then $T \circ f$ is polynomially bounded. Further suppose that A and B are domains. Now the situation is different. Even if f is AP-time computable and T is AP with respect to B, $T \circ f$ is not necessarily AP with respect to A. The problem is that T may be small on average but very large on the range of f. This problem should not arise if f is to be used as a reduction between randomized decision problems. Thus, one may want to say that a function f from A to B reduces A to B if

(R1) f is AP-time with respect to A, and

(R2) For every AP function T on B, the composition $T \circ f$ is AP with respect to A.

Before we proceed, let us make a slight generalization by allowing f to be partial provided that it is defined at every element of positive probability in A. We stipulate that $T(fx) = \infty$ if f is undefined at x. The requirements (R1) and (R2) remain meaningful. For brevity, we will say that such an f is an *almost total* function from A to B.

Unfortunately, the requirement (R2) is difficult to use and a convenient sufficient condition for (R2) is needed. Since we allow ∞ as a possible value of T, (R2) implies

(R0) If $\mathbf{P}_A(x) > 0$ then $\mathbf{P}_B(fx) > 0$.

For, suppose that $\mathbf{P}_A(a) > 0$ but $\mathbf{P}_B(fa) = 0$. Then the function $T(y) = [$ if $y = f(a)$ then ∞ else $0]$ is AP whereas the function $T \circ f$ has the infinite value at a point of positive probability and therefore is not AP.

The next two definitions lead to a more tractable formulation of (R2).

Definition. Let A and B be two domains with the same universe U and the same size function. Then B *dominates* A, symbolically $A \leq B$, if there exists an AP function g on A such that $\mathbf{P}_A(x) \leq g(x) \cdot \mathbf{P}_B(x)$ for all x in U.

This concept was discussed in [Gu1] under the name "weak domination"; the term "domination" was restricted in [Le, Gu1] to the case when g is polynomially bounded. Notice that $A \leq B$ if and only if the ratio $\mathbf{P}_A(x)/\mathbf{P}_B(x)$ is AP with respect to A. It is supposed that $0/0 = 0$ which is consistent with our previous convention $0 \cdot \infty = 0$.

Example. Let BS1 be the uniform domain of nonempty binary strings where the size of a string is its length. Order binary strings lexicographically (more exactly, first by length and then lexicographically). Let BS2 be the domain of nonempty binary strings where the size of a string is its length and the probability of a string x is proportional to the default probability of the number of x in the lexicographical order. It is easy to check that the two domains dominate each other. This would not be true if the default probability of n were n^{-2} as in [Le].

Lemma 3.1 [Gu 1, Section 1]. Let A and B be two domains with the same universe U such that $A \leq B$. For every function T from U to $\bar{\mathcal{R}}^+$, if T is AP with respect to B then it is AP with respect to A.

Definition. Suppose that A, B are domains and f is an almost total function from A to B. *B dominates A with respect to f*, symbolically $A \leq_f B$ if the ratio

$$\mathbf{P}_A[f^{-1}(fx)]/\mathbf{P}_B(fx)$$

is AP with respect to A.

Corollary. If f is one-to-one then $A \leq_f B$ if and only if $\mathbf{P}_A(x)/\mathbf{P}_B(fx)$ is AP with respect to A.

Theorem 3.1. Suppose that A, B are domains and f is an almost total function from A to B such that the function $x \mapsto |fx|_B$ is AP with respect to A. Then (R2) holds if and only if $A \leq_f B$. Moreover, let $\mu(x) = \mathbf{P}_A(x)$ and let ν be the restriction of $\mathbf{P}_B$ to $\{y : \mu[f^{-1}(y)] > 0\}$. Then the condition (R2) is equivalent to each of the following conditions:

(D1) $A \leq_f B$.

(D2) There exists a function g from B to $\bar{\mathcal{R}}^+$ such that

- For all $y \in B$, $\mu[f^{-1}(y)] \leq g(y) \cdot \nu(y)$, and
- $g \circ f$ is AP with respect to A.

(D3) A is dominated by the domain A' obtained from A by replacing μ with the probability distribution

$$\mu'(x) = [\mu(x)/\mu[f^{-1}(fx)]] \cdot \nu(fx).$$

(D4) There is a probability distribution $\mu''(x)$ on the universe of A such that

$$\sum_{x, f(x)=y} \mu''(x) = \nu(y)$$

for every element y of B and the domain A'', obtained from A by replacing μ with μ'', dominates A.

Remark. (D4) is an older definition of domination with respect to a given function; see [Gu1].

Proof. Without loss of generality, we may suppose that A coincides with its subdomain A^+ comprising elements of A of positive probability. Then f is a total function.

To prove that (D1) implies (D2), set $g(y) = \mu[f^{-1}(y)]/\nu(y)$. To prove that (D2) implies (D3), notice that if (D2) holds then $g \circ f$ witnesses that A' dominates A. It is obvious that (D3) implies (D4). The implication (D4) $\rightarrow$ (R2) is proved in [Gu1] in the special case when all values of T are finite. We reduce the general case to the special case. Suppose (D4) and let μ'' be an appropriate witness. Let T be an AP function on B and $S(x) = T(fx)$. Then $\nu[\{y : Ty = \infty\}] = 0$, hence $\mu''[\{x : Sx = \infty\}] = 0$, and therefore $\mu[\{x : Sx = \infty\}] = 0$. Let $T'y = [\text{ if } Ty < \infty \text{ then } Ty \text{ else } 0]$. Obviously, T' is AP with respect to B and all values of T' are finite; hence the function $S'x = T'(fx)$ is AP with respect to A. But $Sx = S'x$ on every x of positive probability. Hence S is AP with respect to A. Thus, (R2) holds.

Finally, we suppose (R2) and prove (D1). Let $R = \{y : \mu[f^{-1}(y)] > 0\}$. By (R0), $\nu(y) > 0$ whenever $y \in R$. Choose $T(y) = [\text{ if } y \in R \text{ then } \mu(f^{-1}y)/\nu(y) \text{ else } 0]$. The function T is linear on average with respect to B:

$$\sum_y T(y)\nu(y)/|y| = \sum_{y \in R} (\mu[f^{-1}y]/\nu(y)) \cdot \nu(y)/|y| \leq \sum_{y \in R} \mu[f^{-1}y] = 1.$$

By (R2), $T(fx)$ is AP with respect to A. Hence (D1) holds. QED

Remark. The proof of the implication (R2) → (D1) does not use the fact that $|fx|_B$ is AP with respect to A. This hypothesis is used in the part of the proof of (D2) → (R2) cited from [Gu1].

Now we are ready to define the notion of (deterministic) domain reduction.

Definition. An almost total function f from a domain A to a domain B *deterministically reduces* A to B if it has the following two properties:

Efficiency: f is AP-time with respect to A.

Domination: B dominates A with respect to f.

We say that a domain A *deterministically reduces* to a domain B if some almost total function f deterministically reduces A to B.

Theorem 3.2. If f deterministically reduces a domain A to a domain B and g deterministically reduces B to a domain C then $g \circ f$ deterministically reduces A to C. Thus the deterministic reducibility relation is transitive.

In essence, Theorem 3.2 is not new [Gu1], but Theorem 3.1 allows us to give a simpler proof.

Proof. To prove that C dominates A with respect to $g \circ f$, use Theorem 3.1. It remains to check that $g \circ f$ is AP-time computable with respect to A. The time to compute $g(fx)$ splits into two parts: The time to compute $f(x)$ from x, and the time $t(x)$ to compute $g(y)$ from $y = f(x)$. Since f is AP-time with respect to A, it suffices to prove that $t(x)$ is AP. We know that the time $T(y)$ needed to compute $g(y)$ from y is AP with respect to B. Obviously, $t(x) = T(fx)$. By Theorem 3.1, $t(x)$ is AP. QED

4. Randomized decision problems

A *randomized decision problem*, in short an RDP, Π may be defined as a domain D together with a function χ, the characteristic function of Π, from D to $\{0, 1\}$. Any element x of D is an *instance* of Π, and the corresponding question is whether $\chi(x) = 1$. The problem Π is AP-*time decidable* if the characteristic function χ is AP-time computable.

Definition. Let Π_1 and Π_2 be RDPs with domains D_1, D_2 and characteristic functions χ_1, χ_2 respectively. An almost total function f from D_1 to D_2 *deterministically reduces* Π_1 to Π_2 if f is a reduction of D_1 to D_2 and f satisfies the following additional requirement:

Correctness: For every instance x of Π_1 of non-zero probability, $\chi_1(x) = \chi_2(f(x))$.

Theorem 4.1. If f deterministically reduces Π_1 to Π_2 and g deterministically reduces Π_2 to Π_3 then the composition $g \circ f$ deterministically reduces Π_1 to Π_3.

Proof. Use Theorem 3.2. QED

Theorem 4.2. If f deterministically reduces Π_1 to Π_2 and Π_2 is AP-time decidable then Π_1 is AP-time decidable.

Proof. By Theorem 3.1, f satisfies the property (R2). QED

Deterministic reductions were used to establish the completeness (for an appropriate class of RDPs) of some natural randomized decision problems [Le, Gu1, BCGL, Gu2].

5. Randomizing domain reductions

According to Section 3, a deterministic reduction of a domain A to a domain B is an AP-time computable almost total function f from A to B such that B dominates A with respect to f. The notion of deterministic reduction is generalized in this section by allowing the algorithm that computes f to flip a coin. For simplicity, only fair coins will be flipped.

Define a *randomizing Turing machine*, in short an RTM, to be a Turing machine that can flip a fair coin. Formally, this may mean that the machine has an additional read-only tape containing a random sequence of zeroes and ones. Thus, an RTM can be seen as a deterministic Turing machine with two input tapes. Our RTMs are transducers, i.e., they compute functions. We say that an RTM *halts* on an input x and a sequence r of coin flips if it reaches a special halting state; if the machine is stuck because the sequence r happened to be too short, it does not halt on (x, r).

Call a set X of binary strings *prefix-disjoint* if no element of X is a prefix of another element of X. A prefix-disjoint set is called a *barrier* if every infinite sequence $b_1 b_2 \ldots$ of bits has a prefix in X. By König's lemma, every barrier is finite.

Lemma 5.1. For every barrier X, $\sum_{r \in X} 2^{-|r|} = 1$.

Proof. Associate the real interval $[0.a_1 \ldots a_k 000\ldots, 0.a_1 \ldots a_k 111\ldots)$ of length 2^{-k} with any binary string $a_1 \ldots a_k \in X$. These intervals partition the interval $[0, 1)$. QED

If M is an RTM, let $\mathcal{B}_M(x)$ be the collection of finite sequences r of coin flips such that the computation of M on (x, r) halts using all bits of r. It is easy to see that $\mathcal{B}_M(x)$ is always prefix-disjoint. We say that M *halts* on x if $\mathcal{B}_M(x)$ is a barrier.

There is a good justification for us to restrict attention to machines that halt on every input: We deal with problems of bounded complexity, e.g., NP problems; given sufficient time, the reducing algorithm can simply solve the problem that it is supposed to reduce. For consistency with preceding sections, we will make a slightly more liberal restriction on our randomized Turing machines. It will be supposed that inputs come from a certain domain and the machine halts on every input of positive probability. More formally, a domain A is an *input domain* for a randomized Turing machine M if every $x \in A$ is a legal input for M and M halts on every $x \in A$ with $\mathbf{P}_A(x) > 0$.

Definition. A *dilator* for a domain A is a function δ that assigns a prefix-disjoint set $\delta(x)$ of binary strings to each $x \in A$ in such a way that if $\mathbf{P}_A(x) > 0$ then $\delta(x)$

is a barrier. Two dilators α and β for A are *equivalent* if $\alpha(x) = \beta(x)$ whenever $\mathbf{P}_A(x) > 0$.

If M is an RTM with input domain A then $\mathcal{B}_M(x)$ is a dilator for A.

Definition. Let A be a domain with a universe U. If δ is a dilator for A, then the *δ-dilation* of A is the domain $A * \delta$ such that:

- The universe of $A * \delta$ is the cartesian product of U and the set $\{0,1\}^*$ of all binary strings.
- $|(x,r)| = |x|$.
- If $r \in \delta(x)$ then $\mathbf{P}_{A*\delta}(x,r) = \mathbf{P}_A(x) \cdot 2^{-|r|}$, and otherwise $\mathbf{P}_{A*\delta}(x,r) = 0$.

The second clause here ensures that the notions like AP refer to the size of the actual input x, not the random string r. Use Lemma 5.1 to check that $\mathbf{P}_{A*\delta}$ is indeed a probability distribution. Even though $A * \delta$ may have infinitely many elements of a given size n, only finitely many of them have positive probability. Thus, $A * \delta$ is indeed a domain though its size function is not conventional. Notice that equivalent dilators give the same dilation. We will ignore the distinction between equivalent dilators.

Definition. Let A be a domain with a universe U. A *random function* on A is a partial function on $U \times \{0,1\}^*$ such that the function

$$\delta_f(x) = \{r : \ f \text{ is defined at } (x,r)\}$$

is a dilator. Two random functions f and g on A are *equivalent* if they have equivalent dilators and they coincide on all elements of the dilated domain which have positive probability.

In terms of Section 3, a random function f is an almost total function on $A * \delta_f$. Every RTM M with input domain A computes a random function on A which will be called F_M. The corresponding dilator is $\mathcal{B}_M$. We will ignore the distinction between equivalent random functions.

One may object to the term "random function" on the grounds that a random function on A should be a randomly chosen function on A. We fashioned the term "a random function" after well accepted terms like "a real function". A real function on a set A assigns real numbers to elements of A. A random function on a domain A can be seen as a function that assigns random objects to (almost all) elements of A.

Definition. A random function T from a domain A to $\bar{\mathcal{R}}^+$ is AP if the following function from $A * \delta_T$ to $\bar{\mathcal{R}}^+$ is AP:

$$(x,r) \mapsto [\text{ if } T \text{ is defined at } (x,r) \text{ then } T(x,r) \text{ else } \infty].$$

Definition. A randomized Turing machine M with input domain A is AP-*time* if the computation time of M is AP.

A randomized Turing machine M with input domain A can be viewed as a deterministic machine with input domain $A * \mathcal{B}_M$. It is easy to see that M is AP-time if and only if the corresponding deterministic machine is AP-time.

Definition. A random function f on a domain A is AP-*time* if there exists an AP-time randomized Turing machine M that computes f, i.e., A is an input domain for M and $f = F_M$.

We proceed to define a composition of random functions.

Definition. Suppose that f is a random function from a domain A to a domain B and g is a random function on B. The *composition* $g \circ f$ of f and g is the random function h on A such that, for every $x \in A$,

- $\delta_h(x) = \{rs : \ r \in \delta_f(x) \text{ and } s \in \delta_g(f(x,r)),$
- if $r \in \delta_f(x)$ and $s \in \delta_g(f(x,r))$ then $h(x,rs) = g(f(x,r),s)$.

Lemma 5.2. The composition of random function is associative.

Proof. Suppose that f is a random function from a domain A to a domain B, g is a random function from B to a domain C, and h is a random function from a C to a domain D. It is easy to see that both $h \circ (g \circ f)$ and $(h \circ g) \circ f$ are equivalent to a random function k on A such that if $x \in A^+$ then $\delta_k(x)$ comprises strings rst with $r \in \delta_f(x)$, $s \in \delta_g(f(x,r))$, $t \in \delta_h(g(f(x,r),s))$ and $k(x,rst) = h(g(f(x,r),s),t)$. QED

Definition. Let f be a random function from a domain A to a domain B. Then B *dominates* A with respect to f, symbolically $A \leq_f B$, if B dominates $A * \delta_f$ with respect to f.

Theorem 5.1. Let f be a random function from a domain A to a domain B such that the random function $x \mapsto |f(x,r)|$ is AP with respect to A. Then the following statements are equivalent:

- $A \leq_f B$,
- For every AP random function T from B to $\bar{\mathcal{R}}^+$, the composition $T \circ f$ is AP.

Proof. Let $\alpha = \delta_f$ so that f is an almost total function on $A * \alpha$. Let β range over dilators for B and γ be the dilator for A such that

$$\gamma(x) = \{rs : \ r \in \alpha(x) \text{ and } s \in \beta(f(x,r))\}.$$

Let x range over elements of A, r range over $\alpha(x)$ and s range over $\beta(f(x,r))$. Define an almost total function $g(x,rs) = (f(x,r),s)$ from $A*\gamma$ to $B*\beta$; $g(x,t)$ is undefined unless t has the form rs. The function $|g(x,t)|$ is AP on $A*\gamma$. For, let ε witness that $|f(x,t)|$ is AP on $A*\alpha$; then

$$\sum_{x r s} |g(x,rs)|^{\varepsilon} \cdot \mathbf{P}_{A*\gamma}(x,rs)/|x| = \sum_{x r} |f(x,r)|^{\varepsilon} \cdot \mathbf{P}_{A*\alpha}(x,r)/|x| < \infty.$$

If $\delta_T = \beta$ then $\delta_{T\circ f} = \gamma$. The composition $T \circ f$ of the random function f from A to B and a random function T on B is the composition $T \circ g$ of the (deterministic) almost total function g from $A * \gamma$ to $B * \beta$ and the (deterministic) almost total function T on $B * \beta$. It suffices to prove that, for every β, the following statements are equivalent:

1. $A * \alpha \leq_f B$,
2. $A * \gamma \leq_g B * \beta$,
3. For every almost total AP function T on $B * \beta$, $T \circ g$ is an almost total AP function on $A * \gamma$.

(2) and (3) are equivalent by Theorem 3.1. It remains to prove that (1) and (2) are equivalent. It is easy to see that if a function π witnesses (1) in the sense of Theorem 3.1(D2) then any function ρ from $A * \gamma$ to $\bar{\mathcal{R}}^+$ such that $\rho(x, rs) = \pi(x, r)$ witnesses (2) in the sense of Theorem 3.1(D2). To prove the other direction, fix a function S that assigns an element of $\beta(x, r)$ to each pair (x, r). If a function ρ witnesses (2), then any function π from $A * \alpha$ to $\bar{\mathcal{R}}^+$ such that $\pi(x, r) = \rho(x, rs)$, where $s = S(x, r)$, witnesses (1). QED.

The restriction on the size $|f(x, r)|$ is not needed to deduce the first statement of the Theorem 5.1 from the second one.

Definition. A random function f from a domain A to a domain B *reduces* A to B if it has the following two properties:

Efficiency: f is AP-time computable, and

Domination: B dominates A with respect to f.

We say that a domain A *reduces* to a domain B if some random function f reduces A to B.

Theorem 5.2. The reducibility relation is transitive.

Proof. Suppose that a random function f reduces a domain A to a domain B, and a random function g reduces B to a domain C. By Theorem 5.1, C dominates A with respect to the composition $h = g \circ f$. It remains to check that h is AP-time computable with respect to A. Let x range over A, r range over $\delta_f(x)$, $y = f(x, r)$ and s range over $\delta_g(y)$. The time to compute $h(x, rs)$ splits into two parts: The time to compute $f(x, r)$ from (x, r), and the time $t(x, rs)$ to compute $g(y, s)$ from (y, s). Since f is AP-time with respect to A, it suffices to prove that t is AP with respect to A. We know that the time $T(y, s)$ needed to compute $g(y, s)$ from (y, s) is AP with respect to B. Obviously, $t(x, rs) = T(y, s)$. Viewing t and T as random functions, we have $t = T \circ f$. By Theorem 5.1, $t(x)$ is AP. QED

6. Randomizing reductions of problems

It turns out to be useful to weaken the correctness property of (randomizing) reductions of decision and search problems [VL, BCGL, IL]. Here is one possible definition of reductions of RDPs.

Definition. Let Π_1, Π_2 be randomized decision problems with domains A, B and characteristic functions χ_1, χ_2 respectively. A random function f from A to B is a *reduction* of Π_1 to Π_2 if f reduces A to B and satisfies the following additional requirement:

Correctness There exists a number $a > 1/2$ such that, for every $x \in A^+$, the probability of the event $\chi_1(x) = \chi_2(f(x,r))$ is at least a.

The notion of AP-time decidability does not fit well the notion of randomizing reductions with a correctness guarantee $a < 1$. If we combine in the obvious way a randomizing reduction of Π_1 to Π_2 and a decision algorithm for Π_2, the result is a randomizing algorithm for Π_1 which is not guaranteed to be correct all the time. Accordingly, we generalize the notion of AP-time decidability. Say that a randomizing Turing machine M solves an RDP Π with a correctness guarantee a if, for every $x \in D^+$, the probability that M computes $\chi(x)$ is at least a.

Definition. A randomized decision problem Π is RAP-*time decidable* if there exists a, $1/2 < a \leq 1$, such that some randomizing Turing machine solves Π with correctness guarantee a.

Lemma 6.1. Let $1/2 < a < b < 1$ and let Π be a randomized decision problem. If there exists a randomizing AP-time Turing machine that solves Π with correctness guarantee a then there exists a randomizing AP-time Turing machine that solves Π with correctness guarantee b.

Proof. Consider Bernoulli trials with probability a for success in a single trial and let k be the least number such that the probability of $> k/2$ successes in k trials is $\geq b$. Given an instance x of Π, repeat the a-correct procedure k times and output the majority answer (in the case of a tie, output any answer). QED

Remark. The situation is even better for search problems, where one needs only one successful attempt [VL, BCGL, IL]. The inequality $a > 1/2$ can be replaced by an inequality $a > 0$ in the case of search problems.

Theorem 6.1. If f reduces an RDP Π_1 to an RDP Π_2 and Π_2 is RAP-time decidable then Π_1 is RAP-time decidable.

Proof. Suppose that the correctness guarantee of f is a_1 and the correctness guarantee of the given RAP-time decision procedure for Π_2 is a_2. By Lemma 6.1, we may assume that a_2 is sufficiently large so that $a_1 a_2 > 1/2$. Consider the randomizing algorithm which, given an instance x of Π_1, first applies f to x, and then – if and when some instance $y = f(x)$ of Π_2 is obtained – applies the a_2-correct decision algorithm to y; of course, the two subcomputations use independent sequences of coin flips. This composite algorithm solves Π_1 with correctness guarantee $a_1 a_2$. QED

Unfortunately, our partially correct reductions of RDP's do not compose in a satisfactory way: If f reduces Π_1 to Π_2 with correctness guarantee a_1 and g reduces Π_2 to Π_3 with correctness guarantee a_2 then the correctness guarantee of $g \circ f$ is only $a_1 a_2$ which may be well below 1/2. (This difficulty does not arise for search problems.) The repetition technique for boosting the probability of correctness, which we applied to randomizing decision (and search) algorithms above, is not directly applicable to our many-one randomizing reductions. Repeating a randomizing reduction k times results in k outputs in the target domain, not one as in the definition of reduction. In other words, such a repetition is a version of Turing (or truth-table) reduction, not a many-one reduction. For simplicity, we spoke about constant correctness guarantees. This restriction can and should be relaxed [VL, BCGL, IL]. We hope to address elsewhere the issues arising from this.

Some randomized decision and search problems complete for RNP with respect to partially correct reductions can be found in [VL, BCGL]. Partially correct reductions play an important role in [IL].

Appendix. On deterministic domain reductions

Return to the motivation of deterministic domain reductions in the beginning of Section 3. The fact that functions T were allowed to have the value ∞ simplified the situation somewhat. It enabled us to derive (R0) from (R2). In this appendix, we give a version of Theorem 3.1 covering the case when functions T have only finite values. We use the notation of Section 3.

Let A, B be domains and f be a function that assigns an element of B to every element of A (including elements of zero probability). Consider the following version of the requirement (R2):

(R3) For every AP function T on B that takes only finite values, the composition $T \circ f$ AP on A.

Obviously, (R3) does not necessarily imply (R0). It turns out that (R3) does not necessarily imply (D1) either. Here is counterexample. Pick any domain A and any element $a \in A$ of positive probability. Let B be the domain such that (i) B has the same universe and the same size function as A, and (ii) $\mathbf{P}_B(a) = 0$ is and $\mathbf{P}_B(x)$ is proportional to $\mathbf{P}_A(x)$ for $x \neq a$. Finally, let f be the identity function. Obviously, (R3) holds and (D1) fails.

Let $\mu(x) = \mathbf{P}_A(x)$, $\nu(y) = \mathbf{P}_B(y)|\text{range}(f)$, $R = \{y : \nu(y) > 0\}$, $E = \{y : \nu(y) = 0 \text{ but } \mu[f^{-1}(y)] > 0\}$, and consider the following version of deterministic domain domination:

(D0) E is finite and there exists a function g from B to $\bar{\mathcal{R}}^+$ such that

- For all $y \in R$, $\mu[f^{-1}(y)] \leq g(y) \cdot \nu(y)$, and
- $g \circ f$ is AP with respect to A.

Theorem. If $|f(x)|$ is AP on A then the condition (D0) is necessary and sufficient for (R3).

Proof. First suppose (R3). Then E is finite: Otherwise, choose T such that T is zero outside E and $T(fx)$ is not AP, and get a contradiction. The desired $g(y) =$ [if $y \in R$ then $\mu[f^{-1}y]/\nu(y)$ else 0]. To check that $g \circ f$ is AP with respect to A, notice that g is AP with respect to B and use (R3).

Next suppose (D0) and let T be an AP function on B taking only finite values. Let m be the maximal value of the function $S(x) = T(fx)$ on E. For any ε,

$$\sum_{x \in E} (Sx)^{\varepsilon} \mu(x)/|x| \leq \sum_{x} m^{\varepsilon} \mu(x) < \infty.$$

Therefore, we have to prove only that, the restriction of S to $f^{-1}(R)$ is AP with respect to A. Without loss of generality, we can suppose that E is empty. Then (D0) implies (D2). By Theorem 3.1, (D2) implies (R2). By (R2), S is AP with respect to A. QED

References

[BCGL] Shai Ben-David, Benny Chor, Oded Goldreich and Michael Luby, "On the Theory of Average Case Complexity", Symposium on Theory of Computing, ACM, 1989, 204–216.

[Gu1] Yuri Gurevich, "Average Case Complexity", J. Computer and System Sciences (a special issue on FOCS'87) to appear.

[Gu2] Yuri Gurevich, "Matrix Decomposition Problem is Complete for the Average Case", Symposium on Foundations of Computer Science, IEEE Computer Society Press, 1990, 802–811.

[GJ] Michael R. Garey and David S. Johnson, "Computers and Intractability: A Guide to the Theory of NP-Completeness", Freeman, New York, 1979.

[IL] Russel Impagliazzo and Leonid A. Levin, "No Better Ways to Generate Hard NP Instances than Picking Uniformly at Random", Symposium on Foundations of Computer Science, IEEE Computer Society Press, 1990, 812–821.

[Le] Leonid A. Levin, "Average Case Complete Problems", SIAM Journal of Computing, 1986.

[VL] Ramarathnam Venkatesan and Leonid Levin, " Random Instances of a Graph Coloring Problem are Hard", Symposium on Theory of Computing, ACM, 1988, 217–222.

From Prolog Algebras Towards WAM– A Mathematical Study of Implementation

Egon Börger

Dipartimento di Informatica
Corso Italia 40
56100 Pisa, Italia
boerger@dipisa.di.unipi.it

Dean Rosenzweig

University of Zagreb
FSB, Salajeva 5
41000 Zagreb, Yugoslavia
fax: +38–41–514535

ABSTRACT. This is the first part of a work presenting a natural and transparent albeit entirely mathematical description of Warren's 1983 abstract machine for executing Prolog. We derive the description from Börger's 1990b phenomenological description of the language, refining Prolog algebras stepwise, proving conservation of correctness at each step.

INTRODUCTION

The dynamic algebra approach to semantics, being mathematical and abstract yet transparent and close to programmer's operative understanding, has so far been successfully used for descriptions of Modula 2 (Gurevich & Morris 1988), Occam (Gurevich & Moss 1990), Prolog (Börger 1990a, b) and Prolog III (Börger & Schmitt 1991). There seems to be no reason, however, to limit it to phenomenological description of programming languages.

We undertake to demonstrate its suitability for describing implementation methods as well, particularly those which rely on a virtual machine concept. The benefits, obviously accruing from a precise mathematical description of implementation methods, are often perceived by implementors as wishful thinking, in view of heavy formalism and/or mathematics involved in semantical methodologies, and combinatorial explosion thereof upon application to non–toy problems. Dynamic algebra approach, however, seems to hold some promise of changing this situation.

We present here what seems to be not only a natural and transparent albeit entirely mathematical description of a standard implementation method—Warren's 1983 abstract machine for executing Prolog—but also its derivation cum correctness proof by stepwise refinement from a phenomenological description of the language. One is almost tempted to speak of a rational reconstruction of the WAM. It seems to us that the mathematical beauty of the WAM gets revealed in the process.

By Prolog we do not mean here something else, such as Horn clauses, but essentially the full language as emerging from the draft standard proposal DFPS 1990, with its most important extralogical constructs—call, metacall, cut, asserta, assertz, retract—and derive an 'implementation' incorporating the usual optimizations (early discarding of environments and choicepoints, determinacy detection, virtual copying of dynamic code, environment trimming). This work is intended to proceed beyond the abstraction level of the WAM as presented by Warren 1983—for instance, to a set of 200 instructions used for ZGLOG 386 Prolog compiler by Zglog Ltd of Zagreb, providing a framework for unambiguous and compact internal documentation for that compiler (debugger and garbage collector included) and its further development.

We proceed by setting up successive dynamic algebras, starting from Prolog algebras of Börger 1990b. Each of them is a full Prolog machine; we isolate however different orthogonal components and optimizations of the WAM and make them explicit by elaborating them into more concrete representations separately, one at a time, leaving the rest (as) 'abstract' (as before), which corresponds nicely to compiling different structures of a Prolog program into (pseudo) WAM code. (Operational semantics of) Prolog could be seen as decomposed into conjunctive, disjunctive and unificational components, corresponding to WAM handling of continuations, choicepoints and term representation, reflected by compilation of the structure of clauses, predicates and terms. We proceed in that order. It could have been partially permuted, by elaborating the structure of predicates before that of clauses. We would not however elaborate the structure of terms first (like Aït–Kaci 1990 in an informal tutorial setting, who proceeds by successive partial Prologs unlike our successive partial elaborations of full Prolog). It seems that the conjunctive and disjunctive components of Prolog and WAM deserve an abstract explication, unburdened at first by details and constraints of term representation. Our explication should thus essentially apply to WAM–based implementation of other logic programming languages with the same and–or structure as Prolog, such as Prolog III or Protos–L (Semle 1989)*.

The paper is organized as follows.

In Section 1 we adapt Gurevich's 1988 notion of dynamic algebra.

In Section 2 we refine Börger's 1990b phenomenological description of Prolog by elaborating the conjunctive component, as reflected by compilation of clause structure into (pseudo) WAM code. It is here that we introduce the last call optimization.

In Section 3 we assume compilation of predicate structure into code, including early discarding of choicepoints and allowing for determinacy detection. Virtual copying of dynamic code (Lindholm & O'Keefe 1987) is included at this stage.

A Section 4, assuming compilation of term structure, thus filling the gap to a usual WAM picture, is left for a sequel to this paper (Börger & Rosenzweig 1991).

The intention of proceeding really stepwise, splitting even steps 2 and 3 into three substeps each, is to have nothing to prove—preservation of correctness (in sense of PROOF of Section 1) at each refinement step is intuitively obvious and follows formally

* **Added in proof.** In February 1991 we became acquainted with the formidable correctness proof for the WAM wrt a form of SLD resolution for Horn clauses by Russinoff 1989. It seems that his conceptual framework would not be easily extendible to account for extralogical predicates, database operations etc.

by perfectly straightforward inductive verification. A major benefit of the method seems to be freedom to choose precisely the right level of abstraction for a rigorous discussion of an implementor's fine point.

Although the paper starts from basic principles, the reader is assumed to have a general understanding of Prolog as a programming language. Acquaintance with Börger 1990b and/or Warren 1983 or Aït–Kaci 1990 tutorial would be helpful, though technically not required. The view of Prolog adopted is that of draft standard proposal DFPS 1990—conjunction and disjunction are transparent for the cut, while call is not, with a resatisfiable retract. Adopting alternative (clear) views, along the lines of Börger 1990a, would be perfectly straightforward and we do not care to discuss it here.

1. DYNAMIC ALGEBRAS

The notion of **dynamic algebra** (Gurevich 1988) allows to capture the intuition of a finite system (abstract machine) with certain structure, evolving in time according to specified rules. Evolution of the system may encompass creation and destruction of objects.

More formally, it involves a (many sorted, for our purposes first order) finite signature of **universes** (possibly empty, growing and shrinking in time) and **functions** (partial, redefinable at any time—constants are nullary functions), together with fixed **transition rules** of form

if *condition*
then
  *update*1
  ⋮
  update$k$,

where *condition* is a boolean expression the truth of which (in a given algebra) triggers simultaneous execution of all the updates in the rule, yielding a new algebra of the same signature. All dynamic algebras in this paper are going to be deterministic, in the sense of at most one transition rule being applicable (without conflicting updates). Rules in which variables (ranging over some of the universes) appear are treated in the usual way, as schemata.

Updates may be of two kinds—**function updates** of form

$$f(t1, \ldots, tn) := t,$$

where everything in sight is a term of the signature, meaning a redefinition of function

f at one point (so that our constants are variable), or **universe updates** of form

$$\begin{array}{ll} \text{EXTEND} & D1 \text{ by } temp(1,1),\ldots,temp(1,t(1)) \\ & \vdots \\ & Dl \text{ by } temp(l,1),\ldots,temp(l,t(l)) \\ \text{WITH} & F1 \\ & \vdots \\ & Fk \\ \text{ENDEXTEND}, & \end{array}$$

where $D1,\ldots,Dl$ are universes, l a number, $t(1),\ldots,t(l)$ terms, and $F1,\ldots,Fk$ function updates in which names $temp(i,j)$ may occur, with parameters i, j. To apply such a universe update to a given dynamic algebra means to

- evaluate $t(1),\ldots,t(l)$ to numbers $n1,\ldots,nl$;
- create a new object $temp(i,j)$ of Di for each $1 \leq i \leq l$, $1 \leq j \leq ni$;
- simultaneously execute the function updates $F1,\ldots,Fk$ for each $1 \leq i \leq l$, $1 \leq j \leq ni$.

This kind of update is called **universe extension***—we shall not need here another kind of universe update, **universe contraction**, allowing destruction of objects. Thus,

DEFINITION. A **dynamic algebra** of a given signature is a pair (A,T) consisting of a finite many sorted partial first order algebra A and a finite set T of transition rules of the signature.

In the rest of this section we shall develop a few examples of (classes of) dynamic algebras, which might assist the reader's understanding of dynamic algebra approach to semantics, serving at the same time as building blocks for subsequent sections.

EXAMPLE 1. A **predecessor algebra** is $(D; top, bottom; pred)$, where

- D is a universe
- $top, bottom : D$ are nullary functions
- $pred : D \rightarrow D$ a unary function,

---

* For the reader who cares about full precision, a few technicalities:

In practice it is convenient to allow names of form $temp(u,v)$ to occur in function updates of a universe extension, where u, v are (arithmetical) terms which may contain parameters i, j. The updates are then understood as being executed only for those values of i, j for which u, v also fall within range, $1 \leq u \leq l$, $1 \leq v \leq t(u)$.

In certain obvious cases below (some occurrences of) parameters i, j will be suppressed from our notation.

Simultaneous extension of several universes is easily reduced to the case of extending only one universe at a time, following a suggestion of Gurevich 1991, by extending a universal domain, simulating extension of subdomains by updating characteristic functions.

which may be depicted as follows.

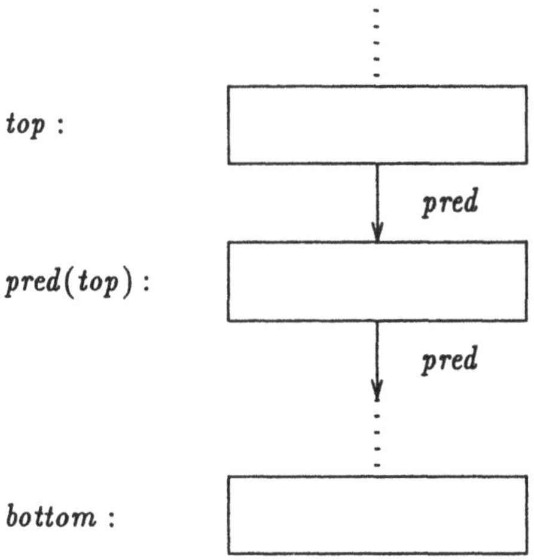

In our applications predecessor algebras will occur as subalgebras; in such a situation compound terms might play the role of top or bottom.

A typical example of predecessor algebra, which will be the basis for our Prolog algebras, is

$$(STATE; \mathit{currstate}, \mathit{nil}; \mathit{choicepoint})$$

with 'backtracking' rule

if *currstate* $\neq$ *nil*
then *currstate* := *choicepoint*(*currstate*).

In presence of such a rule—with possibly enlarged condition—the condition *currstate* = *nil* is therefore a natural termination criterion, in accordance with the above picture.

That picture, however, tells more than is in general implied by the definition. In fact, by reversing the arrows we can think of a predecessor algebra as a successor algebra; an example is a counter or clock

$$(INDEX; \mathit{vindex}, 0; \mathit{succ})$$

which we shall use for variable renaming, with *vindex* representing a fresh variable renaming, together with a function

$$\mathit{rename} : TERM \times INDEX \to TERM.$$

A typical update will be of form *vindex* := *succ*(*vindex*).

If we wish to care about resource bounds, we might introduce a *maxindex* : *INDEX*, assuming *succ*(*maxindex*) undefined. If we then want to enlarge the resource, we can do that explicitly by the following rule:

$$\begin{array}{l}\textbf{if } vindex = maxindex\\ \textbf{then}\\ \quad \text{EXTEND } INDEX \text{ by } temp\\ \quad \text{WITH}\\ \qquad maxindex := temp\\ \qquad \text{succ}(vindex) := temp\\ \quad \text{ENDEXTEND.}\end{array}$$

EXAMPLE 2. Predecessor algebras often come decorated with one or more decorating functions

$$dec : D \rightarrow DECORATION$$

which may be seen as filling the boxes of our picture with some content from *DECORATION*s. This leads us to define a **linked list algebra** as

$$(LOCATION, VALUE; top, bottom; -, val)$$

where

$$(LOCATION; top, bottom; -)$$

is a predecessor algebra, and

$$val : LOCATION \rightarrow VALUE$$

a decorating function.

By restricting the form of permissible updates, as well as introducing possible additional assumptions about the linking function '–', we can impose different disciplines of access to and modification of a linked list algebra, yielding the notions of a stack, buffer or queue, circular buffer, ... algebra.

An example is a **stack algebra**, where the underlying predecessor algebra will be modifiable by updates of form

$$\begin{array}{l}\text{PUSH}(v) \equiv \text{EXTEND } LOCATION \text{ by } temp\\ \qquad\qquad\quad \text{WITH}\\ \qquad\qquad\qquad top := temp\\ \qquad\qquad\qquad temp- := top\\ \qquad\qquad\qquad val(temp) := v\\ \qquad\qquad\quad \text{ENDEXTEND}\end{array}$$

$$\text{POP} \equiv top := top-.$$

In the sequel we shall often refer to certain linked list algebras as 'stacks', although their updates might differ slightly from the above.

The *VALUE* universe of a linked list algebra will often be suppressed, being present only implicitly, as a superset of the (cartesian product of) codomain(s) of (may be several) decorating function(s). The signature will in such a case take the form

$$(LOCATION; top, bottom; -; val_1, \ldots, val_n)$$

together with codomains of the decorating functions val_i, $1 \leq i \leq n$.

EXAMPLE 3. A **pointer algebra** is a linked list algebra with an additional successor function '+' inverse to '−', i.e.

$$(LOCATION, VALUE; currentpointer, end; +, -; val)$$

with the integrity constraint $l{+}{-} = l = l{-}{+}$ for each location $l \neq end$.

If a stack algebra is seen as embedded in a pointer algebra, the above PUSH update can be realized without universe extension, simply updating *currentpointer*, i.e. by

$$\begin{aligned} \text{PUSH}(v) \equiv\ & currentpointer := currentpointer+ \\ & val(currpointer+) := v. \end{aligned}$$

It must be noted, however, that in a stack algebra, information once popped is never destroyed by subsequent pushing (since each time a new element is created). If a stack is embedded in a pointer algebra, popped information may be overwritten by subsequent pushing, and thus irretrievably lost.

EXAMPLE 4. For many domains D we shall use the associated **list domain**, with universe D^* and the usual list functions

$$fst : D^* \to D \qquad rest : D^* \to D^*$$
$$length : D^* \to INTEGER \qquad nth : D^* \times INTEGER \to D$$

relying on Prolog notation for lists. A simple example are the clause analyzing functions

$$\begin{aligned} hd &: CLAUSE \to TERM \\ bdy &: CLAUSE \to TERM^*. \end{aligned}$$

When D is a domain containing Prolog goals, we shall sometimes take the liberty of suppressing the obvious function which transforms a list to an iterated conjunction, replacing recursively [|] by ','.

EXAMPLE 5. The following example of a pointer algebra will play a role in Prolog algebras of the next section.

$$(CODEAREA, INSTR; p, bottom; +, -; code),$$

where universe *INSTR* contains all "instructions" of form *Unify*(H), *Call*(G), *Proceed*, with H, G arbitrary Prolog terms (*INSTR* will be tacitly enlarged in subsequent sections, to contain any further instructions we choose to introduce, such as *Allocate*, *Deallocate*, *Execute*(G), ...).

We assume a Prolog clause to be 'compiled' into pseudo WAM code by a function

$$compile : TERM \to INSTR^*$$

such that

$$compile(H \text{:-} G1, \ldots, Gn) = [Unify(H), Call(G1), \ldots, Call(Gn), Proceed],$$

and 'loaded' into *CODEAREA* by a function

$$load : INSTR^* \rightarrow CODEAREA,$$

which is best explained by setting requirements on its left inverse

$$unload : CODEAREA \rightarrow INSTR^*$$

such that

$$unload(ptr) = \textbf{if} \;\; code(ptr) = Proceed \;\; \textbf{then} \; [Proceed] \;\; \textbf{else} \;\; [code(ptr) \mid unload(ptr+)]$$

for any $ptr : CODEAREA$, and

$$unload(load(List)) = List.$$

for any $List : INSTR^*$.

Representing the database operations will also require a function

$$decompile : CODEAREA \times INDEX \rightarrow TERM$$

of which we assume that

$$decompile(load(compile(Clause)), Vi) = rename(Clause, Vi)$$

for any Prolog clause $Clause : CLAUSE \subseteq TERM$ and index $Vi : INDEX$. The *decompile* function could have been introduced more stepwise, using *unload* and *rename*, but it is the above form that we shall need.

It might be remarked that the bottom of this pointer algebra (and of other linked list algebras occurring as subalgebras of our Prolog algebras) will play a role in our treatment of overflow and and garbage collection, which we intend to develop in sequels to this paper. Here we shall use the above pointer algebra only as a one way pointer structure ('−' will namely not be used until, in Börger & Rosenzweig 1991, we come to deal explicitly with term representation).

2. PROLOG ALGEBRAS WITH COMPILATION OF CLAUSE STRUCTURE

In this section we introduce our notion of Prolog algebra, elaborating the conjunctive component of Prolog. Clauses will be compiled into sequences of instructions, 'abstract' in the sense of (still) containing goals in form of Prolog terms, as in Example 5 of Section 1. Clauses of a Prolog program will be represented by pointers to 'code' (instruction sequences), together with variable indexes and substitutions, and predicates will be represented as lists of pointers to (compiled and loaded) clauses. The notion of correctness, connecting our framework with Prolog algebras of Börger 1990b, will come forward naturally from the homomorphism of PROOF 2.1 below.

In 2.1 the compilation is very simple minded, yielding a *Unify* instruction for the clause head, and a sequence of *Call* instructions for the body, terminated by a *Proceed* instruction. A state of Prolog computation will be represented by a backtracking stack and a stack of continuations (environments). An enviroment will be implicitly allocated for every call by a selection rule.

In 2.2 every clause will explicitly allocate and deallocate an environment, with *Allocate* and *Deallocate* instructions. Selection rule will thus be alleviated of that burden, while the job of goal success rule will be split between new rules *Deallocate* and *Proceed*, corresponding explicitly to instructions executed.

In 2.3 early discarding of environments (last call optimization) is introduced, calling forth a new instruction *Execute*. Environments will now be allocated only when needed, and deallocated as soon as possible.

At every step the new dynamic algebra will be introduced as a refinement of the previous one, almost obviating the need for a correctness proof.

2.1. Simple minded compilation of clauses

SIGNATURE. The core of a **Prolog algebra** is given by two linked list algebras (we shall refer to them as 'stacks', although their transition rules differ slightly from those of a stack algebra)

$$(STATE; currstate, nil; b; g, s, p, vi, ct, e)$$
$$(ENV; e(currstate), nil; ce; p', vi', ct'),$$

where, for a given element of $STATE$,

g	yields a goal from $GOAL \subseteq TERM$
s	a substitution from $SUBST$
p	a program pointer from $CODEAREA$
vi	a variable renaming index from $INDEX$
ct	a cutpoint from $STATE$
e	environment (continuation) from ENV,

and p', vi', ct' assume, for an environment, the same roles and codomains as their homonyms p, vi, ct for a state. The state stack is linked by a function b (for backtracking), while the environment is linked by ce (for continuation environment). The environment stack at present represents just the continuation, i.e. the sequence of goalsequences to be executed once the current goalsequence succeeds, and the state stack holds alternative states for the case that current goal fails. The current goalsequence is represented by $p(currstate)$, $vi(currstate)$, $s(currstate)$, while $g(currstate)$, the current instantiated goal, serves just for unification. To be more precise, the current goalsequence is obtainable by decompiling and instantiating, as

$$subres(decompile(p(currstate), vi(currstate)), s(currstate)),$$

cf. below.

In the signature of a Prolog algebra we meet some old friends, the pointer algebra

$$(CODEAREA, INSTR; p(currstate), bottom; +, -; code)$$

of Example 5, together with associated functions *compile*, *decompile*, *load*, and the counter algebra

$$(INDEX; vindex, 0; succ)$$

from Example 1 of Section 1. Note that, while *vindex* is 'global', i.e. a constant, the program pointer $p(currstate)$ is given by a unary function, i.e. is accessible only from a state.

The universe *SUBST* comes with substitution applying and composing functions

$$subres : TERM \times SUBST \rightarrow TERM$$
$$\circ : SUBST \times SUBST \rightarrow SUBST$$

and a unification algorithm, abstractly encoded by a function

$$unify : TERM \times TERM \rightarrow SUBST + \{nil\},$$

which we assume to yield the unifying substitution of two *TERM*s, or *nil* if none exists.

Additionally, a Prolog algebra contains a universe *PROGRAM*, which comes with the following functions

$$db : PROGRAM$$
$$procdef : GOAL \times PROGRAM \rightarrow CODEAREA^*$$
$$insert : INSTR^* \times PROGRAM \rightarrow PROGRAM$$
$$delete : CODEAREA \times PROGRAM \rightarrow PROGRAM$$

of which we assume that

- *db* represents the current database;
- $procdef(Goal, Db)$ yields the sequence of pointers to compiled clauses for the *Goal* from *Db* (this sequence must contain pointers to at least all those clauses with heads unifiable with the *Goal*);
- $insert(List, Db)$ inserts $List = compile(Head :- Body)$ into *Db* in such a way that $procdef(Head, insert(List, Db))$ finds the pointer $load(List)$ inserted into $procdef(Head, Db)$ at the appropriate place;
- $delete(CodePointer, Db)$ deletes the code sequence pointed at by *CodePointer* from *Db*, in such a way that, if $code(CodePointer) = Unify(Head)$, then the list $procdef(Head, delete(CodePointer, Db))$ does not contain *CodePointer* any more.

Insert should be understood as a paradigmatic representative of a family of *insert* functions, coming with corresponding *assert* rules below, parameterized by a notion of 'appropriate place', with usual cases being '*a*' and '*z*'. We deliberately keep the structure of the universe *PROGRAM* abstract; that will enable us to refine our algebra later with minimal modification to the signature (for the initiated, we might add *abolish*, *predicate_set* and modules by imposing additional structure on *PROGRAM*).

Finally, we have to add constants *error* : *ERRORCONDITION* and *stop* : $\{0, 1, -1\}$, so that a value of 0 indicates absence of error or stopping conditions, a stop value of 1 (-1) success (failure) of a Prolog query, and a function *is_user_defined* : $TERM \rightarrow BOOL$, which recognizes those terms whose functors may be user defined predicates.

To summarize, we could depict a Prolog algebra as follows.

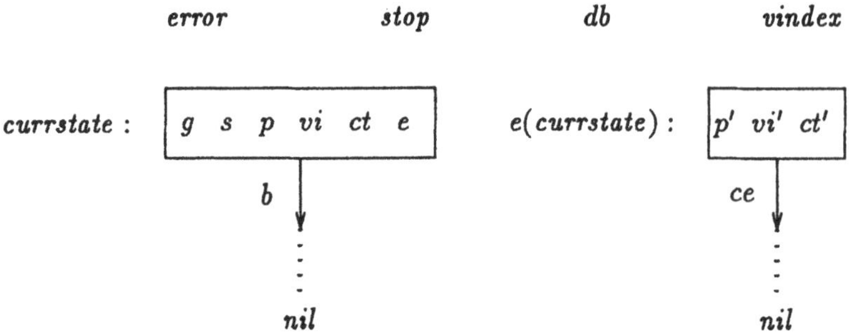

ABBREVIATIONS

$g \equiv g(currstate)$ $\quad p \equiv p(currstate)$ $\quad s \equiv s(currstate)$
$vi \equiv vi(currstate)$ $\quad ct \equiv ct(currstate)$
$e \equiv e(currstate)$ $\quad b \equiv b(currstate)$

$r\_code \equiv rename(code(p), vi)$
$i\_code \equiv subres(r\_code, s)$
(where an instruction containing a term is taken to be renamed or instantiated by renaming or instantiating that term).

ALLOCATE *temp* IN *updates* ENDALLOC $\equiv$ EXTEND *ENV* by *temp*
WITH $p'(temp) := p+$
$vi'(temp) := vi$
$ct'(temp) := ct$
$ce(temp) := e$
updates
ENDEXTEND

Such an update creates, on top of the current environment, a new one for the continuation which points to the instruction following the currently executed one, and saves the current variable renaming index and cutpoint.

ALLOCATE *temp*
PUSH $temp(1), \ldots, temp(l)$ IN
updates
ENDPUSH

$\equiv$ EXTEND *ENV* by *temp*
STATE by $temp(1), \ldots, temp(l)$
WITH
$p'(temp) := p+$
$vi'(temp) := vi$
$ct'(temp) := ct$
$ce(temp) := e$

$currstate := temp(1)$
$b(temp(i)) := temp(i+1)$
$b(temp(l)) := b$

updates
ENDEXTEND

(PUSH $temp(1), \ldots, temp(l)$ IN *updates* ENDPUSH will stand for the obvious simplification of the above, dropping all mention of ENV and its elements).

$$Call\_code(G) \equiv load([Call(G), Proceed])$$
$$Conjunction\_code(G, H) \equiv load([Call(G), Call(H), Proceed])$$
$$Retract\_code(Ptr, Vi) \equiv load([Unify(decompile(Ptr, Vi)), Erase(Ptr), Proceed])$$

$$\text{OK} \equiv (error = 0 \quad \& \quad stop = 0)$$
$$all\_done \equiv (code(p) = Proceed \quad \& \quad e = nil)$$
$$backtrack \equiv currstate := b$$

We will adopt a form 'LET $s = t$' in the usual way, to introduce s as a notation for t.

INITIAL STATE

A Prolog query $G1, \ldots, Gn$ with a program $[C_1, \ldots, C_m]$ will initialize a Prolog algebra with following values.

$error = 0$	$stop = 0$	$db = db0$
$vindex = succ(0)$	$g = G1$	$s = Id$

$p = load([Call(G1), \ldots, Call(Gn), Proceed])$

$vi = 0$	$ct = nil$	$b = nil$

$p'(e) = load([Proceed])$

$vi'(e) = 0$	$ct'(e) = nil$	$ce(e) = nil$

with *Id* being the identity substitution. Initialization of *vindex* ensures that it will be fresh for all states considered later on. (Initializing e with *nil* might be a more natural choice; the reason for the above initialization is the wish to facilitate the equivalence proof of 2.2). We have yet to explain *db0*. If its universe *PROGRAM* were defined as $CODEAREA^*$, we would like to say that

$$db0 = [load(compile(C_1)), \ldots, load(compile(C_m))].$$

Since we chose not to say so, we have to pay the price of our generality, and to require of *db0*:

For any goal $G : TERM$, if $[C_{i1}, \ldots, C_{ik}]$ is the subsequence of clauses for G's predicate, then $procdef(G, db0)$ will return a list of pointers $[p_{i1}, \ldots, p_{ik}]$ such that for any $Vi : INDEX$,

$$decompile(p_{ij}, Vi) = rename(C_{ij}, Vi), \quad 1 \leq j \leq k.$$

RULES

query success

if OK
  & *all_done*
then
  $stop := 1$

query failure

if OK
  & $currstate = nil$
then
  $stop := -1$

goal success

```
if    OK
    & code(p) = Proceed
    & NOT (all_done)
then
    p := p'(e)
    ct := ct'(e)
    vi := vi'(e)
    e := ce(e)
```

backtracking

```
if    OK
    & i_code = Call(G)
    & is_user_defined(G)
    & procdef(G, db) = [ ]
then
    backtrack
```

selection

```
if    OK
    & i_code = Call(G)
    & is_user_defined(G)
    & procdef(G, db) ≠ [ ]
then
    LET cl_list = procdef(G, db)
    LET l = length(cl_list)
        ALLOCATE temp
        PUSH temp(1), ..., temp(l) IN
            g(temp(i)) := G
            s(temp(i)) := s
            p(temp(i)) := nth(cl_list, i)
            vi(temp(i)) := vindex
            ct(temp(i)) := b
            e(temp(i)) := temp
        ENDPUSH
        vindex := succ(vindex)
```

Unify

```
if    OK
    & r_code = Unify(G)
    & unify(g, G) : SUBST
then
    s := s o unify(g, G)
    p := p+
```

```
if    OK
    & r_code = Unify(G)
    & unify(g, G) := nil
then
    backtrack
```

call

```
if     OK
    &  i_code = Call(call(G))
then
    ALLOCATE temp IN
        p := Call_code(G)
        ct := b
        e := temp
    ENDALLOC
```

cut

```
if     OK
    &  i_code = Call(!)
then
    p := p+
    b := ct
```

true

```
if     OK
    &  i_code = Call(true)
then
    p := p+
```

fail

```
if     OK
    &  i_code = Call(fail)
then
    backtrack
```

conjunction

```
if     OK
    &  i_code = Call((G, H))
then
    ALLOCATE temp IN
        p := Conjunction_code(G, H)
        e := temp
    ENDALLOC
```

disjunction

```
if     OK
    &  i_code = Call((G1; G2))
then
    ALLOCATE temp
    PUSH temp(1), temp(2) IN
        s(temp(i)) := s
        p(temp(i)) := Call_code(Gi)
        vi(temp(i)) := vi
        ct(temp(i)) := ct
        e(temp(i)) := temp
    ENDPUSH
```

assert

```
if     OK
    &  i_code = Call(assert(G))
    &  is_user_defined(G)
then
    p := p+
    db := insert(compile(G), db)
```

retract_selection

if OK
& $i\_code = Call(retract(H, B))$
& $is\_user\_defined(H)$
& $procdef(H, db) \neq [\ ]$
then
LET $cl\_list = procdef(H, db)$
LET $l = length(cl\_list)$
ALLOCATE $temp$
PUSH $temp(1), \ldots, temp(l)$ IN
$g(temp(i)) := H :- B$
$s(temp(i)) := s$
$p(temp(i)) := Retract\_code(nth(cl\_list, i), vindex)$
$vi(temp(i)) := vindex$
$e(temp(i)) := temp$
ENDPUSH
$vindex := succ(vindex)$

retract_failure

if OK
& $i\_code = Call(retract(H, B))$
& $is\_user\_defined(H)$
& $procdef(H, db) = [\ ]$
then
backtrack

Erase

if OK
& $code(p) = Erase(Ptr)$
then
$p := p+$
$db := delete(Ptr, db)$

NOTES

- **query rules** are obvious, $currstate = nil$ means that backtracking has found no more alternative states to backtrack to; **goal success** rule passes control to continuation, popping environment stack, retaining substitution accumulated so far;
- **selection rule** copies the relevant part of the database to the state stack, setting up a new state for each alternative, loading and updating global *vindex* to ensure unique renaming; the copying ensures 'logical view' of database operations (cf. Lindholm & O'Keefe 1987);
- **Unify rules** are not used here for the unifiability predicate of Prolog, as in Börger 1990b, where this predicate is used to impose unification of clause head; they could of course easily be used to describe that predicate;
- **call** is not transparent for the *cut*, adhering to DFPS 90—it could be made transparent by simply dropping the *ct* update; the correctness of our rule at this level depends on composition of substitutions being idempotent;
- **assert** is paradigmatic for *asserta*, *assertz* and any other *assert* we might desire, relying on *insert* to select appropriate place;
- **retract** is resatisfiable, creating an alternative state for each candidate clause—*retract_selection* is just a rewriting of *selection* rule (with $p(temp(i))$ pointing not

to code for execution of clauses, but for their unification), and *retract_failure* of *backtracking*; cutpoint *ct* is not updated, being irrelevant for *Retract_code*; **Erase** is auxiliary to *retract*, performing the actual deletion.

To keep the paper within reasonable limits, we do not consider here the rules for the remaining built–in predicates of Prolog (as proposed by DFPS 1990), which are described by extensions of Prolog algebras in Börger 1991; as one can see from there, there is nothing of particularly 'logical' character in predicates for manipulating files, terms, arithmetics and input–output.

PROOF (of correctness wrt Börger 1990b)

Let *RESSTATE*, *DECGLSEQ*, *GOAL*, *decglseq*, *choicepoint* be the universes and functions of Börger 1990b. We define functions

$$F : STATE \to RESSTATE$$
$$G : ENV \times SUBST \to DECGLSEQ$$

by simultaneous recursion (note that the elements of *RESSTATE* are characterized by their decoration through *decglseq* and by their *choicepoint* value)

$$F(nil) = nil$$
$$G(nil) = nil$$
$$decglseq(F(St)) = [\langle goal\_seq(p(St), vi(St), s(St), g(St)), F(ct(St))\rangle \mid G(e(St), s(St))]$$
$$choicepoint(F(St)) = F(b(St))$$
$$G(E, S) = [\langle goal\_seq'(p'(E), vi'(E), S), F(ct'(E))\rangle \mid G(ce(E), S)]$$

$$\text{for} \quad STATE : St \neq nil \neq E : ENV, \quad S : SUBST,$$

where the auxiliary functions

$$goal\_seq : CODEAREA \times INDEX \times SUBST \times GOAL \to GOAL$$
$$goal\_seq' : CODEAREA \times INDEX \times SUBST \to GOAL$$

are defined by cases:

$$goal\_seq(Ptr, Vi, S, G) = \begin{cases} true & \text{if } code(Ptr) = Proceed \\ (G = H, decompile(Ptr+, Vi)) & \text{if } code(Ptr) = Unify(H) \\ subres(decompile(Ptr, Vi), S) & \text{otherwise,} \end{cases}$$

and $goal\_seq'(Ptr, Vi, S)$ is defined likewise, using just the first and the third clause of *goal_seq*.

Since our rules correspond 1–1 to rules of Börger 1990b (up to a trivial decomposition of his *stop* rule to our *query success* and *query failure*), it is straightforward to verify that F preserves initial and stopping states, and commutes with homonymous rules. Hence (by induction) an initial state *s_init* eventually reaches a stopping *s'_stop* iff $F(s\_init)$ eventually reaches $F(s\_stop)$. If there is a nice notion of homomorphism of dynamic algebras, F is an example.

2.2. Explicit allocation and deallocation

Instead of allocating a common environment for every choicepoint in selection rule, every (alternative) clause will explicitly allocate and deallocate, in order to enable later optimizations. We thus introduce *Allocate* and *Deallocate* instructions for that purpose. A clause

$$H :\!- G1, \ldots, Gn$$

thus compiles to

$$[\mathit{Allocate}, \mathit{Unify}(H), \mathit{Call}(G1), \ldots, \mathit{Call}(Gn), \mathit{Deallocate}, \mathit{Proceed}].$$

STATE will get its own continuation pointer *cp*(*currstate*), continuation varindex *cvi*(*currstate*) and continuation cutpoint *cct*(*currstate*) to save continuation info while fussing with environments. We rename the environment codepointer $p'(e)$ to $cp'(e)$, since that suits better its purpose of pointing to continuation, as well as standard WAM usage.

In the sequel $\longleftarrow$ will indicate changes from previous version.

SIGNATURE is retained from 2.1, with renaming p' and additions:

$cp : STATE \rightarrow CODEAREA$ $\qquad$ $cp' : ENV \rightarrow CODEAREA$ $\longleftarrow$
$cvi : STATE \rightarrow INDEX$
$cct : STATE \rightarrow STATE$

so that a Prolog algebra can now be depicted as

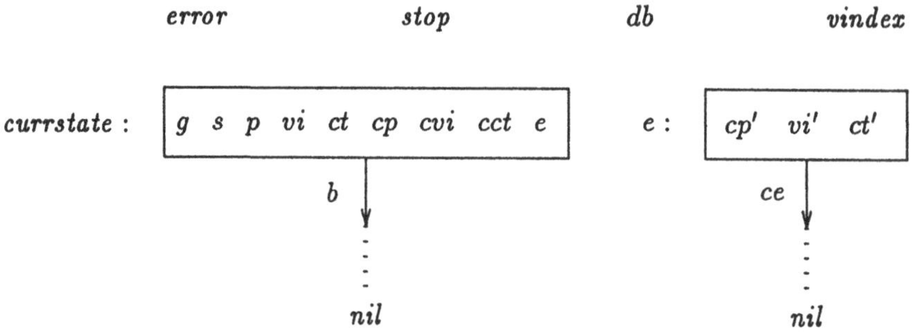

ABBREVIATIONS are retained from 2.1, except for

$cp \equiv cp(currstate) \quad cvi \equiv cvi(currstate) \quad cct \equiv cct(currstate)$ (new)
$all\_done \equiv (code(p) = Proceed \;\; \& \;\; code(cp) = Proceed)$ $\longleftarrow$
$Call\_code(G) \equiv load([Allocate, Call(G), Deallocate, Proceed])$ $\longleftarrow$
$Conjunction\_code(G, H) \equiv load([Allocate, Call(G), Call(H), Deallocate, Proceed])$ $\longleftarrow$
$succeed \equiv p := p+$

$hold\_in\_cont \equiv cp := p+$ $\quad cvi := vi$ $\quad cct := ct$

$pass\_to\_cont \equiv p := cp$ $\quad vi := cvi$ $\quad ct := cct$

INITIAL STATE is essentially retained from 2.1, initializing p, e with

$$\begin{aligned} p &= load([Allocate, Call(G1), \ldots, Call(Gn), Deallocate, Proceed]) \\ e &= nil, \end{aligned}$$

so that code will automatically create the initial environment of 2.1, and the new functions with

$$cp = load([Proceed]) \qquad cvi = 0 \qquad cct = nil.$$

RULES

Rules *query success & failure*, *backtracking*, *Unify*, *cut*, *true*, *fail*, *assert*, *retract_failure*, *Erase* are retained from 2.1 without modification, because they neither influence nor are influenced by the continuation.

New rules *Allocate* and *Deallocate* are added. *Goal success* rule splits into *Deallocate* and *Proceed*, where *Deallocate*, except for popping the environment stack, unloads continuation info into our new 'slots' in *STATE*, to be picked up by *Proceed*. Rules *selection*, *call*, *conjunction*, *disjunction* and *retract_selection* are modified, since they now have to use or affect *cp*, *cvi* and *cct*. The three updates of the allocation part of homonymous rules of 2.1 are namely transferred to updates of *cp*, *cvi*, *cct*, to be saved in *ENV* by subsequent application of *Allocate* rule. Note that we have not modified *Retract_code* of 2.1—since *Unify* and *Erase* do not touch continuation info, *Proceed* will find it exactly as set by *retract* rule.

Proceed (new)

if OK
& $code(p) = Proceed$
& NOT(all_done)
then
pass_to_cont

```
Allocate                                  (new)
  if    OK
     &  code(p) = Allocate
  then
     succeed
     EXTEND ENV by temp WITH
        cp'(temp) := cp
        vi'(temp) := cvi
        ct'(temp) := cct
        ce(temp) := e
        e := temp
     ENDEXTEND

Deallocate                                (new)
  if    OK
     &  code(p) = Deallocate
  then
     succeed
     cp := cp'(e)
     cvi := vi'(e)
     cct := ct'(e)
     e := ce(e)

selection
  if    OK
     &  i_code = Call(G)
     &  is_user_defined(G)
     &  procdef(G, db) ≠ [ ]
  then
     LET cl_list = procdef(G, db)
     LET l = length(cl_list)                         ←
        PUSH temp(1), ..., temp(l) IN
           g(temp(i)) := G
           s(temp(i)) := s
           p(temp(i)) := nth(cl_list, i)
           vi(temp(i)) := vindex
           ct(temp(i)) := b
           cp(temp(i)) := p+                         ←
           cvi(temp(i)) := vi                        ←
           cct(temp(i)) := ct                        ←
           e(temp(i)) := e                           ←
        ENDPUSH
        vindex := succ(vindex)

call
  if    OK
     &  i_code = Call(call(G))
  then                                   ←
     p := Call_code(G)
     hold_in_cont                        ←
     ct := b
```

conjunction

```
if     OK
   &  i_code = Call((G,H))
then                                        ←
   p := Conjunction_code(G,H)
   hold_in_cont                             ←
```

disjunction

```
if     OK
   &  i_code = Call((G1;G2))
then                                        ←
   PUSH temp(1), temp(2) IN
      p(temp(i)) := Call_code(Gi)
      s(temp(i)) := s
      vi(temp(i)) := vi
      ct(temp(i)) := ct
      cp(temp(i)) := p+                     ←
      cvi(temp(i)) := vi                    ←
      cct(temp(i)) := ct                    ←
      e(temp(i)) := e
   ENDPUSH
```

retract_selection

```
if     OK
   &  i_code = Call(retract(H,B))
   &  is_user_defined(H)
   &  procdef(H,db) ≠ [ ]
then
   LET cl_list = procdef(H,db)
   LET l = length(cl_list)
      PUSH temp(1),...,temp(l) IN
         g(temp(i)) := H:-B
         s(temp(i)) := s
         p(temp(i)) := Retract_code(nth(cl_list,i), vindex)
         vi(temp(i)) := vindex
         cp(temp(i)) := p+                  ←
         cvi(temp(i)) := vi                 ←
         cct(temp(i)) := ct                 ←
         e(temp(i)) := e
      ENDPUSH
      vindex := succ(vindex)
```

PROOF

Correspondence to *STATE* of 2.1 is immediate, just disregard *cp*, *cvi* and *cct*. Then notice that execution of

> *selection* followed by *Allocate*, *backtracking* followed by *Allocate*, *call* followed by *Allocate*, *conjunction* followed by *Allocate*, *disjunction* followed by *Allocate*, *retract* followed by *Retract_code*, *Deallocate* followed by *Proceed*

has the same effect as, in 2.1, respectively

> *selection*, *backtracking*, *call*, *conjunction*, *disjunction*, *retract* followed by *Retract_code*, *goal success*.

2.3. Last call optimization

Following observations are obvious:

- *Allocate* and *Unify* may commute, since they update disjoint things;
- *Allocate*, *Deallocate* sequence is a noop, $p := p+$;
- if the last goal knew that it were last, i.e. that continuation info would be found in *cp*, *cvi* and *cct*, we could have *Deallocate*d before it. Hence a new instruction *Execute*(*Goal*), and clauses can be compiled as follows

$$\begin{array}{lll} H & \rightarrow & [Unify(H), Proceed] \\ H\text{:-}G & \rightarrow & [Unify(H), Execute(G)] \\ H\text{:-}G1,\ldots,Gn & \rightarrow & [Allocate, Unify(H), Call(G1),\ldots, Call(G(n-1)), \\ & & \quad Deallocate, Execute(Gn)] \quad (n > 1) \end{array}$$

SIGNATURE and **INITIAL STATE** are those of 2.2.

ABBREVIATIONS are retained from 2.2, except for two (unnecessary but 'efficient') modifications:

$Call\_code(G) \equiv load([Execute(G)])$
$Conjunction\_code(G, H) \equiv load([Allocate, Call(G), Deallocate, Execute(H)])$

RULES are all retained, except that *backtracking* now has to allow for backtracking on *Execute* as well, and all the other rules with *i_code* of form *Call*(...) appearing in the condition, namely

selection, *cut*, *true*, *fail*, *call*, *conjunction*, *disjunction*, *assert*, *retract_failure*, *retract_selection*

obtain their twins

selection_e, *cut_e*, *true_e*, *fail_e*, *call_e*, *conjunction_e*, *disjunction_e*, *assert_e*, *retract_failure_e*, *retract_selection_e*

to be invoked when the respective goal is to be *Execute*d instead of *Call*ed, differing just in looking for continuation in *cp*, *cvi* and *cct* instead of *p+*, *vi* and *ct*. Some *_e* rules will be listed in Section 3.

PROOF consists in above observations together with a straightforward verification that *Call*(*G*), *Deallocate*, *Proceed* has precisely the same effect as *Deallocate*, *Execute*(*G*), whatever *G* might be, *user_defined* or predefined.

3. COMPILATION OF PREDICATE STRUCTURE

Here the disjunctive structure of Prolog predicates will be elaborated, i.e. compiled into code. Hence the whole predicate will be represented as (a pointer to) a sequence of (still abstract) instructions.

The set of backtracking states, created by selection rules of Sections 1, 2, will be replaced in 3.1 by a single reusable choicepoint*. Creation, reusing and discarding of choicepoints will be explicitly determined by code, hence *try_me_else*, *retry_me_else*, *trust_me*, *try*, *retry*, *trust* instructions (close to those) of the WAM. Since (resatisfiable) retract behaves like selection, without loss of generality we postpone its treatment till 3.3, dealing just with 'static' code in 3.1 and 3.2.

In 3.2 we come closer to the usual picture of the WAM as an abstract processor, holding the current state in 'registers' and choicepoints on a stack, instead of a stack of states we have had so far. The stacks of choicepoints and environments will get interleaved on a WAM-like *STACK*, permitting (in Börger & Rosenzweig 1991) a more realistic representation of stacks, by embedding the *STACK* algebra in a pointer algebra.

In 3.3 we resume our treatment of database operations, introducing a variant of 'virtual copying' of Lindholm and O'Keefe 1987. Therefore we introduce a distinction of 'static' predicates from 'dynamic' ones, and *dynamic_else* instructions.

Since current goal is still represented by an instantiated Prolog term g, we decide not to commit at this point to any specific method of 'determinacy detection' (indexing, cf. Warren 1983, Aït-Kaci 1990). We entrust instead the *procdef* function, which gets g as an argument, with selecting the proper sequence of instructions. While in case of 'static' predicates, in 3.1 and 3.2, we can get away with a uniform signature and rules, allowing for (an unspecified method of) determinacy detection abstractly encoded by *procdef*, this does not seem to be possible in the intricate case of determinacy detection for dynamic predicates, in 3.3. We therefore skip the issue at this point, introducing a simplifying hypothesis (about the compiler): no indexing in dynamic code.

3.1. Compiling predicate structure on a stack of states

Here we collapse the set of backtracking states of the selection rule into a single reusable choicepoint (state). While the selection rule of Sections 1, 2 transformed the backtracking stack according to the following picture,

$$\begin{array}{lcl} & & temp(1) \\ state & \longrightarrow & \vdots \\ bstate & & temp(L) \\ & & bstate \quad , \end{array}$$

in 3.1 only one choicepoint is to be created, pictorially

$$\begin{array}{lcl} & & temp \\ state & \longrightarrow & state \\ bstate & & bstate \quad , \end{array}$$

with *state* holding the (pointer to) next alternative clause, and the whole backtracking state of computation. The code must then know when a choicepoint is to be created,

---

* This revitalizes the approach of Börger 1990a, where a *cll* function kept account of alternative clauses. Aiming at the WAM, we could have started from there directly, sidestepping however the intuitive picture of Prolog or-tree, reflected by the selection rule.

reused or discarded. The effect of *backtracking*, which now takes the frugal form of $p := p(b)$, is thus determined by instruction pointed at by $p(b)$, and the burden of optimization, i.e. of deciding not to create a choicepoint if there is just one applicable clause, and to discard it (i.e. to cease using it) when the last clause is entered, can be left to the compiler.

Thus we assume that a Prolog program : $PROGRAM$ has been completely (up to terms) compiled and loaded into $CODEAREA$, and that *procdef* returns a (pointer to a) sequence of form

$$\begin{array}{lllllll} \mathit{fail} & \text{or} & C_k : \ldots & \text{or} & N1 : \mathit{try\_me\_else}(N2) & \text{or} & \mathit{try}(C_1) \\ & & & & \vdots & & \mathit{retry}(C_2) \\ & & & & N2 : \mathit{retry\_me\_else}(N3) & & \vdots \\ & & & & \vdots & & \mathit{trust}(C_n) \\ & & & & Nn : \mathit{trust\_me} & & \\ & & & & \vdots & & (n > 1) \end{array}$$

where 'labels' Nk point to subsequent instructions of the same family, while C_k's point to (code for) clauses like in step 2.3.

We thus allow for 'determinacy detection' without specifying how it is to be done, i.e. without introducing 'switching' instructions (cf. Warren 1983 or Aït–Kaci 1990) or too specific assumptions about the compiler. In view of different methods of determinacy detection present in the literature and in implementations, one may think of a preferred method as being abstractly encoded in the *procdef* function, as indicated above. The chains of *try_me_else*, *retry_me_else* and *trust_me* may thus (but need not) be nested, i.e. such an instruction may be immediately followed by, say, a *try*, *retry*, *trust* chain instead of a clause.

SIGNATURE is that of 2.3, except that procdef now points to code:

$$\mathit{procdef} : GOAL \times PROGRAM \rightarrow CODEAREA.$$

INITIAL STATE is retained as well, except that the exact assumptions about $db0$ ($= db$, since it is here really constant) necessarily depend upon the chosen method of determinacy detection. If it is none, $\mathit{procdef}(G, db)$ should return

$$\mathit{load}([\mathit{fail}])$$

or

$$\mathit{load}([\mathit{compile}(C)])$$

or

$$\mathit{load}([\mathit{try\_me\_else}(N2), \mathit{compile}(C_1), \mathit{retry\_me\_else}(N3), \\ \mathit{compile}(C_2), \ldots, \mathit{trust\_me}, \mathit{compile}(C_n)]),$$

or

$$\mathit{load}([\mathit{try}(P_1), \mathit{retry}(P_2), \ldots, \mathit{trust}(P_n)])$$

where $\mathit{decompile}((Nk)+, Vi) = \mathit{rename}(C_k, Vi)$, or $\mathit{decompile}(P_k, Vi) = \mathit{rename}(C_k, Vi)$ for $k = 2, \ldots, n$.

ABBREVIATIONS are all retained, except for a very significant modification:

$$backtrack \equiv p := p(b).$$

Now we can express even disjunction by code:

$$Disjunction\_code(G, H) \equiv load([try\_me\_else(N), Execute(G), trust\_me, Execute(H)]),$$

where N points to the *trust_me*. Choicepoint handling rules will be alleviated by following abbreviations.

$$\begin{aligned}
store_state_in(temp) \quad \equiv \quad & g(temp) := g \\
& s(temp) := s \\
& vi(temp) := vi \\
& ct(temp) := ct \\
& cp(temp) := cp \\
& cvi(temp) := cvi \\
& cct(temp) := cct \\
& e(temp) := e \\
fetch_state_from_b \quad \equiv \quad & g := g(b) \\
& s := s(b) \\
& vi := vi(b) \\
& ct := ct(b) \\
& cp := cp(b) \\
& cvi := cvi(b) \\
& cct := cct(b) \\
& e := e(b)
\end{aligned}$$

PUSHing will now retain old *currstate*, and state stack will be handled by simple stack algebra updates

```
PUSH_STATE temp IN ≡ EXTEND STATE by temp
    updates            WITH
ENDPUSH                    currstate := temp
                           b(temp) := currstate
                           updates
                       ENDEXTEND
```

POP_STATE $\equiv$ *currstate* $:= b$

In order to avoid boring repetition of similar rules, let us adopt the convention that

```
if    condition
    & cond1 | cond2 | ... | condk
then
    updates
    upd1 | upd2 | ... | updk
```

stands for k obvious conditionals, with at most one alternative in the condition part of the rule, and zero or more in the action part. An empty alternative action means (doing) nothing.

RULES are essentially retained from 2.3, except those for database operations; (the obvious) new version of *backtracking* will be called *Call_failure* | *Execute_failure*, while the job of *selection* will be shared between new *Call_selection* | *Execute_selection* rule and the whole new family of *try*, *retry* and *trust* rules, modifying the *disjunction* rule as well.

Call_selection | Execute_selection (new)

if OK
& $i\_code = Call(G) \mid i\_code = Execute(G)$
& $is\_user\_defined(G)$
& $code(procdef(G, b)) \neq fail$
then
$g := G$
$p := procdef(G, db)$
$vi := vindex$
$ct := b$
$hold\_in\_cont \mid$
$vindex := succ(vindex)$

try_me_else | try (new)

if OK
& $code(p) = try\_me\_else(N) \mid code(p) = try(C)$
then
PUSH_STATE *temp* IN
$store\_state\_in(temp)$
$p(temp) := p+ \mid p(temp) := C$
$p := N \mid p := p+$
ENDPUSH

retry_me_else | retry (new)

if OK
& $code(p) = retry\_me\_else(N) \mid code(p) = retry(C)$
then
$fetch\_state\_from\_b$
$p := p+ \mid p := C$
$p(b) := N \mid p(b) := p+$

trust_me | trust (new)

if OK
& $code(p) = trust\_me \mid code(p) = trust(C)$
then
fetch_state_from_b
$p(b) := p+ \mid p(b) := C$
POP_STATE

disjunction | disjunction_e

if OK
& $i\_code = Call((G; H)) \mid i\_code = Execute((G; H))$
then
$p := Disjunction\_code(G, H)$ ⟵
hold_in_cont |

PROOF (of correctness wrt 2.3)

It seems to us that sheer inspection of the rules should suffice for accepting conservation of correctness wrt 2.3; the choicepoints pushed by selection, insofar as they are ever used, now appear one at a time. If the reader is not convinced, however, here is a more formal argument.

Assume that determinacy is not detected, i.e. that *procdef* of 3.1 points to a *try_me_else*, *retry_me_else*, *trust_me* chain of all clauses for the predicate. Let then *STATE2*, *ENV*2, *currstate*2, *STATE*3, *ENV*3, *currstate*3 be respective universes and constants of 2.3 and 3.1. The critical stages in a sequence of Prolog algebras of 2.1 generated by successive rule applications (in a 'run', cf. Gurevich & Moss 1990) are obviously those when the or–stack has just been modified, i.e. those produced by executing *selection* rule, *disjunction* rule or the *backtrack* update. The corresponding critical stages of 3.1 are those produced by executing *try_me_else* preceeded by *Call* | *Execute*, or one–clause *Call* | *Execute* not to be followed by *try_me_else*, or *retry_me_else* or *trust_me* preceeded by the *backtrack* update. We shall define functions

$$R : STATE3 \rightarrow STATE2$$
$$F : ENV3 \rightarrow ENV2$$

such that applying R to *ct*, *cct*, *b* of a critical *currstate*3, and F to its *e*, reconstructs precisely the corresponding critical *currstate*2. Once the functions are defined, conservation of correctness follows by straightforward induction over critical stages.

Function F is simple, it just applies R to ct', recursively down the environment stack. Function R will be defined by a simultaneous recursion, together with a function

$$V : STATE3 \times CODEAREA \rightarrow STATE2,$$

so that R reconstructs 'real' choicepoints, those pushed in 3.1, while V reconstructs 'virtual' ones, those implicit in code. Minding that states are characterized by their decorations, let

$$R(nil) = nil$$

$$\langle g, s, p, vi, ct, cp, cvi, cct, e\rangle(R(St))$$
$$= \langle g(St), s(St), p(St)+, vi(St), R(ct(St)), cp(St), cvi(St), R(cct(St)), F(e(St))\rangle$$

$$\text{for} \quad nil \neq St : STATE3,$$

(note that cutpoints are always real, and that p of any choicepoint points to a *retry_me_else* or *trust_me* instruction), while b is given by

$$b(R(St)) = \begin{cases} V(St, N) & \text{if } code(p(St)) = retry\_me\_else(N) \\ R(b(St)) & \text{if } code(p(St)) = trust\_me. \end{cases}$$

We define $V(St, N)$ to be the same as $R(St)$, except for

$$p(V(St, N)) = N+$$

$$b(V(St, N)) = \begin{cases} V(St, M) & \text{if } code(N) = retry\_me\_else(M) \\ R(b(St)) & \text{if } code(N) = trust\_me. \end{cases}$$

The rest is induction.

Our vagueness about determinacy detection does not allow us to prove anything about it. It allows us then to assume, for now, that it is done in a correct, i.e. globally equivalent way.

3.2. Compiled predicates 'in registers' with one stack

In this section we come closer to WAM layout, moving *currstate* into 'registers' (constants of a dynamic algebra) in 3.2.1, and allowing for a more realistic representation of stacks, embedding them in a pointer algebra, which could involve overwriting of popped elements, in 3.2.2. We would then have to face the problem of 'environment hiding' (cf. Aït–Kaci 1990), which forces us to interleave the two stacks like in the WAM.

3.2.1. Registers

WAM is usually thought of as an abstract machine holding its state in registers and choicepoints on a stack, while we have so far had a stack of states. Our abbreviations will help us switch to 'registers', constants in this framework, with little notational change. PUSHing not *currstate* but its choicepoint to the stack will do, pictorially

$$\begin{array}{ll} & \\ & bstate \\ state & \vdots \end{array} \quad \longrightarrow \quad \begin{array}{ll} & temp \\ & bstate \\ state & \vdots \end{array}$$

instead of

$$\begin{array}{ccc} & & temp \\ state & \longrightarrow & state \\ bstate & & bstate \\ \vdots & & \vdots \end{array},$$

We must now partially reverse the roles of '*currstate*' and *temp* while PUSHing and popping, and watch out for stopping and failing, because we have no more a *nil* state to backtrack to.

SIGNATURE is the same as before, except that there is no *currstate* any more, and

$$g,\ p,\ s,\ vi,\ ct,\ cp,\ cvi,\ cct,\ e,\ b$$

become official constants of their respective domains (not to be confused by homonymous functions decorating *STATE*). **INITIAL STATE** should be modified accordingly, so that b, e are initialized with *nil*.

ABBREVIATIONS are retained from 3.1 except for two modifications and an addition:

PUSH_STATE *temp* IN *update* ENDPUSH ≡ EXTEND *STATE* by *temp* WITH
$b := temp$
$b(temp) := b$
update
ENDEXTEND

POP_STATE ≡ $b := b(b)$

$$\begin{array}{lcl} \textbf{if } condition & \equiv & \textbf{if} \quad condition \\ \textbf{then} & & \quad \& \ \ b \neq nil \mid b = nil \\ \quad updates & & \textbf{then} \\ \quad backtrack & & \quad updates \mid stop := -1 \\ & & \quad p := p(b) \mid \end{array}$$

(If condition contains a k–fold alternative, backtracking alternative is 'outermost', i.e. the above stands for $2k$ conditionals).

RULES may be written as in 3.1, with their new meanings determined by new signature and abbreviations, except for codepointer updates on PUSHing and POPping.

try_me_else | try

if OK
& $code(p) = try\_me\_else(N) \mid code(p) = try(C)$
then
PUSH_STATE *temp* IN
store_state_in(*temp*)
$p(temp) := N \mid p(temp) := p+$ ⟵
$p := p+ \mid p := C$ ⟵
ENDPUSH

trust_me | trust

if OK
& $code(p) = trust\_me \mid code(p) = trust(C)$
then
fetch_state_from_b
$p := p+ \mid p := C$ ⟵
POP_STATE

PROOF of correctness wrt 3.1 is straightforward.

3.2.2. Environment hiding

Note that correctness of our representation depends critically on a peculiar property of our 'stack' algebras—elements popped off a stack are not irretrievably lost. If something else, except top, had pointed to them, they are there. Were they however lost, say by embedding the 'stack' algebra in a pointer algebra, backtracking somewhere long after having *Deallocate*d, i.e. after having lost an environment, could revive a choicepoint requiring precisely that environment, cf. discussion in Aït–Kaci 1990. In a two-stacks implementation *Deallocate* would thus have to check which environments were accessible from existing choicepoints, withdrawing the and–stack just that far. The usual solution, adopted in the WAM model, is however to interleave the two stacks, with choicepoints protecting older environments from being prematurely overwritten.

We have to introduce a new linked list algebra carried by a domain *STACK*, including *STATE* and *ENV* as subdomains. Our two 'logical' stacks will be interleaved on one 'physical' stack, still accessible by b, e and linked by b, ce. The only modifications required are to pushing (*Allocate*, *try_me_else*, *try*) and popping (*Deallocate*, *trust_me_else*, *trust*) rules, which have to use and update a top_of_stack function

$$tos : STATE \times ENV \rightarrow STACK,$$

which would, in an implementation, be computed by simple comparison of pointers. By the way, it is a nice example of embedding two linked list algebras into a third one.

We shall still PUSH by creating new elements as before. We could have alternatively

(a) held a *tos* register (constant), unlike the WAM model and many implementations, which are squeezed for registers;
(b) taken *STACK* to carry a pointer algebra, using $tos(b, e)+$ instead of newly created *temp* in **ABBREVIATIONS** below, extending the stack by simple function

updates. We postpone that step until, in a sequel to this paper, concrete representation of terms forces us to consider 'memory layout' anyway; the difference is all in abbreviations.

SIGNATURE is exactly as before, with the addition of

$$STACK \supseteq STATE, ENV \qquad\qquad tos : STATE \times ENV \to STACK$$

with

$$(STACK; tos(b,e), nil; -; g, s, p, vi, ct, cp, cvi, cct, e, b, cp', vi', ct', ce)$$

being a linked list algebra, where the decorating functions retain their old typings (being thus partial on *STACK*). **INITIAL STATE** has

$$tos(nil, nil) = nil.$$

ABBREVIATIONS are retained except for new ways of PUSHing and POPping:

PUSH_STATE *temp* IN *updates* ENDPUSH ≡ EXTEND *STATE* by *temp* WITH
$b := temp$
$b(temp) := b$
$tos(temp, e) := temp$ ⟵
$temp- := tos(b, e)$ ⟵
updates
ENDEXTEND

POP_STATE ≡ $b := b(b)$
$tos(b(b), e) := tos(b, e)-$ ⟵

PUSH_ENV *temp* IN *updates* ENDPUSH ≡ EXTEND *ENV* by *temp* WITH
$e := temp$
$ce(temp) := e$
$tos(b, temp) := temp$ ⟵
$temp- := tos(b, e)$ ⟵
updates
ENDEXTEND

POP_ENV ≡ $e := ce(e)$
$tos(b, ce(e)) := tos(b, e)-$ ⟵

RULES are all retained, minding new abbreviations, except that *Allocate*, *Deallocate* should now respectively PUSH_ENV, POP_ENV.

PROOF of correctness wrt 3.2.1 consists in a straightforward reconstruction of old stacks.

3.3. Virtual copying of the database

Here we resume our treatment of database operations, adopting a variant of 'virtual copying' technique by Lindholm & O'Keefe 1987. Clauses (for dynamic predicates) will be decorated with their 'timestamps' indicating 'times' of their birth and death, and *Erasing* will just set death times. In order to retain the 'efficiency' of executing static predicates of 3.2, we shall distinguish 'static' predicates from 'dynamic' ones, with *procdef* returning for the latter a pointer to a code sequence of form

$$
\begin{array}{lll}
\textit{fail} \quad \text{or} & & \textit{dynamic\_else}(B1, D1, N2) \\
 & & \vdots \\
 & N2: & \textit{dynamic\_else}(B2, D2, N3) \\
 & & \vdots \\
 & Nn: & \textit{dynamic\_else}(Bn, Dn, \textit{fail}) \\
 & & \vdots \qquad\qquad\qquad\qquad (n > 0)
\end{array}
$$

where Nk's are codepointers, Bk's (for Birth time) and Dk's (for Death time) integers, Dk's possibly even ∞, cf. *try_me_else*, *retry_me_else*, *trust_me*. This time we do care whether dynamic chains could be nested, introducing the simplifying hypothesis that they could not. *Dynamic_else* instructions are thus immediately followed in *CODEAREA* by code for clauses proper.

To minimize notational changes, let us understand the *is_user_defined* function as recognizing static predicates only, while dynamic ones will be recognized by a new function *is_dynamic*. We introduce a global 'call clock' $cc : INDEX$, to be increased (at least) on every modification of the database (and stored in choicepoints as $cc : STATE \rightarrow INDEX$). Given a dynamic predicate, *Call* | *Execute* will seek a clause which is alive at time *cc*, where

> a clause, decorated with $\textit{dynamic\_else}(B, D, N)$,
> **is alive** at time *Clock* if $B < \textit{Clock} \leq D$,

using a new function *find_live*(*Ptr*, *Clock*), which, given a codepointer *Ptr* and index *Clock*, returns the (pointer to) first *dynamic_else* instruction in the chain starting at *Ptr* which decorates a live clause, or fail if none exists.

$\textit{Dynamic\_else}(B, D, N)$ will use a 'mode register' m, a 'retract register' r (with a role to be explained below) and the *find_live* function to determine what to do. It may, before passing to (immediately following) code,

- create a choicepoint, storing the value of *cc* and the next alternative live clause;
- reuse the choicepoint, updating it to next alternative live clause, relying on stored value of *cc*;
- discard the choicepoint;
- do nothing;

i.e. it may assume any of the roles of the corresponding *try_me_else*, *retry_me_else* or *trust_me* instruction, or none of them (in case of a single live clause).

Asserting will be done by a revised *insert* function, which, as well as *delete*, now obtains a clock argument, so that

- *insert*(*List*, *Db*, *Clock*) loads *List* into *Db* at the appopriate place, preceeded by *dynamic_else*(*Clock*, *Infinity*, *Next*), with *Next* linking the chain;
- *delete*(*Ptr*, *Db*, *Clock*) kills the clause at *Ptr+* by setting its death, in *dynamic_else* at *Ptr*, to *Clock*.

The proximity of *retract* to *selection* in Section 2 is reflected here by *retract* | *retract_e* rule being the same as the dynamic version of *Call* | *Execute*, except for setting the 'retract register' r to 1. The job of old *retract* rule will be effectively done by *dynamic_else*, which, when it sees an r value of 1 (in register or stored in a *choicepoint* as $r : STATE \rightarrow BOOL$), will ultimately pass to appropriate *Retract_code* instead of passing to immediately following clause.

Our simplifying hypothesis has allowed us to simplify slightly the treatment of Lindholm and O'Keefe 1987, by having the work of *dynamic_else* in *First* mode, as described by them, shared between that *dynamic_else* and the preceeding *Call* | *Execute* or *retract*, which obviates the need for two clock registers. We have no need for a separate *dynamic_internal_else* instruction as well.

SIGNATURE is the same as before, with the addition of

$$m : \{First, Next\} \qquad is\_dynamic : TERM \rightarrow BOOL$$
$$cc : INDEX \qquad cc : STATE \rightarrow INDEX$$
$$r : BOOL \qquad r : STATE \rightarrow BOOL$$

$$insert : INSTR^* \times PROGRAM \times INDEX \rightarrow PROGRAM$$
$$delete : CODEAREA \times PROGRAM \times INDEX \rightarrow PROGRAM$$
$$find\_live : CODEAREA \times INDEX \rightarrow CODEAREA,$$

setting cc in **INITIAL STATE** to 0.

ABBREVIATIONS are all retained from 3.2, except that *dynamic_else* in *First* mode stores state to a choicepoint using a new update *store_dynamic_state_in*, which extends *store_state_in* by storing cc and r as well. *Retract_code*(*Ptr*, *Vi*) will now *Unify* decompiled *Ptr+*, since *Ptr* points to preceeding *dynamic_else*.

RULES are all retained, with the addition of dynamic version of *Call* | *Execute* and *dynamic_else*, modifying the database rules (of Section 2). *Erase* should now increase cc.

dyn_Call_failure | dyn_Execute_failure | retract_failure | retract_failure_e

if OK
& $i\_code = Call(H) \mid i\_code = Execute(H) \mid i\_code = Call(retract(H, B)) \mid$
$i\_code = Execute(retract(H, B))$
& $is\_dynamic(H)$
& $code(find\_live(procdef(H, db), cc)) = fail$
then
backtrack

dyn_Call_selection | dyn_Execute_selection | retract_selection |
retract_selection_e

if OK
& $i\_code = Call(H) \mid i\_code = Execute(H) \mid i\_code = Call(retract(H, B)) \mid$
$i\_code = Execute(retract(H, B))$
& $is\_dynamic(H)$
& $code(find\_live(procdef(H, db), cc)) \neq fail$
then
$g := H \mid g := H \mid g := H : -B \mid g := H : -B$
$p := find\_live(procdef(H, db), cc)$
$ct := b \mid ct := b \mid \mid$
$vi := vindex$
$hold\_in\_cont \mid \mid hold\_in\_cont \mid$
$vindex := succ(vindex)$
$m := First$
$r := 0 \mid r := 0 \mid r := 1 \mid r := 1$

Dynamic_else

if OK
& $code(p) = dynamic\_else(B, D, N)$
& $m = First$
& $code(find\_live(N, cc)) \neq fail$
& $r = 0 \mid r = 1$
then
$p := p+ \mid p := Retract\_code(p, vi)$
$m := Next$
PUSH_STATE *temp* IN
$p(temp) := find\_live(N, cc)$
$store\_dynamic\_state\_in(temp)$
ENDPUSH

if OK
& $code(p) = dynamic\_else(B, D, N)$
& $m = Next$
& $code(find\_live(N, cc(b))) \neq fail$
& $r(b) = 0 \mid r(b) = 1$
then
$p := p{+} \mid p := Retract\_code(p, vi)$
fetch_state_from_b
$p(b) := find\_live(N, cc(b))$

if OK
& $code(p) = dynamic\_else(B, D, N)$
& $m = Next$
& $code(find\_live(N, cc(b))) = fail$
& $r(b) = 0 \mid r(b) = 1$
then
$p := p{+} \mid p := Retract\_code(p, vi(b))$
fetch_state_from_b
POP_STATE

if OK
& $code(p) = dynamic\_else(B, D, N)$
& $m = First$
& $code(find\_live(N, cc)) = fail$
& $r(b) = 0 \mid r(b) = 1$
then
$p := p{+} \mid p := Retract\_code(p, vi(b))$
$m := Next$

assert | assert_e

if OK
& $i\_code = Call(assert(G)) \mid i\_code = Execute(assert(G)$
& $is\_dynamic(G)$
then
$p := p{+} \mid pass\_to\_cont$
$db := insert(compile(G), db, cc)$
$cc := succ(cc)$

PROOF of correctness wrt 3.2.

Note that a *dynamic_else* instruction always finds the right value of m, i.e. there is no way for backtracking to have the net effect of resetting it from *Next* to *First*.

It is now straightforward to adapt the proof of 3.1 (in two copies, one for the dynamic version of *Call* | *Execute* and one for *retract*) to *dynamic_else*: the choicepoint reconstructing functions R, V of that proof should now obtain cc and use *find_live* in order to reconstruct only the choicepoints corresponding to live clauses.

Put together with PROOFs of previous sections, this constitutes a proof of correctness of our final Prolog algebra wrt to that of Börger 1990b.

Added in proof. After this paper was written the ISO WG17 decided to change from the logical view to "a view which allows both the logical and immediate view" (see North & Scowen 1990). This however affects our treatment in Section 3 only with respect to the notion of a clause being *alive* at time *Clock* and early discarding of choicepoints. Three more notions of *alive* have to be allowed, namely $D = \infty$ ("*immediate view*"), $B < Clock$ & $D = \infty$ ("*minimal view*"), $B < Clock$ or $D = \infty$ ("*maximal view*"). All the rest of Section 3 remains unchanged.

Acknowledgements

Part of the work was done while the first author was visiting IBM Wissenschaftliches Zentrum in Heidelberg, and some of the results were reported in a talk the second author was invited to present there, in October 1990. The work of the second author was supported by Fond za znanost Republike Hrvatske. Thanks are also due to Zagrebačka banka, which has in various ways stimulated the second author's interest in the WAM, and to Sanja Singer, who did the difficult job of typesetting the paper.

REFERENCES

Aït–Kaci, K. 1990: The WAM: A (Real) Tutorial, PRL Research Report 5, Digital Equipment Corporation, Paris Research Laboratory

Börger, E. 1990a: A Logical Operational Semantics of Full Prolog. Part I. Selection Core and Control, in: CSL '89. 3rd *Workshop on Computer Science Logic* (Eds. E. Börger, H. Kleine Büning, M. Richter), Springer LNCS 440, pp. 36–64

Börger, E. 1990b: A Logical Operational Semantics of Full Prolog. Part II. Built–in Predicates for Database Manipulations, in: MFCS '90. *Mathematical Foundations of Computer Science* (B. Rovan, Ed.), Springer LNCS 452, pp. 1–14

Börger, E. 1991: A Logical Operational Semantics of Full Prolog. Part III. Built–in Predicates for Files, Terms, Arithmetic and Input–Output. In: *Proc. Workshop on Logic from Computer Science* (Y. Moschovakis, Ed.), Berkeley 1989, Springer MSRI Publications (to appear)

Börger, E. & Rosenzweig D. 1991: WAM Algebras–A Mathematical Study of Implemetation, Part II, in preparation

Börger, E. & Schmitt P. 1991: A Formal Operational Semantics for Languages of Type Prolog III, this volume

DFPS 1990: Deransart, P., Folkjær P., Pique, J–F., Scowen, R. S. : Prolog. Draft for Working Draft 4.0, ISO/IEC JTC1 SC22 WG 17 No **64**, September 1990

Gurevich, Y. 1988: Logic and the Challenge of Computer Science, in: *Trends in Theoretical Computer Science* (E. Börger ed.), Computer Science Press, pp. 1–57

Gurevich, Y. 1991: Evolving Algebras. A Tutorial Introduction, in: *EATCS Bulletin* **43**, February 1991

Gurevich, Y. & Morris, J. M. 1988: Algebraic Operational Semantics and Modula–2, in: CSL '87. 1st *Workshop on Computer Science Logic* (Eds. E. Börger, H. Kleine Büning, M. Richter), Springer LNCS 329, pp. 81–101

Gurevich, Y. & Moss, L. S. 1990: Algebraic Operational Semantics and Occam, in: CSL '90. 3rd *Workshop on Computer Science Logic* (Eds. E. Börger, H. Kleine Büning, M. M. Richter), Springer LNCS 440, pp. 176–192

Lindholm, T. G. & O'Keefe, R. A. 1987: Efficient Implementation of a Defensible Semantics for Dynamic Prolog Code, in: *Proceedings of the Fourth International Conference on Logic Programming*, pp. 21–39

North, N. D. & Scowen, R. S. 1990: Budapest 1990 Meeting ISO/IEC JTC1 SC22 WG 17 No **68**, November 1990

Rusinoff, D. M. 1989: A Verified Prolog Compiler for the Warren Abstract Machine, MCC Techical Report ACT–ST–292–89, Austin, Texas

Semle, H. 1989: Erweiterung einer abstrakten Maschine für ordnungssortiertes Prolog um die Behandlung polymorpher Sorten, IWBS Report 75, IBM Deutschland

Warren, D. H. D. 1983: An Abstract Prolog Instruction Set, Technical Note 309, Artificial Intelligence Center, SRI International

A formal operational semantics for languages of type Prolog III

E. Börger[1]
Dipartimento di Informatica
Università di Pisa

Peter H. Schmitt
Fakultät für Informatik
Universität Karlsruhe

Abstract

We use dynamic algebras introduced by Gurevich in [Gurevich 1988],[Gurevich 1991]to develop a formal semantics for the logical core of constraint logic programming languages of type Prolog III [Colmerauer 1990]. Our specification abstracts away from any particular feature of the mechanism for the resolution of constraints, thus providing a uniform description of constraint logic programming languages which turns out to be a natural refinement of the standard Prolog algebras developed in [Börger 1990]. In particular we show how our method can be used for a precise but simple method to handle specification problems connected to the freeze predicate.

0. Introduction

Recently Y.Gurevich proposed in [Gurevich 1988] a framework for semantics of programming languages which directly reflects the dynamic and resource bounded aspects of computation. This approach is based on (essentially first-order) structures that evolve over time and are finite in the same way as real computers are (so called "dynamic algebras", transition systems working on abstract data types).

Dynamic algebras have been used to give a new form of mathematical semantics for real programming languages, namely for Modula 2 [Gurevich & Morris 1988, Morris 1988], Occam [Gurevich & Moss 1990], Prolog [Börger 1990, a,b] as defined by the emerging standard proposal, and Smalltalk [Blakley1990]. These semantics are operational but different from what we have seen in connection with the Vienna Definition Method in that the possibility of appropriate data type abstraction is explicitly brought into the specification language. This allows to give an operational semantics for full (not toy)

---

[1] The bulk of this work was done when the first author from November 1989 till October 1990 was guest scientist at the Scientific Center of IBM Germany GmbH in Heidelberg, on sabbatical from University of Pisa.

programming languages as they are used in real life and to do this in a simple but complete, an abstract but natural way which supports the intuitive understanding of programs as processes.

We want to show here that also the constraint logic programming paradigm can be naturally described by appropriate dynamic algebras. As representative running system we have chosen Prolog III, developed by Colmerauer's group in Marseille, see [PrologIA 1990]. It turns out that Prolog III algebras can be defined as natural and simple refinements of the Prolog algebras in [Börger 1990 a]. In doing this we deliberately abstract away from any particular feature of the mechanism for the resolution of constraints. This mechanism clearly is crucial for the quality of the realization of every constraints logic programming system; our specification is uniform with respect to the notion of constraint solution.Our approach also shows in a mathematically nice way how more natural the notion of general constraints fits the logic programming paradigm than the merely logical unifiability condition.

Our description of Prolog III is a description of the full language, including all built-in predicates and control features. Since however Prolog III seems to behave with the usual built-in predicates mutatis mutandis in the same way as well known standard Prolog systems we can take the description of these built-in predicates from the case of standard Prolog, see [Börger 1990, a,b]. In this paper we concentrate therefore on a presentation of that part of the system which is concerned with the user defined predicates.

We discuss however in some detail the built-in predicate freeze which was not in Prolog I ,but appears already in Prolog II and other commercial Prolog versions and which we describe here in order to show how the use of dynamic algebras allows a simple but complete mathematical specification of this predicate, avoiding certain problems which seem to have appeared in the present Prolog III implementation.

The paper is organized as follows: After a quick review of dynamic algebras in chapter 1 we describe our Prolog III algebras in chapter 2. The next chapter 3 is devoted to the presentation and explanation of the transition rules for dynamic Prolog III algebras. Chapter 4 contains a discussion of some of the problematic issues of the freeze predicate and a proposal of a formal semantics of freeze.

1. Dynamic Algebras

The basic idea behind the dynamic algebra approach to operational semantics of programming languages - which has been developed by Y. Gurevich in [Gurevich 88] and applied to Modula-2 in [Gurevich & Morris 88] and to Occam in [Gurevich & Moss 90] - is to view an abstract machine as given by a collection of (finite) mathematical structures and a set of transition rules among the latter. For our purposes it suffices to consider first-order structures. The structures are many-sorted, universes may be empty and functions are allowed to be partial. We assume that the universe BOOL = {0,1} of Boolean values is always present, so that we may restrict our attention without loss of generality to algebras. The **transition rules** are of the form:

IF b(x) THEN $U_1(x)$

:

:

$U_k(x)$

where b(x) is a quantifier-free formula in the signature of the algebras under consideration, k is a natural number and $U_1(x),\ldots, U_k(x)$ are **updates** of the types described below. **x** is a sequence of variables ranging over some of the universes of the considered algebras. If **x** occurs in a rule then this rule is treated as a schema. In [Gurevich 88] only parameterfree rules are considered. To execute a transition rule on a given algebra $\mathcal{M}$ means to perform **simultaneously** all the updates $U_1(a),\ldots$ $U_k(a)$ provided that b(a) is true in $\mathcal{M}$. The application of a transition rule on $\mathcal{M}$ results in another algebra of the same signature, the update of $\mathcal{M}$.

Updates U_i are either function updates or universe extensions. A **function update** is an equation of the form: $f(t_1,\ldots,t_n) := t$ where $t_1,\ldots,t_n$, t are terms and f is a function symbol of the signature under consideration. To execute a function update in the algebra $\mathcal{M}$ amounts to evaluating the terms $t_1,\ldots,t_n$, t in $\mathcal{M}$, say with resulting elements $e_1,\ldots,e_n$, e and then assigning e as the new value of f for the argument tupel $(e_1,\ldots,e_n)$. It will be clear from our rules that we will never use any inconsistent set of updates, though formally they are not excluded.

A **Universe extension update** is of the form:

EXTEND U by temp(1) ... temp(t)

WHERE F_1

:

:

F_k

END EXTEND

Here t is a term of the underlying signature and $F_1,\ldots,F_k$ are function updates in which temp(1),temp(j),temp(t) may occur. To apply a universe extension to an algebra $\mathcal{M}$ means to evaluate the term t in $\mathcal{M}$, say with result a number n, to add n new elements temp(1),...,temp(n) to the universe U and then to execute simultaneously for each j, $1 \le j \le n$, the function updates $F_1,\ldots,F_m$.

A universe extension update is thus treated as a rule schema in t.In case t=1 we will simply write temp instead of temp(1).

Note: For the reader who cares about full precision of syntax here is a technicality: in reality it is convenient to allow temp(v) to occur in the function updates $F_1,\ldots,F_k$ where v is an arithmetical term which may contain the parameter j.The updates are executed only for those values of the parameter j for which v also falls within the range $1 \le v \le t$.

We will not use universe contraction rules which were considered in [Gurevich 88]. For PROLOG III their application would in most cases correspond to garbage collection, which need not be part of the formal description of the semantics of a programming language. This practice was already adopted in [Börger 90a].

The general definition of dynamic algebras would allow for non-deterministic transition rules, i. e. in a given algebra the condition for more than one transition rule might be true. Since we are at the level of deterministic "implemention" of Prolog III, our rule set for its description will be deterministic as well. Before we turn to the specific universes of Prolog III-algebras we mention two generic constructions that will be used several times later on. For a universe U we may add another universe called U* and thought of to represent (if the reader wants: a finite subset of) the universe of all finite sequences of elements of U. The following functions will be available on U* with their obvious meaning:

head	:	$U^* \rightarrow U$
tail	:	$U^* \rightarrow U^*$
nil	:	U^*
length	:	$U^* \rightarrow$ INTEGER
component	:	INTEGER $\times U^* \rightarrow U$
[|]	:	$U \times U^* \rightarrow U^*$ or $U \times U \times U^* \rightarrow U^*$ etc.

By overloading notation we will use the same names for these functions for all universes U. For simplicity we do not treat these list functions as dynamic but consider them as predefinied functions which therefore need no update (This is in accordance with the usual mathematical technique for implementing lists).

We will also have occasion to consider for two universes U and W the new universe $U \times W$ consisting of all pairs (a,b) with $a \in U$ and $b \in W$. We assume the total projection functions

first : $U \times W \rightarrow U$

second : $U \times W \rightarrow W$

and a binary construction function $\langle\, , \rangle$, such that $\langle a,b \rangle \in U \times W$ for $a \in U$, $b \in W$ to be available whenever $U \times W$ is included among the universes of Prolog III-algebras and adopt overloading notation as above.

2. Universes and Functions of PROLOG III Algebras

In this section we introduce and explain the basic universes of Prolog III-algebras and the functions defined on them. We try to work out that Prolog III algebras can be defined as a natural extension of standard Prolog algebras, see [Börger 90a].

A Prolog (III) program is a finite sequence of Prolog (III) clauses. The distinguished element (0-ary function):

database : CLAUSE*

represents a fixed Prolog (III) program in our algebra. Clauses for Prolog III usually are written in the form

$$L_1 \leftarrow L_2 \ldots L_k, S$$

where L_i are literals and S is a set of constraints ,i.e. elements of a universe CS.(In standard Prolog S is empty).

We can therefore represent the universe CLAUSE of Prolog III clauses as crossproduct of the universe of standard Prolog clauses with CS. This means that in addition to the functions

clhead : CLAUSE $\rightarrow$ LIT

clbody : CLAUSE $\rightarrow$ LIT*

cllength : CLAUSE $\rightarrow$ INTEGER

which we had already for standard Prolog, we introduce for Prolog III the new function

clconstraint : CLAUSE $\rightarrow$ CS

For the generic clause C$\in$CLAUSE above we thus have:

clhead(C) = L_1 clbody(C) = $[L_1,\ldots,L_k]$

cllength(C) = k clconstraint(C) = S

We assume that the universe CS contains the empty constraint system, which is always solvable , and that we have union and solvability functions for it:

empty : CS solvable : CS $\rightarrow$ BOOL $\cup$: CS $\times$ CS $\rightarrow$ CS

For the present definition of PROLOG III we will not specify furthermore the internal structure of constraint systems. The reader may wish to think of systems of linear equations and inequalities between rational number. But other choices are possible, e. g. systems of equations and inequations between infinite rational trees as in [Colmerauer 1984] or systems of linear equations and inequations between real numbers as in [Jaffar & Michaylov 1987] or systems of equations and inequations between numbers and (possibly infinite) trees, lists and boolean expressions as in .[PrologIA 1990]

The basic part of the Prolog algebras from [Börger 90a] is a predecessor structure with universe STATE, distinguished elements currstate, empty and a predecessor function choicepoint:

currstate : STATE

empty : STATE

choicepoint : STATE $\rightarrow$ STATE

The members of STATE are thought to represent the various stages through which the computation of a Prolog (III)-query evolves. The predecessor function, choicepoint, gives for every state s the next choicepoint, to which the computation will revert after the failure of the attempt to satisfy s.

The information we want to attach to members of STATE may be pictorially represented as:

$goal_1$ $cutpoint_1$	...	$goal_n$ $cutpoint_n$		constraint system

Figure 1: The information associated with states

Formally this was achieved in Prolog algebras by introducing a universe DECGOAL (of goals which are decorated by their cutpoint information) and the function

decglseq : STATE $\rightarrow$ DECGOAL*

to which we add for the case of Prolog III algebras a function

constraint : STATE $\rightarrow$ CS

We think of decorated goals as pairs of goals and STATE elements, as was done in [Börger 90a], i. e.

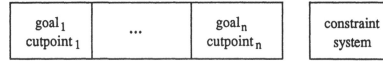

we assume

DECGOAL $\subseteq$ GOAL $\times$ STATE

We also assume GOAL $\subseteq$ LIT*.

As abbreviations we will use:

currdecglseq = decglseq(currstate) : DECGOAL*

currdecgl = head (currdecglseq)) : DECGOAL

curractivator = first(head (currdecgl)) : LIT

currcutpoint = second(head (currdecgl)):STATE

currconstraint = constraint (currstate) : CS

These abbreviations will only be used for brevity. The functions defined by these abbreviations do not belong to the signature of our algebras.

We can carry over from standard Prolog algebras to Prolog III algebras the remaining universes and functions introduced in [Börger 90a], refering to the new interpretation of the universe CLAUSE. This is true in particular for the crucial function

procdef : CLAUSE* $\times$ INDEX $\times$ LIT $\rightarrow$ CLAUSE*

which to a given clause list L, an index i and a literal b associates the definition - with variables "renamed" at level i - of the predicate b in L, i.e. the sequence of all clauses in L which are relevant for the execution of b. This definition abstracts away from any particular mechanism for the selection of these clauses.

(INDEX; varindex,+) is a predecessor structure which is used for abstract renaming to get fresh copies of clauses where necessary.

3.The transition rules

Our Prolog III transition rules are obtained from the standard Prolog transition rules in [Börger 90a] by the following refinements:

- add solvable(currconstraint) = 1 to the condition of (almost all) rules
- add a second backtracking rule corresponding to the case of an unsolvable current constraint system
- extend the selection rule by constraint updates (where the usual unifiability condition is moved from the goal sequence to the accumulated constraint system).

The modified selection rule embodies the characteristic feature of Prolog III that instead of asking for unification of the current goal g with the head h of the next relevant clause before entering the compution of its body, the equation g = h is added as a further constraint to the current constraint system.

We split the stop rule proposed for standard Prolog in [Börger 90a] into two rules corresponding to the two stop situations "no answer found" and "an answer found". The following new stop rule for the first situation is taken unchanged from [Börger 90a].

stop rule

IF still working correctly & currstate = empty
THEN stop := 1

Here still working correctly is the abbreviation for an error and stop handling condition , namely error = 0 & stop = 0.

The next rule - which reflects the successful case of the stop rule in [Börger 90a] - is obtained from there by adding into the condition the solvability property of the current constraint system:

success rule

IF still working correctly & decglseq(currstate) = [ ]
& solvable(currconstraint) = 1
THEN stop := 1 "output solution of currconstraint"

We do not describe here how to output computed solutions.

It is understood that the partial function decglseq is not defined on the state "empty". Thus the equation decglseq(currstate) = [ ] implies in particular currstate $\neq$ empty, therefore the success rule is disjoint from the stop rule.

In the same way we could take from [Börger 90a] the goal success rule mutatis mutandis. For reasons of uniformity we decide to formalize goal success as successful erasing of the whole literal sequence which constitutes the goal (instead of only asking whether its first element, curractivator, evaluates to true). Therefore we have the following:

goal success rule

IF still working correctly & currdecgl = [ ] & solvable(currconstraint) = 1
THEN decglseq(currstate) := tail(currdecglseq)

Our first backtracking rule corresponds to the standard Prolog backtracking rule:

backtracking rule 1

IF still working correctly & solvable(currconstraint) = 1 &
user_defined(curractivator) = 1 &
procdef(database,varindex,curractivator) = [ ]
THEN currstate := choicepoint(currstate)

The second backtracking rule describes the backtracking which is provoked by the unsolvability of the accumulated constraint system:

backtracking rule 2

```
IF   still working correctly  &  solvable(currconstraint) = 0
          THEN      currstate := choicepoint(currstate)
```

In the next transition rule we will use " put temp(i) on top " as an abbreviation for:

```
currstate := temp(1)
choicepoint(temp(i)) := temp(i+1)
choicepoint(l) := choicepoint(currstate)
```

selection rule

```
IF      still working correctly  &  solvable(currconstraint) = 1  &
        user_defined(curractivator) = 1  &
        procdef(database,varindex, curractivator) ≠ [ ]
THEN    LET    CL   procdef(database, varindex,curractivator)
        LET    CutPt         choicepoint(currstate)
        LET    Cont  [tail(currdecgl) | tail(currdecglseq)]
        LET          ℓ       length(CL)
        EXTEND    STATE          by temp(1) ... temp(ℓ)
        WHERE
        put   temp(i) on top
        decglseq(temp(i)) := [<clbody(component(i,CL)),CutPt>  | Cont]
        constraint(temp(i)):=currconstraint ∪ clconstraint(component(i,CL))
                        ∪ { curractivator = clhead(component(i,CL)) }
        END EXTEND
        varindex := varindex+
```

4. The freeze predicate

PROLOG III offers, as already did PROLOG II, as additional feature the built-in predicate "freeze". For a term t and a goal P the effect of the predicate freeze(t, P) is described in the manual [PrologIA 90] as:

> Delays execution of a goal P for as long as the term t is not known.

This is followed by explanations when a term is to be considered as (sufficiently) known. In the simplest case, when t is a variable X, known(X) means that the current constraint system forces X to have a unique value.In general x is (sufficiently) known if the label of the root of (the tree) x is known and if it is known wether x has no or at least one child. In our dynamic algebra formalization below we

will not look into the details of the evaluation of the predicate known(X) but introduce a function "known" from Terms × CS into BOOL.
With the above explanation of the freeze predicate in mind let us consider the following program:

```
start   :-   freeze(X, s1),freeze(X, s2),freeze(X, s3),f(X) = f(a).
s1      :-   write("s1").
s2      :-   write("s2").
s3      :-   write("s3").
```

When the query "start" is posed to the system all three goals s1, s2, s3 will be delayed, the variable X gets instantiated and thus known. What will happen? One might guess, that the goals are activated in the sequence they have been frozen in. Actually running the above Prolog III program produces the print-out:

```
x3
s2
s1 {}
```

Thus the activation seems to proceed by the principle "last frozen in, first activated". Let us consider the slightly more complicated program:

```
start   :- freeze(X, s1),freeze(Y, s2),freeze(Z, s3),f(X,Y,Z) = f(a,b,c).
s1      :-   write("s1").
s2      :-   write("s2").
s3      :-   write("s3").
```

The query "start" produces the output

```
s3
s2
s1 {}
```

as expected.

But replacing f(X,Y,Z) = f(a,b,c) by f(Y,Z,X) = f(b,c,a) we obtain

```
s1
s3
s2 {}
```

further replacing f(X,Y,Z) = f(a,b,c) by f(Z,X,Y) = f(c,a,b) yields the output

```
s2
s1
s3{}.
```

This seems to suggest that the activation of delayed goals in the implemented version of Prolog III depends on the order in which variables get instantiated during the simplification of the constraint system though this is not mentioned in the manual and it is probably not desirable. We feel that here is a need for greater precision: something which can be provided smoothly by the dynamic algebra approach.To illustrate this point we present a proposal for a formal description of the freeze predicate in terms of Prolog III algebras. By this we want above all to illustrate this simple technique by which a designer may formulate in a complete and unambiguous way his own preferred view.

To describe the effect of the freeze command we need one additional universe, called FREEZECOUNTER, FC for short, and the following functions:

freezeflag : BOOL

As long as freezeflag = 0 normal processing of the decorated goal sequence is continued. When freezeflag is turned to 1 the list of frozen goals, which is formally represented by the function

frozengoal : $STATE \times FC \rightarrow LIT^*$,

is searched from top to bottom for the first frozen goal which is executable now and therefore has to be executed. A frozen goal is executable if its guard given by the function

frozenterm : $STATE \times FC \rightarrow TERM$

is known in the current constraint system. Formally we express this by the function:

known : $TERM \times CS \rightarrow BOOL$

As auxiliary structure we need the following "double" predecessor ("pointer") algebra

(FC; bottom, actual,top, prev,next)

where the predecessor, prev, comes with its successor companion ,next, and actual is between bottom and top.

Thus we obtain the following pictorial representation of the extended decoration of states of our Prolog III algebras:

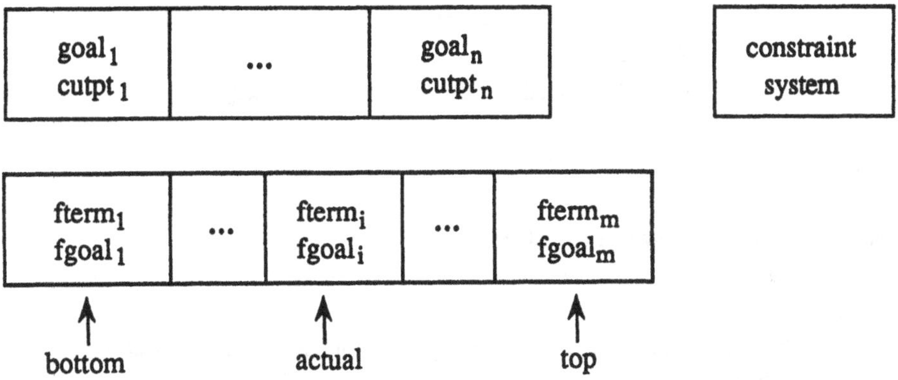

Figure 2: The information associated with states, including FC

freeze rules

```
IF   still working correctly  &  freezeflag = 0  &
          curractivator = freeze(term,goal)
     &  known(term,currconstraint) = 1
THEN
     LET Cont tail(currdecglseq)
     currdecglseq := [insert(goal,tail(currdecgl)) | Cont]
```

Here

insert:GOAL × DECGOAL → DECGOAL

is a function which inserts a given goal in front of a given decorated goal realizing the transparency of freeze for cut, i.e. such that insert(g, <g´,s>) = <gg´,s>.

If a goal g is called by the freeze command and cannot be executed immediately,because its guard term is not known currently, g is put into the list of frozengoals. This is achieved by adding a new top element to the universe FC and updating the function frozengoals and frozenterm accordingly

```
IF   error = 0  &  stop = 0  &  freezeflag = 0  &
          curractivator = freeze(term,goal) &  known(term,currconstraint) = 0
THEN          EXTEND FC    by temp
              WHERE
                    put temp on top
                    frozengoal(temp) := goal
                    frozenterm(temp) := term
              END EXTEND
              currdecglseq := [tail(currdecgl)) | tail(currdecglseq)]
```

In the last rule

put temp on top

is an abbreviation for the following two updates

next(temp) := top

top := temp

Accordingly we have to insert into all of our preceding transition rules the additional condition

freezeflag = 0

An unknown term t can only become known when the constraint system is changed. The only transition rule so far that does change the constraint system is the selection rule. Therefore we have to refine our selection rule by adding the new updates:

freezeflag := 1 and actual := top

These updates provoke the switch of the current computation to the phase where we have to look for frozen goals which may have become executable now.In the following we will formalize a search through the list of frozen goals which looks for the first goal which can be (and will be) defrozen. The reader may supply without difficulty the necessary changes for a more sophisticated treatment of "simultaneous" defreezing of more than one frozen goal. Updating actual by top means that the search starts from the beginning of the frozen list.

frozen rules

The first rule searches through FC, starting at top, skipping each goal with unknown guard.

IF error = 0 & stop = 0 & freezeflag = 1 & actual ≠ bottom &
known(frozenterm(currstate,actual), currconstraint) = 0
THEN actual := next(actual)

The next rule prepares the immediate execution of the first goal g,whose guard has become known now: it puts g in font of currdecgl (conveying to g the very cutpoint information of currdecgl), drops g and its guard from FC by updating the successor of prev(actual) to next(actual)

 is changed to 

and switches freezeflag to 0.

IF error = 0 & stop = 0 & freezeflag = 1 & actual ≠ bottom &
known(frozenterm(currstate,actual), currconstraint) = 1
THEN
LET g frozengoal(currstate,actual)
LET Cont tail(currdecglseq)
currdecglseq = [insert(g,currdecgl) | Cont]
freezeflag = 0
drop actual

Here "drop actual" is short for the update: next(prev(actual)):= next(actual)

The last rule terminates the search for known guards and gives control back to the computation of decglseq.

IF error = 0 & stop = & freezeflag = 1 & actual = bottom
THEN
freezeflag = 0

Bibliography

Blakley,R. 1990: Ph.D.Thesis. University of Michigan at Ann Arbor (in preparation)

Börger, E. 1990: A Logical Operational Semantics of Full PROLOG, Part I. Selection Core and Control, CSL'89.3rd workshop on Computer Science Logic (Eds. E.Boerger, H.Kleine Buening, M. Richter), Springer Lecture Notes in Computer Science ,1989, Vol. 440, pp. 36 -64

Börger, E. 1990a: A Logical Operational Semantics of Full PROLOG,Part II. Built-in Predicates for Database Manipulations, Proc. of MFCS 1990, Springer Lecture Notes in Computer Science, Vol. 452, pp. 1 - 14

Börger, E. 1990b:A Logical Operational Semantics of Full PROLOG,Part III. Built-in Predicates for Files, Terms, Arithmetic and Input-Output Proc. of the workshop on "Logic from Computer Science", Berkely 1989, (Ed. Y. Moschovakis) to appear in MSRI Series, Springer Verlag 1991

Colmerauer, A. 1984: Equations and Inequations on Finite and Infinite Trees. Proc. Int. Conf. on Fifth Generation Computer Systems, Tokyo 1984, p. 85

Colmerauer A.1990: An Introduction to Prolog III. Comm. of the ACM, 33(7), July 1990, pp. 69-90.

Gurevich, Y. 1988: Logic and the Challenge of Computer Science, in: Trends in Theoretical Computer Science (E. Börger, ed.), Computer Science Press, 1988, pp. 1 - 57

Gurevich,Y.1991:Evolving Algebras. A Tutorial Introduction, in., EATCS BULL. FEB. 1991.

Gurevich, Y. & Morris, J. M. 1988: Algebraic Operational Semantics and Modula-2 in: CSL '87 1st Workshop on Computer Science Logic (Eds. E. Börger, H. Kleine Büning, M. Richter) Springer LNCS 329, pp. 81 - 101

Gurevich, Y. & Moss, J. M. 1990: Algebraic Operational Semantics and Occam in: CSL '89 3rd Workshop on Computer Science Logic (Eds. E. Börger, H. Kleine Büning, M. Richter), Springer LNCS 440, pp. 176 - 192.

Jaffar, J. & Michaylov, S. 1987: Methodology and Implementation of a CLP-System. Proc. 4th Int. Conference on Logic Programming, Melbourne 1987

Morris, J. 1988: Algebraic Operational Semantics for Modula 2. Ph.D. Thesis. The University of Michigan at Ann Arbor.

PrologIA 1990: Prolog III, Version 1.1., Reference and User's Manual, Prolog IA, Luminy, Case 919, 13288, Marseille cedex 09, France

Efficiency Considerations on Goal-Directed Forward Chaining for Logic Programs

Wolfram Burgard
Institut für Informatik III, Universität Bonn
Römerstr. 164, D-5300 Bonn

Abstract

This paper presents a linear resolution strategy (GDFC-resolution) for definite logic programs in which goal-directed forward chaining is used to find refutations. GDFC-resolution focuses on relevant facts and rules by means of query independent link clauses. One problem is, that the number of link clauses may be infinite if the program contains recursive rules defined over recursive data structures. We present an approach to generate a more general but finite set of link clauses. We show that GDFC-resolution is sound and complete for definite programs. We discuss the efficiency of GDFC-resolution comparing it with SLD-resolution and present some experimental results.

1 Introduction

Yamamoto and Tanaka presented an approach to transform a definite program to a forward chaining logic program [15]. This technique has been adopted from the bottom-up parsing system BUP, which is implemented in Prolog [11]. In this paper we use a slightly different approach which is equivalent with respect to the declarative semantic [2]. Program 1 is a meta-interpreter for goal-directed forward chaining defined at the goal reduction level which implements this approach.

The first two clauses are also used in the standard three clause meta-interpreter for pure Prolog. The empty goal, represented by the constant true, is true. A conjunction (A,B) is true if A and B are true. The third clause states that B is true if there is a fact A← such that B and A are unifiable. The fourth clause says that B is true if there is a possibly relevant fact A←, for which link(A,B) is true, and A is a subgoal for B. A is a subgoal for B if there is a possibly relevant clause with a leftmost body literal unifying with A and head unifying with B and the remaining body literals are true. Additionally, A is a subgoal for B if there is a possibly relevant clause the leftmost body literal of which is unifiable with A, the remaining body literals are true and the head of the clause is a subgoal for B.

```
fc_solve(true)←

fc_solve((A,B))←
    fc_solve(A),
    fc_solve(B).

fc_solve(B)←
    clause(B,true).

fc_solve(B)←
    clause(A,true),
    link(A,B),
    subgoal(A,B)

subgoal(A,B)←
    clause(B,(A,Body)),
    fc_solve(Body)

subgoal(A,B)←
    clause(C,(A,Body)),
    link(C,B),
    fc_solve(Body),
    subgoal(C,B)
```

Program 1: A meta-interpreter for goal-directed forward chaining

This meta-interpreter for goal-directed forward chaining focuses on relevant clauses calling link goals for link clauses which are generated from the program being interpreted. As already mentioned in [15] these link clauses allow a fast selection of possibly relevant facts and rules without searching all of them. In [1] we showed that the link clauses can be used to implement a necessary condition for the provability of a given goal. Furthermore, we showed how the possibly large number of link clauses can be reduced. The main problem with the link clauses, however, is, that their generation possibly does not terminate if the input program contains left recursive predicates with structured arguments. We will address this problem in the next section.

Additionally we will be concerned with the semantics of this meta-interpreter for goal-directed forward chaining. Based on this interpreter we define a linear resolution strategy (GDFC-resolution) in the third section. We present theoretical results concerning soundness and completeness of GDFC-resolution. In the fourth section we discuss the efficiency of this approach comparing different properties of search trees corresponding to GDFC- and SLD-resolution. The results allow us give a conjecture comparing the average case runtime behaviour of both strategies for propositional programs. Finally we present experimental results confirming this conjecture.

2 Terminating Link Clause Programs

DEFINITION 1 Let P be a definite program. The *set of link clauses* $Link_P$ for P is the set of all clauses link(B_1,B)← modulo variable renaming such that P contains a clause with head B and leftmost body literal B_1. ■

DEFINITION 2 Let P be a definite program. The *link clause program* L_P for P includes $Link_P$. Modulo variable renaming L_P contains a clause link(A,B)θ← for each pair of clauses link(A,C)← and link(D,B)←∈ L_P with θ is an mgu for C and D. ■

As already mentioned, the transitive closure of $Link_P$ may be infinite so that the generation of L_P may not terminate. Consider for example the following program specifying natural numbers:

nat_num(0)←
nat_num(s(X))←nat_num(X)

While $Link_P$ contains only one link clause, L_P is infinite, since it contains infinitely many link clauses of the form

link(nat_num(X),nat_num(s^n(X)))←

There are different approaches to solve this termination problem. The first approach presented in [1, 3] is based on the observation that L_P is finite if P is not left-recursive. Thus one solution is to move appropriate literals of the clause bodies to the leftmost position so that P is not left-recursive. Unfortunately there are programs such as the example given above for which we cannot apply this approach. In this case a second technique presented in [6] may be well suited to solve this problem. The idea is to compute the transitive closure of the link clauses, i.e., L_P at runtime including dynamic size checks to achieve termination. The dynamic size checks are based on termination proofs which may be obtained from the approach presented by Plümer in [12]. For our example this means that we only compute link clauses containing terms with a size smaller than the size of the term given with the input query. This approach, however, can only be applied if termination proofs are available. Furthermore, the dynamic computation of L_P during runtime reduces the efficiency significantly.

In this paper we present a new approach to solve this problem. We simply rename variables in the link clauses of $Link_P$ so that the resulting link clauses are more general but have a finite transitive closure. Again consider the program defining natural numbers. Let us replace X in the left atom by a fresh variable Y which yields the clause

link(nat_num(Y),nat_num(s(X)))←

The transitive closure of this clause is finite, since all generated clauses are a variant of it. Crucial are only such link clauses causing cyclic data flow through structured terms. This data flow, however, can be eliminated by renaming variables.

THEOREM 1 Let P be a definite program with link clauses $Link_P$. L_P becomes finite if a sufficiently large number of variables in $Link_P$ is renamed. ■

Sketch of the proof We introduce migration graphs which are motivated by the migration sets defined in [9]. Migration graphs represent the flow of data between argument positions of atoms contained in the link clauses. Because there are two possible directions of migration in the link clauses, we always have two different migration graphs for one set of link clauses. Then we show that there must be at least one cycle in one migration graph if L_P is infinite. Renaming variables, however, removes arcs from the migration graphs. Hence, if we rename a sufficiently large number of variables, so that both migration graphs are acyclic, then L_P must be finite. ■

An interesting question in this context is, how we can minimize the number of variables which have to be renamed. One approach is to minimize the number of arcs which have to be removed from the migration graphs. This problem exactly is the feedback arc set problem which unfortunately is NP-hard [5]. However, considering corresponding arc-graphs, we can apply approximation algorithms for the feedback vertex set problem [13, 14]. This enables us to reduce the number of variables which have to be renamed.

As already mentioned, the resulting link clause program is more general and therefore has a weaker selectivity than we would have without renaming variables. Whereas a computational overhead may come from the reduced selectivity of the resulting link clause program, we have the great advantage that we can deal with finite link clause programs and that the selection of possibly relevant clauses takes only one step. By all means, in the remainder of this paper we can assume that there is a terminating link clause program L_P^{term} satisfying the following two conditions for each program P and goal G:

1) the refutation of $L_P^{term} \cup \{G\}$ terminates, and

2) for each correct answer θ for $L_P \cup \{G\}$ there is an answer θ' for $L_P^{term} \cup \{G\}$ such that $G\theta'$ is more general than $G\theta$.

3 GDFC-Resolution

In this section we introduce a linear resolution strategy called GDFC-resolution which is based on the meta-interpreter for goal-directed forward chaining defined in the first section. First we need the concept of *subgoal-goal pairs* to represent situations where an atom of the form subgoal(A,B) is selected, i.e., we are solving a goal with an already deduced atom.

DEFINITION 3 A *subgoal-goal pair* is a pair $<A,B>$ where A and B are atoms. A is called the subgoal of the goal B. ■

DEFINITION 4 A *GDFC-goal* is a goal $\leftarrow A_1,\ldots,A_n$ ($n \geq 1$) such that A_i is either an atom or a subgoal-goal pair, for $i=1,\ldots,n$. ■

DEFINITION 5 Let P be a definite program, L_P^{term} a terminating link clause program for P and G be a GDFC-goal $\leftarrow A_1,\ldots,A_i,\ldots,A_n$. G' is derived from G via *GDFC-resolution* w.r.t. L_P^{term} using a clause $C \in P$ and a substitution θ if the following conditions hold:

1) A_i is a selected element (atom or a subgoal-goal pair) in G.

2) Suppose A_i is an atom, and C is a unit clause $B\leftarrow$.

   a) If A_i and B are unifiable with mgu θ, then G' is the goal

   $\leftarrow(A_1,\ldots,A_{i-1},A_{i+1},\ldots,A_n)\theta.$ (Rule 1)

   b) If θ is an answer for $L_P^{term} \cup \{\leftarrow link(B,A_i)\}$, then G' is the goal

   $\leftarrow(A_1,\ldots,A_{i-1},<B,A_i>,A_{i+1},\ldots,A_n)\theta.$ (Rule 2)

3) Suppose A_i is a subgoal-goal pair $<E,F>$. Let C be a non-unit clause $B\leftarrow B_1,\ldots,B_q$ such that E and B_1 are unifiable with mgu θ_1.

   a) If $B\theta_1$ and F are unifiable with mgu θ, then G' is the goal

   $\leftarrow(A_1,\ldots,A_{i-1},B_2,\ldots,B_q,A_{i+1},\ldots,A_n)\theta.$ (Rule 3)

   b) If θ is an answer for $L_P^{term} \cup \{\leftarrow link(B\theta_1,F)\}$, then G' is goal

   $\leftarrow(A_1,\ldots,A_{i-1},B_2,\ldots,B_q,<B,F>,A_{i+1},\ldots,A_n)\theta.$ (Rule 4)

G' is called the *GDFC-resolvent*. ■

Note that Rule 1 or Rule 2 respectively Rule 3 or Rule 4 may be applied non-deterministically, i.e., there are situations where we can either apply Rule 1 or Rule 2 respectively Rule 3 or Rule 4. It is straightforward to see that these rules directly correspond to the last four clauses of Program 1.

As well as for SLD-resolution [8] we can define GDFC-derivations, -refutations, and computed answers in an analogous way. For the sake of brevity we omit these definitions here. Since we need the concept of GDFC-trees to judge the efficiency of this approach, we give its definition below.

DEFINITION 6 Let P be a definite program and G be a definite goal. A *GDFC-tree* for $P \cup \{G\}$ w.r.t. L_P^{term} is a tree satisfying the following conditions:

1) Each node of the tree is a (possibly empty) GDFC-goal.

2) The root node is G.

3) Let $G' = \leftarrow A_1, \ldots, A_i, \ldots, A_n$ ($n \geq 1$) be a node in the tree and suppose A_i is the selected element. Then, for each GDFC-resolvent G" derived from G' using the four derivation rules, G' has a successor node G".

4) Nodes that are the empty clause have no children. ■

THEOREM 2 GDFC-resolution is sound and complete for definite programs. ■

Sketch of the proof By induction on the length n of the SLD-refutation of $P \cup \{G\}$ respectively GDFC-refutation w.r.t. L_P^{term} we show that there is a corresponding GDFC-refutation w.r.t. L_P^{term} respectively SLD-refutation computing the same answer (up to variable renaming). The induction hypothesis is straightforward.

Suppose the computation rule always selects the leftmost element. If we have a GDFC-refutation of $P \cup \{G\}$ with length $n>1$ then the elimination of the leftmost atom of G must have the form which is illustrated in Figure 1. We show that the corresponding SLD-derivation has the form which is illustrated in Figure 2. From the induction hypothesis follows that each conjunction $B_{i,2}, \ldots, B_{i,m_i}$, for $i=2 \ldots r$, can be eliminated in a GDFC-derivation if and only if it can be eliminated in an SLD-derivation. Furthermore, both derivations produce the same answer. $D_1 \ldots D_r$ can be used in a GDFC-refutation if and only if they can be used in an SLD-refutation. Note that SLD-resolution uses them in the reverse order.

This leads to the result that the last goals in both figures are variants. From the induction hypothesis follows that we can find a refutation of $P \cup \{G_k\}$ with GDFC-resolution if and only if we can find one with SLD-resolution. ■

Input clauses	Derived Goals
	$\leftarrow A_1,A_2,\ldots,A_m$
$D_1=C_1$	$\leftarrow(\langle B_1,A_1\rangle,A_2,\ldots,A_m)\theta_1$
$D_2=C_2$	$\leftarrow(B_{2,2},\ldots,B_{2,m_2},\langle B_2,A_1\rangle,A_2,\ldots,A_m)\theta_2$
$Q_{2,1}$	$\vdots$
$\vdots$	$\vdots$
$Q_{2,\lvert Q_2\rvert}$	$\leftarrow(\langle B_2,A_1\rangle,A_2,\ldots,A_m)\theta_{\Sigma_2+2}$
D_3	$\leftarrow(B_{3,2},\ldots,B_{3,m_3},\langle B_3,A_1\rangle,A_2,\ldots,A_m)\theta_{\Sigma_2+3}$
$Q_{3,1}$	$\vdots$
$\vdots$	$\vdots$
D_{r-1}	$\leftarrow(B_{r-1,2},\ldots,B_{r-1,m_{r-1}},\langle B_{r-1},A_1\rangle A_2,\ldots,A_m)\theta_{\Sigma_{r-2}+r-1}$
$Q_{r-1,1}$	$\vdots$
$\vdots$	$\vdots$
$Q_{r-1,\lvert Q_{r-1}\rvert}$	$\leftarrow(\langle B_{r-1},A_1\rangle,A_2,\ldots,A_m)\theta_{\Sigma_{r-1}+r-1}$
D_r	$\leftarrow(B_{r,2},\ldots,B_{r,m_r},A_2,\ldots,A_m)\theta_{\Sigma_{r-1}+r}$
$Q_{r,1}$	$\vdots$
$\vdots$	$\vdots$
$Q_{r,\lvert Q_r\rvert}=C_k$	$\leftarrow(A_2,\ldots,A_m)\theta_{\Sigma_r+r}$

Fig. 1: GDFC-derivation

Input clauses	Derived Goals
	$\leftarrow A_1,A_2,\ldots,A_m$
D_r	$\leftarrow(B_{r,1},B_{r,2},\ldots,B_{r,m_r},A_2,\ldots,A_m)\gamma_1$
D_{r-1}	$\leftarrow(B_{r-1,2},\ldots,B_{r-1,m_{r-1}},B_{r,2},\ldots,B_{r,m_r},A_2,\ldots,A_m)\gamma_2$
$\vdots$	$\vdots$
D_2	$\leftarrow(B_{2,1},B_{2,2},\ldots,B_{2,m_2},\ldots,B_{r,2},\ldots,B_{r,m_r},A_2,\ldots,A_m)\gamma_{r-1}$
D_1	$\leftarrow(B_{2,2},\ldots,B_{2,m_2},\ldots,B_{r,2},\ldots,B_{r,m_r},A_2,\ldots,A_m)\gamma_r$
$R_{2,1}$	$\vdots$
$\vdots$	$\vdots$
$R_{2,\lvert R_2\rvert}$	$\leftarrow(B_{3,2},\ldots,B_{3,m_3},\ldots,B_{r,2},\ldots,B_{r,m_r},A_2,\ldots,A_m)\gamma_{\Sigma_2+r}$
$\vdots$	$\vdots$
$R_{r-1,\lvert R_{r-1}\rvert}$	$\leftarrow(B_{r,2},\ldots,B_{r,m_r},A_2,\ldots,A_m)\gamma_{\Sigma_{r-1}+r}$
$R_{r,1}$	$\vdots$
$\vdots$	$\vdots$
$R_{r,\lvert R_r\rvert}$	$\leftarrow(A_2,\ldots,A_m)\gamma_{\Sigma_r+r}$

Fig. 2: The corresponding SLD-derivation

As a secondary result of this proof is that the same input clauses are used in both refutations, that is the sequences $Q_{i,1},\ldots,Q_{i,|Q_i|}$ and $R_{i,1},\ldots,R_{i,|R_i|}$ in pairs have the same length and contain the same input clauses, for i=2...r.

4 Efficiency of GDFC-Resolution

The problem to find a refutation with linear resolution strategies can also be viewed as a tree-searching problem [4]. The goal is to find one or some or even all of the success branches in the corresponding search tree. Consequently the comparison of different linear resolution strategies can be reduced to the question how fast a solution can be found in corresponding resolution trees. Clearly the efficiency of a strategy depends on the size, i.e., the number of arcs of the tree. The strongest result would be that one of both resolution strategies considered here is always more efficient than the other. The following example, however, demonstrates that this is not true.

EXAMPLE 1 Consider the following program

C_1:	C_2:	C_3:	C_4:	C_5:
$p\leftarrow q,r$	$q\leftarrow s$	$q\leftarrow t$	$s\leftarrow$	$t\leftarrow$
L_1:	L_2:	L_3:	L_4:	L_5:
link(q,p)←	link(s,q)←	link(t,q)←	link(s,p)←	link(t,p)←

Since p is not a logical consequence of this program, SLD-resolution as well as GDFC-resolution w.r.t. L_P has to traverse the complete SLD- resp. GDFC-tree evaluating ←p. Suppose the computation rule is left-first. Then the SLD-tree is smaller than the GDFC-tree. Both trees are contained in Figure 3. Now consider the following program which is

C_1:	C_2:	C_3:
$p\leftarrow q,r$	$p\leftarrow q,t$	$q\leftarrow$
L_1:		
link(q,p)←		

Figure 4 shows that now the SLD-tree is larger than the GDFC-tree. ■

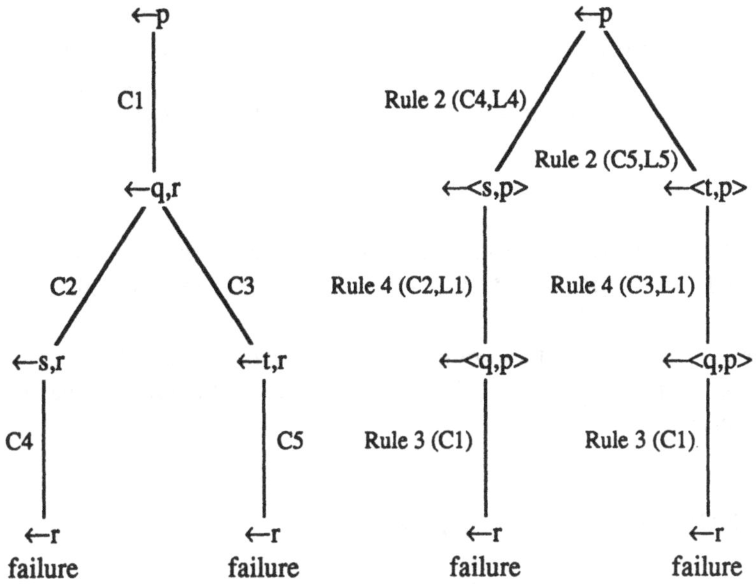

Fig. 3: Corresponding SLD- and GDFC-tree

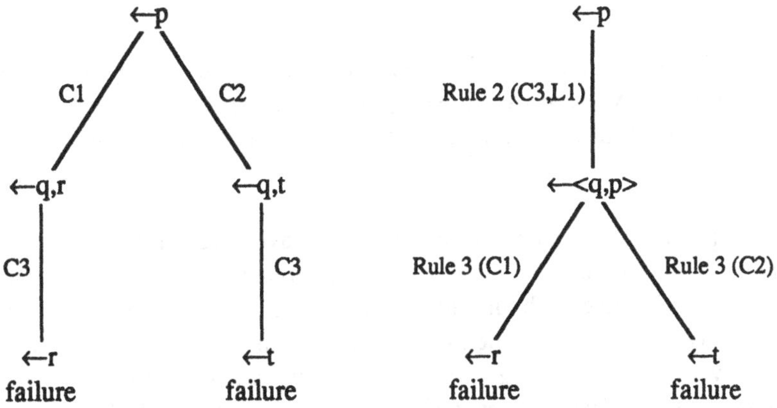

Fig. 4: Corresponding SLD- and GDFC-tree

The example given above shows us that there are situations where SLD-resolution is more efficient than GDFC-resolution and vice versa. This means that none of both strategies is always better than the other. Hence we have to be content with a weaker result. Such a result, for example, would be that one of both strategies in average is more efficient. There are different tree properties we will consider subsequently, namely the length and number of success and failure branches. Clearly, a result saying that one or more of these measures generally are smaller for one strategy gives rise to the anticipation that this strategy in average is better than the other. First we consider the length of corresponding refutations. The following theorem directly follows from the proof of Theorem 2.

THEOREM 3 Let P be a definite program and G be a definite goal. Then corresponding SLD- and GDFC-refutations of $P \cup \{G\}$ w.r.t. L_P^{term} have the same length. ■

Next we will consider the number of success branches in GDFC- and SLD-trees. The following example shows that there are cases where the GDFC-tree contains more success-branches than the SLD-tree.

EXAMPLE 2 Consider the following program

C_1:	C_2:	C_3:
$q(a,a) \leftarrow$	$p(X,Y) \leftarrow q(X,X)$	$p(X,X) \leftarrow q(X,Z)$

L_1:	L_2:
$link(q(X,X), p(X,Y)) \leftarrow$	$link(q(X,Z), p(X,X)) \leftarrow$

Suppose G is $\leftarrow p(X,Y)$. The first link goal called in the refutation is $\leftarrow link(q(a,a),p(X,Y))$. We can use either L_1 or L_2 to compute an answer. Whereas there are only two SLD-refutations of $P \cup \{G\}$, we have four GDFC-refutations of $P \cup \{G\}$ w.r.t. L_P^{term}. ■

The example above demonstrates that we may have redundant proofs caused by different answers to link calls. There are different approaches to solve this problem. First it is reasonable only to use such answers which are not more specific than another. This approach, however, cannot be applied in the example above. One solution is to eliminate redundant derivations by replacing all link clauses with a common instance by their most specific generalization. The most specific generalization can easily be computed [7]. A disadvantage of this transformation is, that it additionally reduces the selectivity of L_P^{term}.

If we consider non-recursive, propositional programs only then this problem of redundant proofs does not arise. Furthermore, the refutation trees are finite, so that the number of success and failure branches is finite too.

THEOREM 4 Let P be a non-recursive, propositional program and G be a goal. The GDFC-tree for P∪{G} w.r.t. L_P contains the same number of success branches as the SLD-tree for P∪{G}. ■

Next we discuss the last two topics of our list concerning failure branches.

EXAMPLE 3 Consider the following program

C_1:	C_2:	C_3:	C_4:
a←b	b←c,d	c←	d←e
L_1:	L_2:	L_3:	L_4:
link(b,a)←	link(c,b)←	link(c,a)←	link(e,d)←

Suppose G is ←a. Since P contains no fact B← such that link(B,d) holds, GDFC-resolution already stops with failure when d is selected. In contrast to SLD-resolution, GDFC-resolution does not use the first and the fourth clause. Both finitely failed trees are contained in Figure 5. ■

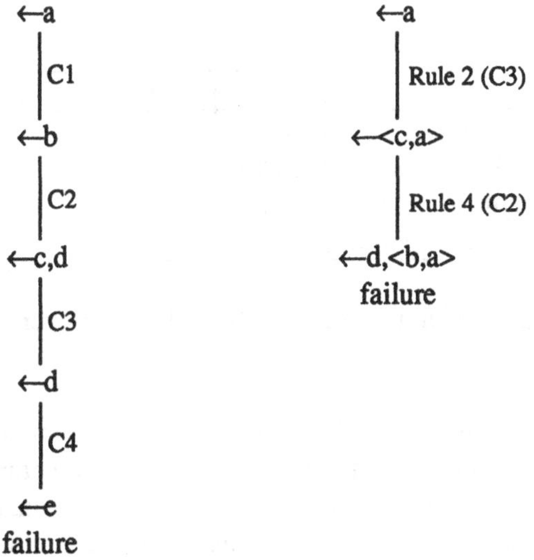

Fig. 5: Corresponding failed SLD- and GDFC-derivations

Example 3 demonstrates, that failed GDFC-derivations may be shorter than corresponding failed SLD-derivations. This observation is confirmed with the following theorem.

THEOREM 5 Let P be a propositional program and G be a goal. Suppose the computation rules always select the leftmost literal. For each failed GDFC-derivation of $P \cup \{G\}$ w.r.t. L_P, there is a failed SLD-derivation of $P \cup \{G\}$ which has at least the same length.■

The next theorem follows from the fact that SLD-resolution possibly needs at least one derivation until it stops with failure, whereas the corresponding failed GDFC-derivation has length zero.

THEOREM 6 Let P be a non-recursive, propositional program and G be a definite goal. There are at least as many failed SLD-derivations of $P \cup \{G\}$ as failed GDFC-derivations of $P \cup \{G\}$ w.r.t. L_P^{term}. ■

Figure 6 sums up the results obtained in this section. If we consider non-recursive, propositional programs, then the number and length of corresponding success branches in GDFC- and SLD-trees are identical. However, number and length of failure branches of GDFC-trees are smaller than those of SLD-trees.

Topic	GDFC $\leftrightarrow$ SLD
Length of success branches	=
Number of success branches	=
Length of failure branches	≤
Number of failure branches	≤

Fig. 6: Comparison of different tree properties

We conclude this section with a discussion concerning implementation aspects. In the context of propositional programs we can use index structures to access possibly relevant clauses in constant time so that there is no computational overhead. Therefore, at least for propositional programs, an implementation of GDFC-resolution can be as efficient as of SLD-resolution. This observation, in common with the results presented above, allows us to anticipate, that GDFC-resolution in average has a better runtime behaviour than SLD-resolution for propositional logic programs.

5 Experimental Results

Our approach to confirm this conjecture by experiments is to generate propositional logic programs randomly and count the number of logical inferences needed to find all solutions, i.e., count the number of arcs of the SLD- resp. GDFC-tree. Our program generator allows us to adjust the following parameters:

- the umber N_{EDB} of predicate symbols defined by a single fact (EDB-symbols),
- the number N_{IDB} of predicate symbols defined by non-unit clauses (IDB-symbols),
- the distribution of the number of rules per IDB-symbol and
- the distribution of the length of the rules.

For different ratios N_{EDB}/N_{IDB} we generated a sample of programs. For each program, we stepwise increased the number of facts and determined the size of the GDFC- and SLD-tree for each IDB-symbol. To avoid that programs are generated containing predicates for which there are too much redundant derivations we included a limit, i.e., we considered only such programs for which the size of all refutation trees is limited. Figure 7 contains the results we obtained with N_{EDB}=40 and N_{IDB}=10 after evaluating about 85.000 goals with a limit of 150.

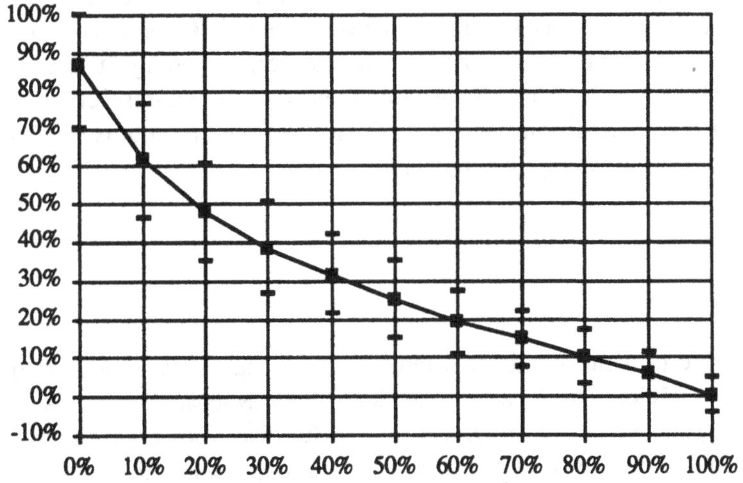

Fig. 7: Relative improvement depending on the number of provable IDB-symbols

The X-axis represents the number of provable IDB-symbols in percent. If $|T_{SLD}|$ is the size of the SLD-tree and $|T_{GDFC}|$ the size of the GDFC-tree, then the Y-axis is the average value of the relative improvement in percent obtained using GDFC-resolution.

The horizontal bars represent the interval of the standard deviation. The relative improvement is computed by the following formula:

$$\text{Relative Improvement} = \frac{|T_{SLD}|-|T_{GDFC}|}{|T_{SLD}|}$$

A quite interesting property of this curve is, that it is largely independent from the ratio N_{EDB}/N_{IDB}, N_{EDB}, N_{IDB} and both distributions (a detailed discussion of all experiments carried out will be presented in [10]). To sum up, we can say that the experimental results are in keeping with the theoretical results suggesting that GDFC-resolution in average is more efficient than SLD-resolution for propositional logic programs.

Conclusions

In this paper we discussed goal-directed forward chaining as a linear resolution strategy denoted by GDFC-resolution. First we solved the problem to obtain a finite transitive closure of the link clauses GDFC-resolution uses to focus on relevant clauses. We showed that GDFC-resolution is sound and complete for definite programs. Furthermore, we compared certain properties of search trees corresponding to GDFC- and SLD-resolution. We showed that the number and size of success branches are the same for propositional programs. Furthermore we presented results saying that an SLD-tree contains at least as many failure branches as the GDFC-tree, and that each failure branch in an SLD-tree is at least as long as the corresponding failure branch in the GDFC-tree. These results give rise to anticipate that GDFC-resolution in average is more efficient than SLD-resolution for propositional logic programs. We presented experimental results confirming this conjecture.

Acknowledgements

I would like to thank Armin B. Cremers who initiated my interest in this area and Lutz Plümer for advice. Thanks also go to Elmar Eder, Michael Hanus, Ralf Hinze, Hans Kleine-Büning, and Stefan Lüttringhaus for fruitful discussions. Furthermore, I want to thank the members of the 'Logic Programming Group' at the University of Dortmund for helpful comments.

References

1. Burgard, W. Generating a Data Structure for Goal-Directed Forward Chaining in Logic Programming. Tech. Rept. 331, Department of Computer Science, University of Dortmund, 1989.

2. Burgard, W. Goal-Directed Forward Chaining: A Linear Resolution Strategy. Tech. Rept. 360, Department of Computer Science, University of Dortmund, 1990.

3. Burgard, W. and Heusch, P. Complexity Results on Generating Link Clauses for Goal-Directed Forward Chaining in Logic Programming. In *Proceedings of the Second PROTOS Workshop, Sils Maria, Switzerland,* Appelrath, H.J., Cremers, A.B., and Herzog, O., 1990.

4. Chang, C.L. and Lee, R.C.T. *Symbolic Logic and Mechanical Theorem Proving,* Academic Press(1973).

5. Garey, M.R. and Johnson, D.S. *Computers and Intractability: A Guide to NP-Completeness,* W. H. Freeman and Company, San Francisco(1979).

6. Heidelbach, M. *Special Problems with Forward-Chaining Logic Programs,* Diploma thesis, In German, University of Dortmund, 1990.

7. Lassez, J.L., Maher, M.J., and Mariott, K. Unification Revisited. In *Deductive Databases and Logic Programming*. Morgan Kaufmann Publishers, Inc., Los Altos, California, pp. 587-626, 1987.

8. Lloyd, J.W. *Foundations of Logic Programming, Second, Extended Edition,* Springer Verlag(1987).

9. Lozinskii, E.L. Evaluating Queries in Deductive Databases by Generating. In *IJCAI 85 Proceedings of the Ninth International Joint Conference on Artificial Intelligence,* Aravind, J., Morgan Kaufmann Publishers, Inc., Los Altos, California, 1985.

10. Magura, N. *An Experimental Study to Estimate the Efficiency of Goal-Directed Forward Chaining*, Diploma thesis, In German, University of Dortmund, 1991.

11. Matsumo, Y., Tanaka, H., Hirakawa, H., Miyoshi, H., and Yasukawa, H. BUP: A Bottom-Up Parser Embedded in Prolog. *New Generation Computing 1*, 2 (1983), 145-158.

12. Plümer, L. *Termination Proofs for Logic Programs,* Springer Verlag, Lecture Notes in Artificial Intelligence, 446(1990).

13. Rosen, B.K. Robust Linear Algorithms for Cutsets. *Journal of Algorithms 3*(1982).

14. Speckenmeyer, E. On Feedback Problems in Digraphs. In *Proceedings of the Fifteenth Workshop on Graphs, Rolduc, Netherlands,* Springer Verlag, 1989.

15. Yamamoto, A. and Tanaka, H. Translating Production Rules into a Forward Reasoning Prolog Program. *New Generation Computing 4*(1986), 97-105.

Decision problems for Tarski and Presburger arithmetics extended with sets. *

D. CANTONE
Dipartimento di Matematica Pura e Applicata
Università di L'Aquila, Italy
V. CUTELLO
International Computer Science Institute, Berkeley CA
J.T. SCHWARTZ
Department of Computer Science
Courant Institute of Mathematical Sciences
New York University

1 Introduction

In this paper we provide decision procedures for the satisfiability problem of the unquantified theory of sets of reals and the unquantified theory of sets of integers, by combining decision methods for set theory (cf. [CFO90] for an extensive bibliography) with the decision algorithms for the classical Tarski real arithmetic (cf. [Tar51] and [Col75]) and Presburger additive theory of integers (cf. [Pre29] and [Pap81]), respectively.

We will also show that the decision problem for the satisfiability problem of the unquantified Presburger arithmetic with sets is $\mathcal{NP}$-complete.

Other related results can be found in [Ble77], [Ble84],[CFOS87], [CO89a] and [CO89b], where real continuous functions and topological constructs are considered.

If universal quantification over numeric variables is allowed, then the satisfiability problem for both Tarski and Presburger arithmetics extended with sets become undecidable. This will be shown in Section 2. Subsequently, our main decidability result will be established. Finally, some possible applications of our decision procedures are discussed.

2 Two undecidability results

It is well known that the decidability problems for Tarski (see [Tar59]) and Presburger (see [Pre29]) arithmetics are solvable (cf. [Tar51] and [Pre29], respectively). In this section we show that if set-valued variables and elementary set constructs are allowed, then decidability is lost even if one restricts consideration to purely universal prenex formulae where quantifiers can apply only to numeric variables (and even if multiplication is not allowed in Tarski theory of real numbers). These undecidability results

*This work has been partially supported by Archimedes S.R.L. - Catania, Italy and by Progetto Finalizzato Sistemi Informatici e Calcolo Parallelo of C.N.R. of Italy, under grant n. 90.00671.69.

will be obtained by showing that multiplication of integers is expressible in Presburger theory with sets and that the set $\mathcal{N}$ of all the integers is expressible in Tarski theory extended with sets, thus reducing the question to the unsolvability of Hilbert's tenth problem (cf. [Mat70]).

Let $\mathcal{TS}_\forall$ denote the quantified additive Tarski arithmetic extended with sets, where only universal quantification on numeric variables is allowed.

For ease of notation we will use capital letters to denote set valued variables and lower case letters to denote numeric variables.

Then $\mathcal{TS}_\forall$ is the propositional combination of purely universal formulae of type $(\forall x_1)\cdots(\forall x_n)\varphi$ with $x_1,\ldots,x_n$ numeric variables and φ propositional combinations of unquantified atoms of type

$$x = 1, x = 0, z = x + y, z = x - y, x < y, x \leq y,$$
$$x \in X, X = Y \cup Z, X = Y \cap Z, X = Y \setminus Z.$$

Numeric variables can range over the set $\Re$ of all real numbers and accordingly set valued variables can range over the subsets of $\Re$.

The syntax of quantified Presburger arithmetic extended with sets, $\mathcal{PS}_\forall$, is the same as that of $\mathcal{TS}_\forall$. However in the case of $\mathcal{PS}_\forall$-theory, we allow numeric variables to range over the set $\mathcal{N}$ of nonnegative integers and set valued variables to range over the subsets of $\mathcal{N}$ and moreover subtraction is modified in such a way that $x - y = 0$ whenever $x < y$.

Remark 2.1 Notice that since we are concerned with satisfiability problems only, it is not restrictive to use predicates in place of the operations $+$, $-$, $\cup$, $\cap$, $\setminus$, since extra free variables can be easily modeled. ∎

Example 2.1 The following formula is a $\mathcal{TS}_\forall$ (and a $\mathcal{PS}_\forall$) formula:

$$x \in S_1 \wedge (\forall x \in S_1)(\forall y \in S_1)(x + y \in S_2)$$

whereas the following is neither a $\mathcal{TS}_\forall$ nor a $\mathcal{PS}_\forall$ formula:

$$S_1 \subseteq S_2 \wedge x \in S_2 \wedge (\forall S_3)(\forall x)(\forall y)(x \in S_2 \vee y \in S_2 \rightarrow (x + y \in S_3 \rightarrow x + y \in S_2))$$

because quantification over set variables in not allowed in $\mathcal{TS}_\forall$ and $\mathcal{PS}_\forall$.

Theorem 2.1 *The satisfiability problem for the theory $\mathcal{PS}_\forall$ is unsolvable.*

Proof. In view of [Mat70], it is enough to show that multiplication of integers is expressible in $\mathcal{PS}_\forall$. Notice that:

- the predicate $y = \min(S)$ with the intended meaning of *y is the minimum element of the set S* can be defined by the following formula:

$$y \in S \wedge (\forall x \in S)(y \leq x)$$

- the set M_y of multiples of y, for $y \geq 1$, can be defined by:

$$0 \in M_y \wedge (\forall x)(x \in M_y \rightarrow x + y \in M_y) \wedge (\forall x)(0 < x < y \rightarrow x \notin M_y)$$
$$\wedge (\forall x)((x > y \wedge x \in M_y) \rightarrow x - y \in M_y)$$

- the predicate $x = lcm(y, z)$ with the intended meaning of *x is the least common multiple of y and z*, where $y, z \geq 1$, can be expressed by:

$$x = \min(M_y \cap M_z)$$

Then the relationship $x = y \cdot z$, for $y, z \geq 1$, can be expressed by the formula

$$v = x + x \wedge v = lcm(y+z, y+z-1) - lcm(y, y-1) - lcm(z, z-1)$$

Indeed, given any integer $x \geq 2$ the numbers x and $x-1$ are relatively prime. It follows that $lcm(x, x-1) = x \cdot (x-1)$. So $v = (y+z) \cdot (y+z-1) - y \cdot (y-1) - z \cdot (z-1)$, yielding $v = 2 \cdot y \cdot z$. ∎

By the preceding theorem, to prove unsolvability of the satisfiability problem for $\mathcal{TS}_\forall$ it is enough to observe that the set $\mathcal{N}$ of nonnegative integers can be characterized in $\Re$ by the following $\mathcal{TS}_\forall$-formula:

$$\begin{aligned} 0 \in \mathcal{N} \quad &\wedge \quad (\forall x)(x \in \mathcal{N} \rightarrow (x = 0 \vee x \geq 1) \wedge x + 1 \in \mathcal{N}) \\ &\wedge \quad (\forall x)((x \geq 0 \wedge x + 1 \in \mathcal{N}) \rightarrow x \in \mathcal{N}). \end{aligned}$$

In summary, this gives

Theorem 2.2 *The satisfiability problem for the theory $\mathcal{TS}_\forall$ is unsolvable.* ∎

3 The unquantified case

In view of the negative results discussed in the previous section, it is interesting to investigate the decidability problems for Tarski and Presburger arithmetics with sets when quantification is not allowed.

In this section we prove two decidability results for the unquantified theories, of reals and integers respectively, extended with sets and with some additional operators and predicates such as min, max, least upper bound, greatest lower bound, interval.

3.1 Unquantified Tarski real arithmetic extended with sets

By $\mathcal{TS}$ we denote the propositional closure of atomic formulae of the form

(=) $x = 0, x = 1, x = y + z, x = y \cdot z$

($\cup, \setminus$) $X = Y \cup Z, X = Y \setminus Z$

($\in$) $x \in S$

($\leq, <$) $x \leq y, x < y$

(lub) $x = lub(S)$

(glb) $x = glb(S)$

(int) $interval(S)$

where, as usual, lower case letters denote numeric variables which can range over real numbers and upper case letters denote set variables which can range over *bounded* sets of real numbers (we will show later that the boundedness restriction can be dropped).

The meaning of the literal $x = lub(X)$ is that *x is the least upper bound of the set X*. Similarly, a literal of type $x = glb(X)$ stands for the predicate *x is the gratest lower bound of X*. Obviously, the predicate $interval(X)$ is true if and only if X is an interval of the real axis. Several other constructs are immediately expressible in the language $\mathcal{TS}$, e.g. $\cap, \subseteq, \{x\}$ (singleton), open_interval(X), closed_interval(X). For instance, $X = \{x\}$ can be expressed by $x = lub(X) \wedge x = glb(X) \wedge x \in X$ and open_interval(X) can be expressed by $interval(X) \wedge x = glb(X) \wedge y = lub(X) \wedge x \notin X \wedge y \notin X$. Also, the relations $<$ and $\leq$ can be easily generalized to sets with the following meaning

$x < X$ stands for $(\forall y \in X)(x < y)$;

$X < Y$ stands for $(\forall x \in X)(\forall y \in Y)(x < y)$;

Observe that Remark 2.1 concerning the use of predicates in place of operations applies as well here.

The decision procedure for TS contained in the theorem below is based on Tarski's algorithm for real geometry (cf. [Tar51]) and therefore is untractable, at least in the general case. (Much work has been done to find faster algorithms for solving systems of polynomial equalities and disequalities (see [DST88]). The best known algorithm is based on the *Cylindrical Decomposition Technique* and runs in time $O(d^{2^n} m^{2^n})$ (see [Col75]) where d is the maximum degree of any polynomial in the system, m is the size of the system, and n is the number of distinct variables. Recently it has been shown that this bound is tight (see [DH88, Wei88]). No tight bound is known when quantifiers are not allowed. Our algorithm will run in nondeterministic double-exponential time.)

Theorem 3.1 *There is a procedure for deciding the satisfiability problem for TS.*

Proof. Let φ be a conjunction of atoms of the above type and let $\mathcal{N} = \{x_1, \ldots, x_m\}$ and $\mathcal{S} = \{S_1, \ldots, S_n\}$ be the collections of numeric and set variables, respectively, occurring in φ.

Consider all possible orderings $\prec$

$$x_{i_1} \prec x_{i_2} \prec \cdots \prec x_{i_m}$$

of the numeric variables in φ, where $\prec$ can stand for either $<$ or $=$.

Once a set of inequalities is fixed it can be checked for consistency with the clauses of type $(=)$ and $(\leq, <)$ in φ by using a simple variant of Tarski's algorithm.

An $O(m!)$ set of candidate orderings is produced and each of these must be checked for consistency with the remaining clauses.

Once an ordering has been chosen, we can identify all variables that are considered equal in the ordering. Let us then assume, without loss of generality, that we have the following ordering:

$$x_1 < x_2 < \cdots < x_k,$$

with $k \leq m$.

At this point, we introduce *explicit* names for the singletons $\{x\}$ and the open intervals (x_i, x_{i+1}) as follows.

- For each x_i we introduce a new set variable S_i^0 and three conjuncts $x_i \in S_i^0 \wedge x_i = lub(S_i^0) \wedge x_i = glb(S_i^0)$.
- For each x_i, x_{i+1} we introduce a new set variable S_i^* and the conjuncts $interval(S_i^*) \wedge x_i = glb(S_i^*) \wedge x_{i+1} = lub(S_i^*) \wedge x_i \notin S_i^* \wedge x_{i+1} \notin S_i^*$.

The left-to-right order in which the sets S_i^0, S_i^* would have to occur in any model of our set of clauses (assuming that these clauses are satisfiable) is :

$$S_1^0, S_1^*, S_2^0, \ldots, S_{m-1}^0, S_{m-1}^*, S_m^0$$

which will be referred to as the *combinatorial order* of these sets.

To handle the purely set theoretic clauses of type $(\cup, \setminus)$ and $(\in)$, we form the *places* of the Venn diagram (see [CFO90]) of the collection of all the set variables $S_1, \ldots, S_n$ occurring in our collection of clauses φ, along with the additional set variables added in the step above, all these constituting a collection $S_1, \ldots, S_\ell$ of set variables.

We recall here that the places are sets of the form:

$$e = (S_{i_1} \cap \ldots \cap S_{i_s}) \setminus (S_{i_{s+1}} \cup \ldots \cup S_{i_\ell}) \qquad (1)$$

where $i_1, \ldots, i_\ell$ is a permutation of $1, \ldots, \ell$.

Every set S_i has a canonical expression as a disjoint union of places, so each place that intersects (the model of) one of the sets S_i is wholly contained in S_i.

Places can be defined syntactically as maps from the set of variables into $\{0,1\}$ (see [CFO90]) in such a way that:

$$\pi(S_i) = 1 \text{ if and only if } e_\pi \subseteq S_i.$$

The collection of places has clearly an exponential bound in the number of set variables. In fact, as in [COP90], it can be shown that only a quadratic number of places are needed (though they are to be chosen nondeterministically).

The clauses of type $(\cup, \setminus), (\in)$ simply constrain the collection of the sets e which can be nonnull. We enumerate all the patterns of nonnull e which are consistent with the $(\cup, \setminus), (\in)$ clauses. These, and the candidate ordering $x_1 < x_2 < \cdots < x_k$ of the variables x_i, must then be tested for consistency with the remaining clauses. This is done by imposing conditions on the places as follows.

(C1) In any model of our set of clauses, each place e would have to be contained in exactly one of the sets S_i^0, S_i^*. We therefore define a map Φ, which we will call *combinatorial assignment* of the places and we put $\Phi(e) = S_i^0$ if $e \subseteq S_i^0$ and $\Phi(e) = S_i^*$ if $e \subseteq S_i^*$. Notice that this will always be known from the expression (1) for e.

(C2) The presence of the conjunct $x_j = glb(S)$ is then equivalent to the condition that no e such that $\Phi(e)$ comes earlier in the combinatorial order than S_j^0 can appear in the decomposition of S.

(C3) The presence of the conjunct $x_j = lub(S)$ is equivalent to the condition that no e such that $\Phi(e)$ comes later in the combinatorial order than S_j^0 can appear in the decomposition of S.

(C4) The presence of a conjunct of type *interval*(S) implies that:

1. there exist x_i, x_j such that every e for which $\Phi(e)$ lies between S_i^0 and S_j^0 in the combinatorial order appears in the decomposition of S (however the two sets S_i^0, S_j^0 may or may not appear in the decomposition of S) and vice versa.
2. The union of all the places e which appear in the decomposition of S_i^* must be the entire open interval (x_i, x_{i+1}). This condition which is less explicitly combinatorial than the others will be called *the first subsidiary condition.*

(C5) The presence of a conjunct $x_i = lub(S)$ (resp. $x = glb(S)$) implies that:

1. either the place S_i^0 appears in the decomposition of S or at least one place e such that $\Phi(e) = S_{i-1}^*$ (resp. $\Phi(e) = S_i^*$) appears in the decomposition of S;
2. in the latter case above one of the places with this property must satisfy $x_i = lub(e)$ (resp. $x_i = glb(e)$); this is again a noncombinatorial condition, which we will call *the second subsidiary condition.*

(C6) Each set S_i^0 must be interpreted as a singleton (*third subsidiary condition*).

The theory is decidable if all essential possibilities can be examined comprehensively. To establish that this is the case, we need just to show that the three subsidiary conditions imply no additional combinatorial constraints.

To this end, we collect for each i all the nonnull places e such that $\Phi(e) = S_i^*$. The combinatorial (i.e. the non-subsidiary) conditions require only that these e be mutually disjoint subsets of the open interval (x_i, x_{i+1}). Hence, they can be modeled by any collection $\sigma_1, \ldots, \sigma_t$ of disjoint subsets of this interval, each of which is everywhere dense in (x_i, x_{i+1}); to ensure that the first subsidiary condition is satisfied we have only to insist that $\sigma_1 \cup \cdots \cup \sigma_t = (x_i, x_{i+1})$. By imposing this condition we obviously ensure that the first and second subsidiary conditions are always satisfied. Finally, we interpret S_i^0 as $\{x_i\}$, proving decidability.

■

Remark 3.1 By applying the ellipsoid method for linear programming, it follows that our procedure runs in nondeterministic polynomial time, at least in the case in which multiplication is absent. In presence of multiplication, the cylindrical decomposition technique gives an upper bound of nondeterministic double-exponential time. ■

Remark 3.2 It is easy to see that a set of clauses can be modeled using unbounded sets if and only if it can be modeled using bounded sets only. Moreover, the extended language in which a predicate *bounded*(S) is present can readily be reduced to the case we have treated by mapping the real axis topologically to a bounded open interval $(x_{-\infty}, x_{+\infty})$. This allows the predicate *bounded*(S) to be expressed as

$$x_{-\infty} \neq glb(S) \wedge x_{+\infty} \neq lub(S).$$

The same remark also proves that allowing set variables to range over bounded sets only is not restrictive. ■

3.2 Presburger unquantified arithmetic extended with sets

In this subsection we will prove a result analogous to that established above, but now extended to apply to Presburger's unquantified theory of integers extended with sets, denoted $\mathcal{PS}$. In particular we will show that the satisfiability problem for $\mathcal{PS}$-formulae is solvable, in fact that it belongs to $\mathcal{NP}$ and is $\mathcal{NP}$-complete. To this purpose we recall that the satisfiability problem for unquantified Presburger arithmetic is already known to be $\mathcal{NP}$-complete (see [GN72], [Pap81], [Sal75]). Solvability of the satisfiability problem is proved by showing that if there is a solution then there is a solution in an a priori exponentially bounded set of integers.

In [FOS80] the decidability of a subset of set theory, denoted by MLS, involving the basic set-theoretic operators $\cup, \cap, \setminus$ and the membership predicate, was proved. Moreover, in [COP90] it was shown that the satisfiability problem for MLS is $\mathcal{NP}$-complete. The proof is based on a technique, *graph modeling*, introduced in [PP90].

Here we combine the two techniques to prove the decidability of the collection of $\mathcal{PS}$-formulae and to establish a complexity bound.

A $\mathcal{PS}$-formula is any propositional combination of atomic formulae of type

(=) $x = 0, x = 1, x = y + z$

($<, \leq$) $x < y, x \leq y$

($\in$) $x \in X$

($\emptyset$) $X = \emptyset$

($\cup, \setminus$) $X = Y \cup Z, X = Y \setminus Z$

(min) $x = \min(X), \min(X) = -\infty$

(max) $x = \max(X), \max(X) = +\infty$

where x, y, z stand for numeric variables and X, Y, Z stand for set variables. Numeric variables can range over the set $\mathcal{Z}$ of integers, whereas set variables can range over the collection of subsets of $\mathcal{Z}$.

Without loss of generality, we can limit consideration to normalized conjunctions of $\mathcal{PS}$, i.e. to conjunctions φ of atoms of the above types, such that for every set variable X occurring in φ either φ contains exactly one atom of type (min) and one of type (max) involving X or φ contains the atom $X = \emptyset$.

Obviously, the intended meaning of an atom of type $\min(X) = -\infty$ (resp. $\max(X) = +\infty$) is that X is unbounded below (resp. above). Therefore, we shall agree that formulae of type $x = \min(X) \wedge X = Y \wedge \min(X) = -\infty$ must be unsatisfiable.

Let φ be a normalized conjunction of $\mathcal{PS}$. We want to derive conditions which are necessary (and sufficient) for φ to be satisfiable. Assume therefore that φ is satisfiable and let M be a model for φ.

Let $\mathcal{N}$ and $\mathcal{S}$ be the collections of numeric and set variables and constants present in φ. Then we can define the equivalence relations $\sim_{\mathcal{N}}$ (over $\mathcal{N}$) and $\sim_{\mathcal{S}}$ (over $\mathcal{S}$) by putting

$$x_i \sim_{\mathcal{N}} x_j \text{ if } Mx_i = Mx_j, \quad x_i, x_j \in \mathcal{N}$$
$$S_i \sim_{\mathcal{S}} S_j \text{ if } MS_i = MS_j, \quad S_i, S_j \in \mathcal{S}.$$

The formula resulting from φ after identification of equivalent variables in it (which, for simplicity, we continue to denote with φ) is *injectively* satisfiable, i.e. it is satisfiable by models which map distinct variables into distinct values.

Notice that $\sim_{\mathcal{N}}$ and $\sim_{\mathcal{S}}$ can be determined in nondeterministic $\mathcal{O}(|\varphi|)$ time, so that we can restrict consideration to the problem of injective satisfiability for normalized $\mathcal{PS}$-conjunctions without affecting the complexity of the resulting decision procedure. Accordingly, we will assume that M is an injective model for φ.

For any pair of set variables $S_i, S_j \in \mathcal{S}$ for which $MS_i \not\subseteq MS_j$ there exists an element a_{ij} such that $a_{ij} \in MS_i \setminus MS_j$. We associate a new numeric variable t_{ij} to each of these elements a_{ij} and put $Mt_{ij} = a_{ij}$. Let T be the collection of these newly introduced numeric variables. Clearly $|T| \leq m(m-1)/2$, where $m = |\mathcal{S}|$.

For any set variable S such that MS is infinite we have:

- if $\min(MS) = -\infty$ there exists an element $b_- \in MS$, such that $b_- < \min(\{Mx : x \in \mathcal{N} \cup T\})$;
- if $\max(MS) = +\infty$ there exists an element $b_+ \in MS$, such that $b_+ > \max(\{Mx : x \in \mathcal{N} \cup T\})$.

We associate new numeric variables u_-, u_+ to each of these elements b_-, b_+ and put $Mu_- = b_-, Mu_+ = b_+$. Let U_- and U_+ be the collections of these newly introduced variables of type u_-, u_+ respectively. Notice that $|U_-| \leq m$ and $|U_+| \leq m$.

It is now possible to define a directed graph $G = (\mathcal{N} \cup \mathcal{S} \cup T \cup U_- \cup U_+, E)$ such that for any two nodes $v_1, v_2 \in \mathcal{N} \cup \mathcal{S} \cup T \cup U_- \cup U_+$ the edge from v_1 into v_2 is in E if and only if $v_1 \notin \mathcal{S}, v_2 \in \mathcal{S}$ and $Mv_1 \in Mv_2$.

The above discussion shows that the following condition must hold if our collection is injectively satisfible:

Condition A: *There exists a directed graph $G = (N, E)$ with N and E as set of nodes and edges respectively, such that*

1. *$N = \mathcal{N} \cup \mathcal{S} \cup Aux \cup Aux_- \cup Aux_+$ where Aux, Aux_-, Aux_+ are three mutually disjoint collections of nodes disjoint from $\mathcal{N}$ and $\mathcal{S}$ and such that $|Aux| \le m(m-1)/2$, $|Aux_-| \le m$ and $|Aux_+| \le m$.*

2. *The in-degree of any node in $\mathcal{N} \cup Aux \cup Aux_+ \cup Aux_-$ as well as the out-degree of any node in $\mathcal{S}$ is zero[1].*

   Moreover, putting $E(S) = \{z : z \to S \text{ is in } E\}$ for all $S \in \mathcal{S}$, we have

3. *$E(\emptyset) = \emptyset$;*

4. *$E(S_i) \neq E(S_j)$ for $S_i \neq S_j$, $S_i, S_j \in \mathcal{S}$;*

5. *$E(S) \cap Aux_+ \neq \emptyset$ if and only if $\max(S) = +\infty$ is in φ;*

6. *$E(S) \cap Aux_- \neq \emptyset$ if and only if $\min(S) = -\infty$ is in φ;*

7. *if $S_i = S_j \cup S_h$ is in φ then $E(S_i) = E(S_j) \cup E(S_h)$;*

8. *if $S_i = S_j \setminus S_h$ is in φ then $E(S_i) = E(S_j) \setminus E(S_h)$;*

9. *if any of the literals $x = \min(S)$, $x = \max(S)$, with $x \in \mathcal{N}$, is in φ then $x \in E(S)$.*

∎

Next we derive a supplementary condition necessary for satisfiability. Let Ξ be the system of linear equalities and disequalities in the set of integer-valued unknowns $\mathcal{N} \cup T \cup U_+ \cup U_-$ consisting of

- all (dis)equalities in φ of type $(=)$ and $(<, \le)$;
- formulae $z_i \le z_j$ for each $z_i, z_j \in \mathcal{N} \cup T \cup U_+ \cup U_-$ such that there exists a literal $z_j = \max(S)$ in φ for which $z_i \in E(S)$;
- formulae $z_i \ge z_j$ for each $z_i, z_j \in \mathcal{N} \cup T \cup U_+ \cup U_-$ such that there exists a literal $z_j = \min(S)$ in φ for which $z_i \in E(S)$;
- formulae $z_i < z_j$ for all $z_i \in \mathcal{N} \cup T \cup U_-$ and $z_j \in U_+$;
- formulae $z_j < z_i$ for all $z_i \in \mathcal{N} \cup T \cup U_+$ and $z_j \in U_-$;
- formulae $z_i \neq z_j$ for any two distinct $z_i, z_j \in \mathcal{N} \cup T \cup U_+ \cup U_-$;

Notice that Ξ is entirely defined by φ and G. Notice also that:

- any inequality of type $z_i \neq z_j$ is equivalent to $z_i > z_j \vee z_i < z_j$;
- any inequality of type $z_i \le z_j$ (resp. $z_i < z_j$) is equisatisfiable with $z_j = z_i + z_i' \wedge z_i' \ge 0$ (resp. $z_j = z_i + z_i' + 1 \wedge z_i' \ge 0$), where z_i' is a newly introduced variable.

Thus, from Ξ we can derive a collection of systems $\Xi_i, i = 1, \ldots, \ell$ where $l \le 2^{|\Xi|}$, such that

- each Ξ_i consists of equations of type $z_i = z_j + z_h, z_i = z_j + z_h + 1, z_i \ge 0$;
- each Ξ_i contains all the variables in Ξ;

[1] We recall here that the in-degree of a node v is equal to $|\{w : w \to v\}|$ whereas the out-degree of v is equal to $|\{w : v \to w\}|$.

- Ξ is solvable if and only if some Ξ_i is solvable, $i = 1, \ldots, l$;
- the restriction of any solution of Ξ_i to the variables of Ξ is also a solution for Ξ, $i = 1, \ldots, \ell$;
- the size of each system Ξ_i is at most twice the size of Ξ, $i = 1, \ldots, \ell$;
- the number of distinct variables in each Ξ_i is at most three times the size of Ξ_i, $i = 1, \ldots, \ell$ (since any (dis)equality in Ξ_i involves at most three distinct variables).

Clearly, the extended model M satisfies Ξ. Thus some Ξ_i must also be satisfiable. Moreover, if we put $K = |\Xi|$, then $|\Xi_i| \leq 2K$ and the number of distinct variables occurring in Ξ_i is at most $6K$. Therefore, from [Pap81] it follows that Ξ_i, and a fortiori Ξ, must have a solution which is bounded in absolute value by $6K \cdot (2K)^{2K+1}$. Notice that $6K \cdot (2K)^{2K+1} < 2^{4K^2+2K+2}$, for all $K \geq 0$. Thus the following condition must hold.

Condition B: *The system Ξ must have a solution in $\{0, \pm 1, \pm 2, \ldots, \pm 2^{4K^2+2K+2}\}$.* ∎

The above discussion can be summarized in the following lemma.

Lemma 3.1 *Conditions A and B are necessary for a normalized conjunction of $\mathcal{PS}$ to have an injective model.* ∎

Next we prove that conditions A and B are also sufficient for a normalized conjunction φ of $\mathcal{PS}$ to be injectively satisfiable.

To this end, let $G = (\mathcal{N} \cup \mathcal{S} \cup Aux \cup Aux_- \cup Aux_+, E)$ be a directed graph such that properties 1-9 of condition A hold. Also, let

$$\alpha : \mathcal{N} \cup Aux \cup Aux_- \cup Aux_+ \mapsto \{0, \pm 1, \pm 2, \ldots, \pm 2^{4K^2+2K+2}\}$$

be a solution of the system Ξ, where Ξ is derived from G and φ as above and $K = |\Xi|$.

To each distinct node $w \in Aux_+ \cup Aux_-$ we associate a distinct prime number p_w such that

$$p_w \geq \max(\{|\alpha(u)| : u \in Aux_+ \cup Aux_-\}).$$

Let us also put

$$I_w = \begin{cases} \{p_w^k : k = 1, 2, 3, \ldots\} & \text{if } w \in Aux_+ \\ \{-p_w^k : k = 1, 2, 3, \ldots\} & \text{if } w \in Aux_- \end{cases}.$$

Then, plainly,

Lemma 3.2 *For any $w, w' \in Aux_+ \cup Aux_-$, if $w \neq w'$ then $I_w \cap I_{w'} = \emptyset$.* ∎

We are now ready to define an injective model M^* for φ.

Put

$$M^*v = \begin{cases} \alpha(v) & \text{if } v \in \mathcal{N} \cup Aux \cup Aux_- \cup Aux_+ \\ \{M^*w : w \in E(v)\} \cup \bigcup_{w \in E(v) \cap (Aux_- \cup Aux_+)} I_w & \text{if } v \in \mathcal{S}. \end{cases} \quad (2)$$

In view of the inequalities $z_i \neq z_j$ in Ξ for every pair of distinct variables z_i, z_j in Ξ, it follows that M^* is injective at least over $\mathcal{N} \cup Aux \cup Aux_- \cup Aux_+$. The injectivity of M^* over $\mathcal{S}$ follows from Lemma 3.2 and condition A4.

Next we prove that M^* satisfies all conjuncts in φ. Obviously M^* satisfies all conjuncts in φ of type $(=)$ and $(<, \leq)$ since it coincides with the solution α of Ξ over numeric variables. Thus it remains to show that literals in φ of type $(\in)$, $(\cup, \setminus)$, (min) and (max) are also correctly modeled by M^*.

Clauses of type ($\in$).

If $x \in S$ occurs in φ, then by condition A9 the node x is in $E(S)$ which by (2) immediately implies $M^*x \in M^*S$.

Clauses of type ($\cup, \setminus$).

If $S_i = S_j \cup S_h$ is in φ, then by condition A7, $E(S_i) = E(S_j) \cup E(S_h)$. Thus from (2) we have $M^*S_i = M^*S_j \cup M^*S_h$.

If $S_i = S_j \setminus S_h$ is in φ, then by condition A8, $E(S_i) = E(S_j) \setminus E(S_h)$. Hence, in view of the injectivity of M^* and from Lemma 3.2 we can conclude that $M^*S_i = M^*S_j \setminus M^*S_h$.

Clauses of type (min).

Let $x = \min(S)$ occur in φ. By condition A9, $x \in E(S)$, so that $M^*x \in M^*S$. Thus to prove that $M^*x = \min(M^*S)$ it only remains to show that $s > M^*x$ for all $s \in M^*S \setminus \{M^*x\}$. Notice that from condition A5, $E(S) \cap Aux_- = \emptyset$. Moreover, if $s \in M^*S \cap I_w$ for some $w \in E(S) \cap Aux_+$, then $s \geq p_w \geq \max(\{|\alpha(u)| : u \in Aux_+ \cup Aux_-\}) > \alpha(x) = M^*x$, since $x < u$ is present in Ξ for every $u \in Aux_+$. On the other hand, if $s = M^*w$, for some $w \in E(S)$, $w \neq x$, then the literals $x \leq w \wedge x \neq w$ would be present in Ξ, so that $M^*x = \alpha(x) < \alpha(w) = M^*w$. Hence, $M^*x = \min(M^*S)$ holds.

Next assume that $\min(S) = -\infty$ occurs in φ. In this case, by condition A5 we have $E(S) \cap Aux_- \neq \emptyset$. Therefore by (2) and the way sets I_w are defined we have at once $\min(M^*S) = -\infty$.

Clauses of type (max).

These can be dealt with in the same way as the clauses of type (min).

Summing up we have shown that M^* is an injective model for φ thus proving

Lemma 3.3 *Conditions A and B are also sufficient for a normalized conjunction of $\mathcal{PS}$ to have an injective model.* ∎

Lemmas 3.1 and 3.3 yield the decidability of the satisfiability prolblem for the theory $\mathcal{PS}$ since conditions A and B are algorithmically verifiable.

It is easy to prove that the satisfiability problem for normalized conjunctions of $\mathcal{PS}$ is $\mathcal{NP}$-complete. Since integer linear programming is $\mathcal{NP}$-complete (cf. [Pap81]), it follows immediately that our problem is $\mathcal{NP}$-hard.

The following nondeterministic polynomial time algorithm for deciding the satisfiability of a given normalized conjunction φ of $\mathcal{PS}$ summarizes the preceding discussion.

Algorithm

STEP 0. Let $\mathcal{N}$ and $\mathcal{S}$ be the collection of numeric and set variables and constants in φ, respectively.

STEP 1. Guess the equivalence relations $\sim_{\mathcal{N}}$ over $\mathcal{N}$ and $\sim_{\mathcal{S}}$ over $\mathcal{S}$ and identify in φ equivalent variables.

STEP 2. Guess a graph $G = (\mathcal{N} \cup \mathcal{S} \cup Aux \cup Aux_- \cup Aux_+, E)$ satisfying conditions A1-A9, with $|Aux| \leq m(m-1)/2$, $|Aux_-| \leq m$, and $|Aux_+| \leq m$, where $m = |\mathcal{S}|$.

STEP 3. Build the system Ξ and guess a solution α for it in the range $\{0, \pm 1, \ldots, \pm 2^{4K^2+2K+2}\}$, where $K = |\Xi|$. (Observe that $K \leq |\varphi| + 3(|\mathcal{N} \cup Aux \cup Aux_- \cup Aux_+)^2$.)

End

It is easily seen that STEPS 1-3 can all be executed in nondeterministic polynomial time in $|\varphi|$. Thus we have

Theorem 3.2 *The satisfiability problem for $\mathcal{PS}$-formulae is $\mathcal{NP}$-complete.* ∎

Remark 3.3 Notice that congruence relations with respect to a fixed constant $n \in \mathcal{N}$ can easily be expressed by $\mathcal{PS}$-formulae. Indeed,

$$x \equiv y\,(n) \quad \text{is equisatisfiable with} \quad x + n \cdot z = y\,,$$

$$\neg x \equiv y\,(n) \quad \text{is equisatisfiable with} \quad \bigvee_{i=1}^{n-1} x \equiv y + i \cdot 1\,(n)\,.$$

∎

In the next section we discuss a few examples of possible applications of the decision procedures for the theories $\mathcal{TS}$ and $\mathcal{PS}$ to theorem proving and verification of correctness of annotated programs.

4 Applications

4.1 The theory $\mathcal{TS}$.

The decidability of the theory $\mathcal{TS}$ can be used to prove certain elementary theorems in real analysis automatically . Below follow some examples.

Example 4.1 *If $S_1 \subseteq S_2$ and $S_1 \neq \emptyset$ then $lub(S_1) \leq lub(S_2)$ and $glb(S_1) \geq glb(S_2)$.*

This can be proved by, first negating the assertion which gives

$$S_1 \neq \emptyset \wedge S_1 \subseteq S_2 \wedge (lub(S_1) > lub(S_2) \vee glb(S_1) < glb(S_2))$$

This splits into the conjunctions

1. $S_1 \neq \emptyset \wedge S_1 \subseteq S_2 \wedge lub(S_1) > lub(S_2)$
2. $S_1 \neq \emptyset \wedge S_1 \subseteq S_2 \wedge glb(S_1) < glb(S_2)$

to which we can apply the decision procedure for $\mathcal{TS}$ to show unsatisfiability. ∎

Other examples of theorems that can be automatically proved include the following (the first being a sort of converse of the above one).

Example 4.2 *If S_1, S_2 are two bounded intervals and $lub(S_1) \leq lub(S_2)$ and $glb(S_1) \geq glb(S_2)$ then $S_1 \subseteq S_2$.* ∎

Example 4.3 If we define *open* and *closed* intervals as follows:

$$open\_interval(S) \leftrightarrow interval(S) \wedge glb(S) \notin S \wedge lub(S) \notin S$$

$$closed\_interval(S) \leftrightarrow interval(S) \wedge glb(S) \in S \wedge lub(S) \in S$$

the following theorems can be automatically proved:

- *if S_1, S_2 are two open (resp. closed) intervals then their intersection (union) is open (resp. closed).*
- *if S_1 is an open interval and S_2 is a closed interval then their intersection (union) can be either closed or open or non-open and non-closed*

∎

4.2 Applications of the theory $\mathcal{PS}$.

A decidability algorithm for the theory $\mathcal{PS}$ can be quite useful in automated proof of correctness of small program fragments. We shall show various examples, written using a SETL-like programming language, to which this observation applies.

Example 4.4 Consider for example the following *Find_min1* procedure to find the minimum of a two element set $\{x_1, x_2\}$.

proc Find_min1;

> **Input:** integers x_1, x_2
> **Output:** $\min(\{x_1, x_2\})$
>
> $\{true\}$
> $m := x_1;$
> **if** $x_2 < m$ **then** $m := x_2;$
> **end if**
> $\{m = \min(\{x_1, x_2\})\}$

end Find_min1

It is an easy matter to see that the verification conditions for the above annotated procedure are:

- $x_2 \geq x_1 \rightarrow x_1 = \min(\{x_1, x_2\})$ and
- $x_2 < x_1 \rightarrow x_2 = \min(\{x_1, x_2\})$,

which can easily be verified by the decision test for $\mathcal{PS}$. Likewise one could automatically test the correctness of the following procedure *Find_min2*, which computes the minimum of a finite set $\{x_1, \ldots, x_n\}$ of integers. (Notice that finite enumerations are expressible in the language $\mathcal{PS}$. Namely, the literal $X = \{x\}$ is equivalent to $\min(X) = \max(X) = x$, and, inductively,

$$\{x_1, \ldots, x_n\} =_{Def} \{x_1, \ldots, x_{n-1}\} \cup \{x_n\}.)$$

proc Find_min2;

> **Input:** a sequence of integers $x_1, x_2, \ldots, x_n$
> **Output:** $\min(\{x_1, x_2, \ldots, x_n\})$
>
> $\{true\}$
> $m := x_1;$
> **for** i **in** $[2 \,..\, n]$ **loop**
> $\{m = \min(\{x_1, \ldots, x_{i-1}\}) \wedge i < n\}$
>   **if** $x_i < m$ **then** $m := x_i;$
>   **end if**
> **end loop**
> $\{m = \min(\{x_1, \ldots, x_n\})\}$

end Find_min2

In this case the verification conditions are

- $m = x_1 \to m = \min(\{x_1\})$,
- $\bigwedge_{i=2,\ldots,n}(m = \min(\{x_1,\ldots,x_{i-1}\}) \wedge x_i < m \to x_i = \min(\{x_1,\ldots,x_i\}))$, and
- $\bigwedge_{i=2,\ldots,n}(m = \min(\{x_1,\ldots,x_{i-1}\}) \wedge x_i \leq m \to m = \min(\{x_1,\ldots,x_i\}))$.

Example 4.5 The preceding examples can be further generalized to the case of a generic set S of integers. In this case to prove the partial correctness of the following procedure *Find_min3* one has to extend the theory $\mathcal{PS}$ with a choice function η characterized only by the following axiom

$$A \neq \emptyset \to \eta A \in A$$

(see [FO87], [Omo84] and [PP90], for other results related to a choice function η.)

proc Find_min3;

Input: a nonempty set S of integers
Output: $\min(S)$

$\{S \neq \emptyset\}$
$A := S$;
take m from A;
$\{m = \min(S \setminus A) \wedge A \subseteq S\}$
while $A \neq \emptyset$ **loop**
  $\{m = \min(S \setminus A) \wedge A \subseteq S \wedge A \neq \emptyset\}$
  take x from A;
  if $x < m$ **then** $m := x$;
  end if
end loop
$\{m = \min(S)\}$
end Find_min3

In this case the verification conditions are the following:

- $S \neq \emptyset \to (\eta S = \min(S \setminus (S \setminus \{\eta S\})) \wedge S \setminus \{\eta S\} \subseteq S)$,
- $(m = \min(S \setminus A) \wedge A \subseteq S \wedge A \neq \emptyset \wedge \eta A < m) \to (\eta A = \min(S \setminus (A \setminus \{\eta A\})) \wedge A \setminus \{\eta A\} \subseteq S)$,
- $(m = \min(S \setminus A) \wedge A \subseteq S \wedge A \neq \emptyset \wedge \eta A \geq m) \to (m = \min(S \setminus (A \setminus \{\eta A\})) \wedge A \setminus \{\eta A\} \subseteq S)$,
- $(m = \min(S \setminus A) \wedge A \subseteq S \wedge A = \emptyset) \to m = \min(S)$.

The decision procedure for the theory $\mathcal{PS}$ can be easily generalized to deal also with the above formulae, under the sole assumption that for all nonempty sets A, $\eta A \in A$. ∎

Acknowledgments. The authors wish to thank an anonymous referee for his/her comments and suggestions.

References

[Ble77] W.W. Bledsoe. Non-resolution theorem proving. *J. Art. Int.*, 9:1–35, 1977.

[Ble84] W.W. Bledsoe. Some automatic proof in analysis. *Contemporary AMS editor*, Automated Theorem Proving after 25 years, 1984.

[CFO90] D. Cantone, A. Ferro, and E.G. Omodeo. *Computable Set Theory*. Oxford University Press, 1990.

[CFOS87] D. Cantone, A. Ferro, E. Omodeo, and J.T. Schwartz. Decision algorithms for some fragments of Analysis and related areas. *Comm. Pure App. Math.*, XL:281–300, 1987.

[CO89a] D. Cantone and E. Omodeo. On the decidability of formulae involving continuous and closed functions. In N.S. Sridharam, editor, *Eleventh Int. Joint Conf. on Art. Intell.*, pages 425–430, 1989.

[CO89b] D. Cantone and E. Omodeo. Topological syllogistic with continuous and closed functions. *Comm. Pure App. Math.*, XLII, n. 8:1175–1188, 1989.

[Col75] G.E. Collins. Quantifier elimination for real closed fields by cylindrical algebraic decomposition. *Proc. 2nd GI Conf. Automata Theory and Formal Languages. Springer Lecture Notes in CS*, 33:134–183, 1975.

[COP90] D. Cantone, E. Omodeo, and A. Policriti. The automation of syllogistic. II. Optimization and complexity issues. *Journal of Automated Reasoning*, 6:173–187, 1990.

[DH88] J.H. Davenport and J. Heintz. Real quantifier elimination is doubly exponential. *J. Symb. Comp.*, 5:29–35, 1988.

[DST88] J.H. Davenport, Y. Siret, and E. Tournier. *Computer Algebra. Systems and algorithms for algebraic computation.* Academic Press Limited, 1988.

[FO87] A. Ferro and E. Omodeo. Decision procedures for elementary sublanguages of set theory. VII. validity in set theory when a choice operator is present. *Comm. Pure App. Math.*, XL:265–280, 1987.

[FOS80] A. Ferro, E. Omodeo, and J.T. Schwartz. Decision procedures for elementary sublanguages of set theory. I. Multilevel syllogistic and some extensions. *Comm. Pure App. Math.*, XXXIII:599–608, 1980.

[GN72] R.S. Garfinkel and G.L. Nemhauser. *Integer programming.* John Wiley & Sons, Inc., New York, 1972.

[Mat70] Y. Matijasevič. Enumerable sets are Diophantine sets. *Soviet Math. Doklady*, 11:354–357, 1970.

[Omo84] E.G. Omodeo. *Decidability and proof procedures for set theory with a choice operator.* PhD thesis, New York University, 1984.

[Pap81] C.H. Papadimitriou. On the complexity of Integer Programming. *Journal of ACM*, 28, N. 4:765–768, 1981.

[PP90] F. Parlamento and A. Policriti. Decision procedures for elementary sublanguages of set theory. XIII. Model graphs, reflection and decidability. *To appear in J. Automated Reasoning*, 1990.

[Pre29] M. Presburger. Über die Vollständigkeit eines gewissen Systems der Arithmetic ganzer Zahlen, in welchem die Addition als einzige Operation hervortritt. In *Comptes-rendus du Premier Congrès des Mathematiciens des Pays Slaves*, pages 192–201,395. Warsaw, 1929.

[Sal75] H.M. Salkin. *Integer programming*. Addison-Wesley Publishing Co., Inc., Reading, Mass., 1975.

[Tar51] A. Tarski. A decision method for elementary algebra and geometry. *Univ. of California Press, Berkeley*, 2nd ed. rev., 1951.

[Tar59] A. Tarski. What is elementary geometry? In L. Henkin, P. Suppes, and A. Tarski, editors, *The axiomatic method with special reference to geometry and physics*, pages 16–29, 1959.

[Wei88] V. Weispfenning. The complexity of linear problems in fields. *J. Symb. Comp.*, 5:3–28, 1988.

A Fast Garbage Collection Algorithm for WAM – based PROLOG

Igor Đurđanović
Zglog Ltd, Hrgovići 47, 41000 Zagreb, Yugoslavia

Abstract

Garbage collection for Warren abstract machine is complicated by:

- three semantically distinct types of pointers, whose types must be preserved as well as their relative positions;
- need for marking used atomic values because of TRAIL compactification;
- capability of structures to grow incrementally, i.e. in the wrong direction.

Prolog data graphs (DAGs, circular structures would interrput the present algorithm) are represented as linear structures by an invertible pointer reversing transform. Reversal of forward pointers is delayed till needed, so that the algorithm can manage with one pass through the workspace. The technique of delayed reversal enables compactification of (two or more) physically separated memory areas which point at each other.

If the workspace contains n used words k of which are active, the complexity of the algorithm is

$$\alpha\, n \;+\; \beta\, k \;+\; OVERHEAD\,,$$

where $OVERHEAD$ depends on the structure of active data graphs involved.

Introduction

Garbage collection algorithms for WAM based Prolog (cf. Warren) must handle several data types:

- atoms
- integers
- lists
- structures
- variables

which can reside in several memory areas.

While atoms, integers and variables are represented with single words (appropriately tagged), lists and structures are represented by tagged pointers pointing at several consecutive words (cf. Warren). Such is the logical nature of Prolog variables that they can get instantiated to anything at any time, creating data structures which grow both ways.

Data structures in general grow on the HEAP, a stack which shrinks on backtracking. Bindings of variables from the HEAP and from a local STACK (which holds arguments, environments and most backtracking information) may then have to be unbound, according to binding records kept on the TRAIL.

In order to preserve the integrity of WAM state, a garbage collector for the HEAP has to meet the following requirements:

- preserve relative positions of active cells or words
- preserve pointer types
- compactify the TRAIL space (to prevent creation of dangling TRAIL pointers) .

Our algorithm is of **mark_&_copy** type, using pointer reversal (cf. Morris) in the marking phase. Reversing forward pointers is delayed till required by the copy phase, which is crucial for performance - it enables the copying to be done in one pass through the data areas. The most important improvement over the algorithm of Appleby et al. consists precisely in this technique of delayed reversal.

In sections **1,2** we describe the respective phases of the algorithm, disregarding forward pointers. The technique of delayed reversal of forward pointers is explained in section **3**. In section **4** we briefly comment TRAIL compactification and compactification of multiple interlinked data areas. In section **5** we discuss complexity of each phase and present some empirical data.

This algorithm has been implemented in ZGLOG/386 Prolog system by Zglog Ltd of Zagreb.

Acknowledgements

I am indebted to Dean Rosenzweig for guidance and criticism, to Egon Boerger for encouragement to present the paper for CSL '90, and to Zagrebačka Banka of Zagreb for encouragement and financial support of the whole ZGLOG project. The suggestions and remarks of anonymous referees have helped to improve the paper.

1 Marking phase

Marking must transform Prolog data graphs to marked linear structures, from which the former are to be restorable. This sets the following requirements for the marking algorithm:

- if several pointers point to the same cell, this fact must be reflected in the marks so that compactification can restore them at the same time correctly;

- the result of compactification may not depend on the time when a cell is put in reversed chain, the position of cell may vary according to marking time;
- if pointers of distinct types point to same object, we will assume object size is of larger type, and have to restore distinct types correctly;
- active atomic values have to be marked as well, to enable detection and removal of dead TRAIL pointers.

We will not describe various possible optimizations for marking cells from state words, TRAIL compactification, heap backtrack pointers handling, early reset of variable etc. because they have been described by Appleby et al.

We shall need some notation. Let

`value(w), adress(w), type(w), mark(w)`

denote the `value`, `adress`, `type`, `mark` of a word `w`. The `mark` is initially `Unmarked`. If `type(w)` is a pointer type, we take `value(w)` to be the word pointed at. If `value(w)` is a functor, we take its type to be atomic, and its mark to be undefined (it is irrelevant).

1.1 Marking algorithm

Algorithm for `mark_word(v)` will proceed according to cases of `type(v)`:

```
Atomic:
    mark(v) := Marked

Pointers:
    if  mark(v) =\= Unmarked
    then
        proceed                 (* end of chain *)

    if  type(v) = Reference         (* unbound *)
        &
        value(v) = adress(v)
    then
        value(v) := MaxAtoms
        type(v)  := Atomic
        mark(v)  := Marked
    else
        mark(v) := Marked       (* premarking *)
        w := value(v)
        if  type(v) = List
        then
            mark_word(next(w))

        if  type(v) = Structure
            &
            type(value(v)) = Functor
        then
```

```
        for all arguments v(i) in structure do
            mark_word(v(i))
        endfor
    else
       mark_word(w)
    if  mark(v) =\= Marked (* is circular *)
    then
        Error(Circular structure)

    reverse(v,w)
```

where `address(next(w))` = `adress(w)` + `Wordsize`.

Marking is terminated by reaching a non-pointer or a previously marked cell. Unbound variables are handled by seting them internaly to `MaxAtoms` (where `MaxAtoms` is a constant representing the highest possible atom number, properly tagged and marked). The copy phase would detect this situation and restore reference pointer. When the marking algorithm sees a pointer, premarking is done to enable detection of circular structures. If the pointer is of type list we will first mark its tail. If the pointer is of type structure and points to a functor (this ensures that the structure is not already marked) we mark all of its arguments, otherwise (i.e. in cases of reference or pointers from head of a list) we recursively continue to mark other active cells pointed from here. The description of the algorithm relies on recursion, which can be eliminated using standard methods (see eg. Field & Harrison). The measurements reported in section **5** use however a recursive implementation.

Algorithm for `reverse(v,w)` will reverse pointers according to the following cases of `type(w)`:

```
Atomic:
Functor:
    value(w) := adress(v)   \   reverse_pointer procedure
    value(v) := value(w)    /
    type(w)  := type(v)
    type(v)  := type(w)       (* they are already Marked *)
```

where we assume (here and in the sequel) that the assignments are done in parallel.

```
Pointer:
    We proceed according to following cases of mark(w):

    Unmarked:
        reverse_pointer
        mark(w) := Marked
        mark(v) := Unmarked

    Marked:
        reverse_pointer
        mark(w) := MultiplyMarked
        mark(v) := End
```

```
MulitplyMarked:
    reverse_pointer
    mark(v) := Internal

Internal:
    reverse_pointer
    mark(w) := InternalMultiplyMarked
    mark(v) := End

InternalMultiplyMarked:
    reverse_pointer
    mark(v) := Internal

End:
    reverse_pointer
    mark(w) := EndMultiplyMarked
    mark(v) := End

EndMultiplyMarked:
    reverse_pointer
    mark(v) := Internal
```

Each mark carries information (type, location) needed at compactification time, which is essential for restoring (the active part of) the original Prolog data graph.

1.2 Examples

Some comments on meanning and usage of marks might help the reader to follow the examples.

- **Marked** (M) means that this cell is active and has to be saved during compactification by simple (de)reversing. If the value of cell was a pointer, it is reversed, as shown on Fig. 1.1 (a – d).
- **MultiplyMarked** (MM) means that several cells point to this cell and it has been revisited by the marking algorithm, This mark indicates the beginning of a subchain (depth increased wrt the previous part of pointer chain), see Fig. 1.2 (a).
- **Internal** (I) decorates a cell which has pointed to an already 'multiply' marked cell with its subchain already created. The **I**–cell becomes an internal part of that subchain, Fig. 1.2 (b, d,e).
- **End** (E) indicates the end of a subchain created at the same time as a 'multiply' marked cell, some situations are shown in Fig. 1.2 (a, c, e).
- **InternalMultiplyMarked** (IM) means this cell was previously marked **internal** in some subchain with more than one cell pointing to it, which results in creation of a subchain at this point. The situation is shown in Fig. 1.2 (c).
- **EndMultiplyMarked** (EM) is created in the same manner as **IM** except that this time the cell was marked as **End**, Fig. 1.2 (f).

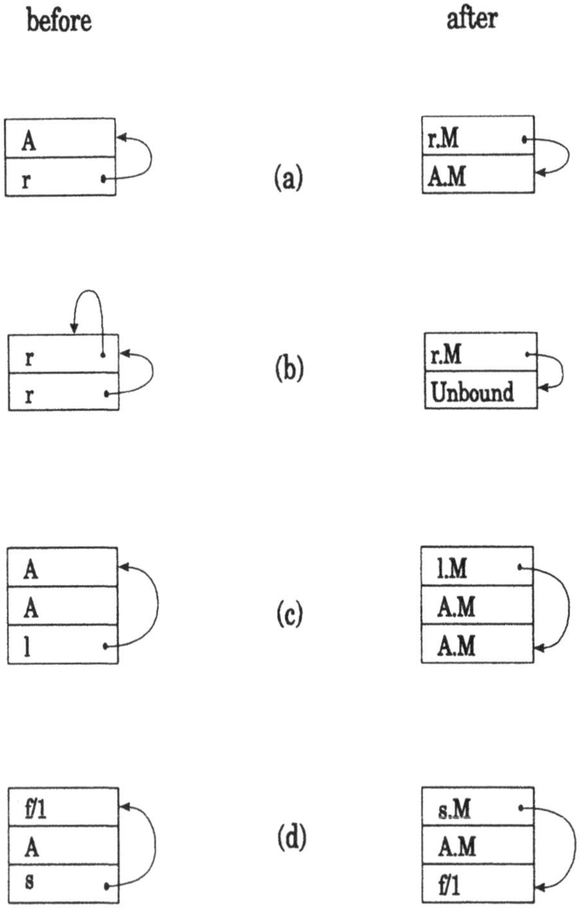

Fig. 1.1 simple marking

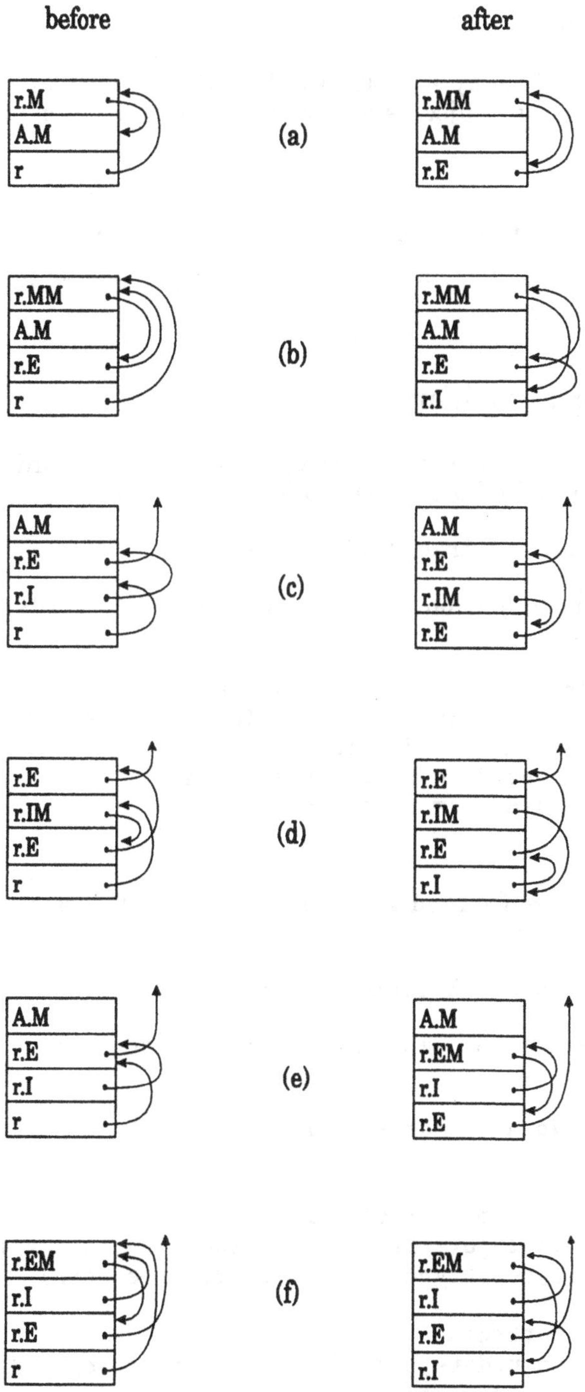

Fig. 1.2 various marking situations

2 Copy phase

Compactification is done by scanning the HEAP from bottom up to the highest used word. Each marked object is relocated to its new position, and its value gets restored.

Let

```
find_marked(w), end_of_chain(w)
```

return first marked word (including w) in HEAP (in upward direction) and last word in reversed chain.

2.1 Copying algorithm

We will use two word variables, one set to current marked word and the other to its new position; initially both are set to bottom of HEAP.

```
marked_word := free_word := HEAP_low
```

Now we succesively copy (from bottom to top) all active words (preserving their relative positions).

```
while adress(free_word) < HEAP_used do
    marked_word := find_marked(marked_word)
    copy(marked_word, free_word)
    marked_word := next(marked_word)
    free_word   := next(free_word)
endwhile
```

Algorithm for copy(marked_word, free_word) will locate the end of chain starting from marked_word,

```
end_word := end_of_chain(marked_word)
```

and then proceed according to case of mark(marked_word):

```
Unmarked:
    free_word := marked_word

Marked:
    type := type(marked_word)
    restore_type(marked_word, free_word, end_word)
    remain_copy(type, marked_word, free_word)

Multiply Marked:
    type := update_chain(marked_word, free_word, end_word)
    restore_type(marked_word, free_word, end_word)
    remain_copy(type, free_word, marked_word)
```

```
InternalyMultiplyMarked:
    type := update_chain(marked_word, free_word, end_word)
    restore_no_type(marked_word, free_word, end_word)
    remain_copy(type, free_word, marked_word)

EndMultiplyMarked:
    type := update_chain(marked_word, free_word, end_word)
    restore_no_type(marked_word, free_word, end_word)
    remain_copy(type, free_word, marked_word)
```

Algorithm for `restore_type(marked_word, free_word, end_word)` has to restore value and type of **marked_word** by taking it from **end_word**, and update value and type of **end_word**:

```
mark(free_word)   := Unmarked
type(free_word)   := type(end_word)
value(free_word)  := value(end_word)
mark(end_word)    := Unmarked
type(end_word)    := type(marked_word)
value(end_word)   := adress(free_word)
```

An unbound variable, replaced previously by **MaxAtom**, can now be restored:

```
if  type(free_word) = Atomic
    &
    value(free_word) = Maxatoms
then
    type(free_word)  := Reference
    value(free_word) := adress(free_word)
```

Algorithm for `restore_no_type(marked_word, free_word, end_word)` has to restore the original value of **marked_word** from **end_word**, and to update the value of **end_word**, preserving type:

```
mark(free_word)   := Unmarked
type(free_word)   := type(marked_word)
value(free_word)  := value(end_word)
mark(end_word)    := Unmarked
value(end_word)   := adress(free_word)
```

Algorithm for `remain_copy(type, free_word, marked_word)` has to copy all arguments of a list or a structure:

```
List:
    marked_word := next(marked_Word)
    free_word   := next(free_word)
    copy(marked_word, free_word)

Structure:
    for all arguments in structure do
        marked_word := next(marked_Word)
        free_word   := next(free_word)
        copy(marked_word, free_word)
    endfor
```

Algorithm for **update_chain(marked_word, free_word, end_word)** has to update all words whose original values were **adress(marked_word)** to new value – **adress(free_word)**, and to return the largest type of updated words (including **type(marked_word)**) (this situation occurs only when a list and a reference pointer point together to a list object, with the reference pointer indicating the head):

```
        tmp_word := marked_word
        type     := type(marked_word)
        deep     := 0
        length   := 0
        IEMM     := 0   (* Internal or End MultiplyMarked flag *)

        repeat
            tmp_word := value(tmp_word)
            length   := length + 1

    (* now we will proceed according to case of mark(tmp_word) *)
            Unmarked:
                if  length = 1
                    &
                    (
                    mark(marked_word) = InternalMultiplyMarked
                    |
                    mark(marked_word) = EndMultiplyMarked
                    )
                    &
                    type < type(tmp_word)
                then
                    type := type(tmp_word)

            Marked:
                proceed

            MultiplyMarked:
                deep := deep + 1

            InternalyMarked:
                if  deep = 0
                    &
                    IEMM = 0
                then
                    if  type < type(tmp_word)
                    then
                        type := type(tmp_word)
                    value(tmp_word) := adress(free_word)
                    length := 0

            InternalyMultiplyMarked:
                if  deep = 0
                then
                    if  IEMM = 0
                    then
```

```
                if  type < type(tmp_word)
                then
                    type := type(tmp_word)
                IEMM := 1
            else
                deep := deep + 1
        else
            deep := deep + 1

    End:
        if  deep = 0
        then
            if  type < type(tmp_word)
            then
                type := type(tmp_word)
            value(tmp_word) := adress(free_word)
            length := 0
        else
            deep := deep - 1

    EndMultiplyMarked:
        if  deep = 0
        then
            if  IEMM = 0
            then
                if  type < type(tmp_word)
                then
                type := type(tmp_word)
                IEMM = 1

    until tmp_word = end_word

    update_chain := type
```

2.2 Comments on compactification

Compactifcation has to copy all active cells to the bottom of HEAP. This is done by scanning the HEAP upwards. Because of reversed pointer chains, the original value of the cell being copied has to be reconstructed, by following the reversed chain. If we encounter any of 'Multiply' Marked cells during chain traversal we will increase depth of chain. Depth is decreased whenever we come to cell marked as **End**. All **Internal** cells at depth zero have to be updated during chain traversal, as well as the last cell in chain.

Depending on the mark we sometimes have to swap both type and value between beginning and end of chain, but sometime type must be preserved. Since a list and a reference pointer may point to the same cell, we have to determine cell size during chain traversal (and use larger size).

The reader may use the copy algorithm to restore original types and values on right hand sides of figures in Section **1**.

3 Forward pointers

So far we have not said anything about forward pointers. Let us see how forward pointers will affect the marking phase. The idea is to delay their reversal by marking them in a special way, so that copy phase can do the actual reversal only when needed (instead of wasting another pass through the HEAP).

3.1 Modifications in the marking algorithm

Algorithm for `mark_word(v)` has to be modified in case of:

```
Pointers:
          ...

        if  mark(v) =\= Marked   (* is circular *)
        then
            Error(Circular structure)

        if  adress(value(v)) > adress(v)    <--- modification
        then
            mark(v) := Forward
        else
            reverse(v,w)
```

Algorithm for `reverse(v,w)`, in case of Pointer, obtains a new subcase of mark(w):

```
    Forward:
        (* do the same as for Atomic *)
```

Fig. 3.1 . illustrates the fact that only list and structure pointers can point forwards, reference pointers always point backwards. It is also shown how such pointers are treated by marking phase.

3.2 Modification in compactification algorithm

Algorithm for `restore_type(marked_word, free_word, end_word)` has to be modified:

```
        ...

    type(end_word)   := type(marked_word)
    value(end_word)  := adress(free_word)

    if   mark = Forward                   <--- modification
    then
        reverse(free_word, value(free_word))

    ...
```

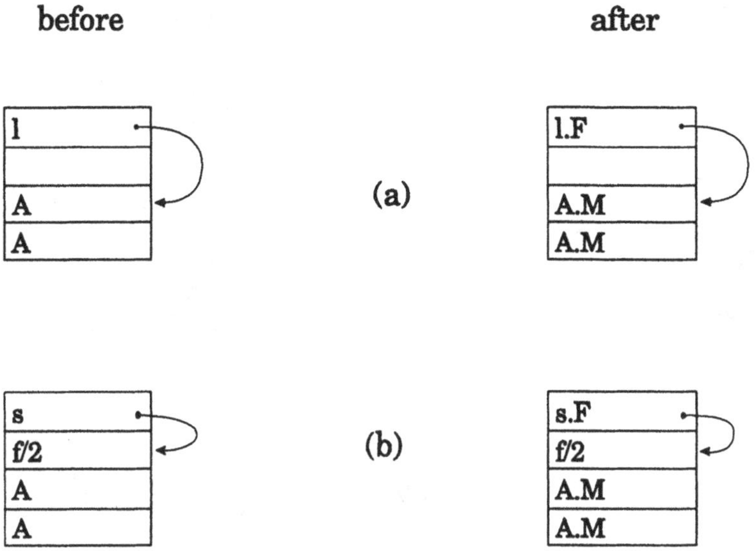

Fig. 3.1 marking forward pointers

Algorithm for `copy(marked_word, free_word)` obtains a new case of mark(marked_word):

```
...
Forward:
    value(free_word):= value(marked_word)
    type(free_word)  := type(marked_word)
    mark(free_word)  := Unmarked
    reverse(free_word, value(free_word))
...
```

Fig. 3.2 shows how forward pointers are handled (reversed) during the copy phase.

4 Other memory areas

4.1 TRAIL

After marking phase is done, TRAIL pointers must be included in marked and reversed chains to enable copy phase to restore them too. All TRAIL pointers that point to unmarked cells have to be removed from TRAIL, because we mark all active objects on HEAP, it is easy to determine dead TRAIL pointers.

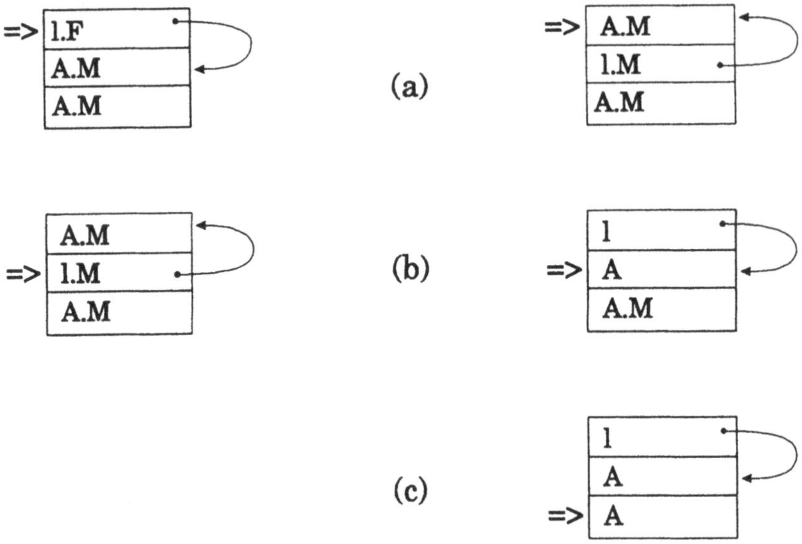

Fig. 3.2 copying forward pointers

4.2 Multiple HEAP areas

The technique of delayed reversal of forward pointers enables us to compactify several physicaly separated data areas which may point to each other (without cyclic structures).

During the marking phase we will not pay any attention to where the pointers point, noting that pointers from LOW to HIGH simply point forward. Thus we just have to compactify LOW area first, and the algorithm will apply delayed reversal automatically. When the HIGH area is compactified, the pointers from the LOW area will thus have been updated automatically.

5 Complexity

5.1 Complexity of marking

During the marking phase we visit each active word exactly once, so the complexity of marking phase depends only on k – number of active words in the HEAP.

We can thus say that, measured in 'word visits',

$$complexity(marking) = k.$$

Delayed reversal of forward pointers reduces time for marking by some incalculable amount, which we can safely ignore, since it is going to be 'paid back' by copying.

5.2 Complexity of copying

During the compactification phase we visit all, say n, words in the HEAP. We also have to follow a reversed chain to its end for every active substructure, what constitutes the $OVERHEAD$, and copy all active words to their new positions (updating ends of reversed chains).

It is hard to obtain a precise realistic estimate of the $OVERHEAD$, since it depends on the structure of Prolog data graphs. It is easy, however, to obtain a worst case estimate, which is likely to exaggerate grossly with respect to real situations.

The worst case is, locally, for a chain of length k_i, obviously the case of a linear (original) structure, which contributes:

$$\frac{k_i^2}{2}$$

to $OVERHEAD$. However, if the data graph resembles a full tree, its contribution will be much closer to k_i. Hence, if the HEAP contains m structures of sizes k_1, ..., k_m with $k_1 + k_2 + \ldots + k_m = k$, the estimates are:

$$k < OVERHEAD \leq \frac{k_1^2}{2} + \ldots + \frac{k_m^2}{2}$$

We can thus say that

$$complexity(compactification) \;=\; n \;+\; OVERHEAD.$$

Since reversals, delayed in the marking phase, have to be performed now, the savings effected in the marking phase have to be repaid, leaving the sum unaltered.

The situation of Fig. 5.1 (b) is very unlikely to happen at any significant depth, since under usual WAM compilation it would require a deeply head - nested list to be explicitly present in the code of a body goal. The situation of Fig. 5.1 a is also anlikely to happen in any significant depth, since it seems to require the chain of unbound heap variables to link recursively at the low (old) end without the high (young) end becoming a garbage at the same time. Both situations seem to require machine - produced Prolog source.

5.3 Discussion

In view of different operations performed in the marking and copying phase, the overall complexity of the algorithm can thus be estimated as:

$$\alpha\, n \;+\; \beta\, k \;+\; OVERHEAD$$

with $OVERHEAD$ very likely to be close to a linear function of k. An analysis of the algorithm of (cf. Appleby) would yield

$$2\,\alpha\, n \;+\; \beta'\, k$$

with $\beta\, k \;+\; OVERHEAD$ unlikely to differ much from $\beta'\, k$.

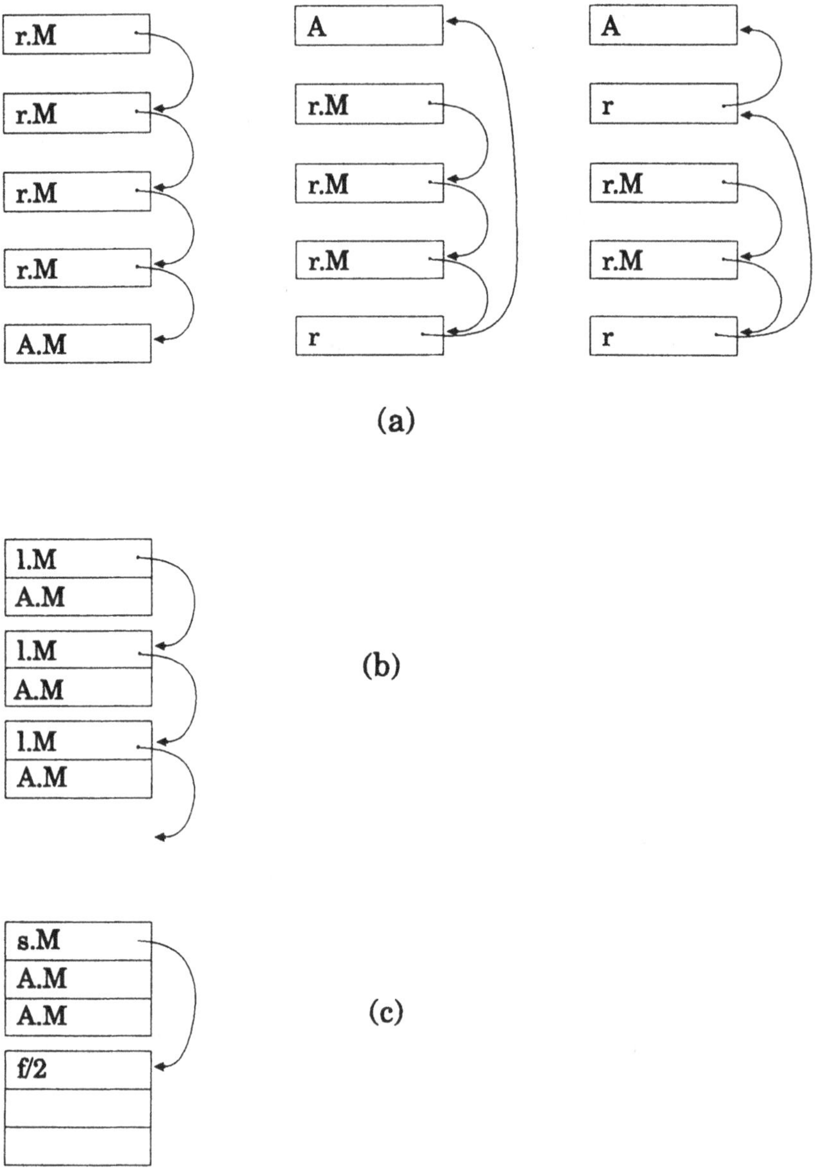

Fig. 5.1 reversed chains

5.4 Experimental data

The following piece of Prolog code was used for generating active and garbage cells on the HEAP and measuring the algorithm performance.

```
gene(0,_,[0])    :- space(_).
gene(A,G,[A|P]) :- A1 is A - 1,
                   gene(G,_),
                   gene(A1,G,P).

gene(0,[0]).
gene(G,[G|T]) :- G1 is G - 1,
                 gene(G1,T).
```

Garbage and data are generated by **gene/2, gene/3**. Garbage collector was invoked by **space/1** predicate at the end of tail recursion loop. On 386/33 we tested various different situations on 1 MB HEAP space (512 KB STACK and 128 KB TRAIL), as shown by table Fig. 5.2 :

A	G	marking	copying
4000	16	0.54	0.69
4000	8	0.54	0.65
4000	4	0.54	0.62
4000	2	0.54	0.58
4000	1	0.54	0.58
8000	8	1.16	1.28
8000	4	1.16	1.11
8000	2	1.16	1.10
8000	1	1.16	1.10
16000	4	2.16	1.78
16000	2	2.16	1.66
16000	1	2.16	1.60

Fig. 5.2 experimental data

The tables show values of variables `A` and `G` (where the amount of active cells is proportional to `A`, and the amount of garbage to `A * G`), and the times of respective phases of garbage collection. Almost linear dependecies, of n (`G`) for the copy phase, and of k (`A * G`) for the marking phase, are obvious. Slight deviations from linearity are due to the quality of time measurements, and to compactification of STACK and TRAIL.

Literature

[Appleby et al.] Appleby, K., Carlsson, M., Haridi, S., Sahlin, D., *'Garbage Collection for Prolog Based on WAM'*. Communications of the ACM, **31**(6), 719-41, (1988).

[Field & Harrison] Field, A.J., Harrison, G.P., *'Functional programming'*. Addison-Wesley Publishers Ltd., (1988).

[Morris] Morris,F.L., *'A time-and space-efficient garbage compaction algorithm'*. Communications of the ACM, **21**(8), 662-5, (1978).

[Warren] Warren, D.H.D., *'An abstract Prolog instruction set'*. Artificial Inteligence Center, SRI International, (1983).

A Resolution Variant Deciding some Classes of Clause Sets

Christian Fermüller
Institut für Praktische Informatik
Resselgasse 3/1/2, 1040 Wien
Technical University Vienna

1 Introduction

The first order decision problem is one of the classical topics in mathematical logic (we refer e.g. to the monographs of Dreben, Goldfarb [1] and Lewis [6]). Moreover, solvable classes are of great importance for database theory, logic programming, and the topical field of non–monotonic logic. Most of the research work has been devoted to classes specified by their prefix type, whereas cases of interest to computer science are frequently formulated in the quantifier free syntax of clause logic and thus demand a characterization of clause sets based on the term structure of their elements. We shall prove the solvability of classes of clause sets that may be characterized as extensions of well known prefix classes.

Traditionally, solvability is proved by model theoretic methods, which hardly lead to practicable decision algorithms. Although Soviet scientists had introduced a proof theoretic approach to the decision problem already in the 1960's, the usefulness of resolution as a decision procedure has only become recognized better with the work of Joyner [3]. In his paper, Joyner only considered the "classical" prefix classes; but recent work by Leitsch, Tammet, Fermüller and others shows how resolution can be applied to decide interesting classes of clause sets with a more general functional structure. Leitsch [5] investigates hyperresolution, Fermüller [2] uses different variants of semantic resolution, whereas in Tammet's paper [12] a special Π–ordering is applied. In this paper we shall introduce a new resolution mechanism, which is based on an A–ordering and is similiar to Joyner's $\mathcal{R}_3$–procedure.

There are various ways to use resolution as a decision procedure. (See, e.g., [3],[9] or [5]). The most obvious strategy to show that a class of clause sets C is decidable by a (refutation complete) resolution variant is to prove that only finitely many resolvents can be generated out of clause sets in C. This can be achieved by demonstrating the following: C is closed with respect to resolution; the resolvents are limited in their length (i.e. the number of literals they contain); and the resolvents never exceed their parents with respect to term depth. To specify our task more accurately we will make use of the following notions:

For a clause set S, let $\mathcal{R}(S)$ denote the set of all resolvents of pairs of clauses in S. Moreover $\mathcal{R}^*(S)$ is defined to be the smallest clause set s.t.

(i) $S \subseteq \mathcal{R}^*(S)$, and,

(ii) for all clause sets X: if $X \subseteq \mathcal{R}^*(S)$ then $\mathcal{R}(X) \subseteq \mathcal{R}^*(S)$.

To prove the decidabilty of a class $\mathcal{C}$ in the way mentioned above we will have to find a resolution variant $\mathcal{R}$, which is complete on clause sets in $\mathcal{C}$, and establish the following facts: For all sets S in $\mathcal{C}$ and all $C, D \in S$

(i) $\mathcal{R}(S) \in \mathcal{C}$,

(ii) if $E \in \mathcal{R}(\{C, D\})$ then $\tau(E) \leq \max(\tau(C), \tau(D))$, and

(iii) there is a constant c s.t.: if $E \in \mathcal{R}^*(S)$ then $|E| \leq c$

where $\tau(C)$ denotes the maximal term depth of a clause C (cf. section 2 below).

In this paper we first investigate a special A–ordering strategy ($\mathcal{R}_{<_d}$) and then introduce a new resolution mechanism $\mathcal{R}_m$, which is based on $\mathcal{R}_{<_d}$. We prove that $\mathcal{R}_m$ provides a decision procedure for an interesting class of clause sets which represents an extension of well known prefix-type classes like the Skolem class, and the Ackermann class.

2 Terminology

We assume the reader to be familiar with the concept of resolution but shortly review the basic definitions for sake of clarity and completeness.

We first define the notions *term*, *atom*, *literal*, *expression* and *clause* formally.

Definition 2.1 *A* term *is defined inductively as follows:*

(i) Each variable and each constant is a term.

(ii) If $t_1, \ldots, t_n$ are terms and f is an n–ary function symbol ($n \geq 1$), then $f(t_1, \ldots, t_n)$ is also a term.

(iii) Nothing else is a term.

Definition 2.2 *If $t_1, \ldots, t_n$ are terms and P denotes an n–ary predicate symbol, then $P(t_1, \ldots, t_n)$ is an* atom. *An* expression *is either a term or an atom. A* literal *is an atom, possible preceded by a negation sign. A* clause *is a finite set of literals. The* empty clause *is denoted by $\Box$.*

Another basic notion is the concept of *substitution*:

Definition 2.3 *Let V be the set of variables and T be the set of terms. A* substitution *is a mapping $\sigma : V \to T$ s.t. $\sigma(x) = x$ almost everywhere. We call the set $\{x/\sigma(x) \neq x\}$* domain *of σ and denote it by $dom(\sigma)$, $\{\sigma(x)/x \in dom(\sigma)\}$ is called* range *of σ ($rg(\sigma)$). By ε we denote the* empty substitution, *i.e. $\varepsilon(x) = x$ for all variables x.*

We shall occasionally specify a substitution as a (finite) set of expressions of the form $(x_i \leftarrow t_i)$ with the intended meaning $\sigma(x_i) = t_i$.

The application of substitutions to expressions is defined as usual.

For the *resolvent* we retain the original definition of Robinson [11], which combines factorization and (binary) resolution.

Definition 2.4 *If C and D are variable disjoint clauses and M and N are subsets of C and D respectively, s.t. $\bar{N} \cup M$ can be unified by the most general unifier (m.g.u.) θ, then $E = (C - M)\theta \cup (D - N)\theta$ is a resolvent of C and D. $\bar{N}$ designates the set of the literals dual to the literals of N. If M and N are singleton sets then E is called binary resolvent of C and D.*

The atom A of the literal $(\bar{N} \cup M)\theta$ is called the resolved atom. *We say that E is* generated by *A. The elements of N and M are the* literals resolved upon.

The critical parts of the proofs in the following sections demand a careful tracing of the unification procedure. For this purpose we review a simple version of the unification algorithm. We first have to introduce some additional terminology, which will prove useful in later sections. We make use of the following naming convention: For variable symbols we use letters from the end of the alphabet (u, v, x, y, z); for constant symbols, letters a, b, c are used; function symbols are denoted by f, g or h; for general terms we use t or s, occasionally augmented by an index; capital letters will denote atoms, literals, clauses or certain sets of expressions.

Definition 2.5 *If t is a term then $\Sigma(t)$ — the number of subterms of t — is defined inductively as follows:*

(i) If t is a variable or a constant, then $\Sigma(t) = 1$,

(ii) if $t = f(t_1, \ldots, t_n)$, then $\Sigma(t) = 1 + \sum_{i=1}^{n} \Sigma(t_i)$.

We call a term t functional *iff it is not variable or a constant (i.e. iff $\Sigma(t) > 1$). A literal or atom is called* functional *iff at least one of its arguments is functional.*

Definition 2.6 *The ith subexpression $SUB(i, E)$ of an expression E is defined inductively as follows:*

(i) $SUB(0, E) = E$,

(ii) if $SUB(k, E) = F(t_1, \ldots, t_n)$, where F is either a n–ary function symbol or a n–ary predicate symbol, then $t_i = SUB(1 + k + \sum_{j=1}^{i-1} \Sigma(t_j), E)$.

We say that a term t occurs in an expression E, if there is an i s.t. $t = SUB(i, E)$. The set of all variables occurring in E is called $V(E)$; if C is a clause, then $V(C)$ is the union over all $V(P_i)$ for all atoms P_i in C. We call an expression constant free *if no constants occur in it.*

Examples. If $E = P(x, f(f(y)))$, then $SUB(0, E) = P(x, f(f(y)))$, $SUB(1, E) = x$, $SUB(2, E) = f(f(y))$, $SUB(3, E) = f(y)$, and $SUB(4, E) = y$; $V(E) = \{x, y\}$.

It can be verified easily that, if $SUB(n, E)$ exists, $SUB(i, E)$ is defined uniquely for all $i \in \{1, \ldots, n\}$.

Definition 2.7 *The leftmost point of disagreement between two different expressions E_1 and E_2 is defined to be the integer k, s.t. $SUB(k, E_1) \neq SUB(k, E_2)$, but $SUB(j, E_1) = SUB(j, E_2)$ for all $j \in \{1, \ldots, k-1\}$.*

We are now able to present a unification algorithm, which finds a m.g.u. θ of two variable disjoint expressions E_1 and E_2, if one exists[1]:

```
begin
    θ := ε;
    i := 0;
    while E₁θ ≠ E₂θ do
        find the leftmost point of disagreement k
        between E₁θ and E₂θ;
        i := i + 1;
        if SUB(k, E₁θ) ∈ V(E₁θ) then
                x_i := SUB(k, E₁θ)
                t_i := SUB(k, E₂θ)
        elseif SUB(k, E₂θ) ∈ V(E₂θ) then
                x_i := SUB(k, E₂θ)
                t_i := SUB(k, E₁θ)
        else {no m.g.u. exists} failure
        endif;
        if x_i occurs in t_i then {no m.g.u. exists} failure
        else θ := θ · {x_i ← t_i}
        endif
    endwhile
    {θ is the m.g.u. of E₁ and E₂}
end.
```

The substitution component $\{x_i \leftarrow t_i\}$ is called the ith *mesh substituent* of θ.

The depth of an expression or a clause is defined as follows:

Definition 2.8 *The* depth of an expression E $\tau(E)$ *is defined by:*

(i) If E is a variable or a constant, then $\tau(E) = 0$,

(ii) if $E = F(t_1, \ldots, t_n)$, where F is either an n-ary function symbol or an n-ary predicate symbol, then $\tau(E) = 1 + \max_{1 \leq i \leq n} \tau(t_i)$.

The depth of a clause C is the maximum of $\tau(P_i)$ for all atoms P_i of C.

---

[1] Remember that any two different m.g.u.s of a set of expressions only differ in the names of the variables.

We not only need to speak about the depth of a term or of a clause, but also about the depth of occurrence of subexpressions of an atom or a clause.

Definition 2.9 *The* depth of occurrence *of the ith subexpression of an expression E $\tau_{SUB}(i,E)$ is defined inductively by:*

(i) $\tau_{SUB}(0,E)=0$

(ii) if $SUB(k,E)=F(t_1,\ldots,t_n)$, where F is either an n–ary function symbol or an n–ary predicate symbol, then $\tau_{SUB}(j,E)=\tau_{SUB}(k,E)+1$ iff $SUB(j,E)$ is an argument of $SUB(k,E)$ (i.e. $SUB(j,E)=t_i$ for some $i\in\{1,\ldots,n\}$).

With $\tau_{MAX}(t,E)$ we denote the maximal depth of occurrence of a term t within an expression E, i.e.

$$\tau_{MAX}(t,E)=\max_{SUB(i,E)=t}\tau_{SUB}(i,E).$$

Examples. If $P_1=P(x,f(f(y)))$, $P_2=Q(f(x))$ and $C=\{P_1,\neg P_2\}$, then $\tau(P_1)=3$, $\tau(P_2)=2$, $\tau(C)=3$, $\tau_{SUB}(0,P_1)=\tau_{SUB}(0,P_2)=0$, $\tau_{SUB}(1,P_1)=1$, $\tau_{SUB}(3,P_1)=2$, $\tau_{MAX}(y,C)=3$.

3 Covering terms and atoms

Throughout the paper we speak of terms and atoms characterized by a special property which we shall call *covering*:

Definition 3.1 *A functional term t is called* covering *iff for all functional subterms s occurring in t we have $V(s)=V(t)$. An atom or literal A is called covering iff each argument of A is either a constant or a variable or a covering term t s.t. $V(t)=V(A)$.*

Examples. $f(x,g(a,f(y,x),y))$ and $g(h(x),a,f(x,x))$ are covering terms; $g(f(x,y),x,h(x))$ is not covering. $P(h(x),x)$ and $P(f(x,f(x,y)),f(y,x))$ are covering atoms; $P(f(a,h(b)),h(c))$, like any ground atom, is covering, too. $P(h(x),y)$ and $P(h(x),f(x,y))$ are not covering.

Covering atoms originate for instance by skolemization of prenex formulas without function symbols when the universal quantifiers precede all existential quantifiers.

Example. The clause $C=\{P(x,y,f(x,y)),Q(f(x,y),y)\}$ corresponds to the formula $\forall x\forall y\exists zP(x,y,z)\vee Q(z,y)$. Clearly, the atoms of C are covering.

We prove some properties of covering atoms that will assist us in establishing the decidabilty results of the following sections.

The following lemma will be used to guarantee a term depth limit for resolvents of certain clauses.

Lemma 3.1 *If θ is a m.g.u. of two covering atoms A and B then $A\theta$ $(=B\theta)$ is covering, too. Moreover $\tau(A\theta)=\max(\tau(A),\tau(B))$.*

Proof. Some additional terminology will result in a more concise formulation of the proof: Let us write $A \simeq B$ if $A = B$ or if B can be obtained from A by substituting some variables of A by constants or other variables. Observe that, if $A \simeq B$, then B is covering whenever A is covering. Moreover we write $A \prec B$ if each functional subterm t of B contains all variables of A (i.e., $V(A) \subseteq V(t)$). We have to trace the process of unification. Let ρ_i denote the concatenation of the mesh substituents applied so far, i.e.

$$\rho_i = \{x_1 \leftarrow t_1\} \cdot \{x_2 \leftarrow t_2\} \cdot \ldots \{x_i \leftarrow t_i\}.$$

We prove the following induction hypothesis by the number of substitution steps:
Either

(IH1) $A \simeq A\rho_i$ and $B \simeq B\rho_i$, or

(IH2) $A \simeq A\rho_i$ and $A\rho_i \prec B\rho_i$, or

(IH3) $B \simeq B\rho_i$ and $B\rho_i \prec A\rho_i$.

(IH1) trivially holds for $i = 0$ as $\rho_0 = \varepsilon$. Let $\{x_{i+1} \leftarrow t_{i+1}\}$ be the next mesh substituent. W.l.o.g. we may assume that x_{i+1} is found in $A\rho_i$ and t_{i+1} in $B\rho_i$ in the point of disagreement between $A\rho_i$ and $B\rho_i$ (otherwise exchange $A\rho_i$ and $B\rho_i$).
We have to consider the following cases:

(1) (IH1) holds:

(1a) t_{i+1} is a variable or a constant: In this case, $A\rho_i \simeq A\rho_{i+1}$ and $B\rho_i \simeq B\rho_{i+1}$. As the relation "$\simeq$" is transitive, (IH1) still holds after applying $\{x_{i+1} \leftarrow t_{i+1}\}$.

(1b) t_{i+1} is a functional term: By (IH1) $B\rho_i$ is covering. Therefore $V(t_{i+1}) = V(B\rho_i)$. This implies that

$$B\rho_i \prec A\rho_{i+1} \; (= A\rho_i \cdot \{x_{i+1} \leftarrow t_{i+1}\}).$$

Now observe that $x_{i+1} \notin t_{i+1}$ (otherwise A and B would not be unifiable). But as $B\rho_i$ is covering this amounts to

$$B\rho_i = B\rho_{i+1}.$$

Summarizing we have shown that (IH3) must hold for $i + 1$.

(2) (IH2) holds:

(2a) t_{i+1} is a variable or a constant: Clearly $A\rho_i \simeq A\rho_{i+1}$ and $A\rho_{i+1} \prec B\rho_{i+1}$. Therefore (IH2) remains valid.

(2b) t_{i+1} is a functional term: This cannot occur because $A\rho_i \prec B\rho_i$ implies that $V(A\rho_i) \subseteq V(t_{i+1})$, and especially, we would have $x_{i+1} \in V(t_{i+1})$. But this contradicts the assumption that A and B are unifiable.

(3) (IH3) holds: The argument is completely analogous to case (2).

Summarizing we have shown that $A \simeq A\theta$ or $B \simeq B\theta$. As A and B are covering it follows that $A\theta(= B\theta)$ is covering, too. Moreover $A \simeq A\theta$ implies $\tau(A) = \tau(A\theta)$. Thus we have proved q.e.d. ■

Two more simple lemmata will be used in the sections below:

Lemma 3.2 *If θ is a m.g.u. of two covering atoms A and B then* $|V(A\theta)| \leq \max(|V(A)|, |V(B)|)$.

Proof. In the proof of lemma 3.1 above we have shown that $A \simeq A\theta$ or $B \simeq B\theta$. The lemma is a direct consequence of this fact. ■

Lemma 3.3 *Let A and B be two covering atoms s.t. $V(A) = V(B)$. For any substitution θ: If $A\theta$ is covering then $B\theta$ is covering, too.*

Proof. $V(A) = V(B)$ clearly implies $V(A\sigma) = V(B\sigma)$ for any substitution σ. If $A\theta$ is covering, any functional term t which replaces a variable x of A has to fulfill the following conditions

(i) t is covering, and

(ii) $V(t) = V(A\theta)$.

Clearly any covering atom B remains covering if a variable of B is replaced by a covering term t, s.t. $V(t) = V(B)$. Therefore $B\theta$ is covering, too. ■

4 An A–ordering strategy as decision procedure

Consider the following class of clause sets:

Definition 4.1 ($\mathcal{E}_1$) *A clause set S belongs to $\mathcal{E}_1$ iff the following holds for all clauses C in S:*

(i) All literals in C are covering, and

(ii) for all literals $L, M \in C$ either $V(L) = V(M)$ or $V(L) \cap V(M) = \emptyset$.

Examples. Let $C_1 = \{P(f(x,y),a),\ Q(y,x,x)\}$, $C_2 = \{P(f(x,f(x,a)),\ Q(z,y,a)\}$, $C_3 = \{P(x,f(a))\}$ and $C_4 = \{Q(x,y,a),\ P(x,x)\}$. Then $\{C_1, C_2\} \in \mathcal{E}_1$, but any clause set containing C_3 or C_4 is not in $\mathcal{E}_1$.

Observe that we do not allow the literals to contain functional ground terms together with variables: In that case we could not prevent the resolvents to be of greater term depth than their parent clauses, in general.

$\mathcal{E}_1$ contains the Ackermann class, characterized by the prefix type $\forall\exists^*$, and even the initially extended form of the Ackermann class (i.e. the prenex class with prefix type $\exists^*\forall\exists^*$. — We refer to Dreben/Goldfarb [1] for formal definitions of these classes).

Also the class of clause sets only containing clauses C s.t. $|V(L)| = 1$, but no functional ground terms occur in L, for all literals L in C is a subclass of $\mathcal{E}_1$.

A closely related class, called $\mathcal{E}^+$, was proved to be decidable by Tammet [12]. $\mathcal{E}^+$ is even slightly more general, because it additionally allows arbitrary ground terms to occur in the clauses. Tammet refers to the completeness of Π–orderings to achieve his result. We demonstrate that a simple A–ordering strategy guarantees a term depth limit for resolvents of clauses in $\mathcal{E}_1$. A limit for the clause length can be achieved by splitting; thus we arrive at a resolution variant which provides a decision procedure for $\mathcal{E}_1$.

The concept of A–orderings was introduced by Kowalski and Hayes [4]. We review their definition:

Definition 4.2 *An A–ordering $<_A$ is a binary relation on atoms s.t.*

(i) $<_A$ *is irreflexive,*

(ii) $<_A$ *is transitive,*

(iii) for all atoms A,B and all substitutions θ: $A <_A B$ implies $A\theta <_A B\theta$, and

(iv) $<_A$ *is compatible with some enumeration of the set of ground atoms (i.e. there is some enumeration $A_0, A_1 \ldots$ of all ground atoms s.t. if $A_i <_A A_j$ then $i < j$).*

Remark. Kowalski and Hayes omit *(iv)*, although it is necessary for their completeness proof, as Joyner [3] has pointed out.

The following A–ordering will prove suitable to provide a term–depth limit for covering literals.

Definition 4.3 *Let A and B be two atoms, then $A <_d B$ iff*

(i) $\tau(A) < \tau(B)$*, and*

(ii) for all $x \in V(A)$: $\tau_{MAX}(x, A) < \tau_{MAX}(x, B)$ (implying $V(A) \subseteq V(B)$).

We first have to show:

Lemma 4.1 $<_d$ *is an A–ordering.*

Proof. Irreflexivity directly follows from the irreflexivity of "$<$"; transitivity holds because both "$<$" and "$\subseteq$" are transitive. Moreover it is obvious that $<_d$ is compatible with any enumeration of ground terms arranging deeper atoms after less deep ones. It remains to show that condition *(iii)* of definition 4.2 holds.

Let A and B be atoms s.t. $A <_d B$ and let $\sigma = \{x \leftarrow t\}$ be a component of any substitution. If $x \notin V(A)$ then $A\sigma = A$; therefore $A\sigma <_d B\sigma$ holds. If $x \in V(A)$ then, by definition of "$<_d$", there is an occurence of x in B which is deeper than any occurrence of x in A. Let y be any variable of $V(t)$; we then have $\tau_{MAX}(y, A\sigma) < \tau_{MAX}(y, B\sigma)$. Applying some substitution θ to an atom has the same effect as applying all components of θ separately. Thus we have proved that $A <_d B$ implies $A\theta <_d B\theta$. ∎

We define $\mathcal{R}_{<_d}(S)$ as follows:

Definition 4.4 *For any clause set S: $E \in \mathcal{R}_{<_d}(S)$ iff E is a resolvent of clauses in S s.t. for no atom B in E: $A <_d B$, where A is the resolved atom.*

Remark. Kowalski and Hayes define the A–ordering strategy by only allowing to resolve on maximal literals of the clauses (with respect to the defined ordering). The definition stated above is more restrictive and still induces a complete resolution strategy, as Joyner [3] has proved.

The following lemma is a consequence of lemmata 3.1 and 3.3:

Lemma 4.2 *If $E \in \mathcal{R}_{<_d}(\{C, D\})$, where $\{C, D\} \in \mathcal{E}_1$, then $\{E\} \in \mathcal{E}_1$ and $\tau(E) \leq \max(\tau(C), \tau(D))$.*

Proof. Let M be the set of literals resolved upon in C and L be a literal in $C - M$; by θ we denote the m.g.u. used to generate E. By definition any two literals of C share either all or none of their variables. We have to consider the following cases:

(1) $V(L) \cap V(M) = \emptyset$: As $\theta(x) = x$ for all $x \in V(C - M)$ this implies $L\theta = L$ and L occurs unchanged in E.

(2) $V(L) = V(L')$ for some $L' \in M$:

 (2a) $L <_d L'$: As "$<_d$" is an A–ordering we have $L\theta <_d L'\theta$ which implies that $\tau(L\theta) < \tau(L'\theta)$. Thus by lemma 3.1 also $\tau(L\theta) < \max(\tau(C), \tau(D))$. Moreover it follows from lemma 3.3 that $L\theta$ is covering.

 (2b) $L \not<_d L'$: By definiton of $\mathcal{R}_{<_d}$ we also know that $L' \not<_d L$. Now observe that in a covering atom A all variables occur somewhere in maximal depth; i.e. for all $x \in V(A)$: $\tau_{MAX}(x, A) = \tau(A)$. This implies $\tau(L) = \tau(L')$ and $\tau(L\sigma) = \tau(L'\sigma)$ for any substitution σ. Moreover lemmata 3.1 and 3.3 again guarantee that $L\theta$ is covering.

By analogy the same holds for the literals in D. Thus we have proved that E fullfills the relevant conditions. ∎

What is still missing in order to conclude the decidability of $\mathcal{E}_1$ is a length limit for the resolvents. Such a limit can be achieved easily if we apply the well known splitting rule (see e.g.[7] and section 5 below). It clearly follows from condition *(ii)* in the definition of $\mathcal{E}_1$ that the set of variables $V(L)$ is the same for all literals L of a clause after splitting. Observe that the number of variables in the resolved atom cannot be greater than the maximum of the number of variables in the atoms resolved upon. This implies that

$$|V(E)| \leq \max(|V(C)|, |V(D)|)$$

for resolvents E of clauses C and D fulfilling the conditions of class $\mathcal{E}_1$. Considering the term depth limit for the resolvents as expressed by lemma 3.1 we arrive at a length limit because, under equivalence w.r.t. renaming of variables, there are only finitely many literals L s.t. $|V(L)|$ as well as $\tau(L)$ are bounded by constants. (As we only consider finite clause sets we may presume that there are only finitely many predicate letters and function symbols). Summarizing the results of this section we have proved the following theorem:

Theorem 4.1 *Class $\mathcal{E}_1$ is decidable; $\mathcal{R}_{<_d}$ combined with the splitting rule provides a decision procedure.*

In the following we investigate a refinement of the $\mathcal{R}_{<_d}$–strategy which will allow us to decide an even more interesting class of clause sets.

5 An extension of the Skolem class

In this section we consider a class which strongly generalizes the extended Skolem class, which is the class of prenex formulas with prefix $\exists z_1 \cdots \exists z_l \forall y_1 \cdots \forall y_m \exists x_1 \cdots \exists x_n$ s.t. each atom of the matrix has among its arguments either

(i) at least one of the x_i, or

(ii) at most one of the y_i, or

(iii) all of $y_1, \ldots y_m$.

Consider the following definition:

Definition 5.1 ($\mathcal{S}^+$) *A clause set S belongs to $\mathcal{S}^+$ iff for all clauses C in S and all literals L in C:*

(i) If t is a functional term occurring in C then $V(t) = V(C)$.

(ii) $|V(L)| \leq 1$ or $V(L) = V(C)$.

Observe that condition *(i)* is equivalent to:

(i') If L is functional, then L is covering and $V(L) = V(C)$.

Class $\mathcal{S}^+$ not only contains the extended Skolem class but also the intially extended Gödel–Kalmár–Schütte class (i.e. the prefix class with quantifier prefix type $\exists^*\forall\forall\exists^*$, cf. [1]) and other classes, like Maslov's class $\mathcal{K}$ (cf. [8],[13]).

$\mathcal{S}^+$ is closely related to $\mathcal{E}_1$. In fact we have (analogously to the result of the last section):

$$E \in \mathcal{R}_{<_d}(\{C, D\}) \qquad \text{implies} \qquad \tau(E) \leq \max(\tau(C), \tau(D))$$

if $\{C, D\} \in \mathcal{S}^+$. The only problem is that in general $\mathcal{R}_{<_d}(S) \notin \mathcal{S}^+$ for clause sets $S \in \mathcal{S}^+$. Atoms may be generated which, besides covering terms, have variables as arguments that do not occur in those terms. Such atoms, of course, are not covering any more.

The following definition is partly due to Joyner [3]. It will allow us to argue more accurately:

Definition 5.2 *An atom or literal A is called* essentially monadic on a term *t iff t is an argument of A and each other argument is either equal to t or a constant. A is called* almost monadic on *t iff t is functional and – besides t and constants – also variables that are not in $V(t)$ are among the arguments of A.*

Examples. $P(g(x), b, g(x))$ is essentially monadic on $g(x)$. $P(f(f(z)), x, a, f(f(z)))$ is almost monadic on $f(f(z))$. $P(x, y, a, x)$ and $Q(f(x), f(y))$ are neither essentially monadic nor almost monadic (on some subterm).

As mentioned above $\mathcal{R}_{<_d}$ provides a term depth limit for the resolvents of $\mathcal{S}^+$. But by resolving such clauses, almost monadic atoms may be generated besides covering ones.

Example. Let $C = \{P(x), Q(x, y)\}$ and $D = \{\neg P(f(z)), Q(g(z), z)\}$. The only $\mathcal{R}_{<_d}$–resolvent of C and D is $E = \{Q(f(z), y), Q(g(z), z)\}$. The first literal of E is almost monadic on $f(z)$; the second literal is covering.

It is interesting to mention that (for class $\mathcal{S}^+$) almost covering atoms are generated by $\mathcal{R}_{<_d}$ only if one of the parent clauses is function free, and the atom(s) resolved upon in this clause contain(s) one variable only, whereas other atoms must have also additional variables as arguments. In all other cases all atoms of a resolvent are covering. It would be an easy task to refine the ordering $<_d$ in a way that it gets sufficiently restrictive to decide $\mathcal{S}^+$ (and many other classes). We would just have to add

(iv) $V(A) \subset V(B)$ *implies* $A <_d B$ *for function free atoms* A *and* B

to the defining conditions for $<_d$ (definition 4.3). Unfortunately the resulting resolution variant is not an A–ordering strategy any more as no ground projection of the ordering exists in general. However the ordering is a so called Π–ordering as described by Maslov [9] and used by Zamov [13], Tammet [12], and others. According to Maslov Π–ordering strategies are always complete, but we do not want to use this result here since no proof has been published so far[2].

We define for any almost monadic atom a corresponding set of essentially monadic atoms.

Definition 5.3 *Let A be almost monadic on some functional term t and K be some set of constants then the* monadization $MON(A, K)$ *consists of atoms that are like A except for replacing each variable that occurs as argument of A (but not of t) by t or some constant in K. More exactly:*

Let $\Sigma_{t,K}$ be the set of all substitutions of the form $\{(x_i \leftarrow t_i)/x_i \in V(A) - V(t)\}$ where $t_i = t$ or $t_i \in K$ then $MON(A, K) = \{A\sigma/\sigma \in \Sigma_{t,K}\}$.

We extend the definition of MON to clauses and clause sets: If all almost monadic atoms A of a clause C are almost monadic on the same functional term t then $MON(C, K) = \{C\sigma'/\sigma' \in \Sigma'_{t,K}\}$ where Σ'_t is the set of substitions $\{(x_i \leftarrow t_i)/x_i \in V(C) - V(t)\}$, s.t. $t_i = t$ or $t_i \in K$.

If C is function free and there is one and only one variable x, s.t. all literals $L \in C$ with $|V(L)| \geq 2$ contain x then $MON(C, K) = \{C\sigma/\sigma \in \Sigma_{x,K}\}$ where $\Sigma_{x,K}$ is the set of substitutions $\{(x_i \leftarrow t_i)/x_i \in V(C) - V(t)\}$, s.t. $t_i = x$ or $t_i \in K$.

In all other cases $MON(C, K) = \{C\}$.

For any clause set S: $MON(S) = \bigcup_{C \in S} MON(C, K_S)$, where K_S is the set of all constants occurring in clauses of S.

---

[2] Very recently T. Tammet (personal communication) seems to have clarified this point by providing accurate definitions and completeness proofs for Π–orderings and related resolution strategies.

Remark. As we only use the monadisation operator in the context of finite clause sets S we shall implicitly assume that the set of constants occurring in S is always used for the monadisation of atoms or clauses. For sake of readability we therefore suppress the second argument and write $MON(A)$ and $MON(C)$ in stead of $MON(A, K)$ and $MON(C, K)$, respectively. Observe that, in this context, $MON(A)$ and $MON(C)$ are always finite.

Examples. For all examples we assume that there is just one constant a. Let $A = P(f(x,y), z)$ then $MON(A) = \{P(f(x,y), f(x,y)),\ P(f(x,y), a)\}$. Let $C = \{Q(u, x, f(x,a)),\ P(u, f(x,a))\}$ then $MON(C) = \{\{Q(f(x,a), x, f(x,a)),\ P(f(x,a), f(x,a))\},\ \{Q(a, x, f(x,a)),\ P(a, f(x,a))\}\}$. For $D = \{P(x,x), Q(u,a,x), P(x,v)\}$ we have $MON(D) = \{\{P(x,x),\ Q(x,a,x)\},\ \{P(x,x),\ Q(x,a,x), P(x,a)\},\ \{P(x,x),\ Q(a,a,x)\},\ \{P(x,x),\ Q(a,a,x), P(x,a)\}\}$.

We may now define a new resolution variant $\mathcal{R}_m$ which is based on $\mathcal{R}_{<_d}$.

Definition 5.4 *For any clause set S:*

$$\mathcal{R}_m(S) = MON(\mathcal{R}_{<_d}(S)).$$

The members of $\mathcal{R}_m(S)$ are called $\mathcal{R}_m$-resolvents.

We state:

Lemma 5.1 *If $\{C, D\} \in S^+$ and $E \in \mathcal{R}_m(\{C, D\})$ then $\tau(E) \leq \max(\tau(C), \tau(D))$.*

Proof. As mentioned above, each atom of an $\mathcal{R}_{<_d}$-resolvent of any clause set in S^+ is either almost monadic or covering. There are two cases:

(1) If E is an $\mathcal{R}_{<_d}$-resolvent of $\{C, D\}$ and does not contain almost monadic atoms: Then E is an $\mathcal{R}_m$-resolvent, too, and $\tau(E) = \max(\tau(C), \tau(D))$ follows directly from lemma 4.2.

(2) Let E be an $\mathcal{R}_{<_d}$-resolvent generated using resolved atom A and let E contain some almost monadic atom B. In this case one of the parent clauses must be function free and the literals resolved upon in that clause may only contain one variable. Moreover B is the instance of an atom in that clause, containing additional variables. It follows that A is essentially monadic on some term t and B is almost monadic on t. Therefore $\tau(B) = \tau(A)$. By definition we have $\tau(B') = \tau(B)$ for all $B' \in MON(B)$. (The same holds for any almost monadic atom of E.) By lemma 3.1 $\tau(A) \leq \max(\tau(C), \tau(D))$. Therefore we conclude that $\tau(E') \leq \max(\tau(C), \tau(D))$ for all $E' \in MON(E)$, which is q.e.d. ∎

In order to prove that the resolvents of clause sets in S^+ are in S^+ again, we make use of the well known splitting mechanism. We want to remark that even without splitting $\mathcal{R}_m$ decides class S^+. We employ the splitting rule to make the proof somewhat simpler.

For any clause set S let $SPLIT(S)$ denote the set of clause sets obtained by splitting all members of S as far as possible. More accurately: $S' \in SPLIT(S)$ iff for all $C \in S$

there exists a clause $C' \in S'$ s.t. $C' \subseteq C$ and $V(C') \cap V(C - C') = \emptyset$ and C' is not further splitable (i.e. for every proper subset C'' of C' we have $V(C'') \cap V(C' - C'') \neq \emptyset$). We call C' *split component of* C.

The splitting rule says that we have to apply resolution to all members of $SPLIT(S)$ to test for the unsatisfiablity of S. (S is unsatisfiable iff $\square \in \mathcal{R}^*(S')$ for all $S' \in SPLIT(S)$).

Lemma 5.2 *If $S \in S^+$, then all members of $SPLIT(\mathcal{R}_m(S))$ are in class S^+, too.*

Proof. Let A be an atom resolved upon in clause $C \in S$ to generate some $\mathcal{R}_{<_d}$-resolvent E. Let θ be the m.g.u. used to get E. (Thus $A\theta$ is the resolved atom). We have to consider the following cases:

(1) $A\theta$ is function free: This only occurs if both parent clauses of E are function free. Therefore E is function free, too.

  (1a) $|V(A\theta)| = 0$: In this case there may be atoms B in E s.t. $|V(B)| > 1$ but $V(B) \neq V(E)$. Any two such atoms are variable disjunct or share all variables. Thus the split components of E fulfill the defining conditions of class S^+.

  (1b) $|V(A\theta)| = 1$: Let x be the only element of $V(A\theta)$. Then for any literal $L \in E$ s.t. $|V(L)| \geq 2$ we have $x \in V(L)$. Thus, by definition of monadisation, $|V(E')| \leq 1$ for all $E' \in MON(E)$. Moreover, all clauses in $MON(E)$ are function free, too. This implies that the $\mathcal{R}_m$-resolvents are in class S^+.

  (1c) $|V(A\theta)| > 1$: In this case E itself is an $\mathcal{R}_m$-resolvent and fulfills the relevant conditions.

(2) $\tau(A\theta) > 1$: W.l.o.g. we just investigate what happens with literals of C when θ is applied.

  (2a) $|V(A)| = 1$: If all literals of C contain at most one variable the defining conditions of class S^+ clearly remain satisfied after applying θ and splitting. The only interesting case arises when there is some $B \in C$, s.t. $|V(B)| > 1$. As A only contains one variable and – by definition of S^+ – no functional ground term, it follows that $A\theta$ is essentially monadic on some functional term t. Therefore $B\theta$ is almost monadic on t. By the definition of the monadisation operator $V(B') = V(t) = V(A\theta)$ for all $B' \in MON(B\theta)$. This holds for all such literals. Therefore $V(A\theta) = V(E')$ for all $E' \in MON(E)$, which implies q.e.d.

  (2b) $|V(A)| > 1$: It follows from the definition of S^+ that A is covering and that $V(A) = V(C)$. For each $B \in C$ we have either

   (i) $V(B) = V(A)$, which implies that $V(B\theta) = V(A\theta)$,

   or

   (ii) B is not functional and contains just one variable. Call this variable x. By lemma 3.3 $\theta(x)$ is either a variable, a constant or a functional term t, s.t. $V(t) = V(A\theta)$.

Since $A\theta$ contains all variables that occur in any functional term of $C\theta$ or $D\theta$ we have $V(A\theta) = V(E')$ for all $E' \in MON(E)$. Therefore the defining conditions of class S^+ remain satisfied for the respective resolvents. ∎

We may now state that $\mathcal{R}_m$ decides S^+:

Theorem 5.1 *S^+ is decidable; $\mathcal{R}_m$ combined with the splitting rule provides a decision procedure.*

Proof. Lemma 5.1 guarantees a term depth limit for $\mathcal{R}_m$–resolvents of clause sets in S^+. Lemma 5.2 shows that the resolvents (at least after splitting) are in S^+ again. It remains to establish a length limit for the resolvents: By lemma 3.2 and the definition of S^+ it follows that the number of variables of an $\mathcal{R}_{<_d}$–resolvent is bounded by the number of variables of its parent clauses. Clearly, this number can never increase through splitting or monadisation. Together with the term depth limit, this guarantees a length limit for the $\mathcal{R}_m$–resolvents. This in turn implies that $\mathcal{R}^*_m(S)$ is finite for any finite clause set S. Thus the theorem follows from the completeness of $\mathcal{R}_m$ on clause sets of S^+ (cf. section 6 below). ∎

6 Completeness of $\mathcal{R}_m$

In order to prove the completeness of $\mathcal{R}_m$, combined with the splitting rule, for class S^+ we make use of the concept of semantic trees. We assume the reader to be familiar with this important device (see e.g. [4] or [10]), but briefly review the terminology.

For any clause set S the *Herbrand base* of S is the set of all ground atoms consisting of predicate symbols, function symbols and constants occurring in S (if there are no constants in S we introduce one). A *semantic tree* of S is a binary tree with elements of the Herbrand base and their negations labeling the edges. Let $A_0, A_1, A_2, \ldots$ be an enumeration of the Herbrand base of S. The two edges leaving the root of the corresponding semantic tree T are labeled A_0 and $\neg A_0$, respectively; and if either A_i or $\neg A_i$ labels the edge entering a node, then A_{i+1} and $\neg A_{i+1}$ label the edges leaving that node. With each node of T we associate a *refutation set*, consisting of the literals labeling the edges of the path from the root down to this node.

A clause C in S *fails* at a node of T if the complement of every literal in some ground instance $C\gamma$ of C is in the refutation set of that node. A node n is a *failure node* for S if some clause of S fails at n but no clause of S fails on a node above n (i.e. a node with a shorter path to the root). A node is called *inference node* if both of its sons are failure nodes. T is *closed* for S if there is a failure node for S on every branch of T.

A completeness proof for a resolution strategy $\mathcal{R}$ applied to clause sets S essentially rests on the following well known fact:

(1) If S is unsatisfiable, every semantic tree T for S is closed.

Our task is to show:

(2) There is a semantic tree T for S s.t. any two clauses which fail immediately below an inference node of T form an $\mathcal{R}$–resolvent which fails at the inference node.

Remark. If a clause C fails at some node n then clearly all split component of C fail at n, too. Thus no problems arise by the application of the spliting rule.

Given *(1)* and *(2)* the completeness of $\mathcal{R}$ follows by well known arguments (see e.g. [4]). Observe that T can be based on an arbitrary enumeration of the Herbrand base. Whereas any enumeration may be used to prove the completeness of Robinsons original resolution strategy, we will have to make a more judiciuos choice of the enumeration.

Theorem 6.1 *Resolution procedure $\mathcal{R}_m$, combined with the splitting rule, is complete for all clause sets S in S^+.*

Proof. For any clause set S in S^+ we define an enumeration $A_0, A_1, \ldots$ of the corresponding Herbrand base s.t. less deep atoms precede deeper ones.

Within atoms of equal depth essentially monadic atoms precede those atoms that are not essentially monadic. More formally we have:

(i) if $\tau(A_i) < \tau(A_j)$ then $i < j$, and

(ii) if $\tau(A_i) = \tau(A_j)$ and $i < j$ then A_i is essentially monadic whenever A_j is essentially monadic.

Observe that "$<_d$" is compatible with all such enumerations. Let T be the semantic tree for S based on such an enumeration. Let C and D be clauses in S failing immediately below, but not at, a node n of T. Let A_k and $\neg A_k$ be the two ground literals labeling the edges leaving the inference node n. By the proof of lemma 4.1 and the completeness of A–ordering strategies with respect to semantic trees (cf. [4]) there exists an $\mathcal{R}_{<_d}$–resolvent E of C and D, s.t. A_k is a ground instance of the resolved atom A, which fails at n. By definiton of S^+ and lemma 3.1 each atom of E either is covering or almost monadic on some term. If for all atoms B of E, B is covering and $|V(B)| \leq 1$ or $V(B) = V(E)$ then E is an $\mathcal{R}_m$–resolvent, too, and the theorem clearly holds.

There are two crucial cases:

(1) The $\mathcal{R}_{<_d}$–resolvent E contains literals that are almost monadic on some term: This may only be the case if A is essentially monadic on some term t, and all atoms of E are either almost monadic on t or covering. We have to show that some clause in $MON(E)$, fails at node n.

Let γ be the substitution s.t. $E\gamma$ is in the refutation set of n. Clearly $A\gamma$ is essentially monadic on $t\gamma$. For any almost monadic atom B of E, $B\gamma$ is of same depth than $A\gamma$ because all functional arguments of A as well as of B are equal to t. Therefore, by definition of the enumeration of the Herbrand base, also $B\gamma$ is essentially monadic. This implies that $B'\gamma = B\gamma$ for some $B' \in MON(B)$. Thus some $E' \in MON(E)$ fails at node n, too.

(2) E is function free and for some literal L in E we have $|V(L)| > 1$ but $V(L) \neq V(E)$ (i.e. condition *(ii)* of the definition of class S^+ does not hold). There are two subcases:

(2a) The resolved atom A is ground: In this case we can split E into components that fulfill the relevant conditions.

(2b) A contains just one variable x: Then all atoms of E with more than one variable contain x, too. By definition of MON all variables of E are replaced by x or by constants. As A is function free and $|V(A)| = 1$ we know that $A\gamma$ is essentially monadic on some term t. Clearly $\gamma(x) = t$. Therefore, by the construction of T, also $L\gamma$ is essentially monadic on t. Observe that, by definition of S^+, for every literal $M \in E$, s.t. $|V(M)| = 1$ there is some literal L, s.t. $V(M) \subset V(L)$ and $x \in V(L)$. Therefore all literals of $E\gamma$ are essentially monadic on t or some constant. It follows, like in case (1), that $E'\gamma = E\gamma$ for some $E' \in MON(E)$. ■

7 Conclusion

In this paper we investigated a special resolution variant, based on an A–ordering, which decides some classes of clause sets. We think that similiar mechanisms may be defined to provide decision algorithms for other interesting classes of clause sets. We shall inquire along this line of arguments into extensions of the Maslov class (i.e. the class of formulas with prefix of type $\forall^*\exists^*$ and at most two literals in each disjunct) in a forthcoming paper.

Acknowledgement

I want to thank Prof.Dr. Alexander Leitsch and the referee for valuable suggestions and comments.

References

[1] DREBEN, B., AND GOLDFARB, W.D., *The Decision Problem*. Addison-Wesley, Massachusetts 1979.

[2] FERMÜLLER, C., *Deciding some Horn Clause Sets by Resolution*. Yearbook of the Kurt Gödel Society 1989, pp. 60-73.

[3] JOYNER, W.H., *Resolution Strategies as Decision Procedures*. J. ACM 23,1 (July 1976), pp. 398-417.

[4] KOWALSKI, R., AND HAYES P.J., *Semantic trees in automated theorem proving* in Machine Intelligence 4, B. Meltzer and D. Michie, Eds., Edinburgh U. Press, Edinburgh 1969, pp. 87-101.

[5] LEITSCH, A., *Deciding Horn Classes by Hyperresolution*. Proc. of the CSL '89, LNCS 440, pp. 225-241.

[6] LEWIS, H.R., *Unsolvable Classes of Quantificational Formulas*. Addison-Wesley, Massachusetts 1979.

[7] LOVELAND, D., *Automated Theorem Proving: A Logical Basis*. North Holland, Amsterdam 1978.

[8] MASLOV, S.JU., *An Inverse Method of Establishing Deducability in the Classical Predicate Calculus*. Soviet Math.–Doklady 5 (1964), pp. 1420-1424 (translated).

[9] MASLOV, S.JU., *Proof-search Strategies for Methods of the Resolution Type*. Machine Intelligence 6, American Elsevier, 1971, pp. 77-90.

[10] NILSON, N.J., *Problem-Solving Methods in Artificial Intelligence*. McGraw-Hill, New York, 1971.

[11] ROBINSON, J.A., *A Machine-oriented Logic Based on the Resolution Principle*. J. ACM 12,1 (Jan. 1965), pp. 23-41.

[12] TAMMET, T., *A Resolution Program, Able to Decide some Solvable Classes*. Proc. of the COLOG '88, LNCS 417, pp. 300-311.

[13] ZAMOV, N.K., *Maslov's Inverse Method and Decidable Classes*. Annals of Pure and Applied Logic 42 (1989), pp. 165-194.

Subclasses of Quantified Boolean Formulas

Andreas Flögel
FB 11-Praktische Informatik
Universität Duisburg
D 4100 Duisburg 1 (Germany)

Marek Karpinski
Institut für Informatik
Universität Bonn
D 5300 Bonn, (Germany)

Hans Kleine Büning
FB 11-Praktische Informatik
Universität Duisburg
D 4100 Duisburg 1 (Germany)

Abstract

Using the results of a former paper of two of the authors [KaKB 90], for certain subclasses of quantified Boolean formulas it is shown, that the evaluation problems for these classes are coNP-complete. These subclasses can be seen as extensions of Horn and 2-CNF formulas.
Further it is shown that the evaluation problem for quantified CNF formulas remains PSPACE-complete, even if at most one universal variable is allowed in each clause.

1 Introduction

It is well known that the evaluation problem for quantified CNF formulas, this means the problem to determine whether a formula $Q_1z_1 \ldots Q_nz_n(\alpha_1 \wedge \ldots \wedge \alpha_m)$ is true, where Q_i is either $\exists$ or $\forall$ and α_j is a propositional clause, is PSPACE-complete [StM 73]. There are also subclasses decidable in polynomial time. In [APT 78] a linear time algorithm for quantified CNF formulas with clauses of at most 2 variables only is given. In [KKBS 87] a cubic time algorithm for quantified Horn formulas was given. This result was improved in [KaKB 90]. It was shown that the evaluation problem for quantified Horn formulas can be decided in $O(rn)$ time, where n is the length of the formula and r is the number of universal variables occuring positive in the formula. In the same paper a complete and sound resolution operation for quantified CNF formulas, called Q-resolution, was introduced. A Q-resolution is an ordinary resolution, where only existential variables can be matched and additionally, each universal variable of a clause, for which no existential variable exists in that clause occuring after the universal variable in the prefix, is omitted.

As an extension of Horn formulas one might think of formulas, where only the parts of the formula, that are built by a certain subset of the atomset are restricted to Horn form (see e.g. [AaBo 79]).
In [KaKB 90] an extension of quantified Horn formulas, where universal literals of a clause can be arbitrary, but the existential literals are in Horn form, was introduced. For this subclass, called quantified extended Horn formulas, it was shown by means of Q-resolution, that the evaluation problem for such formulas with prefix $\forall \ldots \forall \exists \ldots \exists$ is coNP-complete. In this paper our interest is the complexity of quantified extended Horn formulas and quantified extended 2-CNF formulas, that are formulas with at most two existential variables and an arbitrary number of universal variables in each clause.

It is shown that for any fixed number of alternations of $\forall$ and $\exists$ in the prefix the evaluation problem for quantified extended Horn formulas remains coNP-complete. A similar result is shown for quantified extended 2-CNF formulas.

As mentioned above the evaluation problem for quantified CNF formulas is PSPACE-complete. In the last part of this paper it is shown, that this is true, even if at most one universal variable in each clause exists. So there is no improvement of the complexity possible by similar restrictions of the universal variables of a formula.

2 Preliminaries

In what follows quantified CNF formulas are of the form

$$\begin{aligned} \Phi &= \forall x_1 \exists y_1 \ldots \forall x_k \exists y_k (\phi_1 \wedge \ldots \wedge \phi_m), && \text{where} \\ x_i &= \mathrm{x}_{n_{i-1}+1} \cdots \mathrm{x}_{n_i} && \text{and } n_0 = 0, \\ y_i &= \mathrm{y}_{m_{i-1}+1} \cdots \mathrm{y}_{m_i} && \text{and } m_0 = 0. \end{aligned}$$

Further we assume that the variables in a clauses are given in a fixed order. We say a literal L_1 is before a literal L_2, if the variable of L_1 occurs in the order of the prefix before the variable of L_2. We write clauses in the form $(L_1 \vee \ldots \vee L_t)$, where L_{i-1} is before L_i. An *X-literal* resp. Y-literal is a literal of the form x_e or $\bar{\mathrm{x}}_e$ resp. y_e or y_ϵ. A pure *X-clause* is a non-tautological clause consisting of X-literals only. In particular the empty clause is a pure X-clause.
Finally a formula does not contain tautological clauses and each variable of the prefix occurs in some clause.

Definition (1) *A formula $\Phi = \forall x_1 \exists y_1 \ldots \forall x_k \exists y_k (\phi_1 \wedge \ldots \wedge \phi_r)$ is a* quantified extended *Horn Formula resp. 2-CNF formula, if for each ϕ_i the Y-part is a Horn-clause resp. 2-CNF-clause, i.e. the clause ϕ_i omitting all X-literals is a Horn-clause resp. 2-CNF-clause.*

In the following a prefix $\forall \ldots \forall \exists \ldots \exists$ is denoted as Π_1 and Π_{k+1} is defined as $\forall \ldots \forall \exists \ldots \exists \Pi_k$

Definition (2) *For each $k \geq 1$ the class of quantified extended Horn resp. 2-CNF formulas with prefix in Π_k is denoted as Π_k(extended Horn) resp. Π_k(extended 2-CNF).*

Now we introduce our generalized resolution operation, called Q-resolution.

Definition (3)

The formula does not contain tautological clauses and a clause is seen as set of literals, so there are no multiple occurences of the same literal in one clause.

a: *Let α be a pure X-clause, then we replace α by the empty clause. In each other clause we delete all X-literals, that are not before any Y-literal.*

b: *Let α_1 be a clause with Y-literal y_l and α_2 be a clause with Y-literal $\overline{\mathrm{y}_l}$. In both clauses all occurences of X-literals not before any Y-literal are removed. Then the* Q-resolvent *α of α_1 and α_2 is obtained as follows:*

1. *Remove all occurences of y_l and $\overline{\mathrm{y}_l}$ in $\alpha_1 \vee \alpha_2$.*
2. *If the resulting clause contains complementary literals, then no resolvent exists. Otherwise the resolvent is obtained by removing all occurences of X-literals, that are not before any Y-literal occuring in the resulting clause.*

Comparing ordinary resolution with Q-resolution, we see that only literals bounded by existential quantifiers can be matched and universal variables not before an Y-literal will be eliminated.

Example:

$$\forall x_1 x_2 \exists y_1 y_2 \forall x_3((x_1 \vee \overline{x_2} \vee \overline{y_1} \vee \overline{y_2} \vee x_3) \wedge (\overline{x_1} \vee \overline{y_1}) \wedge (\overline{x_2} \vee y_2) \wedge (\overline{x_1} \vee \overline{y_2}))$$

The following Q-resolution steps can be performed:

$$(x_1 \vee \overline{x_2} \vee \overline{y_1} \vee \overline{y_2} \vee x_3),\ (\overline{x_2} \vee y_2) \vdash_{\overline{Q-Res}} (x_1 \vee \overline{x_2} \vee \overline{y_1})$$

$$(\overline{x_1} \vee \overline{y_2}), (\overline{x_2} \vee y_2) \vdash_{\overline{Q-Res}} \sqcup$$

where $\vdash_{\overline{Q-Res}}$ denotes the application of Q-resolution steps.

It is well known that resolution is complete and sound for propositional formulas in CNF, i.e. a formula $\exists y_1 \ldots y_n(\alpha_1 \wedge \ldots \wedge \alpha_r)$ is false iff the empty clause can be derived by resolution applied to $\alpha_1 \wedge \ldots \wedge \alpha_r$.

A similar result for Q-resolution and quantified CNF formulas is shown in [KaKB 90]:

Theorem 2.1 *A quantified CNF formula Φ is false iff $\Phi \vdash_{\overline{Q-Res}} \sqcup$.*

Analogously to the case of unit-resolution for ordinary resolution we can define a Q-unit resolution. A clause ϕ is called a *Y-unit clause* if ϕ contains exactly one Y-literal and arbitrarily many X-literals. A *positive* Y-unit clause is a unit clause with a positive Y-literal.

Definition (4) *The Q-unit-resolution (Q-U-Res) is a Q-resolution, where one of the clauses is a positive Y-unit clause.*

The Q-unit-resolution is useful not only for quantified Horn formulas, but for quantified extended Horn formulas too (see again [KaKB 90]):

Theorem 2.2 *A quantified extended Horn formula Φ is false iff $\Phi \vdash_{\overline{Q-U-Res}} \sqcup$.*

3 The Evaluation Problem for extended Formulas

Theorem 3.1 *The evaluation problem for Π_1(extended Horn) and Π_1(extended 2-CNF) is coNP-complete.*

Proof: That the falsity of an quantified extended Horn formula and an quantified extended 2-CNF formula with prefix $\forall\exists$ can be decided nondeterministically in polynominal time is obvious. Thus the evaluation problem belongs to coNP.

Now we associate to each formula $\alpha = \alpha_1 \wedge \ldots \wedge \alpha_n$ in 3-CNF with variables $x_1, \ldots, x_m$ a quantified CNF formula, which is in Π_1(extended Horn) and in Π_1(extended 2-CNF), such that the propositional formula α is satisfiable iff the quantified extended CNF formula is false.

Let $y_0, \ldots, y_n$ be new variables, then we associate to each clause $\alpha_i = (L_{i_1} \vee L_{i_2} \vee L_{i_3})$ three clauses $\phi_{i_1} = (L_{i_1} \vee \overline{y}_{i-1} \vee y_i)$, $\phi_{i_2} = (L_{i_2} \vee \overline{y}_{i-1} \vee y_i)$ and $\phi_{i_3} = (L_{i_3} \vee \overline{y}_{i-1} \vee y_i)$.

Then we define $\Phi(\alpha) := \forall x_1 \ldots x_m \exists y_0 \ldots y_n\, (y_0 \wedge \overline{y}_n \wedge \phi_{1_1} \wedge \phi_{1_2} \wedge \ldots \wedge \phi_{n_3})$. If $\Phi(\alpha)$ is false then it yields $\Phi(\alpha) \vdash_{Q-Res} \sqcup$. Since for each $i (1 \leq i \leq n)$ one of the clauses ϕ_{i_1} or ϕ_{i_2} or ϕ_{i_3} must be matched, there is some $L_{i_{j_i}}$ in α_i, such that $\{L_{1_{j_1}}, \ldots, L_{n_{j_n}}\}$ does not contain complementary literals. Then, we can define a satisfying truth assignment $\Im$ for α choosing $\Im(L_{i_{j_i}}) = 1$.

In the converse direction let α be satisfiable. Then there is some truth assignment $\Im$, such that $\Im(\alpha) = 1$. Hence, for each i there is some literal $L_{i_{j_i}}$ with $\Im(L_{i_{j_i}}) = 1$. Now we can apply the Q-resolution to $\Phi(\alpha)$ obtaining the empty clause as follows:

$(y_0), (L_{1_{j_1}} \vee \overline{y}_0 \vee y_1) \vdash_{Q-Res} (L_{1_{j_1}} \vee y_1), (L_{2_{j_2}} \vee \overline{y}_1 \vee y_2) \vdash_{Q-Res} (L_{1_{j_1}} \vee L_{2_{j_2}} \vee y_2), (L_{3_{j_3}} \vee \overline{y}_2 \vee y_3) \vdash_{Q-Res} \cdots \vdash_{Q-Res} (L_{1_{j_1}} \vee \ldots \vee L_{n_{j_n}} \vee y_n), \overline{y}_n \vdash_{Q-Res} \sqcup$.

Since the satisfiability problem for 3-CNF is NP-complete and the transformation can be performed in polynominal time, we have proved our desired result.

q.e.d.

For technical reasons we need the following definition:
A quantified CNF formula $\Phi = \forall x_1 \exists y_1 \ldots \forall x_k \exists y_k (\phi_1 \wedge \ldots \wedge \phi_m)$ is called *reduced*, if each variable in the prefix occurs in $\phi_1 \wedge \ldots \wedge \phi_m$ as literal and for each i the quantified CNF formula $\Phi - \{\phi_i\}$ is true.

In the case of the falsity of Φ reduced means, that removing one clause we obtain a formula, which is true.

For each quantified CNF formula $\Phi = \forall x_1 \exists y_1 \ldots \forall x_k \exists y_k (\phi_1 \wedge \ldots \wedge \phi_m)$ a reduced formula Φ^r can be obtained by removing nondeterministically some clauses and removing all variables in the prefix that do not occur in the remaining clauses.
It is easy to see that if Φ is false there is a selection of clauses so that Φ^r is false and reduced and that if Φ is true, Φ^r remains true.

Theorem 3.2 *For each fixed $k \geq 1$ and Φ in Π_k(extended Horn):*
Φ is false iff $\Phi \vdash_{Q-U-Res} \sqcup$ in at most n^k (nondeterministic) Q-unit-resolution steps, where length(Φ) $= n$.

Proof: We know that for a formula Φ, which is false, there is a derivation to the empty clause by means of Q-unit-resolution.
We will prove by induction on k that the number of Q-unit-resolution steps can be restricted to n^k.

For $k=1$ the formula Φ has the form $\forall x_1 \exists y_1 (\phi_1 \wedge \ldots \wedge \phi_m)$.

Obviously at most length(Φ) resolution steps are sufficient, because for a propositional Horn formula β length(β) unit-resolution steps lead to the empty clause.

Now let be $k>1$. We assume $\Phi = \forall x_1 \exists y_1 \ldots \forall x_k \exists y_k (\phi_1 \wedge \ldots \wedge \phi_m)$ is false and reduced. For a clause ϕ and a set S of variables $\phi(S)$ is the clause ϕ omitting all variables not in S.
Since Φ is false and reduced the set of clauses $\Phi(x_1) = \{\phi_1(x_1), \ldots, \phi_m(x_1)\}$ does not contain complementary literals, i.e. for each literal L in $\phi_i(x_1)$ the complement $\bar{L}$ does not occur in $\Phi(x_1)$. Otherwise the empty clause could not be derived by Q-unit-resolution.
Then we can remove the x_1-variables in Φ without effect to the truth of Φ obtaining the formula

$$\Phi' = \exists y_1 \forall x_2 \exists y_2, \ldots \forall x_k \exists y_k (\phi_1' \wedge \ldots \wedge \phi_m')$$

We define

$$\text{Part}(y_1) := \{\phi_i' | 1 \leq i \leq m \text{ and } \phi_i'(y_1) \text{ is not empty}\}$$
$$\text{Rest}\,(y_1) := \{\phi_1', \ldots, \phi_m'\} - \text{Part}(y_1).$$

Since Φ' is reduced for each $\phi \in$ Part(y_1) the subclause $\phi(y_1)$ can be derived applying the Q-unit-resolution with Rest$(y_1) \cup \{\phi\}$.
Thus, Rest$(y_1) \cup \{\phi(x_2, y_2, \cdots, y_k)\} \vdash_{Q-U-Res} \sqcup$ and the prefix is in Π_{k-1}. Applying the induction hypothesis we know that $\phi(y_1)$ can be derived in length $(\text{Rest}(y_1) \cup \{\phi\})^{k-1}$ Q-unit-resolution steps.

Altogether we obtain

$$\sum_{\phi \in \text{Part}(y_1)} \text{length}\,((\text{Rest}(y_1) \cup \phi)^{k-1}$$

Q-unit-resolution steps to generate $\Phi'(y_1)$ and then length $(\Phi'(y_1))$ resolution steps to derive the empty clause. Since Part(y_1) must contain at least two clauses, otherwise Φ' is not reduced and false, we obtain our desired upper bound length $(\Phi')^k$.

q.e.d.

Theorem 3.3 *For each fixed $k \geq 1$ the evaluation problem for Π_k(extended Horn) is coNP-complete.*

Proof: The evaluation problem for Π_k(extended Horn) belongs to coNP, because the falsity can be decided nondeterministically by choosing at most length $(\phi)^k$ Q-unit-resolution steps (see theorem 3.2).
Since Π_1(extended Horn) is coNP-complete we have proved our desired result.

q.e.d.

This result can not be improved by means of Q-resolution only. The following lower bound for quantified extended Horn formulas is shown in [KaKB 90]:

Theorem 3.4 *There exist quantified extended Horn formulas $\Phi_t(t \geq 1)$ of length $18t+1$, which are false and a refutation to the empty clause requires at least 2^t Q-resolution steps.*

The following definition is needed in the proof of the next theorem: A *renaming* is the replacing of all occurences of some literals by their complements. A renaming f is a one-to-one mapping $Lit(\alpha) \rightarrow Lit(\alpha)$, such that for each $L \in Lit(\alpha) : f(L) \in \{L, \neg L\}$ and $f(L) = \neg f(\neg L)$.

Lemma 3.5 : *For each satisfiable formula α in 2-CNF a renaming f can be found in $O(length(\alpha))$ such that $f(\alpha)$ is a Horn formula.*

Proof: Since α is satisfiable and in 2-CNF a satisfying truth assignment $\Im$ for α can be found in $O(length(\alpha))$ [APT 78].
Now the renaming is defined as

$$f(A) = \begin{cases} \bar{A} & \text{if} \quad \Im(A) = 1 \\ A & \text{if} \quad \Im(A) = 0 \end{cases} \qquad \text{for each variable } A \text{ of } \alpha.$$

For at least one literal L_i in each clause $\Im(L_i) = 1$, so we can analyse the following cases with the 5 possible structures of 2-CNF clauses

$$\begin{array}{llll}
1) & f(B \vee C) = (\bar{B} \vee f(C)) & \textit{or} & f(B \vee C) = (\bar{C} \vee f(B)) \\
2) & f(\bar{B} \vee \bar{C}) = (\bar{B} \vee f(\bar{C})) & \textit{or} & f(\bar{B} \vee \bar{C}) = (\bar{C} \vee f(\bar{B})) \\
3) & f(B \vee \bar{C}) = (\bar{B} \vee f(\bar{C})) & \textit{or} & f(B \vee \bar{C}) = (\bar{C} \vee f(B)) \\
4) & f(B) = (\bar{B}) & & \\
5) & f(\bar{B}) = (\bar{B}) & &
\end{array}$$

In each case the result is a Horn-formula.

q.e.d.

Theorem 3.6 *For each fixed $k \geq 1$ the evaluation problem for Π_k(extended 2-CNF) is coNP-complete.*

Proof: Π_1(extended 2-CNF) is known to be coNP-complete, because of theorem 3.1.

Now we will show by induction on k that the complement of the evaluation problem for Π_k(extended 2-CNF) lies in NP.

Let be given $k > 1$. By induction hypothesis there is a nondeterministic algorithm NA_{k-1} deciding whether an arbitrary formula Φ_{k-1} in Π_{k-1}(extended 2-CNF) is false.
We generate for each formula Φ_k in Π_k(extended 2-CNF) nondeterministically in time

$O(\text{length}(\Phi_k))$ two formulas $\Phi_{k-1,1}, \Phi_{k-1,2}$, such that $\Phi_{k-1,1}, \Phi_{k-1,2} \in \Pi_{k-1}$(extended 2-CNF), so that both formulas are false if Φ_k is false and length $(\Phi_{k-1,i}) <$ length (Φ_k) for $i = 1, 2$.

Let be given

$$\Phi_k = \forall x_1 \exists y_1 \ldots \forall x_k \exists y_k (\phi_1 \wedge \ldots \wedge \phi_m)$$

Remember that $\phi(S)$ with a set S of variables is the clause ϕ omitting all variables not in S and $\Phi(S) = \{\phi(S) | \phi \in \Phi\}$.

Step 1: [Building a reduced formula]

If Φ_k is false we nondeterministically obtain a reduced and false formula Φ'_k. If Φ_k is true the generated formula remains true. In the following whenever we observe that the resulting formula is true or not reduced we halt and nothing is known.

If a formula Φ'_k is reduced and false then each clause is necessary to derive the empty clause. Then $\Phi'_k(x_1)$ does not contain complementary literals.

Step 2: [Deletion of x_1, renaming of y_1]

If $\Phi'_k(x_1)$ contains no complementary literals, then remove all x_1-variables in Φ'_k obtaining a formula

$$\Phi''_k = \exists y_1 \forall x_2 \ldots \forall x_k \exists y_k (\phi_1 \wedge \ldots \wedge \phi_r).$$

In case of complementary literals in $\Phi'_k(x_1)$ Φ'_k is not reduced or true, so we halt.

Let be $\Phi''_k(y_1)_i$ the set of clauses in $\Phi''_k(y_1)$ with exactly i literals.

For a non-empty set $\Phi''_k(y_1)_2$ the formula Φ is false, if $\Phi''_k(y_1)_2$ is not satisfiable. In case of satisfiability there is some renaming, i.e. literals L may be replaced by $\overline{L}$ simultanously, such that the resulting formula $\Phi''_{k,H}(y_1)$ is a Horn formula. Since $\Phi''_k(y_1)_2$ contains 2-clauses only a renaming can be found in $O(\text{length}(\Phi''_k(y_1)_2))$ time (see Lemma 3.5).

Now in Φ''_k we replace $\Phi''_k(y_1)$ by the obtained Horn formula $\Phi''_{k,H}(y_1)$. The resulting formula is denoted as Φ'''_k.

Step 3: [Simplification of $\Phi'''_k(y_1)$]

If $\Phi'''_k(y_1)_2$ is empty then only 1-clauses occur in $\Phi'''_k(y_1)$ and $\Phi'''_k(y_1)_1$ must contain complementary literals, otherwise we halt, because Φ'''_k is not reduced or true.
If $\Phi'''_k(y_1)_2$ is not empty and assumed to be satisfiable we can simplify $\Phi'''_k(y_1)$ as follows:

All X-variables in the clauses of $\Phi'''_k(y_1)_2$ are not before any Y-variable.

Each clause in $\Phi'''_k(y_1)_2$ is an implication of the form $(y_{i_1} \leftarrow y_{i_2})$ or a negated clause of the form $(\bar{y}_{i_1} \vee \bar{y}_{i_2})$ for some $y_{i_1}, y_{i_2} \epsilon y_1$. Each clause ϕ_s in Φ'''_k with exactly one literal $\overset{(-)}{y}_i \in y_1$ can be seen as a sort of unit-clause of the form $(\overset{(-)}{y}_i \vee \phi_s(x_2 \ldots y_k))$. A resolution of such a clause $(\overset{(-)}{y}_i \vee \phi_s(x_2 \ldots y_k))$ with a clause in $\Phi'''_k(y_1)_2$ results in a clause $(\overset{(-)}{y}_j \vee \phi_s(x_2 \ldots y_k))$ with $y_j \in y_1$. For all clauses $\phi_s = (\overset{(-)}{y}_i \vee \phi_s(x_2 \ldots y_k)) \in \Phi'''_k$ with exactly one y-variable

$y_i \in y_1$ all resulting clauses $(\overset{(-)}{y}_j \vee \phi_s(x_2 \ldots y_k))$ from resolution with clauses in $\Phi_k'''(y_1)_2$ can be produced in linear time by means of unit resolution. If we add these clauses to Φ_k''', we can choose nondeterministically two clauses $\phi_s = (y_i \vee \phi_s(x_2 \ldots y_k))$ and $\phi_t = (\bar{y}_i \vee \phi_t(x_2 \ldots y_k))$ with $y_i \in y_1$. All other clauses containing variables from y_1 will be removed.

If Φ_k''' is false than there is some selection such that the resulting formula $\Phi_k^{\prime v}$ is false and if Φ_k''' is true $\Phi_k^{\prime v}$ remains true.

The obtained formula $\Phi_k^{\prime v}$ has the form
$\exists y_1 \forall x_2 \ldots \forall x_k \exists y_k((y_1 \vee \phi_1(x_2 \ldots y_k)) \wedge (\bar{y}_1 \vee \phi_2(x_2 \ldots y_k)) \wedge \phi_3 \wedge \ldots \wedge \phi_t)$,
where $\phi_3, \ldots, \phi_t$ does not contain the variable y_1.

We can divide the formula $\phi_k^{\prime v}$ into two formulas $\Phi'_{k-1,1}$ and $\Phi'_{k-1,2}$ with free variables as follows:
$\Phi'_{k-1,i} = \forall x_2 \ldots \exists y_k : ((y_1^{\epsilon_i} \vee \phi_i(x_2 \ldots y_k)) \wedge \phi_3 \wedge \ldots \wedge \phi_t)$,
where $y_1^{\epsilon_1} = y_1$ and $y_1^{\epsilon_2} = \bar{y}_1$.

If $\Phi_k^{\prime v}$ is false and reduced then $\Phi'_{k-1,1}$ resp. $\Phi'_{k-1,2}$ must imply y_1 resp. $\bar{y}_1$. Then it remains to decide whether

$$\Phi_{k-1,i} = \forall x_2 \ldots \exists y_k(\phi_i(x_2 \ldots y_k) \wedge \phi_3 \wedge \ldots \wedge \phi_t)$$

are false.

Now we can apply our nondeterministic algorithm NA_{k-1} with $\Phi_{k-1,i}(i = 1, 2)$.

Adding the above steps to NA_{k-1} we obtain a nondeterministic algorithm NA_k, which decides in polynominal time, whether a formula Φ_k in Π_k(extended 2-CNF) is false.

q.e.d.

Instead of restricting the existential part of the formulas to Horn-clauses resp. 2-clauses, one may also consider restrictions of the universal variables of the clauses. The following theorem shows, that from such a restriction, no gain in complexity will result.

Theorem 3.7 *The evaluation problem for quantified CNF formulas, for which each clause contains at most one universal variable is PSPACE-complete.*

Proof: Let be given any quantified CNF formula $\Phi = \forall x_1 \exists y_1 \ldots \forall x_k \exists y_k(\phi_1 \wedge \ldots \wedge \phi_m)$, where $x_i = x_{n_{i-1}+1} \ldots x_{n_i}$, $1 \leq i \leq k$ and $n_0 = 0$.

Now we introduce new existential variables $z_1, \ldots, z_{n_k}$ and define

$$\Phi^* = \forall x_1 \exists z_1 \forall x_2 \exists z_2 \ldots \forall x_{n_1} \exists z_{n_1} \exists y_1 \ldots \ldots \forall x_{n_{k-1}+1} \exists z_{n_{k-1}+1} \ldots \forall x_{n_k} \exists z_{n_k} \exists y_k$$
$$\left(\bigwedge_{1 \leq i \leq n_k} (x_i \vee \bar{z}_i) \wedge (\bar{x}_i \vee z_i) \right) \wedge \left(\bigwedge_{1 \leq j \leq m} \phi_j[x_1/z_1, \ldots, x_{n_k}/z_{n_k}] \right),$$

where $\phi_j[x_1/z_1, \ldots, x_{n_k}/z_{n_k}]$ is the clause obtained by replacing in ϕ_j each occurence of x_i resp. $\bar{x}_i$ by the variable z_i resp. $\bar{z}_i$.

As easily can be seen Φ is true iff Φ^* is true.

Φ^* contains clauses with at most one universal variable only. Thus we have proved our desired result, because the evaluation problem for quantified CNF formulas is PSPACE-complete and our transformation can be performed in polynominal time.

q.e.d.

4 Conclusion

We have shown that the evaluation problem for the subclasses Π_k(extended Horn) and Π_k(extended 2-CNF) of quantified Boolean formulas for any fixed $k \geq 1$ is coNP-complete. We hope these results will help to settle the computational complexity of the evaluation problem for quantified extended Horn resp. 2-CNF formulas, that still remains open.

References

[AaBo 79] S. O. Aanderaa, E. Börger: *The Horn complexity of Boolean functions and Cook's problem*, Proc. 5th Scand.Logic Symp. (Eds.B.Mayoh, F.Jensen), 1979, 231-256

[APT 78] B. Aspvall, M. F. Plass and R.E. Tarjan: *A Linear-Time Algorithm for Testing the Truth of Certain Quantified Boolean Formulas*, Information Processing Letters, Volume 8, number 3, 1978, 121-123

[DoGa 84] W. F. Dowling and J. H. Gallier: *Linear-Time Algorithms For Testing The Satisfiability Of Propositional Horn Formulae*, J. of Logic Programming, 1984, 267-284

[GJ 79] M. R. Garey and D. S: Johnson: *Computers and Intractability, A Guide to the Theory of NP-Completeness*, W. H. Freemann and Co., New York, 1979

[Ha 85] A. Haken: *The intractability of resolution*, Theoret. Comput. Sci. 39, 1985, 297–308

[KaKB 90] M.Karpinski, H.Kleine Büning: *Resolution for Quantified Boolean Formulas,* submitted for publication

[KKBS 87] M. Karpinski, H. Kleine Büning and P.H. Schmitt: *On the computational complexity of quantified Horn clauses*, Lecture Notes in Computer Science 329, Springer-Verlag, 1987, 129-137

[StM 73] L. J. Stockmeyer and A. R. Meyer: *Word problems requiring exponential time*, Proc. 5th Ann. ACM Symp. Theory of Computing, 1973, 1-9

Algorithmic Proof with Diminishing Resources Part 1

D M Gabbay*
Department of Computing
Imperial College, 180 Queen's Gate
London SW7 2BZ.
e-mail: dg@doc.ic.ac.uk
Tel: 071 589 5111

Abstract

We present goal directed computation prcedures for classical, intuitionistic and linear implication. The procedure allows for using assumptions at most once. Completeness is proved and proof theoretic results such as interpolation are indicated.

1 Introduction

This paper examines the following theme. Assume we are given a consequence relation $\vdash$ of the form $A_1, \ldots, A_n \vdash B$ and an algorithmic proof system $\mathbf{S}_\vdash$ for deciding (either in a recursive of RE manner) whether for given $A_1, \ldots, A_n, B$ the relation $A_1, \ldots, A_n \vdash B$ holds. Is it possible to perturb $\mathbf{S}_\vdash$ into a proof system $\mathbf{LS}_\vdash$ which is still complete for $\vdash$, and such that each A_i is used in the proof at most once?

The above problem requires clarification. We need to define the notion of "used" in the proof system $\mathbf{S}_\vdash$. This will probably not be too difficult to do. We also note that $\vdash$ can be any logic, and hence the further restriction imposed on $\mathbf{S}_\vdash$ may be only sound, but not complete. Further, it is not at all clear whether such a restriction has a corresponding logical meaning. The resulting system $\mathbf{S}^1_\vdash$, which is obtained from $\mathbf{S}_\vdash$ by restricting the use of assumptions to at most once, may not be a logic at all. Furthermore, for two proof systems for the same consequence relation, say $\mathbf{S}_\vdash$ and $\mathbf{T}_\vdash$, the corresponding systems obtained by the above restriction, namely $\mathbf{S}^1_\vdash$ and $\mathbf{T}^1_\vdash$ may not be the same. These matters will become clearer later in the paper. To consider some examples, assume that $\vdash$ is the implicational fragment of Girard's linear logic (see section 5) then any $\mathbf{S}_\vdash$ will probably do as an $\mathbf{LS}_\vdash$. For other logics, for example intuitionistic implication, our theme would seek to compensate for the lack of completeness of $\mathbf{S}^1_\vdash$, by adding new rules to the system $\mathbf{S}^1_\vdash$ to obtain $\mathbf{S}^1_\vdash*$ which is complete. We thus want to systematically study the trade off between bounded resource and new clever deduction rules.

Of course we must be careful not to replace the resource by a rule which will eventually bring the resource back. So some measure of complexity of resources must be available in the system and during the algorithmic proof execution the total complexity (or weight) of the available resource must be continually reduced. When no more resource is available and the algoirthmic proof execution stops then we decide whether to consider the "run" as success or failure.

There are several good reasons for considering our "diminishing resource" theme. The first is that it may provide efficiency and implementational advantage without losing the character of the algorithmic system. Algorithmic systems are proof theoretic and are put forward for conceputal as well as computational reasons. Many of them are non-deterministic and allow for looping and non-termination when implemented naively. The natural response for eliminating loops which is conceputally clean is to provide for a historical loop checker. This is nice but computationally expensive. There are probably other alternatives which are likely to be very specific to the particular algoirthmic system at hand but may change the character of the system beyond recognition. The presence of additional side effects having to do with the efficiency considerations may overshadow the original system itself, which will sacrifice clarity and adaptability. The diminishing resource theme allows in principle for efficiency while retaining the character of the algorithmic system.

The second advantage is conceputal. Diminishing resources is a natural theme, (as we cannot expect to be using a resource too many times) and because of that can be applied to practically any system. This allows us to develop a methodology which can be meaningful and intuitive. We know that a system $\mathbf{S}_\vdash$ which is complete, may become only sound (and not necessarily complete) when perturbed into a system $\mathbf{S}^1_\vdash$ of diminishing resource. We also know how to compensate. The difference between $\mathbf{S}_\vdash$ and $\mathbf{S}^1_\vdash$ is that there may be occasions in $\mathbf{S}^1_\vdash$ where a resource is about to be used again and is blocked. By looking historically at what happened in the system when the resource was used in the first time we may be able to find the additional rules which compensate.

*I am grateful to J Hudelmaier for critical comments. See his paper for similar results in connection with N-Prolog. I also benefitted from a discussion with Roy Dyckhoff

The above description of our theme is vague and non-mathematical. The best approach is to *just do* what we described for intuitionistic and classical logics and then discuss general definitions and concepts. Thus the next section will describe our goal directed sound and complete algorithmic proof system for the $\rightarrow$ propositional fragment of intuitionistic logic [7], then impose the restriction that each assumption can be used at most once, which destroys completeness, and then show that a new rule, the ***bounded restart rule***, can restore completeness. If we add a variation rule, the ***(unbounded) restart rule***, we get completeness for classical logic.

We start with the $\rightarrow$ fragment because it is more instructive to introduce our theme for the relatively simple fragment of $\rightarrow$ only. Note that it is obvious that we should deal with multisets of assumptions, rather than sets, because of the very nature of the underlying concepts of limited resource. The multisets, for example $\{A, A\}$ allows us to use A twice and is different from $\{A\}$. $\cup$ will denote multiset union.

2 An Algorithmic Proof System for Intuitionistic Implication

We begin with a formal definition of the intuitionistic system.

Definition 2.1 *Let $Q = \{q_1, q_2, \ldots\}$ be a set of atomic propositions and let $\rightarrow$ be implication. We recursively define the notions of a well formed formula A, the multiset $\mathbf{B}[A]$ called the body of A, and the atom $\mathbf{H}[A]$ called the head of A.*

(a) any atom q is a well formed formula. $\mathbf{H}[q] = q$ and $\mathbf{B}[q] = \emptyset$.

(b) If A and B are formulas then also $A \rightarrow B$ is a formula, with $\mathbf{H}[A \rightarrow B] = \mathbf{H}[B]$ and $\mathbf{B}[A \rightarrow B] = \{A\} \cup \mathbf{B}[B]$.

We can now give an algorithmic computation for deciding whether $\mathbf{P} \vdash ?A$ holds or not. $\mathbf{P}$ is a set of well formed formulas called the database and A is the goal. We define the computation in terms of a computation tree. The computation is based on the two Heuristic rules **HR1** and **HR2**, which reflect our intuitive understanding of the computation.

[HR1:] $\mathbf{P} \vdash A \rightarrow B$ if $\mathbf{P}, A \vdash B$

[HR2:] $\mathbf{P} \vdash q$, for q atomic if for some $A \in \mathbf{P}, \mathbf{H}[A] = q$ and for all $B \in \mathbf{B}[A], \mathbf{P} \vdash B$.

In view of the relevance of the $\mathbf{B}$ function to the computation, it is convenient to represent all formulas in a ready for computation form.

Definition 2.2 *An RC-clause (ready for computation clause) is defined by induction as follows:*

1. *Any atomic q is an RC-clause, with head $\mathbf{H}[q] = q$ and body $\mathbf{B}[q] = \emptyset$.*
2. *If A_i are RC-clauses and q is atomic then $B = \{A_i\} \rightarrow q$ is an RC-clause with head $\mathbf{H}[B] = q$ and body $\mathbf{B}[B] = \{A_i\}$.*

Note that if we had conjunctions in the language then any RC-clause A would be equivalent to $\wedge\mathbf{B}(A) \rightarrow \mathbf{H}(A)$, in intuitionistic logic.

Our computation procedure is defined on RC-clauses. $\mathbf{P}_0$ denotes a multiset of RC-clauses, referred to as data and $\mathbf{G}_0$ denotes a multiset of RC-clauses referred to as goals.

We follow consistently our convention of writing A instead of the singleton $\{A\}$.

Definition 2.3 *[Goal directed computation, locally linear goal directed compuation and linear goal directed computation] We present three notions of computation by defining their computation trees. These are:*

1. *The goal directed computation for intuitionistic logic.*
2. *Locally linear goal directed computation*
3. *Linear goal directed computation.*

The goal directed computation tree is defined by clauses (a) - (e), (f1, (f2) and (g) below. The locally linear computation is obtained by replacing clause (f2) by (f3). The linear goal directed-computation is obtained by replacing clause (f1) by (f3), clause (f2) by (f4) and clause (d) by (d).*

Let $\mathbf{P}_0$ be a database and let $\mathbf{G}_0$ be a goal. A quadruple $(T, \leq, 0, V)$ is said to be a successful computation tree of the goal $\mathbf{G}_0$ from $\mathbf{P}_0$ if the following conditions hold:

(a) *$(T, \leq, 0)$ is a finite tree; T is its set of nodes, 0 is its root, and for s, t in $T, t \leq s$ means that $t = s$ or t lies closer to the root than s.*

(b) *V is a labelling function on T. For each $t \in T, V(t)$ is a pair $(\mathbf{P}, \mathbf{G})$, where $\mathbf{P}$ is a database and $\mathbf{G}$ is a goal.*

(c) *$V(0) = (\mathbf{P}_0, \mathbf{G}_0)$. We say the clauses of $\mathbf{P}_0$ are* put in the data at node 0.

(d) *If $V(t) = (\mathbf{P}, \mathbf{G})$ and $\mathbf{G} = \{A_1, \ldots, A_n\}$, with A_i RC-clauses then the node t has exactly n immediate successors in the tree, say $t_1, \ldots, t_n$ with $V(t_i) = (\mathbf{P}_i, A_i)$, where $\mathbf{P}_i = \mathbf{P}$.*

(d) For the case of linear computation we require that* $V(t_i) = (\mathbf{P}_i, A_i)$, *with* $\cup\mathbf{P}_i = \mathbf{P}$.

(e) If $V(t) = (\mathbf{P}, \mathbf{G})$, *and* $\mathbf{G} = \mathbf{A} \to q$, *then t has exactly one immediate successor in the tree, say s, with* $V(s) = (\mathbf{P} \cup \mathbf{A}, q)$. *We say that the clauses in* $\mathbf{A}$ *were* put in the data at node t.

(f) If $V(t) = (\mathbf{P}, q)$, *with q atomic, then exactly one of the following holds.*

(f1) t has no immediate successors and $q \in \mathbf{P}$. *We say* $q \in \mathbf{P}$ *is* used at node t.

(f2) t has exactly one immediate successor s in the tree with $V(s) = (\mathbf{P}, \mathbf{A})$ *and with* $(\mathbf{A} \to q) \in \mathbf{P}$. *In this case we say that* $\mathbf{A} \to q$ *is* used at node $t \in T$.

(f3) For the case of linear computation require that t has no immediate successors and $\{q\} = \mathbf{P}$.

(f4) For the case of locally linear computation we require that $V(s) = (\mathbf{P}', \mathbf{A})$, *with* $\mathbf{P} = \mathbf{P}' \cup \{\mathbf{A} \to q\}$.

(g) Note that it is possible in case (f2) or (f3) that $q \in \mathbf{P}$. *Nevertheless the computation goes through* $\mathbf{A} \to q$. *Further note that the databases involved need not be finite. In a successful computation tree only a finite number of clauses from the database will be used in the sense mentioned above.*

We use the notation $\mathbf{P}?A$ to denote the problem of finding a successful computation tree of A from $\mathbf{P}$. We will indicate explicitly in which computation. We write $\mathbf{P}?A = 1$ to denote the fact that such a tree exists. $\mathbf{P}?A = 0$ would then be used for finite failure but the notion of finite failure will have to be precisely defined. $\mathbf{P}?q =?$ can be used for loops. We are not defining these concepts now but as we are operating in a positive fragment it makes sense to add various notions of negation, among them falsity $\perp$ and failure.

We now have to give the connection between our intuitionistic consequence relation and the goal directed computation.

Definition 2.4 *1. With every wff A of intuitionistic logic, we associate a unique RC-clause* A^* *as follows.*

(a) $(q)^* = q$, *for q atomic.*

(b) If A is not atomic, it has the form $A_1 \to (A_2 \to \ldots \to (A_n \to q)\ldots)$ *with q atomic. Then* $A^* = \{A_1^*, \ldots, A_n^*\} \to q$.

2. Conversely, with every RC- clause A we associate a formula of intuitionistic logic $A\sharp$ *as follows:*

(a) $q\sharp = q$, *for q atomic.*

(b) $(\{B_1, \ldots, B_n\} \to q)\sharp = B_1\sharp \to (\ldots \to (B_n\sharp \to q)\ldots)$.

Where the ordering $B_1 \ldots B_n$ is arbitrary. Notice that if $\{C_1, \ldots, C_n\} = \{B_1, \ldots, B_n\}$ is another listing of the set then $B_1\sharp \to (\ldots \to (B_n\sharp \to q)\ldots)$ is intuitionistically equivalent to $C_1\sharp \to (\ldots(C_n\sharp \to q)\ldots)$.

Further note that we can assume $A^*\sharp = A$ for all A. From now on we can allow ourselves to use $\{A_1, \ldots, A_n\} \to q$ ambiguously, either as an RC-clause or as an abbreviation for $A_1 \to \ldots \to (A_n \to q)\ldots)$.

Theorem 2.5 $\mathbf{P} \vdash A$ *iff* $\mathbf{P}^*?A^* = 1$, *in the goal directed computation, (ie there exists a finite successful computation tree).*

Proof. See reference N-Prolog part II. ∎

Example 2.6 *In the examples each line follows from the previous line by applying a rule, in the goal directed computation.*

1. $(a \to b) \to a?a$
   $(a \to b) \to a)?a \to b$
   $(a \to b) \to a, a?b$
   fails

2. $(c \to a) \to c, c \to a?a$
   $(c \to a) \to c, c \to a?c$
   $(c \to a) \to c, c \to a?c \to a$
   $(c \to a) \to c, c \to a, c?a$
   $(c \to a) \to c, c \to a, c?c$
   succeeds

3. $(a \to b) \to b?(b \to a) \to a$
   $(a \to b) \to b, b \to a?a$
   $(a \to b) \to b, b \to a?b$
   $(a \to b) \to b, b \to a?a \to b$
   $(a \to b) \to b, b \to a, a?b$
   $(a \to b) \to b, b \to a, a?a \to b$

$(a \to b) \to b, b \to a, a?b$
loops

4. $(a \to x) \to a, (b \to x) \to b, (a \wedge b \to x)?a$
$(a \to x) \to a, (b \to x) \to b, a \wedge b \to x?a \to x$
$(a \to x) \to a, (b \to x) \to b, a \wedge b \to x, a?x$
$(a \to x) \to a, (b \to x) \to b, a \wedge b \to x, a?a \wedge b$
The left conjunct succeeds so we continue with the right conjunct ?b.
$(a \to x) \to a, (b \to x) \to b, a \wedge b \to x, a?b \to x$
$(a \to x) \to a, (b \to x) \to b, a \wedge b \to x, a, b?x$
$(a \to x) \to a, (b \to x) \to b, a \wedge b \to x, a, b?a \wedge b$
success

3 Resource Boundedness; Locally Linear Goal Directed Computation

We now want to adopt the policy of using each assumption at most once. First we want to motivate the advantages of resource boundness.

Example 3.1 Historical Loop Checking
Consider the data and query below:

$$q \to q?q.$$

Clearly the algorithm has to know it is looping. The simplest way of loop checking is to record the history of the computation. This can be done as follows: Let $\mathbf{P}?(G, \mathbf{H})$ *represent the query of the goal* G *from data* $\mathbf{P}$ *and history* $\mathbf{H}$. $\mathbf{H}$ *is a list. The computation rules become:*
Historical Rule for →
$\mathbf{P}?(A \to B, \mathbf{H}) = 1$
if $[\mathbf{P} \cup \{A\}?(B, \mathbf{H} * ((\mathbf{P}, A \to B))) = 1$ *and* $(\mathbf{P}, A \to B)$ *is not in* $\mathbf{H}]$
where * *denote concatentation of lists. Thus* $(\mathbf{P}, A \to B)$ *is appended to* $\mathbf{H}$.
Historical Rule for atoms
$\mathbf{P}?(q, \mathbf{H}) = 1$, *for* q *atomic,*
if for some $B = (x_1 \to (x_2 \to \ldots (x_n \to q) \ldots)$ *in* $\mathbf{P}$ *we have that for all* i
$\mathbf{P}?(x_i, \mathbf{H} * ((\mathbf{P}, q)) = 1$ *and that* $(\mathbf{P}, q)$ *is not in* $\mathbf{H}$.
Thus the computation for our example above becomes:

$$(q \to q)?(q, \varnothing)$$

$$(q \to q)?(q, ((q \to q, q)))$$

fail.

Note that the historical loop checking conjunct in rule for → *is redundant as the loop will be captured in* rule for atom. *Also note that in this case we made the decision that looping means failure. The most general case of loop checking may first detect the loop and then decide that under certain conditions looping implies success. Thus a rule where looping is considered success can be of the form*

$$\mathbf{P}?A = 1 \text{ if } \mathbf{P}?B \text{ and not } (\mathbf{P}, B) \in \mathbf{H}.$$

This type of loop checking, although effective, is very expensive on resources. Many copies of the database are required. One can optimise it a bit by listing in the history **H** only the additions to the original database, but this is expensive as well, as the additions to the data are of the same order of magnitude as the data.

Trying to think constructively, in order to make sure the algorithm terminates, let us ask ourselves what is involved in the computation. There are two parameters; the data and the goal. The goal is reduced to an atom via the rule for → and the looping occurs because a data item is being used by the rule for atom again and again. Our aim in the historical loop checking is to stop that. Well, why don't we try a modified rule for atoms which can use each item of data only once? This way, we will certainly run out of data and the computation will terminate. We have to give a formal definition of the computation, where we keep track on how many times we use the data.

Let us adopt the point of view that each database item can be used at most once. Thus our rule for atoms becomes (LLR stands for locally linear rule):

LLR for atoms: $\mathbf{P}, \mathbf{A} \to q?q = 1$ if $\mathbf{P}?\mathbf{A} = 1$

The item $\mathbf{A} \to q$ is thus thrown out, as soon as it is used. This condition is clause (f3) of definition 2.3.

The question is now do we retain completeness? Are there examples for intuitionistic logic where items of data essentially need to be used locally more than once?

Example 3.2 $c \to a, (c \to a) \to c?a$
The clause $(c \to a)$ has to be used twice in order for a to succeed.

This example can be generalised.

Example 3.3 *Let* $A_0 = c$
$A_{n+1} = (A_n \to a) \to c$.
Consider the following data and query:
$P_n = \{A_n, c \to a\}?a$
The clause $c \to a$ has to be used n times.

Example 3.4 *Another example is*
$a \to (b \to c), a \to b, a?c$
here a has to be used twice globally, but not locally on each branch of the computation. Thus the locally linear computation succeeds.

Let us do the full computation:

$$0 : \{((a \to (b \to c)), (a \to b), a\}?c$$

$$t : \{(a \to b), a\}?a \qquad s : \{(a \to b), a\}?b$$

$$s_1 : \{a\}?a$$

It is obvious that "a" is used twice "globally" in the entire computation tree but not more than once "locally", on each path.

Example 3.5 *Note that the above examples show that locally linear goal directed computation does* not *give rise to a consequence relation. The cut rule is violated, if we understand cut as defined before.*

In 3.4, a has to be used twice. Hence certainly

$$a, a \vdash (a \to b) \to [(a \to (b \to c)) \to c]$$

in globally linear logic where each assumption can be used only once. Also $a \vdash a$ holds. Therefore, by cut we should get

$$a \vdash (a \to b) \to [(a \to (b \to c)) \to c]$$

which, of course, does not hold. Thus, the cut rule

$$\frac{\mathbf{P} \vdash A; \mathbf{P}, A \vdash B}{\mathbf{P} \vdash B}$$

does not hold.

However, if we write the cut rule as

$$\frac{\mathbf{P}_1 \vdash A; \mathbf{P}_2, A \vdash B}{\mathbf{P}_1, \mathbf{P}_2 \vdash B}$$

The rule does hold.

For $\mathbf{P}_1 = \mathbf{P}_2 = \mathbf{P}$, we can understand $\mathbf{P}$, $\mathbf{P}$ as two copies of $\mathbf{P}$.[1]

Example 3.6 *A simpler example is the following database:*
$(A_1 \to a) \to c$
$(A_n \to a) \to c$
$A_1 \to (A_2 \to \ldots (A_n \to c) \ldots)$

[1] The notion of a consequence relation can also be modified, so that the reading of union $\cup$ and intersection $\cap$ refer to multiset union and intersection. These are:
Reflexivity: $\mathbf{P} \vdash \mathbf{R}$ if $\mathbf{P} \cap \mathbf{R} \neq \emptyset$
Monotonicity: $\mathbf{P} \vdash \mathbf{R}$ implies $\mathbf{P} \cup \mathbf{Q} \vdash \mathbf{R}$
Cut: $\mathbf{P}_1 \vdash A, \mathbf{R}_1$ and $\mathbf{P}_2, A \vdash \mathbf{R}_2$ imply $\mathbf{P}_1 \cup \mathbf{P}_2 \vdash \mathbf{R}_1 \cup \mathbf{R}_2$.

$c \to a$
the query is "?a".
$c \to a$ has to be used $n+1$ times globally to ensure success, but locally it is needed only twice.

We can combine 3.6 and 3.3 to show a database and a query where a clause in the query needs to be used at least n times globally and at least m times locally. 3.6 is a case of $m = 2n = m$.

The above examples show that we do not have completeness for locally linear goal directed computations. We thus have to first check what we are getting, and second check whether we can compensate for the use of the locally linear rule (ie for throwing out the data) by some other means. We need to make our studies in a general framework where formulas can be "used" a fixed number of times (eg only once). The appropriate notations and concepts must be developed. Such logics we call **BR-Logics** or logics of bounded resource. One needs an appropriate notation and an appropriate definition of the notion "used". In our computation above, we know intuitively what "used" means, and our resource bounded principle was to allow each RC-clause to be "used" at most once. The obvious generalisation of the notion is to annotate a formula with a natural number which indicates how many times the formula can be used. That is why we found it convienient to use the notation of multisets.

The reader is warned that the locally linear system of 2.3 is not the same as the linear system. First we do not require in the locally linear system that all assumptions must be used, (conditin (f3) of definition 2.3), we only ask that they be used no more times than specified. A more serious difference is that we do our "counting" of how many times a formula is used separately on each path of the computation and not globally for the entire computation. The linear goal directed system is global, (condition (d*) of definition 2.3).

For example, the query $\{A, A \to (A \to B)\}?B$ will succeed in our locally linear computation because A is used once on each of two parallel paths. It will not be accepted in the linear system because A is used globally twice. This is ensured by condition (d*) of 2.3. We shall see in section 5 that the linear computation of 2.3 corresponds to the implicational fragment of Girard linear logic [5].

We are happy with the locally linear system because of our modest motivation which is the optimisiation our computation procedures and not the development of a general conceptual resource allocation framework.[2]

Example 3.7 *This example shows that the notion of diminishing resource depends on the computation.*

1. *Consider*

   (a) $(c \to a) \to c$.

   (b) $c \to a$

   To show a either by forward modus ponens or by backward goal directed reasoning, we need to use $c \to a$ twice.
   We compare with resolution. The translated database is $\{c, \neg c \vee a, \neg a \vee c\}$.
   To show a we use clauses only once.

2. *Consider the database*

   $$\{\neg a \vee d \vee \neg c, a, \neg a \vee c\}$$

   To show d using resolution we need to use a twice. The same database written with $\wedge$ and $\to$ is:

   (a) $a \wedge c \to d$

   (b) $a \to c$

   (c) a

   To show d using backward reasoning, we use a once on each path, twice globally.

BR-logics, the logics of bounded resource, are extensively studied in [3]. Here we deal only with the aspects that have to do with our algorithmic proof system.

Let us now go back to the notion of locally linear computation. We saw that we do not have completeness. There are examples where formulas need to be used several times. Can we compensate? Can we at the same time throw data out once it has been used and retain completeness for intuitionistic logic by adding some other computation rule? The answer is yes. The rule is called ***(Linear) bounded restart rule*** and is used in the context of the notion of locally linear computation with history.

Let us examine more closely why we needed in example 3.3 the clause $c \to a$ several times. The reason was that from other clauses, we got the query "?a" and we wanted to use $c \to a$ to continue to the query "?c". Why was not $c \to a$ available ?; because $c \to a$ has already been used. In other words, ?a as a query, has already been asked and $c \to a$ was used. This means that the next query after "?a" in the history was "?c".

If **H** is the history of the atomic queries asked, then somewhere in **H** there is "?a" and immediately afterwards "?c".

[2] We study general resource problems in my book on Labelled Deduction Systems [3]. We have a chapter on what we call *Control Derivation Logics*. These are logics where the proofs are controlled either by limiting resource or by giving order and preferences on the use of the inference rules or a combination. Thus this class of logics contains two pure subclasses, the pure *Resource Logics* and the pure *Control Inference Logics*.

We can therefore compensate for the re-use of $c \to a$ by allowing ourselves to go back in the history to where "?a" was, and allow ourselves to ask all queries that come afterwards. Let us see what happens to our example 3.3

$(c \to a) \to c$	[1]	$?a$
$c \to a$	[1]	

We use the second clause to get

$(c \to a) \to c$	[1]	$?(c,(a))$
$(c \to a)$	[0]	

we continue

$(c \to a) \to)c$	[0]	$?(c \to a,(a,c))$
$(c \to a)$	[0]	

we continue

$(c \to a) \to c$	[0]	$?(a,(a,c))$
$(c \to a)$	[0]	
c	[1]	

The "1"("0") annotate the clause to indicate it is active (inactive) for use. The history can be seen as the right hand column of past queries.

We can now ask any query that comes after an "a" in the history, hence

$(c \to a) \to c$	[0]	
$(c \to a)$	[0]	$?(c,(a,c,a))$
c	[1]	

Success.

The previous example suggests the following new computation with bounded restart rule.

Definition 3.8 *[Locally Linear Computation with Bounded Restart; Locally Linear Computation with Restart] Let* $\mathbf{M}_0$ *be a database and let* $\mathbf{G}_0$ *be a goal. We define two computations.* Locally linear computation with bounded restart, and locally linear compuatation with restart. *The first is defined with clauses (a) - (e), (f1) - (f3), (g) and the second with clauses (a) - (e), (f1), (f2), (f4) and (g). In the computation with bounded restart the history* $\mathbf{H}_t$ *is a seqauence, while in the computation with restart* $\mathbf{H}_t$ *is a set.* $\mathbf{M}_t$ *and* $\mathbf{G}_t$ *are multisets in both computations.*

A quadruple $(T, \leq, 0, V)$ *is said to be a successful computation tree of the goal* $\mathbf{G}_0$ *from* $\mathbf{M}_0$ *with initial history* $\mathbf{H}_0$ *iff the following conditions hold.*

(a) $(T, \leq, 0)$ *is a finite tree; T is its set of nodes, 0 is its root and for* $s,t \in T, t \leq s$ *means* $t = s$ *or t lies closer to the root than s.*

(b) V is a labelling function on T. For each $t \in T, V(t)$ *is a triple* $(\mathbf{M}_t, \mathbf{G}_t, \mathbf{H}_t)$ *where* $\mathbf{M}_t$ *is the current database at* $t, \mathbf{G}_t$ *is the current goal and* $\mathbf{H}_t$ *is the current history of previous goals.*

(c) $V(0) = (\mathbf{M}_0, \mathbf{G}_0, \mathbf{H}_0)$. *We say the clauses in* $\mathbf{M}_0$ *were put in the data at node 0.*

(d) If $V(t) = (\mathbf{M}_t, \mathbf{G}_t, \mathbf{H}_t)$ *with* $\mathbf{G}_t = \{A_1, \ldots, A_n\}$ *then the node t has exactly n immediate successors in the tree, say* $t_1, \ldots, t_n$, *with* $V(t_i) = (\mathbf{M}_t, \{A_i\}, \mathbf{H}_t)$

(e) If $V(t) = (\mathbf{M}_t, \mathbf{G}_t, \mathbf{H}_t)$ *and* $\mathbf{G}_t = \{(\mathbf{A} \to q)\}$ *then t has exactly one immediate successor s in the tree with* $V(s) = (\mathbf{M}_t \cup \mathbf{A}, q, \mathbf{H}_t)$. *We say that the new additional RC-clauses in* $\mathbf{M}_t \cup \mathbf{A}$, *were put in the data at node s.*

(f) If $V(t) = (\mathbf{M}_t, \{q\}, \mathbf{H}_t)$, *for q atomic then exactly one of the following holds:*

(f1) t has no immediate successors and $q \in \mathbf{M}_t$. *We say that the RC-clause q is used at this node t.*

(f2) t has exactly one successor s in the tree, and for some $(\mathbf{A} \to q), \in \mathbf{M}_t$ *we have:*

(a) $V(s) = (\mathbf{M}_s, \mathbf{A}, \mathbf{H}_t * (q))$, *where * is concatenation of sequences, or set union for the case* $\mathbf{H}_t$ *are sets.*

(b) $\mathbf{M}_t = \mathbf{M}_s \cup \{(\mathbf{A} \to q)\}$

We say that the clause $(\mathbf{A} \to q)$ *is used once at the node t.*

(f3) Bounded Restart rule

t has exactly one successor s in the tree, and $V(s)) = (\mathbf{M}_t, \{q_1\}, \mathbf{H}_t * (q))$, *provided for some* $\mathbf{H}_1, \mathbf{H}_2, \mathbf{H}_3$ $\mathbf{H}_t = \mathbf{H}_1 * (q) * \mathbf{H}_2 * (q_1) * \mathbf{H}_3$ *where* $\mathbf{H}_i$ *can be empty.*

(f4) Restart rule

If the proviso of (f3) is cancelled, we get the restart rule, namely $V(s) = (\mathbf{M}_t, \{q_1\}, \mathbf{H}_t * (q))$, *with* $q_1 \in \mathbf{H}_t$. *In this case we can simplify the entire definition and allow* $\mathbf{H}_t$ *to be* a set, *the history of previous goals. Thus* $\mathbf{H}_t * (q)$ *should be replaced by* $\mathbf{H}_t \cup \{q\}$.

(g) We write $\mathbf{P}?(\mathbf{G}, \mathbf{H}) = 1$ *iff there exists a successful computation tree of* $\mathbf{G}$ *from* $\mathbf{P}$ *with initial history* $\mathbf{H}$. *the computation involved will be specified in the context.*

We still have the problem of completeness..

Consider the query: $\mathbf{P} = \mathbf{P}_0, x_1 \to (x_2 \to q)[1]?q$.

This continues as two parallel computations (a) and (b) which should both succeed where

(a) $\mathbf{P}_0, x_1 \to (x_2 \to q)[0]?x_1$ and

(b) $\mathbf{P}_0, x_1 \to (x_2 \to q)[0]?x_2$

Computation (a) may proceed and end up with the query $\mathbf{P}_1?q$ with history $(q, x_1, \ldots, q)$. Our rule allows us to continue to ask $\mathbf{P}_1?x_1$.

However, if we really had the clause itself available to us we would split the computation again into

$$P_1?x_1 \qquad \text{and} \qquad \mathbf{P}_1?x_2$$

to be again continued in parallel.

So we are basically chopping out $\mathbf{P}_1?x_2$, in our present algorithm

Does this affect the outcome? We do have the original branch (b), namely $\mathbf{P}_0, x_1 \to (x_2 \to q)?x_2$. If (b) succeeds, then we know $\mathbf{P}_1?x_2$ succeeds. If (b) fails, then $\mathbf{P}_1?x_2$ may still succeed because $\mathbf{P}_1 \supseteq \mathbf{P}_0$ may contain more data, but that makes no difference because the original $\mathbf{P}?q$ fails anyway, as (b) must succeed.

This requires proper study and we shall give the appropriate proof later on.

Example 3.9 *1. The following example shows the problem we might have. Consider schematically an original query ?c which splits to two queries as in fig 1:*

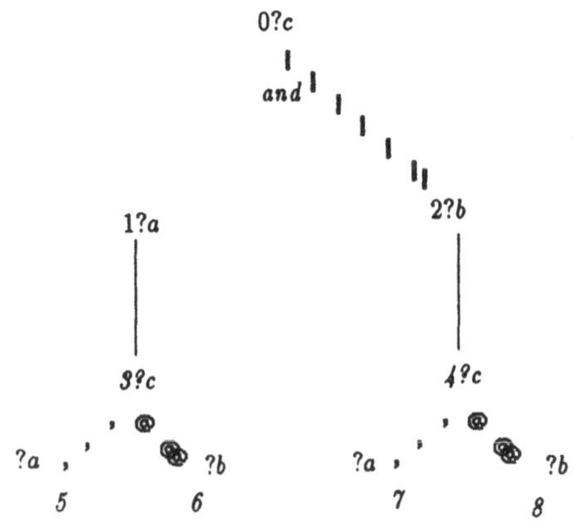

Figure 1:

We numbered the end nodes 5, 6, 7, 8. We might have a situation where we do not pursue ?b at node 6 because it is already asked above at node 2 and at the same time we do not ask ?a at node 7 because ?a is already asked above at node 1. We might think that we never compute nodes 1 and 2 properly, and that we will get wrong results.

If we carry out the computation fully on the left branch (i.e. node 1) then at node 7 we can rely on the fact that ?a has already been computed at node 1.

2. Consider the following example:

$$(((a \to c) \to a) \to a) \to ((((b \to c) \to b) \to b) \to c)?c$$

The computation splits in two, fig 2

The figure shows only the data added at each node. If we follow the bounded restart rule we get success. Let us check what happens if we continue with ordinary rule for atoms and try to succeed without bounded restart. From the rule for atoms, using the original data we get fig 3:

The nodes a?b and b?a can further be computed and they succeed.

We now have a rule which allows us to ask any previous query ?a instead of the current query ?q provided ?q was asked earlier than ?a.

We now have two natural problems to consider, the problem of the efficiency of the proposed computation and the problem of the logical meaning of the computation.

We begin with the problem of the logical meaning and start with an example. Consider the following two databases:

$$\mathbf{P}_1 = \{(c \to a) \to b\} \text{ and } \mathbf{P}_2 = \{c, a \to b\}$$

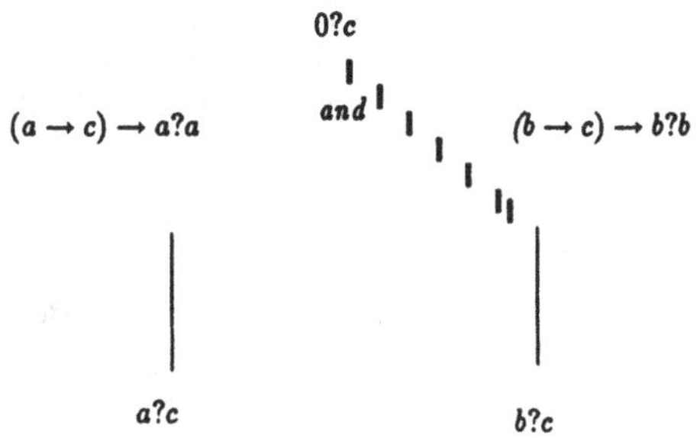

Figure 2:

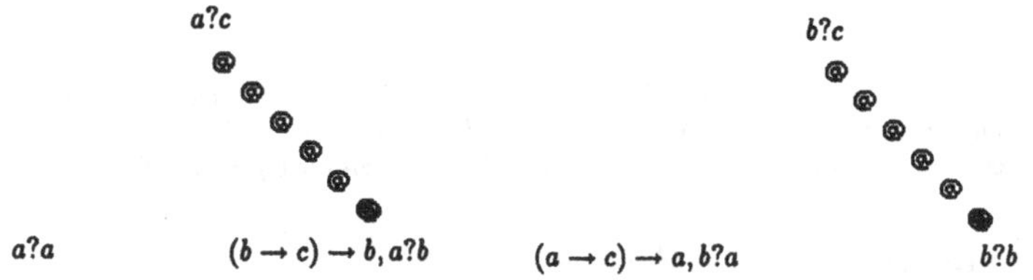

Figure 3:

If we ask the query $?b$ from each database, using the linear bounded restart rule, we get that the query is reduced to

$$\{c\}?(a,(b))$$

Thus we see that one cannot reconstruct the original database from the history. This problem has bearing upon soundness. Given a query $\mathbf{P}?(q,\mathbf{H})$, which succeeds, we can ask what is the logical meaning of the success of the computation, what is being proved? Can we say that for some intuitionistic wff $\varphi(\mathbf{P},q,\mathbf{H})$, embodying in it the logical content of the computation, we have $\vdash \varphi(\mathbf{P},q,\mathbf{H})$?

Ie, we want to find a φ which satisfies the following:

$$\mathbf{P}?(q,\mathbf{H}) = 1 \text{ iff } \vdash \varphi(\mathbf{P},q,\mathbf{H})$$

To achieve that, φ will have to satisfy:

1. $\vdash \varphi(\mathbf{P},q,\varnothing)$ iff $\mathbf{P} \vdash q$
2. $\vdash \varphi(\mathbf{P} \cup \{\mathbf{A} \to q\}, q, \mathbf{H})$ if
   $\vdash \varphi(\mathbf{P},\mathbf{A},\mathbf{H} * (q))$
3. $\vdash \varphi(\mathbf{P} \cup \{x\}, q, \mathbf{H})$
   if $x = q$ or x is in $\mathbf{H}$, bounded by q.
4. $\vdash \varphi(\mathbf{P},\mathbf{A} \to q,\mathbf{H})$ iff $\vdash \varphi(\mathbf{P} \cup \mathbf{A}, q, \mathbf{H})$
5. $\vdash \varphi(\mathbf{P},\{A,B\}\mathbf{H})$ iff $\vdash \varphi(\mathbf{P},A,\mathbf{H})$ and $\vdash \varphi(\mathbf{P},B,\mathbf{H})$

Lemma 3.10 *The followng hold in intuitionistic logic:*

1. *If* $\mathbf{P}, A, q \to x \vdash q$
   Then $\mathbf{P}, (A \to q) \to x \vdash A \to q$
2. $(A \to x) \to A, (B \to x) \to B \vdash (A \wedge B \to x) \to (A \wedge B)$.

Proof.

1. Note that with A in the data, $A \to q$ is equivalent to q and hence $\mathbf{P}, A, q \to x \vdash q$ implies $\mathbf{P}, A, (A \to q) \to x \vdash q$, which implies $\mathbf{P}, (A \to q) \to x \vdash A \to q$.
2. We use the completeness theorem 2.5 and example 2.6, part 4.

■

We now introduce a new rule, the Restart Rule, which is a natural more convenient rule. It is a variation of the Bounded Restart Rule obtained by cancelling any restrictions and simply allowing to ask any earlier atomic query. If we allow that, we will be simplifying our computation a lot. We need not keep history as a sequence but only as a *set* of previous queries. The rule becomes

Restart Rule:

$\mathbf{P}?(q, \mathbf{H}) = 1$ if $\mathbf{P}?(a, \mathbf{H} \cup \{q\}) = 1$, for any $a \in \mathbf{H}$.

The formal definition of *locally linear compuation with restart* is definition 3.8

Remark 3.11 *Note that in definition 3.8, the history is that of atomic queries only. This is sufficient for completeness. There are logics where restart cannot be restricted to the atomic case, see [7]*

Example 3.12 *[Computation using the restart rule]*

$\{(a \to b) \to)a\}?(a, \emptyset)$

use the clause and throw it out and get:

$\emptyset?((a \to b), \{a\})$

$a?(b, \{a, a \to b\})$

restart

$a?(a, \{a, a \to b, b\})$

success.

This example shows that we are getting a stronger "logic"? (we don't know that we get a logic here yet) than intuitionistic logic. The above should fail in the intuitionistic computation. What are we getting?

The surprising answer is that we are getting classical logic. This claim has to be properly proved, of course.

4 Completeness Theorems for the Restart Rules

We will now prove completeness of the restart rule for classical logic for the computation where we do not cancel any data formulas, as described in 3.8.

As the argument is involved, we would like to state the computation again explicitly in terms of computation trees.

Definition 4.1 $(T, \leq, 0, V)$ *is said to be a goal directed computation tree with the Restart Rule and no cancellation of data if all the conditions (a) - (e), (f1), (f2) of 2.3 are satisfied together with the following extra clause:*

(g3) t has exactly one immediate successor s in the tree and $V(s) = (\mathbf{P}, \mathbf{G}_0)$.

Theorem 4.2 $\mathbf{P} \vdash A$ *in classical implicational logic iff* $(\mathbf{P}^*, A^*)$ *has a successful computation tree with the Restart Rule (as defined in 4.1 above).*

For the definition of P^*, A^**, see 2.4. For the definition of classical logic, take truth table tautological validity.*

Proof. *To prove this theorem we need a series of definitions and lemmas, 4.3 to 4.6 Note that* $\mathbf{P}?\mathbf{G} = 1$ *is understood according to 2.3 in the proofs below, in 4.4 and 4.5.*

To prove the theorem note that from 4.3 and 4.5 we get that $\mathbf{P} \vdash A$ *in classical logic iff* $\mathbf{P}^* \cup Cop(A^*)?A^* = 1$. *However* $\mathbf{P}^*?A^* = 1$ *with restart iff* $\mathbf{P}^* \cup Cop(A^*)?A = 1$. ■

Definition 4.3 *Let* $\mathbf{A}$ *be a goal. Then the complement of* $\mathbf{A}$*, denoted by* $Cop(\mathbf{A})$ *is the following set of clauses:*

$Cop\ (\mathbf{A}) = \{\mathbf{A} \to x \mid x$ *any atom of the language*$\}$

Lemma 4.4 *For any database* $\mathbf{P}$ *and goal* $\mathbf{G}$ *such that* $\mathbf{P} \supseteq Cop\ (\mathbf{G})$*, and for any goal* $\mathbf{A}$*, conditions (a) and (b) below imply condition (c):*

(a) $\mathbf{P} \cup \mathbf{A}?\mathbf{G} = 1$

(b) $(\mathbf{P} \cup Cop\ (\mathbf{A}))?\mathbf{G} = 1$

(c) $\mathbf{P}?\mathbf{G} = 1$

Proof. Since $\mathbf{P} \cup$ Cop $(\mathbf{A})?\mathbf{G} = 1$ and computations are finite, only a finite number of the elements of Cop $(\mathbf{A})$ are used in the computation. Assume then that

(b1) $\mathbf{P}, (\mathbf{A} \to x_1), \ldots, (\mathbf{A} \to x_n)?\mathbf{G} = 1$

We use the cut rule. Since $\mathbf{P} \supseteq Cop(\mathbf{G})$ we have $\mathbf{G} \to x_i \in \mathbf{P}$ and hence by the computation rules $\mathbf{P} \cup \mathbf{A}?x_i = 1$. Now by cut on (b1) we get $\mathbf{P}?\mathbf{G} = 1$. ■

Theorem 4.5 *For any* $\mathbf{P}$ *and any* A*, (a) is equivalent to (b) below:*

(a) $\mathbf{P} \vdash A$ *in classical logic*

(b) $(\mathbf{P}^* \cup Cop\ (A^*))?A^* = 1$

Proof.

1. Show (b) implies (a):
   Assume $(\mathbf{P}^* \cup Cop(A^*))?A^* = 1$
   Then by the soundness of our computation procedure we get that $\mathbf{P}^* \cup \text{Cop}\,(A^*\sharp) \vdash A$ in intuitionistic logic, and hence in classical logic, where $A^*\sharp$ are defined in 2.4.
   Since the *proof* is *finite* there is a finite set of the form $\{A \to x_i, \ldots, A \to x_n\}$ (note that $A^*\sharp = A$) such that
   (a1) $\quad \mathbf{P}, (A \to x_1), \ldots (A \to x_n) \vdash A$ (in intuitionistic logic).
   We must also have that $\mathbf{P} \vdash A$, in classical logic, because if there were an assignment h making $\mathbf{P}$ true and A false, it would also make $A \to x_i$ all true contradicting (a1).
   The above concludes the proof that (b) implies (a).

2. Show that (a) implies (b)
   We prove that if $\mathbf{P}^* \cup \text{Cop}\,(A^*)?A^* \neq 1$
   then $\mathbf{P} \not\vdash A$ in classical logic.
   Let $\mathbf{P}_0 = \mathbf{P}^* \cup \text{Cop}\,(A^*)$.
   We define a sequence of databases $\mathbf{P}_n, n = 1, 2 \ldots$ as follows:
   Let $\mathbf{B}_1, \mathbf{B}_2, \mathbf{B}_3, \ldots$ be an enumeration of all goals of the language.
   Assume $\mathbf{P}_{n-1}$ has been defined and assume that $\mathbf{P}_{n-1}?A^* \neq 1$. We define $\mathbf{P}_n$:
   If $\mathbf{P}_{n-1} \cup \mathbf{B}_n?A^* \neq 1$, let $\mathbf{P}_n = \mathbf{P}_{n-1} \cup \mathbf{B}_n$. Otherwise from 4.4 we must have:
   $\mathbf{P}_{n-1} \cup \text{Cop}\,(\mathbf{B}_n)?\mathbf{A}^* \neq 1$.
   and so let $\mathbf{P}_n = \mathbf{P}_{n-1} \cup \text{Cop}\,(\mathbf{B}_n)$.
   Let $\mathbf{P}' = \cup n \mathbf{P}_n$
   Clearly $\mathbf{P}'?A^* \neq 1$.
   Define an assignment of truth values h on the atoms of the language by
   $h(x) =$ **true** iff $\mathbf{P}'?x = 1$.

Lemma 4.6 *For any* $B, h(B) =$ **true** *iff* $\mathbf{P}'?B^* = 1$

Proof. By induction on B.

(a) For atoms this is the definition

(b) Let $B = B_1 \to (B_2 \to \ldots (B_n \to q) \ldots)$ and let $B^* = \mathbf{C} \to q$. We thus have to check the case of $(\mathbf{C} \to q), q$ atomic, with $\mathbf{C} = \{B_1, \ldots B_n\}$.

(b1) If $\mathbf{P}'?(\mathbf{C} \to q) = 1$, then if $\mathbf{P}'?\mathbf{C} = 1$ then by 4.4 also $\mathbf{P}'?q = 1$. This means, by the induction hypothesis that if $h(B_i) =$ **true** for all i then also $h(q) =$ **true**, hence $h(B) =$ **true**.

(b2) If $\mathbf{P}'?(\mathbf{C} \to q) \neq 1$, then $\mathbf{P}' \cup \mathbf{C}?q \neq 1$. Hence by definition $h(q) =$ **false**, since certainly $\mathbf{P}'?q \neq 1$.
   If $h(B_i) =$ **true** for all i, then we are finished, since this makes $h(B) =$ **false**. We now show that indeed $h(B_i)$ $=$ **true** for all i by showing that $h(B_1) =$ **false** leads to a contradiction.
   Assume that $h(B_1) =$ **false**.
   Then by the induction hypothesis $\mathbf{P}'?\mathbf{C} \neq 1$.
   Thus certainly $\mathbf{C} \not\subseteq \mathbf{P}'$ (if $\mathbf{C} \subseteq \mathbf{P}'$ then $\mathbf{P}'?\mathbf{C} = 1$). We have that $\mathbf{C} = \mathbf{B}_n$, for some n in the enumeration of wffs and so since $\mathbf{C} \subseteq \mathbf{P}'$, by construction $\text{Cop}\,(\mathbf{C}) \subseteq \mathbf{P}'$, and especially $(\mathbf{C} \to q) \in \mathbf{P}'$. Thus $\mathbf{P}'?(\mathbf{C} \to q) = 1$, which contradicts the assumption of our case (b2).
   Thus case (b) is proved and 4.6 is proved.

■

We can now prove direction 2 of 4.5. Since $\mathbf{P}'?A^* \neq 1$, we get $h(A) =$ **false**. Thus h is an assignment of truth values such that for any $B^* \in \mathbf{P}$), and certainly for any $B^* \in \mathbf{P} \cup Cop(A^*), h(B) =$ **true** (since $B^* \in \mathbf{P}'$ implies $\mathbf{P}'?B^* = 1$) and $h(A) =$ **false**. This means that $\mathbf{P} \cup Cop(A^*)\sharp \not\vdash A$ in classical logic. Thus (a) of 4.5 implies (b), as we showed that not (b) implies not (a).
This proves 4.5. ■

We now examine completeness for locally linear computation.

Lemma 4.7 Soundness and completeness for classical logic of the notion of locally linear compuation with restart *(defined in 3.8)*
$\mathbf{P}\,?\,(G, \mathbf{H}) = 1$ *iff* $\mathbf{P} \vdash G \vee \bigvee \mathbf{H}$ *in classical logic.*

Proof.

a. We prove soundness by induction on the length of the computation.

1. *Length 1*
   In this case $G = q$ is atomic and $q \in \mathbf{P}$.
   Thus $\mathbf{P} \vdash q \vee \bigvee \mathbf{H}$.
2. *Length* $k+1$
   (a) $G = A_1 \rightarrow (A_2 \rightarrow \ldots \rightarrow (A_n \rightarrow q)\ldots)$
   Then $\mathbf{P}?(G, \mathbf{H}) = 1$ if $\mathbf{P} \cup \{A_i\}?(q, \mathbf{H}) = 1$ and hence $\mathbf{P} \cup \{A_i\} \vdash q \vee \bigvee \mathbf{H}$ by the induction hypothesis and hence $\mathbf{P} \vdash G \vee \bigvee \mathbf{H}$.
   (b) $G = q$ and for some $B = B_1 \rightarrow (\ldots \rightarrow (B_n \rightarrow q)\ldots) \in \mathbf{P}, (\mathbf{P} - \{B\})?(B_i, \mathbf{H} \cup \{q\}) = 1$ for $i = 1, \ldots, n$.
   By the induction hypothesis $\mathbf{P} - \{B\} \vdash B_i \vee q \vee \bigvee \mathbf{H} i = 1 \ldots n$.
   However in classical logic $\bigwedge_{i=1}^{n}(B_i \vee q) \equiv (B_1 \rightarrow (B_2) \rightarrow (B_n \rightarrow q)\ldots) \rightarrow q$. Hence $\mathbf{P} - \{B\} \vdash (B \rightarrow q) \vee \bigvee \mathbf{H}$ and by the deduction theorem $\mathbf{P} \vdash q \vee \bigvee \mathbf{H}$
   (c) The restart rule was used, ie for some $a \in \mathbf{H}$

$$\mathbf{P}?(a, \mathbf{H} \cup \{q\}) = 1$$

   Hence $\mathbf{P} \vdash a \vee q \vee \bigvee \mathbf{H}$ and since $a \in \mathbf{H}$ we get $\mathbf{P} \vdash q \vee \bigvee \mathbf{H}$

b. We prove completeness by induction on the structure of the goal and the size and complexity of the database.

1. The case of G atomic and the database $\mathbf{P}$ are all atoms is clear, because $\mathbf{P} \cap (\{G\} \cup \mathbf{H}) \neq \emptyset$.
2. For the case G not atomic of the form $\mathbf{P} \rightarrow q$ we use the deduction theorem for classical logic and the induction hypothesis.
3. For the case $G = A \wedge B$, we have $\mathbf{P} \vdash (A \wedge B) \vee \bigvee \mathbf{H}$
   iff $\mathbf{P} \vdash (A \vee \bigvee \mathbf{H}) \wedge (B \vee \bigvee \mathbf{H})$
   iff $\mathbf{P} \vdash A \vee \bigvee \mathbf{H}$ and $\mathbf{P} \vdash B \vee \bigvee \mathbf{H}$
   and again we use the induction hypothesis.
4. Assume $\{\mathbf{A}_i \rightarrow x_i\} \vdash q \vee \bigvee \mathbf{H}, q$ atomic. Clearly $\{x_i\}$ and $\{q\} \cup \mathbf{H}$ must have an element in common, otherwise we can define a countermodel. Assume it is x_1. Thus $x_1 \in \mathbf{H}$. If A_1 is non existent, then the computation succeeds by the Restart rule. Otherwise,

$$\{A_i \rightarrow x_i\} \vdash q \vee \bigvee \mathbf{H}$$

   iff in classical logic

$$\{A_2 \rightarrow x_2, A_3 \rightarrow x_3, \ldots\} \vdash A_1 \vee q \vee \bigvee \mathbf{H}.$$

   By the induction hypothesis,

$$\{A_2 \rightarrow x_2, \ldots\}?(A_1, \mathbf{H} \cup \{q\}) = 1$$

   We can now show that

$$\{A_1 \rightarrow x_1, A_2 \rightarrow x_2, \ldots\}?(q, \mathbf{H}) = 1$$

   as follows:
   Use restart with $x_1 \in \mathbf{H}$ and ask

$$\{A_1 \rightarrow x_1, A_2 \rightarrow x_2, \ldots\}?(x_1, \mathbf{H} \cup \{q\})$$

   Unify with $A_1 \rightarrow x_1$ and ask

$$\{A_2 \rightarrow x_2, \ldots\}?(A_1, \mathbf{H} \cup \{q\} \cup \{x_1\})$$

   which succeeds by the induction hypothesis (remember that $x_1 \in \mathbf{H}$).

■

Lemma 4.8 Soundness and completeness of locally linear computation with bounded restart for intuitionistic logic. *For the computation of 3.8 we have (1) if and only if (2):*

1. $\mathbf{P}?(G, (x_n, \ldots, x_1)) = 1$

2. $\mathbf{P}, G \rightarrow x_1, x_1 \rightarrow x_2, \ldots, x_{n-1} \rightarrow x_n \vdash G$

Proof.
Proof of soundness, (1) implies (2):
By induction on the length of the computation leading to success.

1. If $G \in \mathbf{P}, G$ atomic, then clearly (1) holds.

2. If $G = A_1 \wedge A_2$, then the computation succeeds with A_1 and succeeds with A_2. By the induction hypothesis, we have:
$$\mathbf{P}, A_i \to x_1, x_1 \to x_2, \ldots, x_{n-1} \to x_n \vdash A_i$$
for $i = 1, 2$.
We want to show:
$$\mathbf{P}, A_1 \wedge A_2 \to x_1, x_1 \to x_2, \ldots, x_{n-1} \to x_n \vdash A_1 \wedge A_2$$
The above holds because of lemma 3.10, namely
$$((A \to x) \to A) \wedge ((B \to x) \to B) \vdash ((A \wedge B \to x) \to A \wedge B).$$

3. Case $G = A \to q$
We need to show that
$$\mathbf{P}, A, q \to x_1, \ldots, x_{n-1} \to x_n \vdash q$$
implies
$$\mathbf{P}, (A \to q) \to x_1, x_1 \to x_2, \ldots x_{n-1} \to x_n \vdash A \to q$$
This follows from lemma 3.10 applied to the pair $A, q \to x$.

4. Case of bounded restart:
We have to show that if
$$\mathbf{P}, x_i \to q, q \to x_i, \ldots, x_{i-1} \to x_i, \ldots, x_{i+k} \to q, \ldots, \vdash x_i$$
then
$$\mathbf{P}, q \to x_1, \ldots, x_{i-1} \to x_i, \ldots, x_{i+1} \to q, \ldots, \vdash q$$
however the two are equivalent since $x_i \to q$ is in the data, as x_i is bounded by q.

Proof of completeness

1. We first prove that if $\mathbf{P} \vdash G$ then $\mathbf{P}?(G, \varnothing) = 1$ and then use this fact to show that (2) implies (1).
It is sufficient, in view of the deduction theorem, to assume that $G = q$ is atomic.
Assume that $\mathbf{P} \vdash q$, then by theorem 2.5.
 (*) $\mathbf{P}?q = 1$
 where the computation in (*) is the goal directed computation of 2.3.
 We prove from (*) that
 (**) $\mathbf{P}?(q, \varnothing) = 1$.
 We do this by transforming a computation tree of (*) into a computation tree of (**).

First note that if $\mathbf{P}'$ is any database such that $\mathbf{P}'$? G=1 in the goal directed computation of (2.3) then some RC-clauses may be used ("used" is defined in 2.3,f1 and 2.3,f2) more than once. We can let $\mathbf{P}''$ be like $\mathbf{P}'$ except that some clauses are duplicated according to the number of times they are used in the successful computation of $\mathbf{P}'?G = 1$. For this $\mathbf{P}''$, we have $\mathbf{P}''?G = 1$, through a computation which uses each clause at most once. Further $\mathbf{P}' \equiv \mathbf{P}''$ in intuitionistic logic.
We can regard $\mathbf{P}''$ as a multiset and this will allow for several copies of the same clause B to be members or we can label the various copies of the clause B in some manner.
Thus, for example, $\mathbf{P}' = \{c \to a, (c \to a) \to c\}?a = 1$ in intuitionistic logic. The clause $c \to a$ is used twice. $\mathbf{P}''$ will be the multiset $\{c \to a, c \to a, (c \to a) \to c\}$.
The query $?a$ will succeed through a computation which uses every clause just once. However, it does use two copies of $c \to a$, each copy is used once. $\mathbf{P}''$ however, is obtained only by duplicating elements of $\mathbf{P}'$. This is not sufficient for our purpose. It may be the case that we have a database of the form $B = [(c \to a) \to [((c \to a) \to c) \to a)] \to q]$ and the query is $?q$.
We have that the query $?q$ reduces to the query $\{c \to a, (c \to a) \to c\}?a$ and now we have the case that $c \to a$ needs to be duplicated.
Duplicating B does not help. We have to duplicate inside B itself, ie $c \to a$ can be duplicated inside B by writing B as $(c \to a) \to ((c \to a) \to ((c \to a) \to c) \to a)))$.
Assume now that $\mathbf{P}'?G = 1$, through a tree $(T, \leq, 0, V')$ in the sense of 3.8. Consider the multiset $\mathbf{P}''$ which duplicates clauses in $\mathbf{P}'$ as needed. We can transform the labelling function V' into a labelling function V'' in the sense of 3.8 for a linear computation of $\mathbf{P}''?G = 1$.
The transformation is straightforward. The duplication has to be done more carefully however. Consider $(T, \leq, 0, V')$. Part of the 2.3 is the notion of a wff A being *put into the database at a node t*. We also have a notion of the wff A being *used at a node s*. It is therefore possible to count for any A put in the database at a node t, how many times it has been used at nodes s below t. (ie at nodes s with $t \leq s$). Let m be the maximal number that any A was used in the entire tree. This means that if we duplicate any Am times whenever it was

first put in, then no copy of any A needs to be used more than once. We duplicate m times systematically in the manner described above. We thus can get a computation $(T, \leq, 0, V'')$ that has the additional property that the maximal number of copies are put in right at the start and these copies are the ones used at the nodes. So if at lower nodes more copies are put in, they are not needed and not used. Call such a V'' "rich". Of course, since V'' represents a linear computation, each time a copy of a clause is used, it is thrown out, and a history is recorded. Note that in the tree $(T, \leq, 0, V'')$, the restart rule is not used, since it is not originally used in $(T, \leq, 0, V')$.

Coming back to our successful computation $\mathbf{P}?q$, we can assume there exists a $\mathbf{P}^*$ which is a duplicate extension of $\mathbf{P}$ such that $\mathbf{P}^*?q = 1$ through a tree $(T, \leq, 0, V^*)$ according to 3.8. Our aim is to transform this tree to a tree $(T_0, \leq \restriction T_0, 0, V_0)$ with $T_0 \subseteq \mathrm{T}$ which is also a linear computation tree (3.8), which uses only clauses from $\mathbf{P}$. This will prove the theorem. The proof is by induction on the number of "bad" pairs of nodes in the tree, namely, nodes that "use" copies of the "same" formulas. We drop some nodes out of T and modify V^* and change it to V_0 and thus obtain the new tree.

Take a node $t \in T$ in the computation tree of $\mathbf{P}^*?(q, \varnothing)$, namely in $(T, \leq, 0, V^*)$

This node is t and the tree label is $V^*(t) = (\mathbf{P}''_t, \{x\}, \mathbf{H}_t)$

The original computation tree continues through the use of a clause of the form $B = (C_1 \rightarrow (C_2 \rightarrow, \ldots, (C_k \rightarrow x) \in \mathbf{P}$ with $C_i = C^i_1 \rightarrow (C^i_2 \rightarrow, \ldots, (C^i_{j(i)} \rightarrow y_i)$.

There are, therefore, successor points to t in the tree $t_1, \ldots, t_k$ and with $t \leq t_i, i = 1 \ldots k$ and successor points to t_i (if C^i_j indeed exist) of the form $t_i \leq s_i$ such that the following holds;

$V^*(t_i) = (\mathbf{P}''_t - \{B\}, \{C_i\}, \mathbf{H}_t \cup \{x\})$

$V^*(s_i) = ((\mathbf{P}''_t - \{B\}) \cup \{C^i_1, \ldots, C^i_{j(i)}\}, \{y_i\}, \mathbf{H}_t \cup \{x\})$

We distinguish several cases:

(a) $B \in \mathbf{P}$ and this node t is the first node where a copy of B is used. In this case we do nothing.

(b) $B \in \mathbf{P}$ and at some higher node $t^* \nleq t$ another copy B' of B was used. In this case t^* will have as its immediate successors nodes $t^*_1, \ldots, t^*_k s^*_1, \ldots, s^*_k$ and the node t is below one of the s^*_i, say below s^*_1. We thus have: $V^*(t^*_i) = (\mathbf{P}''_t - \{B'\}, \{C_i\}, \mathbf{H}_t \cup \{x\})$
$V^*(s^*_i) = ((\mathbf{P}''_{t^*} - \{B'\} \cup \{C^i_1, \ldots, c^i_{j(i)}\}, \mathbf{H}_t \cup \{x\})$. See fig 4.

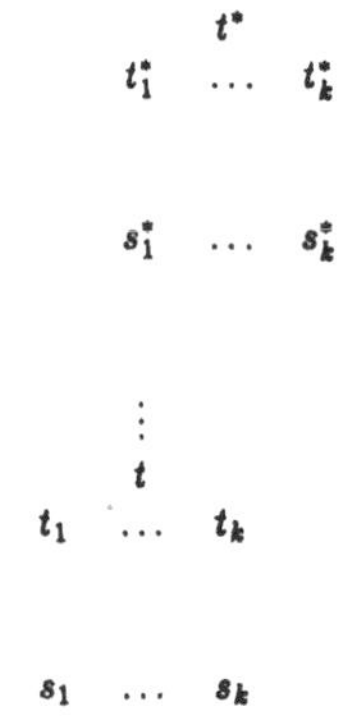

Figure 4:

Notice two important facts:

Fact 1: The clauses $C^i_1, \ldots, C^i_{j(i)}$ are put in $\mathbf{P}_{s^*_1}$ and therefore are redundantly put into $\mathbf{P}_{s_1}$. They are already there. Note that we assume our tree is "rich" and any clause (such as C^i_k) is put in with enough copies "first time".

Fact 2: y_1 is put in the history $\mathbf{H}_s$ at all points s which are immediate successors of s^*_1. Therefore $y_1 \in \mathbf{H}_t$.

In view of the above two facts, we can drop the point t_1 from the tree and also drop all points $t_2, \ldots, t_k$ and all subtrees below $t_1, \ldots, t_k$ respectively including $s_2, \ldots, s_k$. We modify $V^*(s_1)$ and $V^*(s)$ for all s such that $s_1 \leq s$ by *taking out* from $\mathbf{P}_s$ the extra copies of $C^i_1, \ldots, c^i_{j(i)}$ which we put into $\mathbf{P}_{s_1}$. We justify the new junction $t \leq s_1$ by the bounded restart rule in view of fact 2.

We thus get a correct locally linear computation tree with a smaller number of "bad" pairs of nodes.

This inductive proof shows that there exists a tree with no "bad" pairs at all. It follows from the construction that only wffs arising from subformulas of $\mathbf{P}$ are used and these are used at most once. We can therefore take

V_0 to be the following function

$$V_0(t) = (\mathbf{P}^0_t, \mathbf{G}^0_t, \mathbf{H}^0_t)$$

where $\mathbf{P}^0_t, \mathbf{G}^0_t, \mathbf{H}^0_t$ are obtained from $\mathbf{P}^*_t, \mathbf{G}^*_t, \mathbf{H}^*_t$ respectively by substituting "truth" for any duplicate not used in the transformed computation.
It is clear, since there are no "bad" pairs that $\mathbf{P}^0_t \subseteq \mathbf{P}$ for all t. This means that $\mathbf{P}?(q, \varnothing) = 1$.

2. We now prove completeness, namely show that (2) implies (1).
Let $\mathbf{P}' = \mathbf{P} \cup \{q \to x_1, (x_1 \to x_2), \ldots, x_{n-1} \to x_n\}$ then if $\mathbf{P}' \vdash q$ then $\mathbf{P}?(q, (x_n, \ldots, x_1)) = 1$.
By (1) above, we have that

$$\mathbf{P}, q \to x_1, x_1 \to x_2, \ldots, x_{n-1} \to x_n?(q, \varnothing) = 1$$

Let $(T, \leq, 0, V)$ be the computation tree of the above computation as in definition 3.8. For each node $t \in T$ we have $V(t) = (\mathbf{M}_t, \mathbf{G}_t, \mathbf{H}_t)$. Let $\mathbf{X}$ be the sequence $(x_m, \ldots, x_1)$. Let $\mathbf{H}'_t = \mathbf{X} * \mathbf{H}_t$. Let $V'(t) = (\mathbf{M}_t, \mathbf{G}_t, \mathbf{H}'_t)$. Then $(T, \leq, 0, V')$ is a success tree for $\mathbf{P} \cup \{q \to x_1, \ldots, x_{n-1} \to x_n\}?(q, \mathbf{X}) = 1$.
All uses of the bounded restart rule in $(T, \leq, 0, V')$ are still valid because we have appended $\mathbf{X}$ at the beginning of $\mathbf{H}_t$.
Observe that we can use the bounded restart rule to justify any node of the form as in figure 5

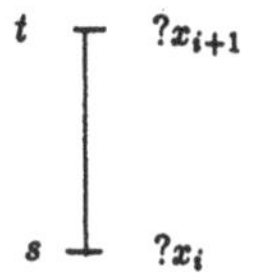

Figure 5:

because $\mathbf{X}$ is appended first in $\mathbf{H}'_t$. Thus if we let $\mathbf{M}'_t = \mathbf{M}_t \cup \{q \to x_1, x_1 \to x_2, \ldots, x_{n-1} \to x_n\}$ for $t \in T$, we get that $(T, \leq, 0, (\mathbf{M}', \mathbf{G}, \mathbf{H}'))$ is also a correct computation tree for $\mathbf{P}, \{q \to x_1, q \to x_1, \ldots, x_{n-1} \to x_n, x_{n-1} \to x_n\}?(q, \mathbf{X})$.
In this computation tree the clauses of $\{q \to x_1, \ldots, x_{n-1} \to x_n\}$ are *never* used. Thus these can be taken out of the database and the computation tree above justifies $\mathbf{P}?(q, \mathbf{X}) = 1$.

■

Lemma 4.9 *In the notation of 4.7, for the case of classical logic, let* $A = A_1 \to (A_2 \to \ldots \to (A_n \to q)\ldots), B = B_1 \to (B_2 \to \ldots \to (B_m \to q)\ldots)$
Then (a) is equivalent to (b):

(a) $\mathbf{P} \cup \{A\}?(B_i, \mathbf{H} \cup \{q\}) = 1$ *for* $i = 1, \ldots m$.
(b) $\mathbf{P} \cup \{B\}?(A_i, \mathbf{H} \cup \{q\}) = 1$ *for* $i = 1, \ldots, n$.

Proof. From 4.7 and conditions (a) and (b) are equivalent, respectively to (a') and (b') below:

(a') $\mathbf{P}, A \vdash B_i \vee q \vee \bigvee \mathbf{H}$
$i = 1, \ldots, n$.
(b') $\mathbf{P}, B \vdash A_i \vee q \vee \bigvee \mathbf{H}$
$i = 1, \ldots, n$.

By classical logic rules (a') and (b') are equivalent to (a'') and (b'') respectively:

(a'') $\mathbf{P}, A \vdash (\bigwedge_{i=1}^m B_i) \vee q \vee \bigvee \mathbf{H}$
(b'') $\mathbf{P}, B \vdash (\bigwedge_{i=1}^n A_i) \vee q \vee \bigvee \mathbf{H}$

Which are equivalent to (a''') and (b''') respectively:

(a''') $\mathbf{P}, A \vdash ((\bigwedge_{i=1}^m B_i \to q) \to q) \vee \bigvee \mathbf{H}$
(b''') $\mathbf{P}, B \vdash ((\bigwedge_{i=1}^n A_i \to q) \to q) \vee \bigvee \mathbf{H}$

Both (a'') and (b'') are equivalent to (c) below by the deduction theorem for classical logic.

(c) $\mathbf{P}, A, B \vdash q \vee \bigvee \mathbf{H}$

(c) is equivalent by 4.7 to
$\mathbf{P}, A, B?(q, \mathbf{H})$.
This proves 4.9. ∎

Lemma 4.10 *In the computation* $\mathbf{P}?(q, \mathbf{H})$ *with restart, no backtracking is necessary. q can unify with any clause* $(A_1 \to \ldots \to (A_n \to q)\ldots) \in \mathbf{P}$ *and success or failure does not depend on the choice of the clause.*

Proof. Follows from 4.9 above. ∎

Example 4.11 *The parallel to 4.9, 4.10 does not hold for the intuitionistic case. In the notation of 4.8, for the case of intuitionistic logic, let* $A = A_1 \to \ldots \to (A_n \to q)\ldots)$, $B = B_1 \to \ldots \to (B_m \to q)\ldots)$
Then (a) is not necessarily *equivlaent to (b):*

(a) $\mathbf{P} \cup \{A\}?(B_i, \mathbf{H} * (q)) = 1$
for $i = 1, \ldots, m$

(b) $\mathbf{P} \cup \{B\}?(A_i, \mathbf{H} * q) = 1$
for $i = 1, \ldots, n$.

Let

$$\mathbf{P} = \{a\}$$

$$A = a \to q$$

$$B = (c \to b) \to q$$

Then

$$\mathbf{P} \cup \{A\}?(b, (q)) \text{ is } \{c, a, a \to q\}?(b, (q))$$

while

$$\mathbf{P} \cup \{B\}?(a, (q)) \text{ is } \{a, (c \to b) \to q\}?(a, (q))$$

The first computation fails (we cannot restart here) and the second computation succeeds. The soundness direction of 4.8 does give us that $\{c, a, a \to q, b \to q\}$ *does not prove b in the first case and it does prove a in the second case.*

We thus see that in the intuitionistic computation $\mathbf{P}?(q, \mathbf{H})$, *with bounded restart, backtracking is certainly necessary. q can unify with any clause* $(A_1 \to \ldots \to (A_n \to q)\ldots)$ *and success or failure may depend on the choice of clause.*

5 Futher Results

This section is a preview of part 2 of this paper. There are three main topics yet to be addressed.

1. Extending the paradigm to predicate logics.
2. Extending the paradigm to neighbouring logics including the family of linear logics.
3. A study of the connection with Gentzen formulations of the various logics. Obtaining some proof theoretic applications (interpolation, complexity bounds) out of the approach.

We briefly discuss each point

1 Predicate Logics

There is no problem in principle to extend the results to quantification. We are naturally led to constraint logic. The conceptual paradigm of diminishing resources manifests itself in the case of quantifiers in following form. A clause $\forall x[A(x) \to B(x)]$ is taken as a conjunction of all its substitution instances. When the clause is used through a substitution Θ (say the goal is $?B(a)$) then it cannot be used again for that same substitution.

Thus during the computation the database will have clauses annotated with constraints. For example $\{A(x) \to B(x); x \neq a\}$. This clause can be used for a goal $?B(y)$ provided the substitution Θ does not instantiate y to be a. Thus the goal during the computation is also annotated, for example when we use the clause above for the query $?B(y)$, we unify and continue with $?A(y), y \neq a$ and add annotation to the clause $\{A(x) \to B(x); x \neq a, x \neq y\}$. As y is further instantitated the constraint on the clause will be futher refined. The exact computational details will be worked out in part 2.

It is not clear at this stage whether there are computational benefits to be gained from the theme of diminishing resource in the predicate case. It is clear, however, that the connection with constraints certainly has conceptual advantages.

2 Linear and other related logics

The connection with linear logic is important, as linear logic is an example of diminishing resource. There are several versions and connectives involved in linear logic. These will be studied in the sequel as well as the addition of other connectives (such as disjunction) to the intuitionistic case. We will define now precisely what implicational version of linear logic our current computations relate to.

Definition 5.1 *1. Consider the language with $\rightarrow$ only, and consider the Hilbert system defined by the following axiom schemas and the rule of modus ponens:*

(a) $A \rightarrow A$

(b) $(A \rightarrow B) \rightarrow ((C \rightarrow A) \rightarrow (C \rightarrow B))$

(c) $(A \rightarrow (B \rightarrow C)) \rightarrow (B \rightarrow (A \rightarrow C))$

The set of theorems of the above system yields the theorems of implicational linear logic. Two more axioms will yield implicational intutionistic logic

(d) $(A \rightarrow (B \rightarrow C)) \rightarrow ((A \rightarrow B) \rightarrow (A \rightarrow C))$

(e) $A \rightarrow (B \rightarrow A)$.

One more axiom will yield implictional classical logic.

(f) $((A \rightarrow B) \rightarrow A) \rightarrow A$

2. The above Hilbert formulation defines the notion of $\vdash A$. We now define the consequence relation $\Vdash$ between multisets and single formulas by:

$$\{A_1, \ldots, A_n\} \Vdash B \text{ iff (def) } \vdash A_1 \rightarrow (A_2 \rightarrow \ldots \rightarrow (A_n \rightarrow B) \ldots)$$

Note that by axiom (c) the definition does not depend on the order of A_i. Thus we get $\Vdash$ for linear or intutionistic or classical implication, depending what $\vdash$ is.

The locally linear computation with bounded restart of definition 3.8 corresponds to (ie is sound and complete for) the intuitionistic consequence. The locally linear computation with restart of definition 3.8 corresponds to classical consequence.

I do not have a corresponding system for the locally linear computation of 3.8.

Turning now to definition 2.3, what does linear computation correspond to? The answer is linear consequence of 5.1. This is not formally proved in this paper,but is easy to do. It will be dealt with in part 2.

Conjunctions $\wedge$ and falsity $\perp$ are easy to add to intutionistic and classical logic and the extension of the computation procedure to these connectives is trivial. In case of linear logic there is a variety of possibilities of additional connectives. This will be dealt with in part 2. See also [3]. We can mention at this stage that the cross $A \otimes B$ can be added to the Hilbert system of linear loigc (of def 5.1) with the following two axioms:

$$\vdash A \rightarrow (B \rightarrow A \otimes B)$$

$$\vdash (A \rightarrow (B \rightarrow C)) \rightarrow (A \otimes B \rightarrow C).$$

The linear computation of definition 2.3 can be extended by the following clause to accommodate the cross:

$$\mathbf{P}?A \otimes B = 1 \text{ iff } \mathbf{P} = \mathbf{P}_1 \cup \mathbf{P}_2 \text{ and } \mathbf{P}_1?A = 1 \text{ and } \mathbf{P}_2?B = 1.$$

Remember that $\mathbf{P}_i$ are multisets. In intuitionistic logic $A \otimes B$ becomes $A \wedge B$.

We conclude my mentioning that various modifications of the restart rule and their connection with intermediate logics are studied in [7].

Gentzen Formulations

There are related Gentzen formulations and already some results of Hudelmaier improving computational complexity. We can sketch here some obvious immediate proof theoretic results that our theme yields.

Lemma 5.2 *1. If $\{A_i \rightarrow q_i\} \vdash q$ then for some $A_i, \{A_i \rightarrow q_i\} \vdash A_i$.*
This is a strong subformula property.

2. Interpolation property for databases **P**:
If $\mathbf{P} \vdash B$ then for some C in the common langauge $\mathbf{P} \vdash C$ and $\vdash C \rightarrow B$.

Proof.

1. Follows from the goal directed computation of 2.3 and the completeness theorem.
2. To show interpolation we use (1) and induction on the complexity of the computation.

Length 1
Clearly B is atomic and $B \in \mathbf{P}$ then the interpolant is B.
Length $n+1$
Assume $\mathbf{P} = \{A_j \rightarrow P_j\}$ and $B = \wedge(B_k \rightarrow y_k) \rightarrow q$. Then we are given that $\mathbf{P} = \{A_j \rightarrow p_j, B_k \rightarrow y_k\} \vdash q$. The goal directed computation of q must succeed. We distinguish two cases:

1. q unifies with some y (and hence $y_m \vdash q$).
   In this case $\mathbf{P} \vdash \wedge(B_k \rightarrow y_k) \rightarrow B_m$. B_m is proved with a shorter computation. By the induction hypothesis there exists an interpolant C such that $\mathbf{P} \vdash C$ and $C \vdash \wedge(B_k \rightarrow y_k) \rightarrow B_m$. Since $B_m \rightarrow y_m$ is one of the conjuncts and $y_m \vdash q$ we get that $C \vdash B$.

2. q unifies with some p_m (and hence $p_m \vdash q$).
   In this case either $p+m = \bot$ or q is in the common language. We have

$$\wedge(B_k \to y_k) \vdash \wedge \mathrm{P} \to A_m$$

   Hence there exists an interpolant I such that $\vdash \wedge(B_k \to y_k) \to I$ and $\vdash I \to (\wedge \mathrm{P} \to A_m)$. Let $r = \bot$ if $p_m = \bot$ and $r = q$, otherwise then $\vdash (I \to r) \to (\wedge(B_k \to y_k) \to r)$ and $\wedge \mathrm{P}, I \vdash A_m$.
   Since $\mathrm{P} \vdash A_m \to p_m$ and $p_m \vdash q$, we get $\mathrm{P}, I \vdash r$ and so $\mathrm{P} \vdash I \to r$. Thus our interpolant is $I \to r$.

If $B = B_1 \wedge B_2$ and the interpolant for B_i is C_i then the interpolant for B is $C_1 \wedge C_2$.
This completes the induction and theorem 5.2 is proved. Notice that the interpolant can be constructed effectively from the computation and different computations may give rise to different interpolants. ■

References

1. J Hudelmaier, 'Decision Procedure for Propositional N-Prolog', in *Extensions of Logic Programming,* (Ed) P Schröeder Heister, Springer Verlag 1990.

2. D M Gabbay, Theory of Algorithmic Proof, A chapter in *Handbook of Logic in Theoretical Computer Science,* Volume 1, S Abramsky, D M Gabbay and T S E Maibaum (Eds), Oxford University Press, to appear.

3. D M Gabbay, *Labelled Deductive Systems,* Oxford University Press, 1992, in preparation.

4. D M Gabbay, U Reyle, N-Prolog I, *Journal of Logic Programming,* Vol 1, 1984, pp 319-355.

5. D M Gabbay, N-Prolog, Part II, *Journal of Logic Programming,* Vol 2, 1985, pp 251-283.

6. J-Y Girard, Linear Logic, *Theoretical Computer Science* 50, 1987, pp 1-102.

7. D M Gabbay, F Kriwaczek, A family of goal directed theorem provers based on conjunction and implication, to appear in *Journal of Automated Reasoning.*

CUTTING PLANE VERSUS FREGE PROOF SYSTEMS

Andreas Goerdt
Universität-GH-Duisburg
Fachbereich Mathematik/Praktische Informatik
Lotharstraße 65, D-4100 Duisburg
Germany
Net address: hn281go@unidui.uucp

ABSTRACT

The cutting plane proof system for proving the unsatisfiability of propositional formulas in conjunctive normalform is based on a natural representation of formulas as systems of integer inequalities. We show: Frege proof systems p-simulate the cutting plane proof system. This strengthens a result in [5] , that extended Frege proof systems (which are believed to be stronger than Frege proof systems) p-simulate the cutting plane proof system. Our proof is based on the techniques introduced in [2].

INTRODUCTION

The systematic study of the complexity of proof length in various proof systems for propositional logic (like resolution [3] , tableaux calculus [11] , Frege systems (or Hilbert style systems, these are the usual systems with modus ponens)) was initiated by Cook/Reckhow in [6] . One motivation for this is the following: As propositional proof systems are just nondeterministic algorithms for the coNP-complete language of unsatisfiable propositional formulas (or equivalently tautologies) the NP $\neq$ coNP assumption implies the existence of hard examples for any proof sytem. Hard examples are infinite families of formulas having only proofs of superpolynomial size in the system considered. It is an interesting problem, to show the existence of hard examples for increasingly powerful proof systems, without assuming NP $\neq$coNP. Even for the relatively weak resolution proof system the existence of hard examples was proved by Haken only in 1985 [7]. Ajtai [1] showed this result for bounded depth Frege systems (which are stronger than resolution). Both authors use a family of formulas encoding the pigeonhole principle (saying that a total mapping from a set with n+1 elements to a set with n elements is not injective). The existence of hard examples for Frege systems with unbounded formula depth (or for stronger systems) is a well-known open problem [4]. Buss [2] proved that Frege systems allow for short (i.e. polynomial) proofs of the pigeonhole principle.

The cutting plane proof systems is based on the representation of propositional formulas in conjunctive normalform as systems of integer inequalities (see example 1.3 below). The only paper investigating the complexity of the cutting plane system is [5], where the following results are shown: The cutting plane proof system p-simulates (see definition 1.1 below) resolution. The cutting plane proof system has short proofs for the formulas encoding the pigeonhole principle. Extended Frege system p-simulate the cutting plane system. The existence of hard examples is mentioned as an open problem. Our result that (not extended) Frege systems p-simulate the cutting plane system shows that finding hard examples for Frege systems is at least as difficult as finding hard examples for the cutting plane system. So perhaps one should try to find hard examples for the cutting plane system before looking at Frege systems. An application of our result is the following: It is possible to prove that a family of formulas has short Frege proofs by constructing short cutting plane proofs. As the cutting plane system is in some respects easier to handle than Frege systems, cutting plane proofs can be simple where Frege proofs are complex. We demonstrate this for a family of formulas introduced in [9], not known to have short Frege proofs before. (Here also [10] might be of interest.)

The second motivation of our paper comes from the fact, that propositional proof systems are the basis of algorithms for the satisfiability problem. In particular, work has been done on applying techniques from linear integer programming (like cutting planes) to obtain algorithms for the satisfiability problem (see for example [8]). Here we contribute to an analysis of the principles underlying these algorithms and to a comparison of these algorithms to the more common logic based ones.

To prove our result, we have to simulate the parts of arithmetic available in the cutting plane system in Frege systems. For this we use the techniques introduced in [2]. As the situation here is more general than in [2], we have to simulate more of arithmetic (multiplication, negative numbers, subtraction) in Frege systems.

In section 1 we start with some basics and give an example of a cutting plane derivation. In sections 2 and 3 we present our simulation of cutting plane proofs by Frege proofs. Finally we give the application of our result. Due to space restrictions we have omitted some proofs. In [12] you can find an appendix of this paper containing all proofs missing here.

1 BASICS

For undefined notions of propositional logic we refer the reader to § 2 of [2]. In particular, by Frege proof system we mean the Frege system presented there. By the size of a proof in any propositional proof system we mean the length of the proof written out as a string over a binary alphabet. Note that the size of a proof also accounts for the size of the formulas occurring in the proof whereas the number of proof steps of a proof does not.

1.1 Definition ([6], definition 1.5)

Let $\mathfrak{P}$ and $\mathfrak{Q}$ be two propositional proof systems. System $\mathfrak{P}$ p-simulates system $\mathfrak{Q}$ iff there is a polynomial $p(x)$ such that for any proof Q of the formula F in system $\mathfrak{Q}$ there is a proof P of F (or a formula corresponding to F) in $\mathfrak{P}$ such that size $P \leq p(\text{size } Q)$.

□

From corollary 2.4 in [6] we know that any two Frege systems p-simulate each other, hence it does not really matter that we base this paper on the Frege system specified in [2].

A literal is a propositional variable or a negated propositional variable. Clauses are disjunctions of literals and a formula in conjunctive normalform is a conjunction of clauses. The cutting plane proof system is based on the representation of clauses as inequalities between integers: The representation $R(x)$ of the positive literal x just is x, viewed as an integer variable. The representation $R(\bar{x})$ of the negative literal $\bar{x}$ is $1-x$. A clause $L_1 \vee \ldots \vee L_n$ is represented as

$$R(L_1 \vee \ldots \vee L_n) = (R(L_1) + \ldots + R(L_n) \geq 1).$$

For example $R(x \vee \bar{y} \vee \bar{z}) = (x + 1-y + 1-z \geq 1)$ which is equivalent to $x-y-z \geq 1-2 = -1$. A formula in conjunctive normalform $\mathfrak{C} = C_1 \wedge \ldots \wedge C_n$, where the C_i are clauses is presented by a sequence of inequalities:

$$R(\mathfrak{C}) = R(C_1), \ldots, R(C_n) \text{ and}$$

$$x \geq 0,\ -x \geq -1 \text{ (saying } 0 \leq x \leq 1)$$

for each occurring variable x. The formula $\mathfrak{C}$ is satisfied by an assignment of truth values to the variables iff $R(\mathfrak{C})$ is satisfied by the corresponding assignment (true $\sim$ 1, false $\sim$ 0) of integers to the variables. In the sequel we assume that inequalities representing clauses are always in the normalform as above: Each summand on the left hand side is an integer multiple of a variable, the right hand side is an integer.

The formulas of the cutting plane proof system are inequalitis of the kind: $v_0 x_0 + \ldots + v_n x_n \geq P$ where the x_i are integer variables and $v_i, P \in \mathbb{Z}$.

The cutting plane proof system has three kinds of proof rules:

<u>Addition</u>

$$\frac{\Sigma\, v_i x_i \geq P \qquad \Sigma\, w_i x_i \geq Q}{\Sigma\, (v_i + w_i) x_i \geq P+Q}$$

<u>Mulitplication</u>

For $c \in \mathbb{Z}^+$:

$$\frac{\Sigma\, v_i x_i \geq P}{\Sigma\, (c \cdot v_i) x_i \geq c \cdot P}$$

<u>Division</u>

For $c \in \mathbb{Z}^+ \setminus \{0\}$, $v_i = c \cdot w_i$ for all i, not all $v_i = 0$, $Q = \lceil \frac{1}{c} \cdot P \rceil$:

$$\frac{\Sigma\, v_i \cdot x_i \geq P}{\Sigma\, w_i \cdot v_i \geq Q},$$

The condition not all $v_i = 0$ is no real restriction. It ensures that we cannot divide by a number which is not contained in the formulas. This could increase the size of a proof without being visible from the formula size. In [5] the c's are explicitly included in the proof.

These rules are correct in the sense that for all integers $x_1,\dots,x_n$ for which the premises hold the conclusion holds. Note that the rounding step in the division rule is correct because the left hand side always is an integer ($\lceil \frac{1}{c}\cdot p \rceil = \mathrm{Min}\{x \mid x \in \mathbb{Z},\ x \geq \frac{1}{c}\cdot p\}$).
A cutting plane proof of a formula in conjunctive normalform consists in the derivation of a contradictory inequality $0 \geq m$ with $m > 0$ from the inequalities representing the formula. Note that our notion of cutting plane proof differs slightly from that in [5], but both notions p-simulate each other and thus can be considered as equivalent in the present context.

1.2 Theorem (see [5])

Let F be a system of inequalities representing a propositional formula in conjunctive normalform. The system F is contradictory iff a contradictory inequality $0 \geq m$ is derivable with the cutting plane proof system from F. This shows: The cutting plane proof system is sound and complete. □

1.3 Example

The formula $(x \oplus y) \oplus (y \oplus z) \oplus (z \oplus x)$ where $x \oplus y = (\bar{x} \wedge y) \vee (x \wedge \bar{y})$ (exclusive or) is unsatisfiable because each variable occurs exactly twice and $\oplus$ is commutative and associative and $x \oplus x$ is false. A stronger formula in conjunctive normalform is:

$$\underbrace{(x \vee y) \wedge (\bar{x} \vee \bar{y})}_{x \oplus y} \wedge (y \vee z) \wedge (\bar{y} \vee \bar{z}) \wedge (z \vee x) \wedge (\bar{z} \vee \bar{x})$$

The formula is represented by the inequalities:

$$\begin{array}{ll} x + y \geq 1, & -x - y \geq -1 \\ y + z \geq 1, & -y - z \geq -1 \\ z + x \geq 1, & -z - x \geq -1 \\ x,y,z \geq 0 & -x,-y,-z \geq -1\ . \end{array}$$

By addition we can derive $2x + 2y + 2z \geq 3$, $-2x - 2y - 2z \geq -3$.
Adding these inequalities would only give $0 \geq 0$. But, dividing by 2 and rounding we get $x + y + z \geq 2$, $-x - y - z \geq -1$ and addition gives $0 \geq 1$.

□

1.4 Outline of the simulation

(a) For the rest of this paper we assume that we have fixed a cutting plane proof of the formula $\mathfrak{C} = C_1 \wedge \dots \wedge C_n$ over the N variables $x_0,\dots,x_{N-1}$. W.l.o.g. we assume that N is a power of 2 and that the inequalities of this proof are written in the form

$$v_0x_0 + \dots + v_{N-1}x_{N-1} - (w_0x_0 + \dots + w_{N-1}x_{N-1}) \geq P$$

where $v_i, w_i \in \mathbb{Z}^+$ and $v_i > 0 \Rightarrow w_i = 0$, $w_i > 0 \Rightarrow v_i = 0$. The reason for this form is that we want the positive and negative summands of the left hand side of an inequality to be separated. We fix the following quantities:

S = smallest power of 2 > the size of the proof,

$T = S^4$,

LT = log T (log x = $\log_2$ x),

LN = log N.

Note that S and T are polynomial in the proof size.

(b) In the sequel we will show how to simulate the given cutting plane proof of $\mathfrak{C}$ by a Frege proof whose size is polynomially bounded in the size of the cutting plane proof. Assume the cutting plane proof is the sequence of inequalities $I_1, \ldots, I_M$ with $I_M = (0 \geq m)$ and $m > 0$. Our simulating Frege proof will have the structure

$$\mathfrak{C} \rightarrow \mathrm{Rep}(I_1)$$
$$\mathfrak{C} \rightarrow \mathrm{Rep}(I_2)$$
$$\vdots$$
$$\mathfrak{C} \rightarrow \mathrm{Rep}(I_M) \rightarrow \text{false}$$
$$\mathfrak{C} \rightarrow \text{false},$$

where Rep(I) (to be defined in 2.6) is a propositional formula representing the inequality I. This Frege proof shows that $\mathfrak{C}$ is contradictory. □

2 PROPOSITIONAL REPRESENTATION OF INEQUALITIES

We show how to represent the inequalities of our cutting plane proof by propositional formulas. This requires parts of arithmetic to be formulated within propositional logic. The definition of addition and iterated addition in propositional logic follows [2].

We interpret vectors of length T of propositional formulas, like $\vec{F} = (F^0, \ldots, F^{T-1})$ as natural numbers: With respect to an assignment of truth values to the occurring propositional variables $\vec{F}$ represents the number $\sum_{i=0}^{T-1} 2^i \cdot m_i$, where $m_i = 1$ if F^i evaluates to true and $m_i = 0$ otherwise. Vectors of length T+1 like $\vec{F} = (F^0, \ldots, F^T)$ are interpreted as signed numbers, where F^T denotes the sign: If F^T is true, the sign is positive, otherwise negative. Let $m \in \mathbb{N}$ with $m = \sum_{i=0}^{T-1} 2^i \cdot m_i$, $0 \leq m_i \leq 1$.

The propositional presentation of m is $\bar{m} = (F^0, \ldots, F^{T-1})$ where F^i = true (= $x_0 \vee \bar{x}_0$) if $m_i = 1$ and F^i = false otherwise. In the vectors $\overline{+m} = (F^0, \ldots, F^{T-1}, G)$ and $\overline{-m} = (F^0, \ldots, F^{T-1}, H)$ G is true and H is false and the F^i are as before. We attach a sign to a vector $\vec{F}$ of length T by defining

$$+\vec{F} = (F^0, \ldots, F^{T-1}, \text{true})$$

and $$-\vec{F} = (F^0, \ldots, F^{T-1}, \text{false}) .$$

2.1 Definition

(a) Let $\vec{F}_0$, $\vec{F}_1$ be vectors of length T of propositional formulas. The formula $\vec{F}_0 \leftrightarrow \vec{F}_1$ is defined as

$$\bigwedge_{0\leq i\leq T-1} (F_0^i \leftrightarrow F_1^i) .$$

The formula $\vec{F}_0 > \vec{F}_1$ as

$$\bigvee_{0\leq i\leq T-1} ((F_0^i \wedge \neg F_1^i) \wedge \bigwedge_{i<j\leq T-1} (F_0^j \leftrightarrow F_1^j)) .$$

The formula $\vec{F}_0 \geq \vec{F}_1$ as

$$\vec{F}_0 > \vec{F}_1 \vee \vec{F}_0 \leftrightarrow \vec{F}_1 .$$

(b) Let F_0, F_1 be vectors of length T+1 of propositional formulas. The absolute value of $\vec{F}_0$ is given by omitting the sign:

$$|\vec{F}_0| = (F_0^0,\dots,F_0^{T-1}).$$

The formula $\vec{F}_0 \leftrightarrow \vec{F}_1$ is given by

$$|\vec{F}_0| \leftrightarrow |\vec{F}_1| \wedge (F_0^T \leftrightarrow F_1^T) .$$

The formula $\vec{F}_0 \geq \vec{F}_1$ is defined as

$F_0^T \wedge \neg F_1^T$ ($\vec{F}_0$ is positive, $\vec{F}_1$ negative)

$\vee$

$F_0^T \wedge F_1^T \wedge |\vec{F}_0| \geq |\vec{F}_1|$ ($\vec{F}_0$ and $\vec{F}_1$ are positive)

$\vee$

$\neg F_0^T \wedge \neg F_1^T \wedge |\vec{F}_1| \geq |\vec{F}_0|$ ($\vec{F}_0$ and $\vec{F}_1$ are negative).

□

2.2 Definition

Let $\vec{F}$, $\vec{F}_0$, $\vec{F}_1$ be vectors of length T of propositional formulas.

(a) The vector of length T of propositional formuals

$$\mathrm{Add}(\vec{F}_0,\vec{F}_1) = (\mathrm{Add}^0(\vec{F}_0,\vec{F}_1),\dots, \mathrm{Add}^{T-1}(\vec{F}_0,\vec{F}_1))$$

is given by

$$\mathrm{Add}^0(\vec{F}_0,\vec{F}_1) = F_0^0 \oplus F_1^0$$

$$\mathrm{Add}^i(\vec{F}_0,\vec{F}_1) = F_0^i \oplus F_1^i \oplus \bigvee_{0\leq j<i} (F_0^j \wedge F_1^j \wedge \bigwedge_{j<k<i} (F_0^k \oplus F_1^k))$$

for i with $1 \leq i \leq T-1$. The number $\mathrm{Add}(\vec{F}_0,\vec{F}_1)$ is the sum modulo 2^T of $\vec{F}_0$ and $\vec{F}_1$. $\bigvee_{0\leq j<i}(\dots)$ is true iff we have a carry from position j into position i.

If s bounds the size of the components of $\vec{F}_0$ and $\vec{F}_1$, then the size of the formulas $\mathrm{Add}^i(\vec{F}_0,\vec{F}_1)$ is bounded by $c\cdot s\cdot T^2$ where c is a constant independent of s and T. The size of $\mathrm{Add}(\vec{F}_0, \vec{F}_1)$ is bounded by $d\cdot s\cdot T^3$ for a constant d.

(b) We define $\neg \vec{F} = (\neg F^0,\ldots,\neg F^{T-1})$. The vector $Suc(\vec{F}) = (Suc^0(\vec{F}),\ldots,Suc^{T-1}(\vec{F}))$ is given by:

$Suc^i(\vec{F}) = ((\bigwedge_{0\le j<i} F^j) \rightarrow \neg F^i) \wedge ((\bigvee_{0\le j<i} \neg F^j) \rightarrow F^i)$, which is equivalent to $((\bigwedge_{0\le j<i} F^j) \wedge \neg F^i) \vee ((\bigvee_{0\le j<i} \neg F^j) \wedge F^i)$. $Suc(\vec{F})$ is the successor of $\vec{F}$ modulo 2^T.

(c) The vector of length T $Sub(\vec{F}_0,\vec{F}_1) = (Sub^0(\vec{F}_0,\vec{F}_1),\ldots,Sub^{T-1}(\vec{F}_0,\vec{F}_1))$ is defined by:

$$\begin{aligned} Sub^i(\vec{F}_0,\vec{F}_1) = {} & \vec{F}_0 \ge \vec{F}_1 \wedge Suc^i(Add(\vec{F}_0,\neg \vec{F}_1)) \\ \vee \ & \vec{F}_0 < \vec{F}_1 \wedge Suc^i(Add(\neg \vec{F}_0,\vec{F}_1)) \end{aligned}$$

for i with $0 \le i \le T-1$. $Sub(\vec{F}_0,\vec{F}_1)$ is the absolute value of the difference of $\vec{F}_0$ and $\vec{F}_1$. Subtraction is based on the 2-complement presentation of binary numbers.

□

Now we can define addition taking account of the signs.

2.3 Definition

Let $\vec{F}_0$, $\vec{F}_1$ be vectors of length T+1 of propositional formulas. The vector of length T+1 of formulas

$$AddSi(\vec{F}_0,\vec{F}_1) = AddSi^0(\vec{F}_0,\vec{F}_1),\ldots,AddSi^T(\vec{F}_0,\vec{F}_1))$$

is given by:

$$\begin{aligned} AddSi^i(\vec{F}_0,\vec{F}_1) = {} & (F_0^T \leftrightarrow F_1^T) \wedge Add^i(|\vec{F}_0|,|\vec{F}_1|) && \text{(the same signs)} \\ \vee \ & (F_0^T \oplus F_1^T) \wedge Sub^i(|\vec{F}_0|,|\vec{F}_1|) && \text{(different signs)} \end{aligned}$$

for $0 \le i \le T-1$. The sign is computed by

$$\begin{aligned} AddSi^T(\vec{F}_0,\vec{F}_1) = {} & (F_0^T \leftrightarrow F_1^T) \wedge F_0^T \\ \vee \ & (F_0^T \oplus F_1^T) \wedge (|\vec{F}_0| \ge |\vec{F}_1| \wedge F_0^T \\ & \qquad \vee \\ & \qquad |\vec{F}_0| < |\vec{F}_1| \wedge F_1^T) \ . \end{aligned}$$

□

In our simulating Frege proof we will compute the values of the left hand sides of the inequalities of the cutting plane proof. This requires two N-fold iterations of addition: one iteration for the positive summands, one for the negative summands. Unfortunately, an N-fold iteration of Add gives us (cf. 2.2(a)) only the bound $c^N s\cdot T^{2\cdot N}$ on the size of the components of the resulting vector. As the number of variables N cannot be bounded by a constant independent of S, this formula size is superpolynomial in S. Even if we organize the iterated additions in a binary tree of depth LN we get a size of at least $T^{2\cdot LN}$ which still is not polynomial in S. This unpleasant growth is due to the computation of the carry bits. "Carry-save addition" allows for iterating addition without superpolynomial growth of formula size. Carry-save addition was used for counting with small formulas by Buss.

2.4 Definition

Let $\vec{F}_0$, $\vec{F}_1$, $\vec{F}_2$, $\vec{F}_3$ be vectors of length T of propositional formulas, we abbreviate $\vec{F} = (\vec{F}_0,\vec{F}_1,\vec{F}_2)$, $\vec{F}' = (\vec{F}_0,\vec{F}_1,\vec{F}_2,\vec{F}_3)$.

(a) The two vectors of length T of propositional formulas

$$\mathrm{SuCSum}(\vec{F}_0,\vec{F}_1,\vec{F}_2) \quad \text{and} \quad \mathrm{CaCSum}(\vec{F}_0,\vec{F}_1,\vec{F}_2)$$

(the sum value of the carry-save sum and the carry value of the carry-save sum) are given by:

$$\mathrm{SuCSum}^i(\vec{F}) = F_0^i \oplus F_1^i \oplus F_2^i$$

for $0 \le i \le T-1$ and

$$\mathrm{CaCSum}^0(\vec{F}) = \text{false},$$

$$\mathrm{CaCSum}^i(\vec{F}) = (F_0^{i-1} \wedge F_1^{i-1}) \vee (F_0^{i-1} \wedge F_2^{i-1}) \vee (F_1^{i-1} \wedge F_2^{i-1})$$

for $1 \le i \le T-1$.

In $\mathrm{SuCSum}(\vec{F})$ the sum values are computed and in $\mathrm{CaCSum}(\vec{F})$ the carries. Adding $\mathrm{SuCSum}(\vec{F}$ and $\mathrm{CaCSum}(\vec{F})$ gives us the sum of $\vec{F}_0$, $\vec{F}_1$ and $\vec{F}_2$ modulo 2^T.

If s bounds the size of the components of the $\vec{F}_j$, $0 \le j \le 2$, then the formulas $\mathrm{SuCSum}^i(\vec{F})$ and $\mathrm{CaCSum}^i(\vec{F})$ are bounded by $c \cdot s$, where c is a constant independent of T and s.

(b) One iteration of the definitions in (a) allows us to combine four numbers into two: The two vectors of length T of propositional formulas

$$\mathrm{SuCSAdd}(\vec{F}') \quad \text{and} \quad \mathrm{CaCSAdd}(\vec{F}')$$

(the sum value of the carry-save addition and the carry value of the carry-save addition) are given by:

$$\mathrm{SuCSAdd}(\vec{F}') = \mathrm{SuCSum}(\mathrm{SuCSum}(\vec{F}),\ \mathrm{CaCSum}(\vec{F}),\ \vec{F}_3)$$

$$\mathrm{CaCSAdd}(\vec{F}') = \mathrm{CaCSum}(\qquad \text{"} \qquad).$$

Summing $\mathrm{SuCSAdd}(\vec{F})$ and $\mathrm{CaCSAdd}(\vec{F})$ gives us the sum of $\vec{F}_0$, $\vec{F}_1$, $\vec{F}_2$, $\vec{F}_3$ modulo 2^T. Concerning the size estimates of the formulas the same property as in (a) holds. □

Now we can add $\le T$ many numbers with only polynomial growth (in T) of formula size.

2.5 Definition

Let $n \le T$ be a power of 2 and let $\vec{F}_0,\dots,\vec{F}_{n-1}$ be vectors of length T of propositional formulas. Let $l = \log n$. We abbreviate $\vec{F} = (\vec{F}_0,\dots,\vec{F}_{n-1})$. For i,j with $0 \le i \le l$ and $0 \le j \le \frac{1}{2^i}\cdot n-1$ we define the following vectors of length T of formulas:

$$\mathrm{CaItAdd}_{i,j}(\vec{F})$$

standing for the carry at position i,j of the iterated addition of $\vec{F}$ and

$$\mathrm{SuItAdd}_{i,j}(\vec{F})$$

standing for the sum at position i,j of the iterated addition of $\vec{F}$. It is helpful to see these vectors of formulas in a binary tree of height l:

$$\begin{array}{ccccccccc}
Su_{o,o} & Su_{o,1} & Su_{o,2} & ,\ldots, & Su_{0,n-2}, & Su_{0,n-1} & Ca_{0,0}\ Ca_{0,1} & \ldots & Ca_{0,n-1} \\
Su_{1,0} & & Su_{1,1} & \ldots & Su_{1,\frac{1}{2}n-1} & & Ca_{1,0} & \ldots & Ca_{1,\frac{1}{2}n-1} \\
 & Su_{2,0} & & \ldots & Su_{2,\frac{1}{4}n-1} & & & \vdots & \\
 & & & \vdots & & & & & \\
 & & & Su_{l,o} & & & & Ca_{l,o} &
\end{array}$$

If the vector of numbers to be added, $\vec{F}$, is clear from the context, we use the abbreviation $Su_{i,j}$ for $\mathrm{SuItAdd}_{i,j}(\vec{F})$ and $Ca_{i,j}$ for $\mathrm{CaItAdd}_{i,j}(\vec{F})$.

The vector $Su_{i,j}$ and $Ca_{i,j}$ are defined by induction on i:

$$Su_{0,j} = \vec{F}_j, \quad Ca_{0,j} = \bar{0}$$

for $0 \le j \le n-1$, and

$$Su_{i+1,j} = \mathrm{SuCSAdd}(Su_{i,2j}, Ca_{i,2j}, Su_{i,2j+1}, Ca_{i,2j+1})$$

$$Ca_{i+1,j} = \mathrm{CaCSAdd}(\qquad \text{"} \qquad)$$

for $1 \le i+1 \le l$ and $0 \le j \le \frac{1}{2^{i+1}} \cdot n-1$.

For i,j with $0 \le i \le l$ and $0 \le j \le \frac{1}{2^i} \cdot n-1$ we define the intermediate values of the iterated addition

$$\mathrm{ItAdd}_{i,j}(\vec{F}) = \mathrm{Add}(Su_{i,j}, Ca_{i,j}).$$

The final value of the iterated addition is given by:

$$\mathrm{ItAdd}(\vec{F}) = \mathrm{ItAdd}_{l,o}(\vec{F}) .$$

Inductively on i we get that $\mathrm{ItAdd}_{i,j}(\vec{F})$ is the sum module 2^T of all $\vec{F}_k$ with $2^i.j \le k \le 2^i \cdot (j+1)-1$.

The size of the components of $\mathrm{ItAdd}_{i,j}(\vec{F})$ is bounded by $b \cdot T^d \cdot s$ if s bounds the size of the components of $\vec{F}$. The constants b and d are independent of T and S.

□

2.6 Propositional representation of inequalities

Let $\vec{F}_0$ be a vector of length T of propositional formulas and let F be a single propositional formula. We define

$$\vec{F}_0 \wedge F = (F_0^0 \wedge F, F_0^1 \wedge F, F_0^2 \wedge F ,\ldots, F_0^{T-1} \wedge F).$$

The vector $\vec{F}_0 \wedge F$ evaluates to $\bar{0}$ if F is false and to $\vec{F}_0$ otherwise. Let

$$I = \Sigma\, v_h \cdot x_h - \Sigma\, w_h \cdot x_h \geq P$$

be an inequality of our cutting plane proof. We abbreviate

$$\vec{v}\cdot\vec{x} = (\bar{v}_0 \wedge x_0, \ldots, \bar{v}_{N-1} \wedge x_{N-1}),$$
$$\vec{w}\cdot\vec{x} = (\bar{w}_0 \wedge x_0, \ldots, \bar{w}_{N-1} \wedge x_{N-1}).$$

The inequality I is represented by the propositional formula Rep(I) which is defined as

$$\text{AddSi}(\, +\text{ItAdd}(\vec{v}\cdot\vec{x}),\ -\text{ItAdd}(\vec{w}\cdot\vec{x})) \geq \bar{P}.$$

(We assume P signed.) As $\Sigma\, v_h$ and $\Sigma\, w_h$ certainly are $< 2^T$ no overflow occurs in computing the iterated additions and the formula Rep(I) holds wrt. an assignment of truth values of the x_h iff I holds wrt. the corresponding assignment of 1,0 to the x_h. Note that the size of Rep(I) is polynomial in T.

We will use the following abbreviations:

$$\begin{aligned}
\text{Pos}_{i,j}(I) &= \text{ItAdd}_{i,j}(\vec{v}\cdot\vec{x})\\
\text{Neg}_{i,j}(I) &= \text{ItAdd}_{i,j}(\vec{w}\cdot\vec{x})\\
\text{Pos}(I) &= \text{Pos}_{LN,0}(\vec{v}\cdot\vec{x}) = \text{ItAdd}(\vec{v}\cdot\vec{w})\\
\text{Neg}(I) &= \text{Neg}_{LN,0}(\vec{w}\cdot\vec{x}) = \text{ItAdd}(\vec{w}\cdot\vec{x})\\
\text{LHS}_{i,j}(I) &= \text{AddSi}(+\text{Pos}_{i,j}(I), -\text{Neg}_{i,j}(I))\\
\text{LHS}(I) &= \text{AddSi}(+\text{Pos}(I), -\text{Neg}(I)),
\end{aligned}$$

with LHS standing for left-hand-side. □

We collect some lemmas preparing for our simulation in the next section.

2.7 Lemma

Let $\vec{q}, \vec{q}_0, \vec{q}_1, \vec{p}, \vec{p}_0, \vec{p}_1$ be vectors of the appropriate length of propositional variables. The following formulas are derivable by short Frege proofs.

(a) $\text{Add}(\vec{q},\bar{0}) \leftrightarrow \vec{q}$ and $\vec{p} < \text{Add}(\vec{q},\bar{1}) \rightarrow \vec{p} \leq \vec{q}$.

Basic properties of adding $\bar{0}$ and $\bar{1}$.

(b) Let

$$\text{Prem} = (\vec{q}_0 \leftrightarrow \vec{p}_0) \wedge (\vec{q}_1 \leftrightarrow \vec{p}_1)$$

then

$$\text{Prem} \rightarrow (\text{AddSi}(\vec{q}_0,\vec{q}_1) \leftrightarrow \text{AddSi}(\vec{p}_0,\vec{p}_1))$$

and

$$\text{Prem} \rightarrow ((\vec{q}_0 \geq \vec{q}_1) \leftrightarrow (\vec{p}_1 \geq \vec{p}_1)) .$$

Invariance wrt. equivalence.

(c) Let

$$\text{Prem} = \bigwedge_{i=0,1} \neg q_1^{T-1} \wedge \neg p_i^{T-1}$$

then

$$\text{Prem} \rightarrow (\text{AddSi}(\text{AddSi}(\vec{q}_0,\vec{q}_1), \text{AddSi}(\vec{p}_0,\vec{p}_1)) \leftrightarrow \text{AddSi}(\text{AddSi}(\vec{q}_0,\vec{p}_0) \text{AddSi}(\vec{q}_1,\vec{p}_1))) .$$

The formula corresponds to (a+b) + (c+d) = (a+c) + (b+d). Prem makes sure that no overflow occurs when performing the inner additions.

(d) Let

$$\text{Prem} = \vec{q}_0 \geq \vec{q}_1 \wedge \vec{p}_0 \geq \vec{p}_1 \wedge \neg\, p_0^{T-1} \wedge \neg\, q_0^{T-1}$$

then

$$\text{Add}(\vec{q}_0,\vec{p}_0) \geq \text{Add}(\vec{q}_1,\vec{p}_1) .$$

Monotonicity of addition.

The proofs are omitted. Essentially the short Frege proofs are obtained by making the appropriate case distinctions. Statement (d) can be found as lemma 2.6 in [2].

□

2.8 Lemma

Let $\vec{q}_0$, $\vec{q}_1$, $\vec{q}_2$, $\vec{q}_3$ be vectors of length T of propositional variables. The formula

$$\text{Add}(\text{Add}(\vec{q}_0,\vec{q}_1), \text{Add}(\vec{q}_2,\vec{q}_3)) \leftrightarrow \text{Add}(\text{SuCSAdd}(\vec{q}_0,\vec{q}_1,\vec{q}_2,\vec{q}_3), \text{CaCSAdd}(\vec{q}_0,\vec{q}_1,\vec{q}_2,\vec{q}_3))$$

is derivable by a Frege proof of size polynomial in T. The formula shows that carry-save addition is equivalent to addition.

This is lemma 4 in [2].

□

2.9 Lemma

Let $n \leq T$ be a power of 2 and let $\vec{q}_0,\ldots,\vec{q}_{n-1}$ be vectors of length T of propositional variables. Let $l = \log n$. We abbreviate $\vec{q} = (\vec{q}_0,\ldots,\vec{q}_{n-1})$. For $0 \leq j \leq n-1$ the formulas

$$\text{ItAdd}_{0,j}(\vec{q}) \leftrightarrow \vec{q}_j$$

are provable by a Frege proof of $O(T^3)$. Let $1 \leq i \leq l$ and $0 \leq j \leq \frac{1}{2^i} \cdot n-1$, then the formulas

$$\text{ItAdd}_{i,j}(\vec{q}) \leftrightarrow \text{Add}(\text{ItAdd}_{i-1,2j}(\vec{q}), \text{ItAdd}_{i-1,2j+1}(\vec{q}))$$

are provable by a Frege proof of size polynomial in T.

The formula shows that the $\text{ItAdd}_{i,j}(\vec{q})$ can be obtained from previously computed ItAdd's by normal addition instead of using carry save addition.

Proof

The first claim follows from definition 2.5 using the proof constructed in 2.7(a). (Note: As we have no substitution rule like from F(x) infer F(G), where x is a propositional variable in Frege systems we can use a result like F(x) is derivable only in the following sense: To get F(G) we repeat the whole proof of F(x) with G instead of x.) The second claim follows by using the proof from lemma 2.8.

□

2.10 Lemma

Let

$$I = \Sigma\, v_h \cdot x_h - \Sigma\, w_h \cdot x_h \geq P$$

be an inequality of our cutting plane proof. The following formulas are derivable by polynomial size Frege proofs:

(a) For j with $0 \leq j \leq N-1$

$$Pos_{0,j}(I) \leftrightarrow \bar{v}_j \wedge x_j\ ,$$

for i,j with $1 \leq i \leq LN$ and $0 \leq j \leq \frac{1}{2^i} \cdot N-1$

$$Pos_{i,j}(I) \leftrightarrow Add(Pos_{i-1,2j}(I),\ Pos_{i-1,2j+1}(I))$$

and analogously for Neg instead of Pos.

(b) $Pos(I) \leq \overline{2^S}$, $Neg(I) \leq \overline{2^S}$, $LHS(I) \leq \overline{2^S}$.

(c) For all j with $0 \leq j \leq N-1$

$$LHS_{0,j}(I) \leftrightarrow \overline{(v_j - w_j)} \wedge x_j\ ,$$

for all i,j with $1 \leq i \leq LN$ and $0 \leq j \leq \frac{1}{2^i} \cdot N-1$

$$LHS_{i,j}(I) \leftrightarrow AddSi(LHS_{i-1,2j}(I),\ LHS_{i-1,2j+1}(I)).$$

Proof

(a) The claim follows by applying lemma 2.9 .

(b) The short proofs are obtained by doing some computing with the constants. This must give the result because the proof size is $\leq S$.

(c) The claim follows with an application of 2.7(c) and (b) of this lemma.

□

3 THE SIMULATION

We show that each proof step of the cutting plane proof can be simulated by a polynomial size Frege proof.

3.1 Theorem

Adding two inequalities can be simulated by a Frege proof of polynomial size.

Proof

We simulate the following derivation step of our cutting plane proof:

$$I_1 = \Sigma\, v_{1,h}\cdot x_h - \Sigma\, w_{1,h}\cdot x_h \geq P_1$$

$$I_2 = \Sigma\, v_{2,h}\cdot x_h - \Sigma\, w_{2,h}\cdot x_h \geq P_2$$

$$I_3 = \Sigma\, v_{3,h}\cdot x_h - \Sigma\, w_{3,h}\cdot x_h \geq P_3$$

where $P_3 = P_1 + P_2$ and the $v_{3,h}$, $w_{3,h}$ are defined as follows:
Let

$$\sigma_h = (v_{1,h} - w_{1,h}) + (v_{2,h} - w_{2,h})$$

be the sum of the components of x_h in I_1 and I_2, then

$$v_{3,h} = \begin{cases} \sigma_h & \text{if } \sigma_h > 0 \\ 0 & \text{otherwise} \end{cases} \qquad w_{3,h} = \begin{cases} \sigma_h & \text{if } \sigma_h < 0 \\ 0 & \text{otherwise} \end{cases} .$$

This expresses the fact that I_3 is the sum of I_1 and I_2. We show how to get from

$$\mathfrak{C} \rightarrow \text{Rep}(I_1) \text{ and } \mathfrak{C} \rightarrow \text{Rep}(I_2)$$

to

$$\mathfrak{C} \rightarrow \text{Rep}(I_3$$

by a polynomial size Frege proof.

First, we derive by a short Frege proof that the left-hand side of I_3 is the sum of the left-hand sides of I_1 and I_2, that is we derive the formula

$$\text{AddSi}(\text{LHS}(I_1), \text{LHS}(I_2)) \leftrightarrow \text{LHS}(I_3).$$

The Frege proof proves the intermediate formulas $F_{i,j}$ for $0 \leq i \leq LN$ and $0 \leq j \leq \frac{1}{2^i} \cdot N-1$ given by

$$\text{AddSi}(\text{LHS}_{i,j}(I_1), \text{LHS}_{i,j}(I_2)) \leftrightarrow \text{LHS}_{i,j}(I_3).$$

Note that $F_{LN,0}$ is the formula we want to prove and that there are only $O(N^2)$ many $F_{i,j}$.

The $F_{0,j}$ are derivable by short proofs as follows easily applying the proof constructed in 2.10(c) as the $\bar{v}_{k,h}$ $\bar{w}_{k,h}$ are fixed vectors of true and false for $k = 1,2,3$.

Now let $i \geq 1$ and assume we have derived the $F_{i-1,j}$. By using the Frege proof constructed in proving lemma 2.10(c) the formula

$$\text{LHS}_{i,j}(I_3) \leftrightarrow \text{AddSi}(\text{LHS}_{i-1,2j}(I_3), \text{LHS}_{i-1,2j+1}(I_3))$$

is derivable by a polynomial size proof.

The rest of the proof is indicated by the following series of propositional equivalences:

$\text{AddSi}(\text{LHS}_{i-1,2j}(I_3),\ \text{LHS}_{i-1,2j+1}(I_3))$

$\leftrightarrow$ 2.7(b) using the $F_{i-1,j}$ already proved

$\text{AddSi}(\text{AddSi}(\text{LHS}_{i-1,2j}(I_1),\ \text{LHS}_{i-1,2j}(I_2)),$

$\quad \text{AddSi}(\text{LHS}_{i-1,2j+1}(I_1),\ \text{LHS}_{i-1,2j+1}(I_2))$

$\leftrightarrow$ 2.7(c), 2.10(b) to eliminate the premise of (2.7(c)

$\text{AddSi}(\text{AddSi}(\text{LHS}_{i-1,2j}(I_1),\ \text{LHS}_{i-1,2j+1}(I_1)),$

$\quad \text{AddSi}(\text{LHS}_{i-1,2j}(I_2),\ \text{LHS}_{i-1,2j+1}(I_2)))$

$\leftrightarrow$ 2.10(c), 2.7(b)

$\text{AddSi}(\text{LHS}_{i,j}(I_1),\ \text{LHS}_{i,j}(I_2))$.

From

$$\mathfrak{C} \rightarrow \text{LHS}(I_1) \geq \bar{P}_1 \ , \quad \mathfrak{C} \rightarrow \text{LHS}(I_2) \geq \bar{P}_2$$

we get by 2.7(d) and 2.10(b) to eliminate the premise of 2.7(d)

$$\mathfrak{C} \rightarrow \text{AddSi}(\text{LHS}(I_1),\ \text{LHS}(I_2)) \geq \text{AddSi}(\bar{P}_1,\bar{P}_2).$$

Now 2.7(b), the formula proved above, and the definition of $\bar{P}_3$ allow us to derive

$$\mathfrak{C} \rightarrow \text{LHS}(I_3) \geq \bar{P}_3$$

by a short proof. □

Simulating the multiplication and division steps requires some preparation.

<u>3.2 Definition</u>

Let $\vec{F}_0$, $\vec{F}_1$ be vectors of length T of propositional formulas

(a) Let $n \geq 0$, then

$$2^n \cdot \vec{F}_i = (\underbrace{\text{false}, \text{false}, .., \text{false}}_{n\text{-times}},\ F_0^0,\ F_0^1,\ F_0^2,\ \ldots,\ F_0^{T-1-n}).$$

The vector of formulas $2^n \cdot \vec{F}_0$ is the multiplication of $\vec{F}_0$ with 2^n modulo 2^T.

(b) Let the vector of length T (of vectors) $\vec{F}$ be given by

$$\vec{F} = (2^0 \cdot \vec{F}_1 \wedge F_0^0,\ 2^1 \cdot \vec{F}_1 \wedge F_0^1,\ 2^2 \cdot \vec{F}_1 \wedge F_0^2,\ \ldots,\ 2^{T-1} \cdot \vec{F}_1 \wedge F_0^{T-1})$$

then the vector of formulas $\text{Mult}(\vec{F}_0,\vec{F}_1)$ is defined by:

$$\text{Mult}(\vec{F}_0,\vec{F}_1) = \text{ItAdd}(\vec{F}).$$

$\text{Mult}(\vec{F}_0,\vec{F}_1)$ is the multiplication of $\vec{F}_0$ and $\vec{F}_1$ modulo 2^T. The formula is the propositional implementation of the school method for multiplication: We add up the appropriate multiples of $\vec{F}_1$, which make up the components of $\vec{F}$.

The intermediate values of $\text{Mult}(\vec{F}_0,\vec{F}_1)$ are given by:

$$\text{Mult}_{i,j}(\vec{F}_0,\vec{F}_1) = \text{ItAdd}_{i,j}(\vec{F})$$

for $0 \le i \le LT$ and $0 \le j \le \frac{1}{2^i} \cdot T-1$.

The size of the $\text{Su}_{i,j}(\vec{F})$ and $\text{Ca}_{i,j}(\vec{F})$ is bounded by $T^c \cdot d \cdot s$ where c and d are constants independent of T and s and s bounds the size of the components of $\vec{F}_0$ and $\vec{F}_1$. In particular the size of the components of $\text{Mult}_{i,j}(\vec{F}_0,\vec{F}_1)$ is bounded by $T^c \cdot d \cdot s$. Thus we have polynomial size formulas for multiplication.

□

We first prove some basic laws for multiplication by short Frege proofs.

<u>3.3 Lemma</u>

Let $\vec{q}_0, \vec{q}_1, \vec{q}_2, \vec{p}_0, \vec{p}_1$ be vectors length T of propositional variables. The following formulas have short Frege proofs:

(a) If

$$\text{Prem} = \vec{q}_0 \leftrightarrow \vec{p}_0 \wedge \vec{q}_1 \leftrightarrow \vec{p}_1,$$

then

$$\text{Prem} \to (\text{Mult}(\vec{q}_0,\vec{q}_1) \leftrightarrow \text{Mult}(\vec{p}_0,\vec{p}_1))$$

Invariance wrt. equivalence.

(b) If

$$\text{Prem} = \vec{q}_1 \ge \vec{q}_2 \wedge \bigwedge_{k=0,1,2} \Big(\bigwedge_{S \le l \le T-1} \neg q_k^l \Big)$$

then

$$\text{Prem} \to \text{Mult}(\vec{q}_0,\vec{q}_1) \ge \text{Mult}(\vec{q}_0,\vec{q}_2).$$

Monotonicity of multiplication. (Prem makes sure that no overflow occurs.)

(c) $\text{Mult}(\vec{p}, \text{Add}(\vec{q}_1,\vec{q}_2)) \leftrightarrow \text{Add}(\text{Mult}(\vec{p},\vec{q}_1), \text{Mult}(\vec{p},\vec{q}_2))$.

Distributivity.

<u>Proof</u>

(a) The formula can be shown by proving the claim for all $\text{Mult}_{i,j}$ instead of Mult using the proof of 2.9 and 2.7(b).

(b) Let for all i,j with $0 \le i \le LT$ and $0 \le j \le \frac{1}{2^i} \cdot T-1$ the formula $F_{i,j}$ be given by:

$$\text{Prem} \rightarrow \text{Mult}_{i,j}(\vec{q}_0,\vec{q}_1) \geq \text{Mult}_{i,j}(\vec{q}_0,\vec{q}_2) \wedge \bigwedge_{2\cdot S+i\leq l\leq T-1} \neg\, \text{Mult}^{l}_{i,j}(\vec{q}_0,\vec{q}_1)\ .$$

The formulas $F_{0,j}$ are derivable using the Frege proofs of 2.9 and 2.7(b) and the premise making a case distinction as to whether $\vec{q}_1 > \vec{q}_2$ or $\vec{q}_1 \leftrightarrow \vec{q}_2$ and in the first case making another T case distinctions as to which bit causes $\vec{q}_1 > \vec{q}_2$.

Assume we have proved $F_{i-1,j}$ for all j. Using the proof constructed in 2.9 we can derive

$$\text{Mult}_{i,j}(\vec{q}_0,\vec{q}_1)$$

$$\leftrightarrow$$

$$\text{Add}(\text{Mult}_{i-1,2j}(\vec{q}_0,\vec{q}_1),\ \text{Mult}_{i-1,2j+1}(\vec{q}_0,\vec{q}_1)).$$

Using $F_{i-1,2j}$ and $F_{i-1,2j+1}$, the proof constructed in 2.7(d), and the definition of addition we prove that

$$\text{Prem} \rightarrow \text{Mult}_{i,j}(\vec{q}_0,\vec{q}_1) \geq \text{Mult}_{i,j}(\vec{q}_0,\vec{q}_2) \wedge \bigwedge_{2S+i\leq l\leq T-1} \neg\, \text{Mult}^{l}_{i,j}(\vec{q}_0,\vec{q}_1).$$

(c) The formula can be proved proving the intermediate formula obtained by substituting $\text{Mult}_{i,j}$ for Mult and using the case of the proof 2.7(c) where all numbers are positive (then the size restriction is not used).

□

<u>3.4 Lemma</u>

Let $\vec{p}$, $\vec{q}_0,\ldots,\vec{q}_{T-1}$ be vectors of length T of propositional variables. The formula

$$\text{Mult}(\vec{p},\ \text{ItAdd}(\vec{q}_0,\ldots,\vec{q}_{T-1}))$$

$$\leftrightarrow$$

$$\text{ItAdd}(\text{Mult}(\vec{p},\vec{q}_0),\ldots,\text{Mult}(\vec{p},\vec{q}_{T-1}))$$

is derivable by a polynomial size Frege proof.

Distributivity for iterated addition.

<u>Proof</u>

A short Frege proof of the above formula is obtained using the intermediate formulas obtained by substituting $\text{ItAdd}_{i,j}$ for ItAdd in the above formula using 3.3(c) and 2.9.

□

3.5 Theorem

The multiplication steps of the cutting plane proof can be simulated by polynomial size Frege proofs.

Proof

We simulate the following derivation step:

$$I = \Sigma\, v_h \cdot x_h - \Sigma\, w_h \cdot x_h \geq P$$

$$J = \Sigma\, (c \cdot v_h) \cdot x_h - \Sigma\, (c \cdot w_h) \cdot x_h \geq c \cdot P\ ,$$

where $c \geq 0$ is an integer.

For simplicity we assume $P \geq 0$, hence LHS(I), LHS(J) ≥ 0. The case $P < 0$ can be reduced to the positive case by changing the signs and the comparison operator using a short Frege proof making some case distinctions.

First we give a short proof of the formula

$$|\mathrm{LHS}(J)| \leftrightarrow \mathrm{Mult}(\bar{c}, |\mathrm{LHS}(I)|)$$

indicated by the following series of equivalences: (We must work with absolute values because of our definition of Mult.)

$\mathrm{Mult}(\bar{c}, |\mathrm{LHS}(I)|)$

$\leftrightarrow$ 2.7(b), definition of AddSi, LHS(I) ≥ 0.

$\mathrm{Mult}(\bar{c}, \mathrm{Sub}(\mathrm{Pos}(I), \mathrm{Neg}(I)))$

$\leftrightarrow$ 3.3(c), 2.7(b)

$\mathrm{Sub}(\mathrm{Mult}(\bar{c}, \mathrm{Pos}(I)), \mathrm{Mult}(\bar{c}, \mathrm{Neg}(I)))$

$\leftrightarrow$ 3.4, 2.7(b), some computation with true, false to get $\overline{c \cdot v_h} \wedge x_h$, $\overline{c \cdot w_h} \wedge x_h$ from $\mathrm{Mult}(\bar{c}, v_h \wedge x_h)$, $\mathrm{Mult}(\bar{c}, \bar{w}_h \wedge x_h)$

$\mathrm{Sub}(\mathrm{ItAdd}((\overline{c \cdot v_0}) \wedge x_0, \ldots, (\overline{c \cdot v_{N-1}}) \wedge x_{N-1})$

$\quad \mathrm{ItAdd}((\overline{c \cdot w_0}) \wedge x_0, \ldots, (\overline{c \cdot w_{N-1}}) \wedge x_{N-1}))$

$\leftrightarrow$ Definition of AddSi, note LHS(J) ≥ 0.

$|\mathrm{LHS}(J)|$.

Using the proof of 2.7(a) and 3.3(b) and 2.10(b) to eliminate the premise of 3.3(b) we get from

$$\mathfrak{C} \rightarrow \mathrm{LHS}(I) \geq \bar{P}$$

the formula

$$\mathfrak{C} \rightarrow \mathrm{Mult}(\bar{c}, \mathrm{LHS}(I)) \geq \mathrm{Mult}(\bar{c}, \bar{P})$$

with the formula proved above, 2.7(a), and some computation with true false we get

$$\mathfrak{C} \to \mathrm{LHS}(J) \geq \overline{c\cdot P}\ .$$

□

3.6 Theorem

The division steps of our cutting plane proof can be simulated by polynomial size Frege proofs.

Proof

Wie simulate the following derivation step of the cutting plane proof:

$$I_1 = \Sigma\, v_{1,h}\cdot x_h - \Sigma\, w_{1,h}\cdot x_h \geq P_1$$

$$I_2 = \Sigma\, v_{2,h}\cdot x_h - \Sigma\, w_{2,h}\cdot x_h \geq P_2$$

where $c \in \mathbb{Z}^+ \setminus \{0\}$ is such that $v_{1,h} = c\cdot v_{2,h}$ and $w_{1,h} = c\cdot w_{2,h}$ and $P_2 = \left\lceil \frac{1}{c}\cdot P_1 \right\rceil$.
Again we only treat the case $P_1 \geq 0$. We only consider division with rounding.
Let

$$P_1 = c\cdot Q+R$$

where $c > R > 0$. Then $P_2 = Q+1$.
We show how to get from

$$\mathfrak{C} \to \mathrm{LHS}(I_1) \geq P_1$$

to

$$\mathfrak{C} \to \mathrm{LHS}(I_2) \geq P_2.$$

The Frege proof proceeds by contradiction. From the assumption

$$|\mathrm{LHS}(I_2)| < \mathrm{Add}(\bar{Q},\bar{1})$$

we get using 2.7(a)

$$|\mathrm{LHS}(I_2)| \leq \bar{Q}.$$

The proofs from 3.3(b) and the proof of 2.10(b) to eliminate the premise of 3.3(b) give us

$$\mathrm{Mult}(\bar{c},\ |\mathrm{LHS}(I_2)|) \leq \mathrm{Mult}(\bar{c},\bar{Q}).$$

As in the proof of 3.5 we get

$$\mathrm{Mult}(c,\ |\mathrm{LHS}(I_2)|) \leftrightarrow |\mathrm{LHS}(I_1)|.$$

We get

$$\mathrm{Mult}(\bar{c},\bar{Q}) < \bar{P}_1$$

by a simple computation with true, false. Using the proof of 2.7(a) we get

$$|\mathrm{LHS}(I_1)| < \bar{P}_1$$

which is easily provable to be equivalent to

$$\neg\,(|\mathrm{LHS}(I_1)| \geq \bar{P}_1).$$

As $\mathrm{LHS}(I_1) \geq \bar{0}$, this contradicts

$$\mathcal{C} \rightarrow \mathrm{LHS}(I_1) \geq \bar{P}_1$$

which we have derived already. Hence, we get

$$\mathcal{C} \rightarrow \mathrm{LHS}(I_1) \geq \bar{P}_2.$$

□

The following theorem finishes our simulation of the cutting plane proof by a polynomial size Frege proof.

3.7 Theorem

The formulas corresponding to the axioms of the cutting plane proof can be derived by short Frege proofs. From $\bar{0} \geq \bar{P}$ where $P > 0$ (the final inequality of the cutting plane proof) we can derive false.

□

3.8 Application

In [9] a propositional formula encoding the principle: Let R be a set of odd cardinality say n, then R cannot be partitioned into a family of disjoint sets each of which has exactly two elements, is formulated as a propositional formula:
We have $\binom{n}{2}$ propositional varaibles $x_{\{i,j\}}$ for $i,j \in \{1,\ldots,n\}$, $i \neq j$. Instead of $x_{\{i,j\}}$ we write x_{ij} (then $x_{ij} = x_{ji}$). The variable x_{ij} indicates that i and j are in the same set of the (assumed) partition.
The formula reads:

$$\bigwedge_{i=1,\ldots,n} \Big(\bigvee_{j\neq i} x_{ij}\Big) \wedge \bigwedge_{i} \Big(\bigwedge_{\substack{j\neq k \\ i\neq j,\, i\neq k}} (\bar{x}_{ij} \vee \bar{x}_{ik})\Big).$$

The first part says that each i is in a set of the partition. The second says that no i is in two sets of the partition. In [9] it is said that these formulas have short extended and symmetric resolution proofs. We can easily strengthen this result showing that the formulas have short Frege proofs by showing that they have short cutting plane proofs. (It is known [6] that extended resolution p-simulates Frege systems.) The direct construction of short Frege proofs for these formulas seems to be much more complex.

From the first half of the formula we get the inequality

$$\sum_{\substack{i=1 \\ j\neq i}}^{n} 2x_{ij} \geq n, \quad \text{hence } \sum x_{ij} \geq \frac{n+1}{2} \quad \text{as n is odd.}$$

From the second half we get

$$\sum_{j=2}^{n} -x_{1j} \geq -1$$

by induction: For $j \neq k$, $1 \neq j$ and $1 \neq k$ we have the inequalities

$$-x_{1j} - x_{1k} \geq -1$$

as induction base. Let $m < n$ and assume we have derived

$$\sum_{j=2}^{m} -x_{1j} \geq -1 \quad \text{and} \quad \sum_{j=3}^{m+1} -x_{1j} \geq -1 \quad .$$

Adding these inequalities gives

$$-x_{12} + \sum_{j=3}^{m} -x_{1j} - x_{1m+1} \geq -2$$

adding

$$-x_{12} - x_{1m+1} \geq -1 \ ,$$

dividing by 2 and rounding gives (finishing the induction step):

$$\sum_{j=2}^{m+1} -x_{1j} \geq -1 \ .$$

We can proceed analogously for all other $i \neq 1$. Adding the n inequalities obtainded in this way gives us

$$\Sigma\ -2x_{ij} \geq -n \quad \text{hence} \quad \Sigma\ -x_{ij} \geq \frac{-n+1}{2}$$

which gives $0 \geq 1$ by adding the inequality derived at the beginning. (Note that this proof is rather similar to the short cutting plane proof of the pigeonhole principle given in [5].)

□

LITERATURE

[1] M. Ajtai, The complexity of the pigeonhole principle, Proceedings of the 29th Symposium on foundations of Computer Science (1988) 346-355.

[2] S. Buss, Polynomial size proofs of the propositional pigeonhole principle, Journ. Symb. Logic 52 (1987) 916-927.

[3] C.-L. Chang, R.C.-T. Lee, Symbolic logic and mechanical theorem proving, Academic Press (1973).

[4] P. Clote, Bounded arithmetic and computational complexity, Proceedings Structure in Complexity (1990) 186-199.

[5] W. Cook, C.R. Coullard, G. Turan, On the complexity of cutting plane proofs, Discr. Appl. Math. 18 (1987) 25-38.

[6] S.A. Cook, R.A. Reckhow, The relative efficiency of propositional proof systems, Journ. Sym. Logic 44 (1979) 36-50.

[7] A. Haken, The intractability of resolution, Theor. Comp. Sci. 39 (1985) 297-308.

[8] J.N. Hooker, Generalized resolution and cutting planes, Annals of Oper. Res. 12 (1988) 217-239.

[9] B. Krishnamurthy, Short proofs for tricky formulas, Acta Informatica 22 (1985) 253-275.

[10] P. Pudlak, Ramsey's theorem in bounded arithmetic, Workshop in Comp. Sci. Logic 1990, Springer LNCS, submitted.

[11] R.M. Smullyan, First-order Logic, Springer Verlag 1968.

[12] A. Goerdt, Cutting plane versus Frege proof systems, Technical report, University of Duisburg.

RAM with compact memory: a realistic and robust model of computation

by Etienne GRANDJEAN, Université de Caen, LIUC 14032 CAEN Cedex FRANCE and J.M. ROBSON, Australian National University, Dept of Computer Science, P.O. Box 4, CANBERRA 2601, AUSTRALIA

ABSTRACT:

An operation op of arity k on $\mathbb{N}$, i.e. a function op: $\mathbb{N}^k \to \mathbb{N}$ is *linear time Turing computable* (for short, LTTC) if it is computable in linear time on a Turing machine (for usual binary or dyadic notation of integers). Let + and *Conc* respectively denote usual addition of integers and concatenation (of their dyadic notations). A RAM which uses only arithmetical operations of a set I is called an I - RAM. An LTTC-RAM is a RAM which only uses LTTC operations.

In the present paper, we use the logarithmic criterion for time measure of RAMs. A RAM works *with polynomially (resp. strongly polynomially) compact memory* if it only uses addresses (resp. addresses and register contents) $\leq t^{O(1)}$ where t is the time of the computation.

Theorem 1. A deterministic LTTC-RAM R with polynomially compact memory is simulated in linear time by a deterministic {+}-RAM (resp. {*Conc*}-RAM) R' with strongly polynomially compact memory.

Theorem 2. A nondeterministic LTTC-RAM R can be simulated in linear time by a nondeterministic {+}-RAM (resp. {*Conc*}-RAM) R' with strongly poynomially compact memory.

Note that Theorem 2 holds for both weak nondeterministic RAMs and strong nondeterministic RAMs, i.e. in case the RAMs have only nondeterministic goto instructions or in case they have an instruction to guess an integer.

If moreover the RAMs R of Theorem 1-2 are *sane* (i.e. R does not use noninitialised registers) then the simulating RAMs R' are sane, too.

We also study and discuss more restrictive notions of compact memory (*linearly compact memory*).

I) Introduction and discussion:

What is a linear time computation? This question depends on two other questions: what is a *good model* of computation? what is a *step*? There are two well-known models of computation: *Turing machine* and *random access machine*. It is generally accepted that Turing machine is not powerful enough to be a model of linear time: it lacks indirect addressing. There is a problem about RAMs: what is the good time measure? uniform measure (one instruction = one step)? logarithmic measure (the execution time of an instruction is the sum of lengths of the integers involved)? Another problem is that the RAM model *is not canonical* : which instructions are allowed? It is generally admitted that multiplication (resp. division) of integers may not be executed by a single RAM instruction and most authors regard the addition (resp. subtraction) of two integers as a primitive instruction. However there are many instructions (shift, concatenation, boolean instructions) which seem to be as simple as addition and are allowed by some authors but are forbidden by many others.

Slot and van Emde Boas (cf. [SlvEB] and [vEB]) recently noted that there is some confusion in the literature about what is space complexity of RAMs (they discuss several definitions and prove some results of invariance or noninvariance). They explain such a confusion as follows: before their papers there was no systematic study of definitions and simulations of the various space measures. We feel that the same argument is applicable to time complexity: there is no confusion in the definitions concerning the time measure but it *lacks* some systematic discussion and some general study. It has been treated in only *scattered* papers.

Cook and Reckhow [CoRe] give the first and the most systematic treatment of RAMs and of their time complexity. Wiedermann [Wi] and, recently, Katajainen and others [KvLP] improve time simulations of RAMs by Turing machines and conversely. Schönhage [Se1] defines storage modification machine, a new model of computation and proves that it simulates several other models within linear time (for unit cost measure). Gurevich [Gu] compares this model with a model defined very early by Kolmogorov and Uspensky [KoUs]. Recently Gurevich and Shelah introduce the class NLT of functions computable in nearly linear time $n(\log n)^{O(1)}$ on "random access computers". A *random access computer* (for short, a RAC) is a restricted RAM model, introduced and discussed by Angluin and Valiant [AnVa], the memory of which depends on the input size: a RAC is a RAM which can only manipulate integers of binary length $\leq c \log n$, where n is the length of the input and c depends on the given machine only; in particular, the addresses of registers used by a RAC are $\leq n^c$. [GuSh] show that the class NLT is very robust: it is invariant if in place of RACs we use Kolmogorov machines or storage modification machines or if we change the set of allowed instructions of RACs. They also show that NLT is invariant if we use "frugal" RACs: a RAC is *frugal* if its visited locations always form an initial segment. The present paper introduces and extensively studies variants of RAC (RAM *with polynomially compact memory* and RAM *with strongly polynomially compact memory*) and variants of frugal RAC (RAM *with linearly compact memory* and RAM *with strongly linearly compact memory*). Note that [GuSh] also study the nondeterministic version of their class NLT.

There is an argument against the RAM model which is implicit in the literature (see the discussion about RACs by [AnVa]) but which, to our knowledge, has not been explicitly presented till now: *in its full generality* the RAM model *is not realistic*! Let us examine the following program

```
BEGIN
     e:=1
     FOR i:=0 to n DO
          BEGIN X(e):=i ; e:=e+e END
END
```

For each i = 0,1,...,n this program stores i into register $X(2^i)$. Note that it uses $O(n^2)$ time (for logarithmic cost measure) but uses the register $X(2^n)$: hence its implementation requires an exponential number of registers (sparsely used)! Of course we feel that this program is badly constructed. Therefore we propose in this paper a restricted definition of a RAM computation.

A RAM R works *with polynomially compact memory* if there is a constant d such that any halting computation of R only uses registers with addresses $O(t^d)$ where t denotes its execution time (for logarithmic cost criterion). This restriction ("polynomially compact memory") is obviously very realistic: we think that most "real" computations on RAMs respect this condition. Moreover this notion gives a very robust model of computation as explained below. Note that there are three differences with the RAC model of [AnVa]. First we do not bound the contents of registers but only their addresses: this difference is not essential because we show (Corollary 6.4 below) that a RAM R with polynomially compact memory is simulated in linear time by a RAM R' *with strongly polynomially compact memory* (it means that R' only uses addresses *and* register contents $O(t^d)$, for a constant d, i.e. R' is a RAC if the time t is polynomial in the length of the input). Another difference is that the function $N(t) = O(t^d)$ which bounds the RAM memory is not a function of the length of the input as in [AnVa] but a function of the execution time t: this allows to define not only time complexity classes of low level (for instance, polynomial or linear time) but also those of high level. The third difference is that [AnVa] use the uniform cost measure but we prefer the logarithmic cost measure because it is more precise and more significant.

An *I-RAM* is a RAM where I is the set of allowed operations (on nonnegative integers). A k-ary operation op (on nonnegative integers) is *linear time Turing computable* (for short, LTTC), if there is a Turing machine which for each k-tuple of integers $x_1,...,x_k$ computes $op(x_1,...,x_k)$ in time

$O(\sum_{i=1,\ldots k} \text{length}(x_i))$

(each integer x_i is represented in dyadic notation; length(x_i) denotes the length of its notation).

Let +, Conc denote the usual addition of two integers and the concatenation of their dyadic notations, respectively (for convenience we identify an integer with its dyadic notation; in particular "zero" is represented by the empty word, denoted 0). Of course + and Conc are LTTC operations. In the present paper we prove the following theorem which shows that {+} and {Conc} are *complete sets* of operations for RAMs with polynomially compact memory.

Theorem 1.1. Let I be a finite set of LTTC operations.
(i) A deterministic I-RAM R with polynomially compact memory is simulated in linear time by a deterministic {+}-RAM (resp.{Conc}-RAM) R' with polynomially compact memory.
(ii) A nondeterministic I-RAM R (without any memory restriction) is simulated in linear time by a non deterministic {+}-RAM (resp.{Conc}-RAM) R' with polynomially compact memory.

(Note that there are both a strong and a weak interpretations of nondeterminism for RAMs: a *strong* or *unbounded nondeterministic* RAM has an instruction to guess an integer; a *weak* or *bounded nondeterministic* RAM has no such an instruction but has only nondeterministic goto instructions ; Theorem 1.1 (ii) holds for both weak and strong interpretations).

In fact our results are slightly stronger than Theorem 1.1. For example, register values of R' are also polynomially bounded and we give a precise polynomial bound in the nondeterministic case.

Theorem 1.1 gives the basis for defining a realistic and robust complexity class for linear time, in both the deterministic and nondeterministic cases.

We have chosen to present our results for the logarithmic cost criterion. However they can be adapted for the uniform cost measure of RAMs (for example in the result similar to Theorem 1.1(i) ,we must add a condition for R: register values of R must be polynomially bounded).

[KvLP] prove that RAMs can simulate Turing machines in linear time (for logarithmic measure) *if we omit* the time to convert a string , i.e. a sequence of bits, into a sequence of integers (and conversely). Note that the classical RAM is an heterogeneous model of computation: its input (resp. output) is a string but its computation works on integers. Schönhage [Se2] has recently studied the time required by the conversion string- integer (which he calls "storing") for on-line random access machines and has proved a nonlinear lower bound for that.

Note that the notion of LTTC operation and then our results about I-RAMs (where I is a set of LTTC operations) seem to be dependent on the (dyadic) representation (of integers) we have chosen. Since the conversions dyadic-binary and binary-dyadic can be computed by finite automata, an operation is LTTC for dyadic notation iff it is LTTC for binary notation (and similarly, for b-adic and b-ary notations for b>2). On the other hand, the careful reader can easily check that all the proofs and results of the present paper still hold if we replace dyadic notation by b-adic notation.

Our results and those of [KvLP] can be regarded as the expression of an intuitive and common idea: the classical RAM model (for example, RAM with addition and subtraction of [CoRe]) has some *completeness property*; it is as powerful as possible for it can simulate efficiently (i.e. in linear time)
- Turing machines (if we omit the conversion string-integer)
- each reasonable RAM with shift, concatenation, boolean operations...(and, more generally, with any linear time Turing computable operation).

More surprisingly, we prove that *addition alone is sufficient* (subtraction is not necessary!). On the other hand, our similar result about {Conc}-RAMs can be regarded as follows: the RAM model where addresses and register contents are *words* on a binary alphabet ({1,2},for instance) and where *concatenation of words* is the only operation (this RAM does not involve numbers!) is exactly*as powerful* as the usual RAM model.

Note that we also study a restriction of the notion of RAM with polynomially compact memory : the RAM *with linearly compact memory* (cf.par.VI) and prove a result similar to Theorem 1.1(i) for this restricted model.

For the deterministic case the complete results are given in par.VI. The hasty reader can omit par.IV and V where we give rigorous but lengthy and technical proofs; however par. III states a simplified result and presents an intuitive (but not rigorous) proof.

Section VII gives the results in the nondeterministic case. The simulation in that case is complicated by the need to simulate the nondeterminism of the simulated machine, especially in the case of unbounded nondeterminism. On the other hand, the simulation is able to utilise nondeterminism in two ways, firstly to guess where information has been previously been stored and secondly to guess a parameter to be used in a hash function.

The results of the present paper and some more have been previously stated without proof in [GrRo,Gr3,Gr4]. Recently, [Wi2] has independently presented related results about time and space of RAMs.

II) Preliminaries

We use the usual notations and definitions in *computational complexity* (see [AHU] and [HoUl]). We use *multitape Turing machines* (for short, TM) where every tape is one-dimensional.

In this paper we will write "integer" in place of "nonnegative integer" (negative integers will not be used). It will be convenient to *identify* each integer n with its *dyadic notation* $\alpha_{l-1}...\alpha_1\,\alpha_0 \in \{1,2\}^l$ defined by

$$n = \sum_{i<l} \alpha_i 2^i$$

(<u>convention</u>: zero is identified with the empty word denoted 0). This identification (integer = dyadic word) will be used in the whole paper. In particular length (n) = l. (The b-adic notation of integers, for b≥2, is defined similarly). Note that we have chosen in this paper not to use the usual binary notation but to work with the dyadic one. Our reasons are as follows:

(i) each integer has *one and only one* dyadic notation (unlike its binary notation which can be padded with 0s to left); this justifies a natural identification of each integer with a word ∈ $\{1,2\}^*$; in particular, number "zero" identifies with the empty word (the binary notation 0 is not absolutely satisfying; why not use the empty word too?)

(ii) some of our sharp results (for example Lemma 5.5, Theorems 6.1 and 6.2,...) no longer hold (or we have not succeeded in proving them) if we use concatenation of binary notations (in place of dyadic notations): moreover the concatenation of binary notations is rather unpleasant: for example, consider 101^0 = 1010 and compare the concatenations 10^11 = 1011 and 10^011 = 10011.

log n will denote the logarithm of n in base 2. In particular length(n) = $\lfloor \log(n+1) \rfloor$

A k-ary operation op on $\mathbb{N}$, i.e. a function op: $\mathbb{N}^k \to \mathbb{N}$, is *linear time Turing computable* (for short, LTTC) if there is a deterministic TM M with k input tapes (for the k arguments) and one output tape, with input-output alphabet {1,2}, such that M computes the function op in linear time

$$O(\text{length}(w_1)+...+\text{length}(w_k))$$

where $w_1,w_2,...,w_k$ are the k arguments.

More particularly, we will use the following LTTC operations:

- *successor, predecessor,* denoted Succ and Pred, respectively

and the following binary LTTC operations:

- *addition, concatenation,* denoted + and Conc, respectively.

We will also denote the concatenation by symbol ^ as above : w^w' = Conc(w,w').

For each word $w = w_0 w_1 ... w_{n-1}$ of length n, the word (with "repetition")

$$w_0\, w_0\, w_1 w_1 ...\, w_{n-1}\, w_{n-1}$$

of length 2n will be denoted rep(w). In particular the binary operation, denoted *Pair*, and defined by

Pair(i,j) = rep(i)^12^j

is clearly a LTTC operation and a *pairing function* (i.e. Pair is an injection: $\mathbb{N}\times\mathbb{N}\to\mathbb{N}$).

We use classical *random access machines* (for short, RAM: see [CoRe] and [AHU]) with *logarithmic cost measure*. However we also introduce the following generalization. Let I be a finite set of LTTC operations (of any arity). An *I-RAM* consists of a finite program which operates on

- a fixed number of *accumulators* denoted u,v,...
- a (potentially infinite) sequence of *registers* denoted X(0), X(1),...

Each register and each accumulator can store any integer (in dyadic notation). The program is a finite sequence of labeled instructions which are given in Figure 1, together with their arguments.

The *execution time* of an instruction is (for logarithmic cost measure) as follows:

- if it is an instruction $u := op(v_1,...v_k)$ for $op \in I$ its time cost is

$$\max [\, 1, (\textstyle\sum_{i=1,\ldots k} \text{length}(v_i)) + \text{length}(op(v_1,...v_k)) \,]$$

- if it is of the form u := C for an integer C its time cost is max(1, length(C))

Instruction	Arguments
u := C , C any integer	None
$u := op(v_1,...v_k)$, op any k-ary operation of I	$v_1, v_2, ... v_k$
u := X(v)	v,X(v)
X(v) := u	u,v
Goto m if u = 0	u
Goto m	None
Halt	None

Note: u (resp. $v, v_1, ..., v_k$) denotes any accumulator.

Figure 1: RAM instructions and their arguments

- otherwise its cost is

$$\max (1, \textstyle\sum_{i=1,\ldots k} \text{length}(x_i))$$

where $x_1,...x_k$ is the list of its arguments.

(For purpose of simplification, we confuse each register or accumulator with the integer it contains: so length(X(i)) denotes the length of the contents of X(i))

Let *cost(Inst)* denote the execution time of an instruction Inst. Clearly if Inst computes an integer x then length(x) $\le$ cost(Inst).

We do not insist on input-output instructions. We can use those introduced by [KvLP] who use one read-only (resp. write-only) one-way input tape (resp output tape) with alphabet {1,2, #} (resp. {1,2}). The instructions are as follows:

input $\lambda_1,\lambda_2,\lambda_\#$. It is executed as follows:

- read the new input symbol i (the input word is of the form w^# , with $w \in \{1,2\}^*$)
- goto label λ_i

output β ($\beta \in \{1,2\}$). It is executed as follows:

-write β on the output tape.

In the present paper, we require all our RAMs to be *sane*, that means :

- no computation indefinitely loops (all terminate on the Halt instruction);
- a register (or accumulator) cannot be used as an instruction argument if it has not been given a value before (we do not require all registers and accumulators to contain zero before the computation; note that the first instruction executed cannot have arguments).

However our results still hold if this assumption is removed.(In just one case the simulating machine may fail to be sane even though the simulated machine is sane. This occurs in the simulation of unbounded nondeterminism by an ND {*Conc*}-RAM and then only for some "unlucky" nondeterministic choices which will lead to the simulation failing immediately).

The *time* of a RAM computation is the sum of the times of its executed instructions.

In the following an accumulator or a register will be denoted generically a *register* .

As in [CoRe] and [KvLP] that we quote:"it will be convenient to use various extensions of the basic RAM instruction set provided that the execution time is adequately measured by the logarithmic cost criterion ... Also in some algorithms it is convenient to have a RAM with k separate memories, denoted $A_0, A_1, \ldots A_{k-1}$ (called "arrays" by [CoRe]), each A_i consisting of a countable sequence of registers indexed

$A_i(0), A_i(1), \ldots$

we call this a *multimemory* RAM. For the sake of readability, we extend the basic instruction set of the RAM with some Pascal-like control structures that have obvious translation ".

Our fundamental model of computation will be the RAM *with polynomially compact memory.*

Definition: A RAM R works *with polynomially* (resp. *strongly polynomially) compact memory* if there is a constant d such that each computation of R uses addresses (resp. addresses and register contents) $O(t^d)$ where t denotes its execution time.

Let us adapt a lemma of [CoRe] and [KvLP].

Lemma 2.1: (i) Every multimemory {+}-RAM (resp. {Conc}-RAM) R can be simulated in O(t) time by an ordinary {+}-RAM (resp. {Conc}-RAM) R'.

(ii) Moreover if R works with polynomially (resp. strongly polynomially) compact memory then R' does too.

Proof: (i) is easy. "The idea is simply to interleave the RAM memories into one" ([KvLP]). If we have k arrays A_i, i<k, register $A_i(j)$ is simulated by register X(i+kj) for I = {+} and by register X(rep(i)^12^j) for I = {Conc}. This new index can be computed in O(log j) time (i is fixed) by using addition (resp. concatenation), which multiplies the time bound by a constant factor.

(ii) is obvious: addresses are linear in the original ones. ∎

A *multidimensional RAM* is a generalization of a multimemory RAM: it has one (or several) multidimensional array(s) of fixed dimensions. For example, a 2-dimensional array A is a 2-dimensional sequence of registers denoted:

A(0,0), A(0,1),...
A(1,0), A(1,1),...
....
A(i,0), A(i,1),....
....

Instructions of Figure 1 and their logarithmic time cost can be easily generalized for multidimensional RAMs.

We will also use *multidimensional* RAMs working *with polynomially (resp. strongly polynomially) compact memory*: they are defined similarly as above.

We will also introduce and study a natural but stronger notion of compact memory:

Definition. A RAM R works *with linearly* (resp. *strongly linearly*) *compact memory* if in each computation of R of time t:

- each address is O(t / log t);
- each register length is O(log t) (resp. each register value is O(t / log t).

Remark. The expression "linearly compact memory" is due to the fact (easily proved) that a RAM which works in time t can only visit O(t / log t) distinct registers and to the fact that the potential memory of a RAM is the product of its maximal address with its maximal register length; with our definition, this product is O((t / log t) log t) = O(t) and then is linear.

We clearly get the following implications:

Proposition 2.2. R is a RAM with strongly linearly compact memory

$\Rightarrow$ R has linearly compact memory

$\Rightarrow$ R has strongly polynomially compact memory

$\Rightarrow$ R has polynomially compact memory.

In order to avoid the hard problem of string-integer conversion which is involved by input-output instructions (see [Se2]), we also use a technical version of RAM computation.

Definition. Let op be a k-ary operation on $\mathbb{N}$ and let R be a multimemory RAM. Let b be a positive integer and $w_1, w_2, ...w_k$ be a k-tuple of integers. We say that R *b-computes* the integer

$$w_0 = op(w_1,...w_k)$$

if R does the following computation:

- at the beginning of the computation a special accumulator contains the value b;
- at the beginning (resp. at the end) of the computation, the k input integers $w_1,w_2,...w_k$ (resp. the output w_0) are stored in k special arrays, denoted $A_1, A_2,...A_k$, respectively (resp. in a special array denoted A_0) with the following conventions: for each i = 0,1,...k let the word w_i be decomposed in

$$w_i = w_i^{n_i} \wedge ... \; w_i^1 \wedge w_i^0$$

where words w_i^j are such that $\text{length}(w_i^j) = b$ for each $j < n_i$ and $0 < \text{length}(w_i^{n_i}) \le b$; then register $A_i(j)$ contains w_i^j for each $j \le n_i$ and $A_i(n_i + 1)$ contains 0 (which marks the end).

III) The deterministic case: a simplified statement of the results and a sketch of proof:

We want to prove the following

Theorem 3.1. Let I be a finite set of LTTC operations. Let R be an I-RAM with polynomially compact memory. Then R is simulated in linear time by a {Conc}-RAM (resp. {+}-RAM) R' with strongly polynomially compact memory (i.e. R' has polynomially compact memory *and* registers of length O(log t)).

The general idea of the proof is to efficiently simulate the Turing machines which compute the I operations by dividing each TM tape into *b-blocks*, i.e. blocks of b consecutive cells, for some fixed b: each b-block will be represented by some register of the RAM. Similarly, each register X(i) of the original RAM R will be represented by a sequence of registers

A(i,0), A(i,1),...

of the simulating RAM R', each one of length b (to contain a b-block) so that register X(i) is the concatenation of registers A(i,0), A(i,1),... in reversal order : A(i,j) will be called a *b-block* of X(i) (or, simply, a *block* of X(i)).

Take $b = \varepsilon \log t$, for a "small" constant ε: then it is possible to precompute tables of b consecutive transitions of the Turing machines involved. It requires polynomial time in $2^b = 2^{\varepsilon \log t} = t^{\varepsilon}$ and therefore o(t) time, for ε is small. A key idea is that the second component j of the address (i,j) of A(i,j), is $\le t / b = O(t / \log t)$ and then its length has the same order O(log t) as $\text{length}(A(i,j)) = \varepsilon \log t$.

Let us explain the ideas of the simulation by simulating two instructions of R.

Let op $\in$ I be a binary operation. We simulate the instruction $u := op(v_1,v_2)$ as follows:

- if $v_1 \geq t^{\varepsilon}$ or $v_2 \geq t^{\varepsilon}$ then simulate the TM which computes op by using b-blocks and the precomputed table of b transitions on b-blocks: b(= ε log t) steps of computation of the TM are simulated in time O(b) by one consultation of the table; hence the simulation of the instruction requires time

$$O(b).O((\text{length}(v_1) + \text{length}(v_2)) / b) = O(\text{length}(v_1) + \text{length}(v_2))$$

- otherwise (i.e. if $v_1 < t^{\varepsilon}$ and $v_2 < t^{\varepsilon}$) use a special precomputed table of operation op for arguments $< t^{\varepsilon}$: the table has $t^{2\varepsilon}$ entries and therefore can be computed in time o(t).

Now let us simulate an instruction (with indirect addressing):

u := X(v)

Recall that the address value v is $< t^d$ (i.e. is polynomial in t since the RAM memory is polynomially compact ; d is a constant integer) and then

$$\text{length}(v) \leq d \log t = (d / \varepsilon)\, b ;$$

hence the value of v is contained in d / ε registers $V(0), V(1), \ldots V(d/\varepsilon - 1)$, each of length $b = \varepsilon \log t$ so that

$$v = V(d/\varepsilon - 1)\wedge \ldots V(1) \wedge V(0).$$

The instruction u:= X(v) is simulated as follows (recall that register u (resp. X(v)) is simulated by a sequence of registers U(0), U(1) ,...(resp. A(v,0), A(v,1),...) of length b) :

```
BEGIN
    v := V(d/ε - 1)^ ... V(1)^ V(0)  {d/ε - 1 concatenations}
    i := 0
    WHILE A(v,i) ≠ 0 DO
        BEGIN u(i) := A(v,i) ; i := i + 1 END
    u(i) := 0
END
```

Clearly the instruction is simulated in linear time.

Theorem 3.1 is proved by a sequence of three lemmas.

Lemma 3.2. Let R be as in Theorem 3.1. There is a 2-dimensional {Conc, Succ, Pred}-RAM R_1 which simulates R in time O(t) with strongly polynomially compact memory.

Sketch of proof. Use the previous informal ideas: precompute the table of the Turing machine associated to each LTTC operation (use the operations Conc and Succ); simulate the RAM R by using for each b-block of R a register of length b (use Conc, Succ, Pred).

The only problem in Lemma 3.2 is that the RAM R_1 is 2-dimensional. We simulate it by a one-dimensional RAM R_2 by using the above pairing function Pair: $\mathbb{N}\times\mathbb{N} \rightarrow \mathbb{N}$ (cf. par. II): each 2-dimensional register A(i,j) of R_1 is simulated by the one-address register X(p) of R_2 where p = Pair(i,j).

Lemma 3.3. Let R_1 be the RAM above. There is a one-dimensional {Conc, Succ, Pred, Pair}-RAM R_2 which simulates R_1 in linear time with strongly polynomially compact memory.

Proof. it is sufficient to note that

$$\text{length}(\text{Pair}(i,j)) = O(\text{length}(i) + \text{length}(j)).$$

Now we only have to get rid of operations Pair, Succ and Pred.

Lemma 3.4. Let R_2 be as above. There is a one-dimensional {Conc, Succ, Pred}-RAM R' which simulates R_2 in time O(t) with strongly polynomially compact memory.

Sketch of proof. Use the same simulation as in Lemma 3.2. The simulating RAM R' is apparently 2-dimensional. However note that the second dimension of the registers is bounded by a constant ; more precisely, there is a constant d such that for each register X(i) of R_2, length(X(i))$\leq$ d log t ; hence register X(i) is simulated by registers A(i,0),

A(i,1),...A(i,d/ε -1) of R' so that

$X(i) = A(i,d/\varepsilon - 1)\wedge \ldots\ A(i,1)\wedge A(i,0)$ and

$\mathrm{length}(A(i,j)) \leq \varepsilon \log t$ for $j < d/\varepsilon$.

So we can regard R' as a multimemory one-dimensional RAM (with d/ε one-dimensional arrays: then use Lemma 2.1). ■

End of the proof of theorem 3.1 (Sketch). We only have to get rid of operations Succ, Pred. Observe that they are essentially applied on addresses of b-blocks : they are integers less than $t / b = t / (\varepsilon \log t) = O(t / \log t)$. From Lemma 5.3. below, the tables of successors and predecessors of integers $O(t / \log t)$ can be precomputed by a {Conc}-RAM (resp. {+}-RAM) within O(t) time. This proves Theorem 3.1 for a {Conc}-RAM.

We obtain the similar result for a {+}-RAM by observing that concatenation is used in only two manners:

- concatenation of one bit to a word of length $\leq \varepsilon \log t$;
- concatenation of a fixed number of words of length exactly $\varepsilon \log t$.

The first use is easily simulated in linear time on a {+}-RAM. The second use is simulated by consulting precomputed tables of multiplication by $t^{\varepsilon}, t^{2\varepsilon}$,...(for $\varepsilon \log t$ - shift, $2\varepsilon \log t$ - shift,... respectively) and adding the results. These tables are precomputed by repeated additions within o(t) time. ■

IV) Some technical lemmas:

Our simulations will use precomputed tables (compare with [KvLP]).

Lemma 4.1. Let op be a LTTC operation. There is a {Conc, Succ, Pred}-RAM R such that for each positive integer b and each k-tuple of integers $w_1,w_2,\ldots w_k$, one of which has length > b,

(i) R b-computes $w = op(w_1,\ldots,w_k)$ in time

$O(L + L/b \log L)$

where $L = \Sigma_{i=1,\ldots k}\ \mathrm{length}(w_i)$. This computation uses some tables (which only depend upon b and op) which are precomputed by R in time 2^{cb}, for a contant c (only depending upon op).

(ii) Moreover integers involved in the computation are bounded by

$2^{cb} + O(L/b)$

and therefore are linearly bounded by the time.

(iii) Moreover predecessor and successor operations are only applied to arguments $O(b + L/b)$

Lemma 4.2. Lemma 4.1 also holds if R is now a {+, Succ, Pred}-RAM.

Proof of Lemma 4.1. As explained in par.III, a basic idea is to precompute in a table all the possible b moves of the r tape Turing machine M which computes operation op. Let Tab(M) denote this table.

Each tape of M is divided into *b-blocks* (i.e. sequences of b consecutive cells) and each b-block is represented by one register of the RAM (also denoted a ***b-block***). Similarly the time of the computation that M executes is divided into ***b-intervals*** (i.e. intervals of b consecutive instants). During a b-interval Δ each tape head can only visit the b-block where it lies at the beginning of Δ and its two adjacent b-blocks: call this a *b-transition* of M.

Therefore table Tab(M) contains for each 3r-tuple of b-blocks (three b-blocks per tape) the result of the next b-transition for this 3r-tuple, i.e.

- the new content of each b-block concerned by the 3r-tuple (including the state and the head position)

- r values $\in \{1,0,-1\}$ indicating for each tape the head movement within the three b-blocks concerned (which determines the new active b-block).

Let us be more precise. Let Q and Σ denote the set of states and the tape alphabet of M, respectively. Each b-block is described by a b length word on the alphabet Alpha = $\Sigma \cup (Q \times \Sigma)$; for instance the word

$$\sigma_0 \sigma_1 ... \sigma_{i-1} (q_j, \sigma_i) \sigma_{i+1} ... \sigma_{b-1}$$

(where $q_j \in Q$ and $\sigma_0, \sigma_1, ... \sigma_{b-1} \in \Sigma$) describes an active b-block where the head is positioned on the ith cell and scans the symbol σ_i.

There is a slight technical problem: we have to encode each b-block by an integer (in dyadic notation) which lies in a register of the RAM. Without loss of generality assume that Alpha has 2^l elements. Let us bijectively associate to each $\varepsilon \in$ Alpha a code :

$$\text{code}(\varepsilon) = \varepsilon_1 \varepsilon_2 ... \varepsilon_l \in \{1,2\}^l$$

Then trivially encode a b-block

$$\beta = \varepsilon_0 \varepsilon_1 ... \varepsilon_{b-1} \in \text{Alpha}^b$$

by

$$\text{code}(\beta) = \text{code}(\varepsilon_0)\text{^}\text{code}(\varepsilon_1)\text{^}...\text{code}(\varepsilon_{b-1}) \in \{1,2\}^{bl}$$

In order to simplify the notation assume that the TM M has only one tape. Each index of table Tab(M) is a dyadic word w of the form

$$w = \text{code}(\beta_{-1})\text{^}\text{code}(\beta_0)\text{^}\text{code}(\beta_1) \in \{1,2\}^{3bl}$$

where β_0 is an active b-block. The table associates to w a 4-tuple which denotes the result of the current b-transition of M:

- the three words $w'_i = \text{code}(\beta'_i)$ for i=-1,0,1 where β'_i is the new value of the block β_i after the b-transition;

- a number $\in \{1,2,3\}$ which indicates the new active block (1,2,3 respectively encode "left move", "no move", "right move").

If M has r tapes then the number of entries of Tab(M) is no more than 2^{3rbl}. The RAM R computes this table by enumerating (lexicographically) all the possible table indexes and by simulating the TM M in a trivial manner (one cell per register: see [AHU, p.31]). For each entry the time of the simulation is polynomial in its length O(b). Hence the whole time to construct the table is $b^{O(1)}.2^{3rbl} \leq 2^{cb}$ for some constant c.

Note that the only operations used in the construction of Tab(M) are successor and predecessor applied to O(b) numbers (for the enumeration of the possible entries in increasing order and also for the simulation of moves of the heads) and concatenation of O(b) length bounded words (for the reconstitution of words w, w'_{-1}, w'_0, w'_1 from their digits).

We also need a table denoted Tab(I / O), for associating to the dyadic words which constitute the k input integers (these are words $\in \{1,2\}^b$, except the last ones which may be shorter) the codes of the blocks which constitute the k input tapes (they include the initial state and the head position; each code belongs to $\{1,2\}^{bl}$). This table is easy to construct (in time 2^{cb}). Conversely a similar table, denoted Tab'(I / O), is precomputed for decoding the output.

The RAM R which b-computes $w = op(w_1,...w_k)$ works as follows:

- encode the input words by consulting Tab(I / O) for each b-block;
- for each b-interval (there are O(L/b) b-intervals) simulate the corresponding b-transition of M by consulting only once Tab(M); (Halt if the final state is encountered)
- decode the output words by consulting Tab'(I / O) for each b-block.

Let us describe more precisely how the RAM R simulates a b-transition of the TM M. R has for each tape of M a special accumulator which contains the number, denoted a, of the active block (a = O(L/b)). The simulation works as follows:

- get the three b-block a, a-1 and a+1 of each tape (similarly indexed in the array which

represents this tape) and concatenate the 3r codes (for the r tapes)
- consult once Tab(M)
- update b-blocks a, a-1 and a+1
- update number a.

The simulation of one b-transition requires time

$$O(b) + O(\log a).$$

Hence the whole simulation requires time

$$O(L/b)\,(O(b) + \log a) = O(L + L/b \log L)$$

(note that L > b). So (i) is proved.

In the simulation Pred and Succ operations are only applied to the argument a which is O(L/b) and concatenation is applied to O(b) length bounded words.

Integers involved in the computation of R are clearly less than $2^{cb} + O(L/b)$. This proves (ii). We have also proved (iii). ∎

Proof of Lemma 4.2. It is absolutely similar to the proof of lemma 4.1. We only have to get rid of the concatenation in the original simulation by using addition. Note that Conc is used in only two manners:

(i) for the construction of tables Tab(M), Tab(I / O) and Tab'(I / O): concatenation of one digit to a word;

(ii) for the concatenation of 3r words each of length bl to get an entry of Tab(M).

Cases (i) and (ii) are treated as follows:

(i) Note that for each integer w and each digit $j \in \{1,2\}$, w^j = w+w+j;

(ii) Note that if u,v are integers such that length(v) = bl then $u\hat{}v = u.2^{bl} + v$. In order to get an efficient simulation, precompute in a table, denoted Tab(shift), for each integer u of length ≤ 3rbl the integer $u' = u.2^{bl}$ with the sequence of instructions:

```
u' := u
FOR i := 1 TO bl DO u' := u' + u'
```

Clearly Tab(shift) can be constructed in time 2^{cb} and the computation of each $u.2^{bl}+v$ with this table requires only time O(length(u)+length(v)). ∎

In Lemmas 4.1 and 4.2 we require at least one of the arguments w_i (i = 1,...k) to have length > b. We can get similar results if this condition is negated.

Lemma 4.3. Let the hypotheses of Lemma 4.1 hold, except that length(w_i)≤ b for each i = 1,2,...k (inputs of the RAM are b, w_1,...w_k each in a special accumulator). Then the conclusion of Lemmas 4.1 and 4.2 hold. More precisely,

(i) there is a {Conc, Succ, Pred}-RAM (resp. {+, Succ, Pred}-RAM) which computes w = op(w_1,...w_k) in O(L) time using tables that R precomputes in time 2^{cb};

(ii) moreover predecessor and successor operations are only applied to arguments O(b).

Proof. The idea is as follows: use a precomputed table of the operation op and a pairing function (on integers) to present this table as a one-dimensional array. Let Pair: NxNa N be the function defined by

Pair(x,y) = rep(x)^12^y

(resp. Pair(x,y) = W(x+y) + y where W(z) = rep(z)^12^1^i where 1^i = 11...1 (repeated i times) and i = length(z)+1).

They are effectively pairing functions: it is obvious for the first function Pair; for the second one, observe that if length(y) < i = length(z)+1, then the dyadic word rep(z)^12 is a prefix of the dyadic word

W(z) + y = rep(z)^12^1^i + y ;

hence we can identify without ambiguity the prefix rep(x+y)^12 in the word

Pair(x,y) = W(x+y) +y

and, consequently, we can recognize the integers x+y and y.

Now let us explain the simulation. Assume for instance that the operation op is binary. Suppose we have constructed a table, denoted Tab(rep), (resp. Tab(W)), which associates to each integer z such that length(z) ≤ b (resp. length(z) ≤ 2b) the integer rep(z) (resp. W(z)), and also a table, denoted Tab(Pair, op), which associates to each integer of the form Pair(x,y) with $x,y \leq b$ the value op(x,y). Then we compute w = op(w_1,w_2) (where length(w_i) $\leq$ b for i = 1,2) as follows:

- read the value rep(w_1) (resp. W(w_1+w_2)) in the table Tab(rep) (resp. Tab(W))
- compute the value Pair(w_1,w_2) = rep(w_1)^12^w_2 by concatenation (resp.= W(w_1+w_2)+w_2 by addition)
- read the result op(w_1,w_2) at the index Pair(w_1,w_2) of table Tab(Pair,op).

This simulation clearly requires time

O(length(w_1)+length(w_2)) = O(L).

It is routine to show that our tables can be computed in time 2^{cb} by using Conc, Succ, Pred (resp. +, Succ, Pred). So (i) is proved. (ii) is obvious. ∎

For purpose of efficiently counting the time of a RAM computation, we will use the following lemma similar to Lemma 3.2.1 of [KvLP] (see also [CoRe]).

Lemma 4.4. Let $T_1,T_2,...T_k$ be positive integers. Assume that each T_i is presented in dyadic notation in an array T(0),T(1),...T(j),... with one digit per register (i.e. T(j) is the jth digit of T_i). Then a {Pred,Succ}-RAM can compute the integer

$$Sum = \Sigma_{i=1,...k} T_i$$

(in a similar presentation) within O(Sum) time.

Proof. First assume that each integer T_i is 1, i.e. the integer Sum is successively 1,2,...k. Each incrementation of Sum is simulated by a sweep over the array denoted SUM where it is stored, beginning at the register SUM(0), till a 1 or the end of the representation is encountered. Each previous digit (which is a 2) is replaced by 1 and

- either the first 1 is replaced by 2
- or (if there is no 1) we write a new digit 1.

Then we conclude as [KvLP] and [CoRe]: the linear time bound is seen as follows; during a count up from Sum = 2^e - 1 to Sum = 2^{e+1} - 1, the length of the sweep is $2^e/2$ times 1, $2^e/4$ times 2, $2^e/8$ times 3, and so forth. Hence the total time is

$$O(2^{e-1}.1.\log 1 + ... + 2^{e-j}.j.\log j + ...+ 2^0.e.\log e) = O(2^e)$$

because the infinite series $\sum_j j.(\log j).2^{-j}$ is convergent; hence a count up from 0 to t = 2^e-1 requires time:

$$O(2^0 + 2^1 + ... + 2^{e-1}) = O(2^e-1) = O(t).$$

This proves the case when each T_i = 1.

For the general case, note that the addition Sum := Sum + T_i is simulated as follows:

```
WHILE Ti ≠ 0 DO
    BEGIN
      Ti := Ti -1; Sum := Sum+1
    END
```

The array SUM is treated as above and the decrementations $T_i := T_i - 1$ can be treated similarly. ∎

V) The simulations:

The simulations are mainly presented in Lemmas 5.1 and 5.5. The other lemmas prepare the proof of the main lemma (5.5) which gives a *dynamic simulation* (the time bound t and the length b of the blocks are not known in advance: moreover b can vary during the simulation). The following Lemma 5.1 presents a *static simulation* (b is known in advance and is the same during the whole simulation; "static" and "dynamic simulation" are intuitive concepts drawn from [KvLPJ]).

Notation. Let R be a RAM with a special accumulator and let $\mathfrak{E}$ be a computation or a subcomputation of R. If at the beginning of $\mathfrak{E}$ the special accumulator contents is b, then $\mathfrak{E}$ is called a *b-execution* of R.

Lemma 5.1. Let R be an I-RAM where I is a finite set of LTTC operations. Let c_1, c_2 be some constants. There is a 2-dimensional {Conc, Succ, Pred}-RAM (resp. {+, Succ, Pred}-RAM) R' such that

if C is a (sub)computation of R of time t using registers of length $\leq L$ and using addresses $\leq N$ where $t,L \leq 2^{c_1 b}$ and $\text{length}(N) \leq c_2 b$, for some integer b, then C is simulated by an O(t) time bounded b-execution $\mathfrak{E}$ of R' where

(i) the first component of any address is O(N);
(ii) the second component of any address is $\leq L/b = O(L / \log L)$;
(iii) each register length is O(b).

This computation $\mathfrak{E}$ use tables (which only depend upon b and R) which are precomputed by R' in time 2^{cb} (for some constant c only depending upon R).

Moreover predecessor and successor operations are only applied to arguments $O(b + L / \log L)$.

Proof. The basic idea of the simulation is very simple. As roughly explained in par.III, each register X(i) (resp. each accumulator u,v,...) of R (its length is $\leq L$) is statically represented by an array of $\lceil L/b \rceil$ registers A(i,0), A(i,1),...(resp. U(0), U(1),... V(0), V(1),...) each of length b (a possible exception: the last one may be shorter) such that at each moment of the simulation

X(i) = ...^A(i,1)^A(i,0) (and similarly for u, v,...).

This intuitively explains conclusions (i-iii) of the lemma. (Note that $c_1 b \geq \log L$ and then $L/b = O(L / \log L)$.

It remains to explain carefully how to do the simulation within O(t) time. We strongly use Lemmas 4.1-4.3. Let R_{op} denote the RAM that Lemma 4.1 (resp. 4.2) associates to an operation $op \in I$. Our RAM R' will use the programs of these RAMs and also the precomputed tables introduced in Lemmas 4.1-4.3 which we will call the *b-tables*. R' first computes the b-tables in 2^{cb} time (for a constant c only depending upon R).

We simulate an instruction of R of the form $u := op(v_1,...v_k)$ in linear time O(l) (where $l = \sum_{i=1,...k} \text{length}(v_i)$) as follows:

- if $\text{length}(v_i) \leq b$ for each i = 1,2,...k then execute the simulation of Lemma 4.3 (in time O(l)
- otherwise work like the RAM R_{op} (in time $O(l + l / b \log l) = O(l)$ since $\log l = O(\log t) = O(b)$).

Note that the condition $\text{length}(v_i) \leq b$ is equivalent to condition $V_i(1) = 0$ (cf. the above representation of registers of R) that R' easily checks.

Now let us simulate an instruction of R of the form u := X(v) . Recall that each address used in computation C is $\leq N$ and then $\text{length}(v) \leq \text{length}(N) \leq c_2 b$. Hence the accumulator v of R is represented by an array (of R') V(0), V(1), ..., V(lmax - 1) of bounded length $lmax \leq c_2$, i.e. V(i) = 0 for each $i \geq lmax$. Let v be an accumulator of R'. Instruction u:= X(v) is simulated as follows:

1) reconstitute the address v: v := V(lmax-1)^...V(1)^V(0)
2) copy the array A(v,0), A(v,1),... into the array U(0), U(1),...

(1) is executed in O(length(v)) time. Let Nblocks denote the number of nonzero registers A(v,i) (i = 0,1,...) i.e. the number of b-blocks of X(v). Clearly

(*) Nblocks = O((log X(v))/ b)

(2) is executed in time O(Nblocks (b + log v + log Nblocks)). We have log v = O(b) and log(Nblocks) = O(log t) = O(b). Hence (from (*)) (2) is executed in O(log X(v)) time and therefore instruction u:= X(v) is executed in linear time

O(log v + log X(v)).

Similarly it is easy to prove that each instruction of the RAM R is simulated by the 2-dimensional RAM R' in linear time. Note that conditions (i-iii) always hold in this simulation.

Successor and predecessor operations are only used:

- for simulating LTTC instructions $u := op(v_1,...v_k)$ where they are applied to arguments O(b) + O(l / b) (cf. Lemmas 4.1-4.3) with $l = \Sigma_{i=1,...k}$ length(v_i) ≤ kL
- for running along the indices ≤ L / b = O(L / log L) of nonzero elements of an array V(0), V(1),... (resp. A(v,0), A(v,1),...).

Hence Succ and Pred operations are applied to arguments O(b + L / log L). ♠

Remark 5.2. By using as a time counter a special array, denoted TIME, where the register TIME(i) contains the ith digit of the time, we can add to the RAM R' of Lemma 5.1 the ability to evaluate the execution time of each instruction of the RAM R (while R' simulates it) and adding it up to the previous time. This requires O(t) time.

Proof. It is a consequence of Lemma 4.4 and of the following. We have seen that each nonzero accumulator u (resp. register X(i)) that the simulated instruction of R manipulates is represented in R' as an array of integers U(0),U(1),...U(k) (resp. A(i,0),A(i,1),...A(i,k)) each of length b, except the last one U(k) (resp. A(i,k)), denoted REMAINDER, such that

0 < length(REMAINDER) ≤ b.

Then the length of the original accumulator (resp. register) of R is

$\underbrace{b+b+...+b}_{k \text{ times}}$ + length(REMAINDER).

The length of each component of the array is first read in a precomputed table, denoted Tab(length) (which stores the length of integers m such that length(m)≤b), and secondly is converted into the array of its digits. Use for that a table denoted Tab(digits) which associates to each integer n such that 0 < n ≤ b the array of its digits. Tab(digits) can be one-dimensional by the following trick: set the 0th, 1st,...i^{th},... digit of n at the address numbered n^2, n^2+1,...n^2+i,... respectively (since length(n)≤n there are zero registers that separate the representations of n and n+1). Use another precomputed table, denoted Tab(square), for storing squares of integers ≤ b. R' precomputes Tab(length), Tab(square) and Tab(digits) within 2^{cb} time. ♠

In order to get rid of Succ and Pred operations, we need to compute their tables dynamically: the idea is first to initialize both tables for the first integers and then to compute them for integers with 2,3,...l,... dyadic digits, by a recurrence on l.

Assume that we have computed the successor table for integers n of length ≤ l (and the symetric predecessor table) and stored it in a one-dimensional array denoted SUCC (resp. PRED) where SUCC(n) = n+1 (resp. PRED(n) = n-1). The procedure denoted SuccPred(l), given in Figure 2, computes the successors of integers of length l+1 (and the symetric table for PRED) on a {Conc}-RAM.

```
PROCEDURE SuccPred(l)
    global variables: one-dimensional arrays SUCC, PRED
    local variables: integers j, n, l_ones, l_twos {which suggestively denote the
integers of length l with l ones and l twos, respectively}
BEGIN
    computes l_ones, l_twos
    n := l_ones
    FOR j := PRED(l_ones) DOWNTO 0 DO
    BEGIN
        SUCC(1^n) := 1^SUCC(n) ; SUCC(2^n) := 2^SUCC(n)
        n := SUCC(n)
    END
    SUCC(1^l_twos) := 2^l_ones ; SUCC(2^l_twos) := 11^l_ones
    Execute the symetric instructions for Pred {i.e. replace in the previous code each
instruction of the form SUCC(p) := s by PRED(s) := p}
END
```

Figure 2: Algorithm to contruct Succ and Pred

Remarks. The loop "for j..." is executed for successive values n = l_ones, l_ones+1,... l_twos - 1 and therefore the procedure exactly computes the successors of integers of length l+1.

The loops involved in computations of l_ones and l_twos and the loop "for j..." implicitly use the Pred operation for integers of length $\leq$ l.

The loop "for j..." is the main time consuming instruction: its internal instruction works in time O(length(n)) = O(l) and is repeated $O(2^l)$ times. Therefore the time bound of SuccPred is $O(l.2^l)$.

Lemma 5.3. There is a {Conc}-RAM (resp. {+}-RAM) with input integer n (in a special accumulator) which computes the successor table for integers of length $\leq$ l (and the symetric predecessor table) within $O(l.2^l)$ time. Its computation only involves integers $O(2^l)$.

Proof. The following program of a {Conc}-RAM uses the above arrays SUC, PRED and the procedure SuccPred of Figure 2.

```
BEGIN
    Initialize the arrays SUCC, PRED for small integers
    FOR j := 1 TO l - 1 DO call PROCEDURE SuccPred(j)
END.
```

It requires time $O(\Sigma_{j=1,...l-1}\, j.2^j) = O(l.2^l)$.

For a {+}-RAM it is easy to construct a variant of the above procedure SuccPred (Hint: use the identities Pred(n+n+1) = n+n and Pred(n+n+1+1) = n+n+1). The remainder of the proof is not modified. ◆

The following easy lemma roughly means that in a RAM computation, one instruction is not much more costly than all the previous ones together.

Lemma 5.4. Let I be a finite set of LTTC operations and R be an I-RAM. There is a contant γ (only depending upon R) such that if $Inst_0, Inst_1,...Inst_q,...$ is the sequence of instructions executed in a computation of R, then for each $q\geq1$

$$\sum_{j\leq q} cost(Instj) \leq \gamma . \sum_{j<q} cost(Instj)$$

Proof. Each integer that each instruction $Inst_q$ manipulates either is still involved in some previous instruction $Inst_j$ ($j < q$) or, in case instruction $Inst_q$ is of the form $u := op(v_1,\ldots v_k)$ for an LTTC operation op, is the result u of the operation and satisfies

$$length(u) = O(\Sigma_{i=1,\ldots k}\, length(v_i))$$

where values v_i have been computed in previous instructions $Inst_j$ ($j < q$). Hence we get $cost(Inst_q) = O(\sum_{j<q} cost(Inst_j))$ and the lemma follows immediately. ◆

We are now ready to prove our main lemma which has a variant (stated with "respectively"); the variant has a similar proof as the original version, excepted some details (stated with "respectively").

Main Lemma 5.5. Let d_1, d_2 be constant integers such that $d_1 < d_2$ and let $N: \mathbb{N} \to \mathbb{R}$ be a function such that

$$n / \log n \leq N(n) \leq n^{d_1} \text{ for each } n.$$

Let I be a finite set of LTTC operations and R be an I-RAM such that in each computation of R of time $t \geq 2$, each address is $\leq N(t)$ (resp. each address is $\leq N(t)$ and each register length is $\leq d_2.\log t$).

Then R is simulated by a 2-dimensional {Conc}-RAM (resp. {+}-RAM) R' in time O(t) so that in the simulation

(i) the first component of any address is O(N(t))

(ii) the second component of each address is $O(t / \log t)$

(resp. O(1): R' is essentially a (multimemory) one-dimensional RAM with addresses O(N(t)))

(iii) each register length is $O(\log t)$.

Remark. Note that if $t \geq 2$ then any integer $x \leq N(t)$ has length $\leq d_2.\log t$.

Proof of the remark. We have

$$length(x) \leq length(t^{d_1}) \leq \log(t^{d_1}+1) \leq \log(2t^{d_1}) \leq \log(t^{d_1+1}) \leq \log(t^{d_2}).$$ ◆

Proof of Lemma 5.5. As mentioned above, a great difficulty in the simulation is that t and, consequently, the block length b are not known in advance (they are not assumed to be constructible) and therefore cannot be precomputed. In order to overcome this difficulty, we use a method rather similar to that of [KvLP]: we combine successive b-blocks pairwise (in the simulation of Lemma 5.1), every time the block length b is too small in comparison with time. This gives successive values for b:

$$p_0, p_1, \ldots p_i = 2^i, \ldots$$

with the respective time bounds for R

$$T_0, T_1, \ldots T_i = 2^{2cp_i}, \ldots$$

The dynamic simulation algorithm is given in Figure 3 (it uses the simulation of each instruction given in the proof of Lemma 5.1 and the counter TIME of Remark 5.2). Note that if c is a constant integer such that all the b-tables (i.e. all the tables Tab(...) we have exhibited till now) can be computed in time 2^{cb}.

Simulation algorithm (first version).
{it uses the following procedure where b is a global variable}

PROCEDURE Double_Block_Length;
BEGIN
(D1): $b := 2b$; recompute the b-tables {it requires 2^{cb} time}
(D2): combine successive blocks paiwise for each register of R which has been visited till this moment
END;

BEGIN {Main}
TIME := 0
b := 1 ; compute the b-tables for b = 1
REPEAT
(*1): FOR each address {resp. address or register value} x used as an argument by the current instruction and such that length(x) > 16dcb where $d=d_1$ {resp. $d=d_2$} DO
REPEAT {*1.1} Double_Block_Length
UNTIL length(x) ≤ 16dcb
(*2): simulate the current instruction of R {as in proof of Lemma 5.1}
(*3): add up its execution time (for R) to TIME {cf. Remark 5.2}
(*4): IF TIME ≥ 2^{2cb} - 1 THEN {*4.1} Double_Block_Length
(*5):
UNTIL a Halt instruction in the simulation
END.

Figure 3

Remarks. The algorithm corresponding to the second version is also given in Figure 3. The only difference with the original one consists of a change in instruction labeled (*1).

The reader may ask why the algorithm calls the procedure Double_Block_Length in two cases denoted {*1.1} and {*4.1}:

- *Case {*4.1}, when TIME is too large*: the reason of the call is the hypothesis $t \leq 2^{c_1 b}$ of Lemma 5.1;
- *Case {*1.1}, when an address (resp. a register value) is too long*: we call the procedure because the hypothesis of Lemma 5.1 stipulates that each address must have length ≤ $c_2 b$ (resp. because the number of blocks of each register must always be less than a fixed constant).

Condition TIME ≥ 2^{2cb} - 1 is easy to check since it is equivalent to the following one: the dyadic representation of TIME has ≥ 2cb digits. (Similarly condition length(x) > 16dcb is easy to check since it holds iff x is represented (in R') by more than 16dc b-blocks.)

Note that TIME is a time counter not for RAM R' but for the original RAM R.

Without loss of generality, assume that the time cost of the first instruction of any computation of R is < 2^{2c} - 1 and that the inequality $\gamma \leq 2^c$ holds for the constant γ of Lemma 5.4.

Now in order to analyze the algorithm of Figure 3, we state and prove a long series of claims.

Claim 5.5.1. Each time we pass label (*5) in the simulation (cf. Figure 3), we have TIME < 2^{2cb} - 1.

Proof. By recurrence on the number q≥1 of simulated instructions of R: $Inst_0$, $Inst_1$,...$Inst_q$,...

- It holds for q=1 (b=1 and TIME < 2^{2c} - 1 by the above remark).

- Let $q \geq 2$. Then $\sum_{j<q-1} \mathrm{cost}(\mathrm{Inst}_j) < 2^{2cb} - 1$ by the recurrence hypothesis. The execution of instrution (*1) may increase b but the inequality still holds (a fortiori) for greater b. When we arrive at instruction (*4) we have two cases:

- if $\mathrm{TIME} = \sum_{j<q} \mathrm{cost}(\mathrm{Inst}_j) < 2^{2cb} - 1$, it is done;
- otherwise, from Lemma 5.4 and an above remark, we get

$$\mathrm{TIME} \leq \gamma.(\textstyle\sum_{j<q-1} \mathrm{cost}(\mathrm{Inst}_j)) < 2^c(2^{2cb} - 1) \leq 2^{4cb} - 1$$

which gives (since we update b by $b := 2b$) $\mathrm{TIME} < 2^{2cb} - 1$ when we arrive at label (*5). This concludes the recurrence and the proof of the claim. ◆

Claim 5.5.2. In any execution of the algorithm of Figure 3, the last call of procedure Double_Block_Length by {*1.1} (if any) is followed by at least three calls of this procedure by {*4.1}.

Proof. Assume that the claim is false, i.e. there is a call of {*1.1} Double_Block_Length among the last three calls. Let b_0 denote the block length after the last execution {*1.1}. It is followed by at most two calls of Double_Block_Length. Hence the final block length, denoted b, is such that $b \leq 4b_0$. Of course the final value of TIME is t. From Claim 5.5.1 we get

$$(*) \quad t < 2^{2cb} - 1 < 2^{8cb_0}$$

A necessary condition for the last call of {*1.1} Double_Block_Length was the following

$$\mathrm{length}(x) > 16dc(b_0 / 2) = 8dcb_0$$

and then $\mathrm{length}(x) \geq 8dcb_0 + 1$ (since $8dcb_0$ is an integer). We have

$$\mathrm{length}(x) = \lfloor \log(x+1) \rfloor \leq 1 + \log(x)$$

and then $\log(x) \geq 8dcb_0$. Hence we get from (*) $x > t^d$ (resp. $\mathrm{length}(x) > d \log t$) which contradicts the hypothesis that each address is $\leq N(t) \leq t^d$ (resp. the hypothesis that each register length is $\leq d \log t$). ◆

Claim 5.5.3. We have $2^{cb} - 1 \leq t < 2^{2cb} - 1$ where b denotes the final block length.
Proof. The second inequality is a particular case of Claim 5.5.1. The first one is a consequence of Claim 5.5.2: the last call of Double_Block_Length is of the form {*4.1} and then requires the condition $\mathrm{TIME} \geq 2^{2c(b/2)} - 1$. ◆

Claim 5.5.4. The cumulated time of construction of the b-tables (for block lengths 1,2,...b/2,b where b is the final block length) is O(t).
Proof. This time is

$$2^{cb} + 2^{cb/2} + 2^{cb/4} + \ldots = O(2^{cb})$$

and then is O(t) (by Claim 5.5.3). ◆

Claim 5.5.5. During the simulation the number of (nonzero) b-blocks of each nonzero register of R is O(TIME / log(TIME)) (resp. O(1)), where TIME is the current value of the time counter.

Proof of the first version, i.e. O(TIME / log(TIME)) b-blocks: We conclude from Claim 5.5.1 that each time we pass label (*5) in the simulation, we have for each nonzero register x :

$$\mathrm{length}(x) \leq \mathrm{TIME} < 2^{2cb} - 1.$$

Hence $2cb > \log(\mathrm{length}(x))$. Let Nblocks denote the number of b-blocks of register x. We have

$$\mathrm{Nblocks} < (\mathrm{length}(x) / b) + 1 = O(\mathrm{length}(x) / \log(\mathrm{length}(x))) = O(\mathrm{TIME} / \log(\mathrm{TIME})).$$ ◆

Proof of the second version, i.e. O(1) b-blocks: Let K denote the maximal number of arguments of an operation $op \in I$. Let γ_0 denote a constant integer such that every operation $op \in I$ can be computed on a TM in linear time $\gamma_0.l$ where l is the sum of the arguments lengths. Without loss of generality, assume that the time cost of the first instruction of any computation of R is $\leq \gamma_0 Kdc$.

It is sufficient to prove the following assertion, denoted (Ass), by recurrence on the number $q \geq 1$ of simulated instructions $Inst_0, Inst_1, \ldots Inst_q, \ldots$ of the computation of R:

(Ass) : When the simulation (of Figure 3) passes label (*5), each register x of R is such that $length(x) \leq 16\gamma_0 Kdcb$ (and then has at most $16\gamma_0 Kdc = O(1)$ b-blocks).

- It holds for q=1 (b=1 and $cost(Inst_0) \leq \gamma_0 Kdc$ and then $length(x) \leq \gamma_0 Kdc$ for each register x).

- Let $q \geq 2$ and assume that $Inst_0, Inst_1, \ldots Inst_{q-1}$ have been simulated. By the recurrence hypothesis, each register length was $\leq 16\gamma_0 Kdcb$ when we passed label (*5) just after the simulation of $Inst_0, \ldots Inst_{q-2}$. There are two cases:

1) If instruction $Inst_{q-1}$ uses no operation $op \in I$, then (Ass) still holds a fortiori after the simulation of $Inst_{q-1}$ (when we pass label (*5)) since its execution introduces no new integer (except eventually an integer C if $Inst_{q-1}$ is of the form $u := C$).

2) If $Inst_{q-1}$ is of the form $u := op(v_1, \ldots v_k)$ then when we arrive at label (*2) (just before the simulation of $Inst_{q-1}$), we get

$length(v_i) \leq 16dcb$, for $i = 1, 2, \ldots k$

because instruction (*1) has been executed. A Turing machine computes $op(v_1, \ldots v_k)$ in time $\leq \gamma_0 (\sum_{i=1,\ldots k} length(v_i))$. Hence

$length(op(v_1, \ldots v_k)) \leq 16\gamma_0 Kdcb$

and then after the simulation of $Inst_{q-1}$ we have

$length(x) \leq 16\gamma_0 Kdcb$

for register x=u and for any other register x. This concludes the proof of (Ass) by recurrence and also the proof of Claim 5.5.5. ∎

<u>Claim 5.5.6</u>. The time of the simulation strictly speaking, i.e. the cumulated time of instructions (*2) in the algorithm of Figure 3, is (at each moment when we arrive at label (*4)) O(TIME) where TIME is the current value of the time counter. In particular it is finally O(t).

Moreover conditions (i-iii) of Lemma 5.5 always hold during the execution of instructions (*2). These instructions only apply predecessor and successor operations to arguments $O(b + TIME / \log(TIME))$ where b is the current block length.

<u>Proof</u>. Let C_b denote the part of the computation C of R which is simulated in (*2) with block length b (note that C_b may be empty for some b).

- Let t_b denote the execution time of C_b (for R). We have $t_b < 2^{4cb} - 1$ (from Claim 5.5.1).

- Let L_b denote the maximal register length during C_b. We have

$L_b \leq TIME < 2^{4cb} - 1$ (from Claim 5.5.1)

(resp. $L_b \leq 16dcb$ because of (*1)).

- During C_b each address has length $\leq 16dcb$ (because of (*1)) and is $\leq N(t)$: then it is no more than the integer

$N_b = \min (2^{16dcb+1} - 2, N(t))$.

The hypotheses of Lemma 5.1 clearly hold for C_b, t_b, L_b, N_b (in place of C, t, L,

N, respectively) with $c_1 = 4c$ and $c_2 = 16dc$. By application of Lemma 5.1 we conclude that C_b is simulated by an $O(t_b)$ time bounded b-execution, denoted E_b, of a 2-dimensional {Conc, Succ, Pred}-RAM or by a 2-dimensional {+, Succ, Pred}-RAM so that:

(i) the first component of any address is $O(N_b) = O(N(t))$;

(ii) the second component is $O(L_b / \log(L_b)) = O(t / \log t)$
(resp. is $\leq L_b / b \leq 16\,dc = O(1)$)

(iii) each register length is $O(b) = O(b_0) = O(\log t)$ (use Claim 5.5.3: here b_0 denotes the final block length).

Moreover each time we pass label (*4), the cumulated time of instruction (*2) is

$$\sum_{b'} \mathrm{time}(E_{b'}) = \sum_{b'} O(t_{b'}) = O(\sum_{b'} t_{b'}) = O(\mathrm{TIME})$$

where the sum $\sum_{b'}$ is taken for each $b' = 1,2,4,\ldots b/2,b$ and b is the current block length.

Lastly Lemma 5.1 states that in the execution E_b, predecessor and successor functions are applied to arguments

$$O(b + L_b / \log(L_b)) = O(b + \mathrm{TIME} / \log(\mathrm{TIME})).$$

This concludes the proof of Claim 5.5.6. ∎

Claim 5.5.7. Each execution of instruction (D2) requires time O(TIME) (where TIME is the current value of the time counter) if the list of registers that RAM R has visited till this moment is given by an oracle. This execution uses predecessor and successor operations on arguments $O(\mathrm{TIME} / \log(\mathrm{TIME}))$ (for $\mathrm{TIME} \neq 0$).

Proof. It is an intuitive consequence of Claim 5.5.6. We clearly see that the work of (D2) is essentially to run along the actual visited memory (addresses and contents of visited registers) of the simulating RAM R'. If an oracle gives the list of visited registers (without any additional time cost), it is obvious that (D2) uses no more time than the computation of R' (i.e. the simulation strictly speaking) till the current instant. By Claim 5.5.6 this time is O(TIME).

Clearly, (D2) uses the successor operation to run along the indices of the blocks which has to be combined and the indices of the resulting blocks. These indices are $O(\mathrm{TIME} / \log(\mathrm{TIME}))$ by Claim 5.5.5. ∎

Remark. A combination of two b-blocks either is a concatenation or, for a {+}-RAM, requires the use of special precomputed b-tables alike the tables Tab(W) and Tab(Pair,op) of the proof of Lemma 4.3.

Claim 5.5.8. Procedure Double_Block_Length is executed log(b) times, where b denotes the final block length.
Proof. Obvious. ∎

Claim 5.5.9. The cumulated time of all the executions of procedure Double_Block_Length is O(t) if the list of registers visited by R is given by an oracle.

Remark. this artificial requirement of an oracle is removed in the final version of our simulation algorithm (cf. Figure 4).

Proof of the claim. Because of Claim 5.5.4, we only have to show that the cumulated time, denoted T(combin), required by instructions (D2) is O(t). The argument is rather subtle. Let TIME_1 (resp. TIME_2, TIME_3) denote the value of counter TIME at the last (resp. last but one, antepenultimate) call of Double_Block_Length. We know from Claim 5.5.2 that the three last calls of this procedure are of the form {*4.1}. Then from Claims 5.5.3 and 5.5.1 we get

$$2^{cb} - 1 \leq t = \mathrm{TIME}_1 < 2^{2cb} - 1, \ \mathrm{TIME}_2 < 2^{cb} - 1, \ \mathrm{TIME}_3 < 2^{cb/2} - 1$$

for the block length before the last but one call (resp. antepenultimate call) is b/2 (resp. b/4). By Claim 5.5.8 there are $\log(b) - 3$ calls of the procedure before the three final calls. At each of these calls we have

$$TIME \leq TIME_3 < 2^{cb/2} - 1.$$

Hence we get (from Claim 5.5.7):

$$T(combin) = O((\log b)\, 2^{cb/2}) + O(TIME_3) + O(TIME_2) + O(TIME_1)$$

$$= O(2^{cb}) + O(t) = O(t) \quad \text{(by Claim 5.5.3).}$$ ◆

Claim 5.5.10. There is a constant integer c' such that at each moment of the simulation (of Figure 3), predecessor and successor operations are only applied to arguments less than c'b or less than

$$f(TIME) = c' \lceil TIME / \log(TIME) \rceil$$

where b (resp. TIME) denotes the current value of the block length (resp. of the time counter $\neq 0$).

Proof. These operations are used:

- for the constructions of b-tables (on arguments O(b): cf. proof of Lemma 4.1)
- for instruction (D2) (on arguments $O(TIME / \log(TIME))$: cf. Claim 5.5.7)
- for instruction (*2) (on arguments $O(b + TIME / \log(TIME))$: cf. Claim 5.5.6)
- for the treatment (*3) of the counter array TIME (its indices are $O(\log(TIME))$). ◆

Remark 5.5.11. In order to avoid using predecessor and successor operations, we will use the following procedure, denoted Construct_Succ_Pred(n) , where n is an integer parameter and SuccPred is the procedure of Figure 2 and Lemma 5.3.

```
BEGIN
    compute the integer l = length(n)
    FOR j := 1 TO l - 1 call procedure SuccPred(j)
END;
```

It constructs the array of successor function, denoted SUCC, for each integer of length $\leq l$ = length(n) (and the symetric array, denoted PRED) and, in particular, for integers $\leq n$: it requires time $O(l.2^l) = O(n \log n)$ on a {Conc}-RAM or on a {+}-RAM (by Lemma 5.3).

The strategy for the construction of arrays SUCC and PRED is the following: each time b is doubled (resp. the counter array TIME has more digits) we compute $n = c'b$ (resp. n = f(TIME)) and call Construct_Succ_Pred(n) which constructs SUCC(i) and PRED(i) for each $i \leq n$ within time $O(n \log n) = O(b \log b)$ (resp. = O(TIME)). This time is negligible in comparison with the time 2^{cb} of construction of the b-tables. (resp. The cumulated time is $O(t + t/2 + t/4 + \ldots) = O(t)$.)

Note that the computation of n and of its length trivially requires polynomial time in length(n) = O(log n); it uses the arrays SUCC(i) and PRED(i) for small arguments $i = O(\log b)$ (resp. $i = O(\log(TIME))$) but it is not a problem since SUCC(i) and PRED(i) have been computed before.

End of the proof of Lemma 5.5. In Claims 5.5.7 and 5.5.9 we have assumed that the list of registers visited by R till some moment can always be exhibited (by an "oracle") without any additional time cost, which is not the case: visited registers may be polynomially scattered in the memory.

Now let us remove this assumption by using the following trick: if successive blocks have to be combined pairwise, we can delay this reorganization for each register x of R till x has to be explicitly reused by R: when register x is reused we combine its blocks pairwise (repeatedly if necessary) till its block length be the current b. Of course the algorithm with postponed combination of blocks consumes no more time than the original version with oracle: in fact it may be faster since a register which is never reused is never reorganized. And the algorithm uses no oracle!

The algorithm of Figure 4 below improves the one of Figure 3 by using the previous trick and Remark 5.5.11.

Simulation algorithm (definitive form)

```
TIME := 0
b := 1
compute the b-tables for b=1 and initialize arrays SUCC, PRED for small integers
REPEAT
    (*0): FOR each register x of R used as an argument by the current instruction DO
              WHILE blocks of x have length < b DO
                  combine successive blocks of x pairwise
    (*1): FOR each address {resp. address or register value} x of R used as an
argument by the current instruction and such that length(x) > 16dcb where d=d1 {resp.
d=d2} DO
          REPEAT
              b := 2b
              recompute the b-tables and the SUCC, PRED arrays for integers ≤ c'b
          UNTIL length(x) ≤ 16 dcb
    (*1'): FOR each register x of R used as an argument by the current instruction DO
              WHILE blocks of x have length < b DO
                  combine successive blocks of x pairwise
    (*2): simulate the current instruction of R {as in proof of Lemma 5.1}
    (*3): add up its execution time (for R) to TIME
    (*3'): IF the number of dyadic digits of TIME has just increased THEN
              compute the SUCC, PRED arrays for integers ≤ f(TIME)
    (*4): IF TIME ≥ 2^(2cb) - 1 THEN
          BEGIN
              b := 2b
              recompute the b-tables and the SUCC, PRED arrays for integers ≤ c'b
          END
    (*5):
UNTIL a Halt instruction in the simulation
```

Figure 4

Note that the condition "blocks of x have length < b" of instructions (*0) and (*1') is easy to check since, if x(0), x(1), x(2),... denotes the array of the blocks of x, it is equivalent to the following:

(length(x(0)) < b) and (length(x(1)) > 0).

This last condition and the condition length(x) > 16dcb of (*1) are efficiently checked by using a new precomputed b-table, denoted Tab(length_less_than_b): it is an array of boolean values which indicates, for each integer n such than length(n) ≤ b whether length(n)<b.

Claims 5.5.6 and 5.5.9 (with Remark 5.5.11) prove that our new simulation works in time O(t) on a {Conc}-RAM or on a {+}-RAM. We have seen in Claim 5.5.6 that conditions (i-iii) of Lemma 5.5 are always respected during the simulation strictly speaking (i.e. instructions (*2)). It is easy to show that they still hold during the whole simulation (note that hypothesis $t / \log(t) \leq N(t)$ is used for (i): clearly, the additional memory used for auxiliary computations consists of a fixed number of one-dimensional arrays of indices $O(\log t)$ or $O(t / \log t)$ or $O((\log t)^2)$; cf. for instance the arrays TIME and SUCC and the table Tab(digits) of Remark 5.2, respectively ; table Tab(length) of Remark 5.2 is a special case: its indices are $O(2^b) = O(2^{cb/2}) = O(t^{1/2})$ by Claim 5.5.3 if we assume c≥2; hence (i) is proved; similarly for (iii): we easily see that each of these arrays only has registers of length O(log t); (ii) holds by Claim 5.5.5). This concludes the proof of Lemma 5.5. ∎

Lemma 5.6. Lemma 5.5 still holds with the following additional condition (which improves (iii)):

(iii') each register value is O(N(t)).

Proof. This refined conclusion is obtained by a more careful analysis of the simulations of par. IV and par. V. There are three points:

1) each b-block of a register of R contains an integer of length $\leq b$;

2) there is a constant c_0 (which only depends upon the set I of LTTC operations, i.e. upon the TMs that compute them in linear time) such that each integer involved in the computations of the b-tables has length $\leq c_0 b$ and then is $O(2^{c_0 b})$;

3) the other (auxiliary) registers of R' (i.e. the registers of arrays TIME, SUCC,... and the accumulators) contain either digits (for example: TIME), or integers $O(t / \log t)$ (for example: SUCC), or integers $O(N(t) + t / \log t) = O(N(t))$ (for example: accumulators which contain addresses).

We only have to examine point (2). Without loss of generality assume that $c \geq 2c_0$; hence the integers involved in the b-tables are $O(2^{cb/2}) = O(t^{1/2})$ (cf. Claim 5.5.3). ∎

VI) The results in the deterministic case:

They are consequences of one form or of both forms of Lemmas 5.5 and 5.6. In the following, letter I will denote any finite set of LTTC operations.

Theorem 6.1. Let $N: \mathbb{N} \to \mathbb{R}$ be a function such that for each n,

$$n / \log n \leq N(n) \leq n^{O(1)}.$$

Let R be an I-RAM such that in each computation of R of time t :

- each address is $O(N(t))$;
- each register length is $O(\log(t))$.

Then R is simulated by an ordinary (i.e. one-dimensional) {Conc}-RAM (resp. {+}-RAM) R' in time $O(t)$ so that in the simulation each address or register value is $O(N(t))$ (in particular, each register length is $O(\log t)$).

Proof. It is a reformulation of the second form of Lemmas 5.5 and 5.6. ∎

In case $N(n) = n / \log n$ we can reformulate Theorem 6.1 as follows:

Corollary 6.2. An I-RAM with linearly compact memory is simulated in linear time by an ordinary {Conc}-RAM (resp. {+}-RAM) with strongly linearly compact memory. ∎

Remark. These RAMs can be compared with the "frugal" RAC of [GuSh] (if we study polynomial time, a *RAC* is a RAM with strongly polynomially compact memory; it is *frugal* if at any moment of its computations, the addresses of its visited registers form an initial segment).

Our main corollary is a consequence of the following

Theorem 6.3. Let $N: \mathbb{N} \to \mathbb{R}$ be a function such that for each n,

$$n / \log n \leq N(n) \leq n^{O(1)}$$

and R be an I-RAM using addresses $\leq N(t)$ in each computation of time t.

Then R is simulated by an ordinary {Conc}-RAM (resp. {+}-RAM) R' in time $O(t)$ so that in the simulation the following conditions (1,2) hold:

1) each address is $O(N(t).t.\log t)$ if $N(t) \geq t \log t$ and otherwise is $O((N(t))^2)$;

2) each register length is $O(\log t)$.

Proof. It uses successively both versions of Lemma 5.5. First we get (from the first version) a 2-dimensional {Conc}-RAM R_1 which simulates R in time $O(t)$ so that

(i) the first component of any address is $O(N(t))$;

(ii) the second component is $O(t / \log t)$;

(iii) each register length is $O(\log t)$.

Let Pair denote the following function : $\mathbb{N} \times \mathbb{N} \to \mathbb{N}$:

$$\text{Pair}(x_1, x_2) = \text{rep}(\text{length}(x_1))\text{^}12\text{^}x_1\text{^}x_2$$

(resp. $\text{Pair}(x_1, x_2) = (x_1+x_2)\text{^}z\text{^}i$ where $z = \max(x_1, x_2)$ and $i = 1$ if $z = x_1$, i.e. if $x_1 \geq x_2$ and $i = 2$ if $z \neq x_1$, i.e. if $x_1 < x_2$).

It is easy to check that both functions Pair are pairing functions (for the second one, note that $z \leq x_1+x_2 \leq 2z$ and consequently $\text{length}(z) \leq \text{length}(x_1+x_2) \leq \text{length}(z)+1$; so length(z) is the quotient of the euclidean division of $\text{length}((x_1+x_2)\text{^}z)$ by 2). We also clearly see that they are LTTC operations.

Using the function Pair, we can encode each ordered pair (x_1, x_2) which constitutes an address of a register of R_1 by the integer $\text{Pair}(x_1,x_2)$. Since $x_1 = O(N(t))$ and $x_2 = O(t / \log t)$ we get

$$\text{Pair}(x_1,x_2) = O(x_1.x_2.(\log x_1)^2) = O(N(t).t.\log t)$$

(resp. $\text{Pair}(x_1,x_2) = O(x_1{}^2+x_2{}^2) = O((N(t))^2)$).

So R_1 becomes (i.e. is simulated by) an O(t) time bounded one-dimensional {Conc,Pair}-RAM R_2 which obeys conditions (1,2) of the theorem.

Now let us apply the second form of Lemma 5.5 to R_2. R_2 is simulated in O(t) time by an ordinary {Conc}-RAM (resp. {+}-RAM) R' which fulfils conditions (1,2). ■

Remark. It would be satisfying to have addresses $O(N(t).t / \log t)$ for R': this depends on an improvement of the pairing functions: we would like to get the optimal result: $\text{Pair}(x,y) = O(x.y)$. This result is proved only in the particular case $N(t) = O(t / \log t)$.

Our main result is the following:

Corollary 6.4. An I-RAM R with polynomially compact memory is simulated in linear time by a {Conc}-RAM (resp. {+}-RAM) R' with strongly polynomially compact memory.

Proof. It is an immediate consequence of Theorem 6.3. ■

Corollary 6.5. Corollary 6.4 still holds if the I-RAM R is multidimensional.

Proof. Use an LTTC pairing function (as above) to encode each tuple (which is an address) with one integer. ■

Compare Corollary 6.5 with the following recent result by one of the present authors [Ro] : each multidimensional {+,-}-RAM (without any condition about memory) is simulated in linear time by a one-dimensional {+,-}-RAM.

There is a natural question: do our results depend on the dyadic notation of integers? The following results essentially give a negative answer.

Corollary 6.6. Theorems and corollaries 6.1-5 still hold if integers are represented in b-adic notation, for each $b \geq 2$.

Proof. Note that our results and proofs can be trivially adapted for b-adic notation (in place of dyadic notation). ■

Remark. Concatenation operation is strongly dependent on the (b-adic, b-ary,...) notation. There are as many different concatenations as different notations for integers. At the opposite, the addition of integers does not depend on their notation. Corollary 6.7 below follows from this fact and from Corollary 6.6 (with Theorem 6.1).

Corollary 6.7. Let $N: \mathbb{N} \to \mathbb{R}$ be a function such that $n / \log n \leq N(n) \leq n^{O(1)}$ and let b,b' be integers ≥ 2. Let I be a set of LTTC operations for b-adic notation of integers.

Let R be an I-RAM using addresses and register values $\leq N(t)$. Then there is a set I' of LTTC operations for b'-adic notation of integers and an I'-RAM R' which simulates R in time O(t) by using addresses and register values O(N(t)).

Moreover the assertion holds for I'= $\{\text{Conc}_{b'}\}$ where $\text{Conc}_{b'}$ denotes the concatenation of b'-adic notation. ∎

<u>Remark</u>. The corollary is interesting even in the particular case where $I = \{Conc_b\}$ and $I' = \{Conc_{b'}\}$.

<u>Corollary 6.8</u>. Corollaries 6.4 and 6.5 still hold for binary notation (resp. b-ary notation, for any $b \geq 2$) of integers.
<u>Proof</u>. Assume that b=2 (for instance). We use the following trick: encode each dyadic notation $w \in \{1,2\}^*$ with a binary notation $\pi(w) \in \{0,1\}^*$; function $\pi:\{1,2\}^* \to \{0,1\}^*$ is the homomorphism of free monoids given by the following conditions on generators: $\pi(1) = 10$ and $\pi(2) = 11$. Of course π is injective (if regarded as a map on integers respectively represented in dyadic and binary notation) and, by definition, respects concatenation. Moreover, since

$$\text{binary\_length}(\pi(n)) = 2\ \text{dyadic\_length}(n), \text{ for each integer } n > 0,$$

we have $\pi(n) = O(n^2)$; therefore if Conc and Conc' respectively denote the dyadic and the binary concatenation of integers, then a {Conc}-RAM with strongly polynomially compact memory can be simulated in linear time (by using π) by a {*Conc'*}-RAM with strongly polynomially compact memory. ∎

VII) The nondeterministic case:

VII.a Various interpretations of time complexity of nondeterministic RAMs

VII.a.i Bounded versus unbounded nondeterminism

There are two reasonable interpretations of what is meant by a nondeterministic RAM with operation set I (ND I-RAM for short). *Weak* or *bounded* nondeterminism allows the choice at nondeterministic steps between a finite number of possibilities; without loss of generality we can allow simply a nondeterministic jump which may jump or not. *Strong* or *unbounded* nondeterminism allows choice between an unbounded number of possibilities; this can be provided by a CHOOSE instruction which places in a specified register an arbitrary (non-negative) integer. Weak nondeterminism is the version described by Aho, Hopcroft and Ullman [AHU] and is what is provided by higher level constructs such as guarded commands [Di].

Clearly a weakly nondeterministic program can simulate strong nondeterminism by guessing the bits of the integer to be chosen. However this simulation is non-linear in time because the strongly nondeterministic instruction which CHOOSEs n, say, takes time $\log n$ whereas the simulation of this instruction would take time $\Omega(\log^2 n)$ if done in the obvious way or $\Omega(\log n \log\log n)$ using a divide and conquer approach. Thus it is not clear whether t time complexity classes defined by ND I-RAMs using bounded and unbounded nondeterminism are the same.

VII.a.ii Which computations should a time bound apply to?

Another divergence of interpretation occurs when we ask what does it mean for an ND I-RAM to obey a time bound t. Does this mean that all computations on any input w should halt in time $t(|w|)$? Or should this bound apply only to accepting computations? Many authors have an even broader interpretation, namely that for each accepted input there should exist at least one accepting computation which halts within time $t(|w|)$.

In the case that t is time-constructible, it is easy to see that these interpretations are equivalent as far as definition of time complexity classes is concerned (because a program can compute $t(|w|)$ and halt all computations which take more time). However in this paper we are making no such assumptions about the functions t defining our complexity classes so the interpretations may yield different classes.

VII.a.iii Initialisation of registers

Many authors assume that the contents of each register of a RAM is zero at the start of any computation. As in the deterministic case all the discussions of this section will be completely transparent to this assumption, that is the simulating RAM will rely on this assumption if and only if the simulated RAM does so.

VII.a.iv Outline of results of this section

The results of this section can be summarised by saying that a nondeterministic I-RAM for any set I of LTTC operations can be simulated in linear time by a ND {+}-RAM (resp ND {Conc}-RAM). This will hold for either interpretation of nondeterministic RAM and for each of the three interpretations of the time bound of the ND I-RAM.

To clarify the important ideas of the proof, we will concentrate on the case of a weakly nondeterministic ND I-RAM and time defined by the existence of an accepting computation (for each accepted input) within that time. This is the case of greatest interest as well as being the easiest one to prove. Sections D and E will briefly describe how to extend the proof to the other cases. From here on, unless otherwise stated, we are discussing weak nondeterminism and are interested in showing the existence of a constant c say such that if the original ND I-RAM has an accepting computation in time t, then the simulating ND I-RAM has an accepting computation in time ct.

The methods of sections III to V extend trivially to the case of weak nondeterminism so that the proof would be easy if we had assumed that the ND I-RAM to be simulated had polynomially compact memory. To prove the result in general, we proceed in two stages, lemmas 7.1 and 7.2, of which the first reduces the maximum address used to a low enough level for the methods of sections III to V to be utilised with minor modifications in the second.

VII.b Avoiding very large addresses

VII.b.i Increasing the operation set

In this section we show that an ND I-RAM, with I a set of LTTC operations, can be simulated in linear time by an ND I'-RAM, with I' a (possibly larger) set of LTTC operations, so that the simulation of a computation taking time t uses addresses bounded by $O(t^{\log t})$. To carry out this simulation we introduce the following four LTTC operations:

(a) + : the usual addition operation,

(b) $\times_2$: a multiplication operation guaranteed to give the correct result provided its operands are powers of 2; for other operands we can take its result as 0,

(c) LOG : when applied to a single operand, say x, LOG returns $\lceil \log_2 x \rceil$ (or zero if x is zero),

(d) < : applied to two operands, x and y say, returns 1 if $x<y$ and zero otherwise.

Lemma 7.1

Let R be an ND I-RAM where I is a finite set of LTTC operations. R is simulated in linear time by R^-, an ND (I $\cup$ {+,$\times_2$,LOG,<})-RAM so that, in simulating a computation of R taking time t, R^- uses addresses $O(t^{\log t})$.

Proof

The proof of this lemma takes up the remainder of section VII.b.

VII.b.ii Tables used by R^-

In discussing the data structures used by R^- at a given stage of the simulation, we will make extensive use of $TIME$ the time taken by R up to this point and of i defined as $\lfloor \log_2 TIME \rfloor$. Addresses less than $2^{(i+1)^2}$ will be called *small* and other addresses *large*.

R^- uses a table $SMALL$ in a straightforward way to represent the current value in every register (of R) with a small address; if x is a small address then $SMALL[x]$ (in R^-) contains the same value as register x (in R) provided any value has ever been assigned to that register.

Three tables of R^-, namely $ADDRESS$, $CONTENTS$ and $LINK$ contain the information about all registers of R with large addresses which have ever been assigned a value. If y is such an address, then there will be some y' such that $ADDRESS[y'] = y$ and $CONTENTS[y']$ is the current contents of register y (of R). The indices (such as y') of all these $ADDRESS$–$CONTENTS$ pairs used up to this point in the simulation will be a contiguous set of integers $[1..maxindex]$ where $maxindex$ is held in a register of R^-. All of the current such pairs (that is those whose addresses are still large) are linked into a list in increasing order of $ADDRESS$ by use of the $LINK$ table; this list contains, as well as pairs for addresses used by R a *dummy* pair for each large address of the form 2^{2^j} which is less than an address actually used; this linked list is handled in the usual way, with a register of R^- containing the index of the first entry ($START$) and the last entry having a zero $LINK$.

Another table $TIME$ holds the binary digits of $TIME$ the time taken so far by R; these digits are held in reversed order so that the least significant is at the lowest address.

Finally two tables hold needed values which do not depend on the course of the simulated computation except that the tables are extended and updated when i increases: *TWOSQUARE* holds the value 2^{j^2} for every j up to $i+1$ and *LESS* $[x]$ holds the boolean value $(x<i)$ for x up to $2i$.

VII.b.iii Simulating one instruction of R

The simulation of an instruction of R is straightforward except for three issues, namely simulating accesses to the registers of R, updating the table *TIME* and updating all the other tables whenever *TIME* passes a power of 2.

(a) simulating an access to a register with address x:

This is done in three steps:

(a1) guess whether x is small or large;

(a2) confirm the guess;

(a3) if x is small

then (a31) simulate the access by using *SMALL*

else (a32) simulate the access using *ADDRESS*, *CONTENTS* and *LINK*.

Of these, (a1) is a simple nondeterministic choice. (a2) is easy when x is guessed to be large: R^- simply compares x with *TWOSQUARE* $[i]$ using the $<$ operation; if x is guessed to be small, R^- will guess a j less than or equal to i and check that $x <$ *TWOSQUARE* $[j+1]$. (To guess a j less than or equal to i, compute the successive powers of 2 using $\times_2$ adding them nondeterministically into j, and halting when a power is not less than i or nondeterministically at any point)

(a31) is very simple: R^- uses *SMALL* $[x]$ in whatever way R would have used x. (a32) involves guessing an *index* not greater than *maxindex* (similarly to the guessing of j described above), checking that this is the correct position for x in the list of *ADDRESS*es (that is either [a: this is exactly the position for x] *ADDRESS* [*index*] $= x$, or [b: x should be inserted between this position and its successor] *ADDRESS* [*index*] $< x <$ *ADDRESS* [*LINK* [*index*]] or [c: x should be inserted after the end of the list] *ADDRESS* [*index*] $< x$ and *LINK* [*index*] $= 0$ or [d: x should be inserted before the start of the list] *START* $= 0$ or $x <$ *ADDRESS* [*START*]), inserting an entry for x into the list if necessary [b, c or d] and accessing the entry found or inserted for x. Note that in case [c] and the case of an empty list (*START* $= 0$) extra dummy entries may need to be made into the list for large addresses 2^{2^j} between the current last in the list and x; the number of these can be computed using *LOG LOG* x and each one can be found from its predecessor by applications of $\times_2$.

(b) updating the table *TIME*:

This is done in three steps also:

(b1) calculate the time for this instruction;

(b2) guess and check the digits of this time;

(b3) add the digits found into those stored in *TIME*.

Each of these is straightforward since the operands of the instruction are known and the *LOG* operation is available for (b1) and + and < for (b2) and (b3).

(c) updating all other tables when *TIME* passes a power of 2:

There are four steps to be considered here:

(c1) decide if *TIME* has passed a power of 2 and if so:

(c2) update *TWOSQUARE*;

(c3) update *LESS*;

(c4) move information from the linked list to *SMALL* for all addresses which have now become small, updating *START* as necessary.

The information needed in (c1) will be available as a result of (b3); (c2) and (c3) are simple; (c4) is straightforward since the organisation of the linked list in *ADDRESS* order means that the information on addresses which have become small is contained in an initial segment of the list.

VII.b.iv The time used by the simulation

The linear time property of the simulation depends on operations (a) (excluding the addition of dummy records into the linked list in (a3)) taking a time which is linear in the time taken by the instruction of R being simulated and all other operations up to time *TIME* taking time linear in *TIME*.

(a) simulating an access to a register x:

(a1) takes time $O(1)$;

(a2): If x is large, this step takes time $O(length(x) + i^2)$ which is $O(length(x))$ since x is large. If x is small it takes time $O(length(x))$ provided that the testing of whether a power (used in the guessing of j) is less than i is done using the array *LESS*.

(a31) clearly takes time $O(length(x))$ since the address of $SMALL[x]$ is $O(x)$.

(a32) excluding the addition of dummy entries to the linked list involves guessing *index* less than *maxindex* and therefore certainly less than *TIME* which can easily be done in time $O(\log^2 TIME)$ which is $O(length(x))$ since x is large. After this there are $O(1)$ instructions involving operands which are either $O(x^2)$ or $O(TIME)$ so that the total time is $O(length(x) + \log TIME)$ which is $O(length(x))$ and so linear in the time of the register access being simulated. Note the reliance on the dummy entries in the list here to ensure that $ADDRESS[LINK[index]]$ is $O(x^2)$.

The remaining time for stage (a), namely the insertion of dummy entries into the list is bounded by the observation that the maximum address that can be generated up to time *TIME* is $O(2^{TIME})$ and that there can never be two dummy entries made for the same address. Thus even if all possible addresses of the form 2^{2^j} had dummy entries made, the total time used for them would be $O(\sum_{j=1}^{\log t} 2^j)$ which is $O(TIME)$.

(b) updating the table *TIME*:

(b1) calculating the time for this instruction (say τ) takes time $O(\tau)$ using the LOG operation;

(b2) computing all powers of 2 less than or equal to τ, nondeterministically adding those guessed to be the digits of τ and checking the result takes time $O(\log^2\tau)$;

(b3) adding the digits of τ into the digits of *TIME* in the table *TIME* takes amortised time $\log\tau$.

Thus stage (b) contributes $O(TIME)$ to the simulation time of R up to simulated time *TIME*.

(c) updating all other tables:

(c1) takes time $O(1)$

(c2) when t passes 2^i, *TWOSQUARE* is extended with a new value of $2^{(i+1)^2}$ which can be computed as $2^{i^2} \times_2 2^i \times_2 2^i \times_2 2$ in time $O(i^2)$. Thus the total time spent in these operations up to time t is $o(t)$;

(c3) when *TIME* passes 2^i, *LESS* is extended up to $2i$ with *FALSE* values and $LESS[i-1]$ is set to *TRUE*. This takes time $o(i^2)$ and so again $o(TIME)$ up to time

TIME.

(c4) moving items from the linked list to *SMALL* takes time linear in the time to insert the same items into the linked list plus $O(i^2)$ to recognise the end of the initial segment of the list to be moved. Thus the total time in this step is linear in the time in other steps plus $o(TIME)$.

This completes the proof that the total simulation time is linear in the time taken by the original machine R and so completes the proof of lemma 7.1.

VII.c Avoiding all large addresses

VII.c.i Overview

In this section we will prove a lemma showing that an ND I-RAM with I a set of LTTC operations can be simulated in linear time by an ND {+}-RAM provided a weak condition holds on the addresses used. The RAM shown to exist by lemma 7.1 will satisfy this address condition completing our proof (for the operation set {+}). The method of proof is essentially that of sections III to V with one new idea, the use of a hash function to reduce the space used by the simulation. In this section (VII.c.i) we state the lemma; the proof will take up the remainder of VII.c.

Lemma 7.2

Let R^- be an ND I-RAM with I a finite set of LTTC operations such that there exists a constant $c < 1$ such that in any computation of R^- taking time t, the length of the maximum address used is $O(t^c)$. Then R^- is simulated in linear time by an ND {+}-RAM R^+ such that for at least one successful simulation of any computation of R^- taking time t, the maximum address used by R^+ is $O(t^2)$.

VII.c.ii The hash function

First we show the existence of a hash function which causes no collisions, has a very simple form and takes on only fairly low values.

Lemma 7.3

In any computation taking time t on a logarithmic cost RAM, there exists a number $p < t^2$ such that every pair X and Y of distinct addresses used satisfy $X \neq Y \ (mod\ p)$.

Proof

Consider an arbitrary X used as an address in the computation. Next consider an arbitrary $Y > X$ also so used (necessarily in an instruction taking time $\geq \log_2 Y$ so that the sum over all these Y of $\log_2 Y$ is at most the sum of the time of all instructions namely t).

$$Y - X \leq Y$$

$$\log_2(Y-X) \leq \log_2 Y$$

$$\sum_{\substack{Y\ used \\ Y>X}} \log_2(Y-X) \leq \sum_{Y\ used} \log_2 Y \leq t$$

$$\sum_{X\ used}\ \sum_{\substack{Y\ used\\ Y>X}} \log_2(Y-X) \le \sum_{X\ used} t$$

$$\le t\,(t^{1/2} + \frac{2t}{\log_2 t})$$

(since there can be at most $\frac{2t}{\log_2 t}$ addresses X used which exceed $t^{1/2}$)

$$\text{so} \prod_{\substack{X,Y\ used\\ Y>X}} (Y-X) \le 2^{t(t^{1/2} + \frac{2t}{\log_2 t})}.$$

The product on the left hand side is the product of the absolute values of all differences between distinct addresses used. But the product of all primes less than t^2 exceeds $2^{t(t^{1/2} + \frac{2t}{\log_2 t})}$ for all $t > 6$. (For t from 7 to 17 this can be checked by inspection. For t greater than 17 it follows by a minor modification of Erdős' proof of Bertrand's postulate (see e.g. [Ar]): the product of the primes less than t^2 is at least $2^{t^2}/(t^2)^{1+\pi(t)}$; using the fact that, for t greater than 17, $\pi(t)$ is less than $t/2$, we see that the base 2 logarithm of this product of primes is more than $t^2 - (2+t)\log_2 t$; the conclusion follows since t^2 already exceeds $t\,(t^{1/2} + \frac{2t}{\log_2 t}) + (2+t)\log_2 t$ when $t = 18$ and clearly the ratio of $t\,(t^{1/2} + \frac{2t}{\log_2 t}) + (2+t)\log_2 t$ to t^2 is decreasing.) Hence for $t > 6$ there is a prime $p < t^2$ which is not a factor of the product $\prod_{\substack{X,Y\ used\\ Y>X}}$ and so no two distinct addresses can be equal modulo p. For $t \le 6$ the result is trivial so the lemma is proved.

When it is necessary to compute this hash function x *mod* p for an argument x in the simulation, x will be represented in the form used in lemma 5.5, namely a sequence of blocks of length b (except possibly the last which can be shorter) of digits of the dyadic representation of x. The computation of the hash function starts by computing the same representation (which we call the *b block dyadic* representation) of x *mod* p. To achieve this five tables are used:

BLOCKVALUE $[i,j,k]$ gives the jth block of the b block dyadic representation of $(i \times 2^{bk})$ *mod* p for all triples i,j,k required, namely i up to the number with dyadic representation of b twos, j up to the number of blocks in the b block dyadic representation of p, k less than the number of blocks in the b block dyadic representation of the largest addresses used. This notionally 3 dimensional array is stored in lexicographic order of triples (k,j,i) so that the elements necessary for evaluating x *mod* p are available using a sequential scan through *BLOCKVALUE* in parallel with the scan through the blocks of x. Since the maximum address used is not known in advance, *BLOCKVALUE* may need to be extended when a new address is used; using k as the most significant field in the ordering of elements ensures that this can be done without any need to move the existing elements.

To handle carries as x *mod* p is being accumulated, another table *MOD* is used. *MOD* $[x]$ for any x up to twice the largest dyadic integer of length b contains the value equal to x modulo 2^b and represented by a dyadic number of length exactly b.

As each block of x has its value modulo p added into the sum, the sum has to be reduced by p if it is now p or greater. To determine if it is p or greater, another table *COMPARE* tells, for any i and j, whether the jth block of the representation of p is equal to or greater than or less than i. Reduction of a number by p is achieved by addition of $2^{b \times l} - p$ (where l is the length in blocks of the representation of p) with the carry out of the last block being ignored; the blocks of this integer are held in a table TWOSCOMP; to handle the last block correctly, another table *SHORTMOD* gives, for each dyadic integer of length b, the smallest dyadic integer equal modulo 2^b (which may have length less than b).

Having computed the b block dyadic representation of $x \ mod \ p$, conversion into a single integer is achieved easily with the aid of a table *INTVALUE*. $INTVALUE[i,j] = i \times 2^{bj}$ for all i of length b or less and all j less than the number of blocks in the representation of p.

If, as will be the case in what follows, the number of b blocks in the representation of p is bounded by a constant, the time required to compute $x \ mod \ p$ in this way is linear in $length(x)$.

VII.c.iii The simulation

The essential idea of the simulation of R^- by R^+ is that p is guessed nondeterministically and the simulation proceeds using the hash function given by this p; if p is found to be unsuitable (because a collision is detected in the hash table), the computation halts. Lemma 7.3 guarantees that there is some p for which the computation will successfully simulate that of R^-.

As in lemma 5.5, a dynamic simulation is necessary with a sequence of assumed values of t until a large enough t is used for the computation to terminate. Each t will have its own p and so its own hash function. When a new p is chosen, all information in the old hash table must be transferred to the new one.

The simulation proceeds essentially as in lemma 5.5 with instructions of R^- being simulated by table look up if all operands are single b blocks and by simulation of the Linear Time Turing Machine otherwise. The major difference is in the representation of the contents of the registers of R^-. Every integer has a *representation* consisting of firstly its length in blocks and secondly the block (if its length is 1) or the index in a table *BLOCKS* of the first block; subsequent blocks are simply contained in the subsequent elements of *BLOCKS*. The fact of R^- having value V at address A is simulated by the representations of V and A being stored at locations $A \ mod \ p$ of two tables called *ADDRESS* and *VALUE* which together constitute the hash table. As new multi-block integers are computed, space is allocated for them in *BLOCKS* in the next free elements of that table.

When R^- reads a value from address A, R^+ will compute $A \ mod \ p$ and check whether $ADDRESS[A \ mod \ p]$ contains the representation of A. If it does, then $VALUE[A \ mod \ p]$ contains the representation of the contents of A; if another representation is found, the simulation halts having detected a collision; (if no representation is found, R^- is accessing a register without having assigned a value to it and R^+ can take appropriate action depending on whether R^- is assumed to be allowed to do this).

When R^- writes a value V to an address A, R^+ simply writes the representations of V and A to elements $A \ mod \ p$ of *VALUE* and *ADDRESS* respectively. It seems dangerous to do this without checking for the possibility that a value was already held there for another address A' equal to A modulo p; however either such a value would never be read subsequently, in which case the overwriting is benign, or the collision will be detected when such a read is attempted.

Next we consider the dynamic nature of the simulation, namely the need to increase the block length and the hash table size from time to time depending on the time taken by the simulated machine R^-. We use a procedure Double_Block_Length analogous to that of lemma 5.5. Our choice of block length b is dictated by a need to have

$TIME^c \times 2^b = o(TIME)$ (where $TIME$ is the simulated time so far and c is the constant in the conditions of lemma 7.2). Accordingly we choose an integer n such that $n > 1/(1-c)$ and ensure that $2^b < TIME^{1/n}$ whenever $TIME > 2^n$ by starting with $b = 1$ and doubling b whenever $TIME > 2^{2bn}$.

Double_Block_Length will choose a new p nondeterministically less than 2^{4bn} for the new b and then construct a new hash table with new tables $VALUE$, $ADDRESS$ and $BLOCKS$. To do this it needs to be able to find all the entries in the hash table. This is made possible by two tables, one dealing with hash table entries whose $ADDRESS$ part is a single b block integer and another for multi-block addresses. $SINGLE$ is a boolean table with one entry for each single block integer, simply saying whether this is an address which has been used. $ENTRIES$ is a list of the representations of all multi-block addresses used, in chronological order of their uses. These two tables make it easy (and efficient) to scan through the old hash table and build the new one. (They can also be used to detect hash function collisions which have been undetected up to this point; this is only necessary if R^- depends on the assumption of registers being initialised to zero.)

VII.c.iv The time of the simulation

There are three aspects of the simulation which differ markedly from that of lemma 5.5 and whose influence on the time taken needs discussion; these are the handling of numbers via their representations, the handling of register accesses via the hash table and the maintenance of the various tables.

For the handling of numbers via their representations, the crucial fact is that for single block numbers the value of the block is available directly whereas for a multi-block number, i say, the number of blocks is $O(length(i)/logb)$ and the time taken to access each block is $O(logTIME)$ since consecutive locations in $BLOCKS$ are used. Thus, since $\log t = O(\log b)$, in either case, the time taken to retrieve the number, given its representation, is $O(length)$.

As for register accesses via the hash table, the time taken to compute the hash function has already been discussed (in section VII.c.ii); the writing or reading in the table is clearly linear in the address and value of the simulated register (note the reliance here on the fact that the hash function has values no greater than the corresponding argument); and the updating of tables $SINGLE$ and $ENTRIES$ is similarly linear (again an index in $SINGLE$ is the same as the corresponding single block register address and indices in $ENTRIES$ and $BLOCKS$ have length $O(\log TIME)$ with the number accessed being $O(length(address)/logb))$.

Of the tables which are set up at each call of Double_Block_Length, the largest is $BLOCKVALUE$ with $O(2^b \times length(p) \times length(maximum\ address\ used))$ elements of $O(b)$ bits. Since we know from the hypothesis of the lemma that $length(maximum\ address\ used) = O(t^c)$ and the decision as to when to call Double_Block_Length ensured that the base 2 logarithm of the greatest b used is $O(t^{1/n})$ with $1/n < 1-c$ and $length(p) = O(logt)$, the sum over all b used of the time to set up $BLOCKVALUE$ is $o(t)$. Similar remarks apply to the other much smaller tables MOD, $COMPARE$, $TWOSCOMP$, $SHORTMOD$ and $INTVALUE$ and also the necessary initialisation of $SINGLE$ to $FALSE$.

Finally the moving of a record from the old hash table to the new one by Double_Block_Length takes time linear in the time taken in inserting the record originally. Certainly some records from early in the computation may be moved up to $\log t$ times but the fact that Double_Block_Length is only called with $TIME$ a power of 2 is already sufficient to ensure that the total work done in all these record moving steps is at most twice what could be required at the last one, namely $O(t)$.

We remark that any successful simulation of a computation taking time t will use space $O(t^4)$ (since Double_Block_Length chooses a value of p less than t^4 where t is the current time) but that there will be at least one such simulation (choosing the least suitable p at the last call of Double_Block_Length) whose space requirement is $O(t^2)$.

VII.d Conclusions for bounded accepting computations

We now restate and prove theorem 1.1 for the case of bounded nondeterminism:

Theorem

Let I be a finite set of LTTC operations. A boundedly nondeterministic I-RAM R is simulated in linear time by a boundedly nondeterministic {+}-RAM (resp. {Conc}-RAM) R^- with polynomially compact memory. Moreover the space used in simulating a computation of R taking time t is $O(t^2)$ (for the best simulation) and $O(t^4)$ for every simulation.

Proof

(i) For the operation set {+} the theorem is a trivial consequence of lemmas 7.1 and 7.2 since $length(t^{\log t}) = O(t^c)$ for *any* positive c.

(ii) For the operation set {Conc}, the proof is completed by the observation that the methods of lemma 5.5 can be used to simulate the ND {+}-RAM given by part (i) by an ND {Conc}-RAM.

VII.e Unbounded nondeterminism

With a little more care we can show the same result in the case of unbounded nondeterminism.

Theorem

Let I be a finite set of LTTC operations. An unboundedly nondeterministic I-RAM R is simulated in linear time by an unboundedly nondeterministic {+}-RAM (resp. {Conc}-RAM) R^- with polynomially compact memory.

Proof

The only added problem is in the simulation of unbounded nondeterministic CHOOSE instructions of R. In the analogue of lemma 7.1, this simulation is straightforward: where R guesses an arbitrary integer to be placed in a given register, R^- guesses an arbitrary integer to be placed in the location simulating that register. The difficulty arises in the analogue of lemma 7.2 where we must simulate the guessing of an arbitrary integer by the guessing of the integers representing its blocks; guessing the integers is easy but it must also be verified that they are valid blocks, that is that they have an appropriate length (b for every block except the least significant and at most b for that); simply using a table of lengths would suffice if we assumed that registers were initialised to zero, but without that assumption we cannot rely on the table until we know that the length is at most b.

In the case of the operation set {+}, this is fairly easy. When a block is guessed as an integer, say x, the simulation also guesses two other integers x^+ and x^-. Now if the length of x was required to be b, the simulation checks that the sum of x and x^+ is equal to the largest integer of length b and that x is equal to x^- plus the smallest integer of length b, halting the simulation if not. (The two integers 111...1 and 222...2 of length b are easily precomputed and equality testing between two integers i and j, on a machine with only a test for zero instruction, can be done by using a special table, $COMP$ say, and setting $COMP[i]$ to 1, and then $COMP[j]$ to zero and finally testing whether $COMP[i]$ is zero.) This forces x to be guessed in the correct range in any successful simulation and all the additions, guesses and tests are done in time $O(b)$ as required. If, on the other hand, the length of x is required to be at most b, bounded nondeterminism will be used to choose an integer l at most b and a similar process will be used to check that x lies between the smallest and largest numbers of length l. Provided a table of size b is maintained which holds FALSE for every index except b, this can be done as required in time $O(length(x))$.

The case of operation set {Conc} is more complex. We have found no way of checking that a guessed number has length b except by a method which will produce "insane" computations, that is computations which read registers to which they have never written a value, where a nondeterministic choice is "wrong". This potentially "insane" method is to keep a table, from one call of Double_Block_Length to the next, of how many times each possible number (up to the largest integer of length b) has been chosen, and also how many numbers have been guessed in total. At the next call of Double_Block_Length it can be checked that the total is indeed the sum of the counts for all "correct" numbers. All the necessary arithmetic can be done using the tables SUCC and PRED discussed in section V, since incrementing or decrementing a number (which cannot exceed 2^{2bn}) takes time $O(b)$ and needs to be done $O(1)$ times for each number guessed of length b. Again numbers of length less than b need to be dealt with separately because a time of $\Omega(b)$ would be too great.

VII.f Other time complexity criteria

Thus far we have exclusively considered the time of the best successful simulation, that is the one which has made all the "correct" nondeterministic choices. This suffices to answer the most interesting questions about nondeterministic machines, namely "Is there a computation on input I which terminates in time t?" or "Is input I accepted in time t?". Now we briefly look at the other computations which are relevant to questions such as "Do all computations on input I terminate in time t?" or "Do all accepting computations on I terminate in time t?".

If we limit our attention to successful simulations, we see that the only newly introduced nondeterminism is the choice of p for the hash function in lemma 7.2. The use of nondeterminism in lemma 7.1 is restricted to directly simulating the nondeterminism of the simulated machine and to guessing indexes which must be guessed right for the simulation to proceed. The fact that p is chosen less than 2^{4bn} when $TIME$ first exceeds 2^{bn} ensures that the time for any successful simulation is at most twice the bound stated for the best. On the other hand the bound of $O(t^2)$ on the space used in simulating a time t computation does not hold for all accepting computations unless the simulation actually checks the final value of p and does not accept if it was too large (greater than kt^2 for a well chosen constant k).

The time bound even applies to unsuccessful simulations (of bounded nondeterminism). This is because we have taken a little care whenever a number was guessed to make sure that there was a known upper bound on the number so that any process of guessing a number bit by bit could halt as soon as a power of 2 exceeded this upper bound. The only places where a "wrong" guess might have taken much more time than the right one were at the guessing of an index in lemma 7.1 and the guessing of p in lemma 7.2. The guessing of p has not been discussed in detail but we have already implicitly assumed that the choice of p less than 2^{4bn} is done by $4bn$ steps of computing a power of 2 by doubling the previous power and nondeterministically deciding whether to add this power into a sum; this will take time $O(b^2)$ and so is negligible compared with the remainder of the time taken by Double_Block_Length. For the guessing of an index in lemma 7.1, we have the upper bound of *maxindex* so that similarly correct or incorrect guesses will take time $O(\log^2 maxindex) = O(\log^2 TIME)$. For unsuccessful simulations the space bound is increased to $O(t^4)$ because the final value of p chosen may be up to the fourth power of *TIME* at that point; this bound could be reduced by choosing a new p more often than calls of Double_Block_Length.

These results cannot be extended to all simulations of an unboundedly nondeterministic machine (unless unbounded nondeterminism can be shown to be no stronger than bounded) because any unbounded guess may take unbounded time under the logarithmic criterion. But for the same reason it is impossible to put any time bound on all computations of a machine using unbounded nondeterminism, so that it is vacuously true, if we are implying unbounded nondeterminism and expect time bounds to apply to all computations, to say that if a problem is solved in time t by an ND I-RAM, then it is solved in time $O(t)$ by an ND {+}-RAM or by an ND {Conc}-RAM.

VIII) Conclusion:

We have achieved some linear time simulations which show that the LTTC-RAM is a very robust model of computation. Of course much of the complexity of the proofs came from the fact that we made no assumption of the existence of an easily computable bound on the time taken by the simulated machine and from the classical convention (see [CoRe]) that the input (resp. output) must be read (resp. written) only bitwise on a one-way tape without reset (resp. rewrite).

It has been shown that the operation set {+} (resp. {*Conc*}) is *complete* for reasonable RAMs in the sense that any *reasonable* RAM program is simulated in linear time by one with operation set {+} (resp. {*Conc*}). Here a reasonable RAM is one whose operations are all LTTC. In the deterministic case we also need to assume that the RAM is reasonable in its use of memory; that is that it has polynomially compact memory. In all cases, the simulating machines obey this condition.

It is of great interest to know whether these results can be extended. In the deterministic case it would be more satisfactory to prove the result without the assumption of polynomially compact memory. In fact the first author [Gr3,Gr4,Gr5] has recently proved that any LTTC-RAM can be simulated in linear time by a {+}-RAM (resp. {*Conc*}-RAM) with polynomially compact memory. However the simulation needs the usual assumption that each register content is initially zero (the simulating RAMs *are not sane*); the proof strongly uses the results of the present paper.

In the nondeterministic case, nondeterministic operations (computed in linear time by a nondeterministic Turing machine) are of interest but whether they can be simulated in linear time except by using unbounded nondeterminism seems closely bounded up with the difficult question of whether there is a real difference between the power of the two types of nondeterminism.

We claim that this LTTC-RAM is the *most realistic* model in which to make statements about the time complexity of problems. Of course a claim such as this can never be proved, but can only be justified by appeals to what appears to be reasonable. Our reasons for believing the claim are firstly that the Turing machine model, apparently weaker, is generally accepted to be too weak, secondly that stronger, that is non LTTC, operations such as multiplication and the floor function [St] are generally accepted to be unreasonably strong, simply because they can apparently not be simulated in linear time by the standard operations, and thirdly that our completeness result shows that the

LTTC-RAM, in many forms which have been or might be proposed, continues to define *the same* complexity classes (up to O()). Certainly the restriction to algorithms with polynomially compact memory is not a major restriction in practice. It is hard to imagine an algorithm whose space is not bounded by a polynomial in its time except for one invented just to demonstrate the existence of such pathological cases. Of course for many assertions about complexity, such as that a given problem is polynomial time, the model used is irrelevant; however the LTTC-RAM does seem to *capture* what most complexity theorists accept as reasonable in considering more precise assertions such as *linear* or *quadratic* time, and to capture it in a pleasingly general way.

For practical reasons, one may prefer the more restrictive notion of *linearly compact memory* (the potential memory is of the order of the execution time): Corollary 6.2 asserts that this notion is also very robust. However we are not sure that all the concrete algorithms in the literature (which of course use polynomially compact memory) which may use multidimensional arrays or other complicated data structures can be implemented with linearly compact memory *and* in linear time: as we have seen above, multidimensional arrays can be simulated on one-dimensional RAMs with polynomially compact memory, *but not* with linearly compact memory (to our knowledge). The linear-time decision algorithm for Horn propositional formulas described in [DoGa] and [Mi] needs the preliminary construction of a table which associates to each propositional variable p the list of Horn clauses where p is negative (i.e. is a premise): we do not know how to implement that both in linear time and with linearly compact memory;

Therefore it seems to be very difficult to determine which of the two robust notions of compact memory (*linearly* or *polynomially*?) is the "good" one for a *very* realistic definition of *linear time*. Maybe an intermediate notion would be the best one. For instance [Gr4] states that a (sane or nonsane) LTTC-RAM using integers $O(t(\log t)^{O(1)})$ can be simulated by a sane {+}-RAM (resp. sane {Conc}-RAM) using integers $O((t/\log t)(\log t)^{\varepsilon})$ for any positive number ε.

Regardless of the question of whether one accepts the LTTC-RAM as the most realistic model, the results proved here will be very useful in proofs that, for instance, some problem has linear time complexity on the standard RAM. The ability to use arbitrary LTTC operations together with the ability to use multidimensional arrays will make many such proofs easier

References.

[AHU] A.V. AHO, J.E. HOPCROFT and J.D.ULLMAN, The design and analysis of computer algorithms, Addison-Wesley, Reading, MA, 1974.

[AnVa] D. ANGLUIN and L. VALIANT, Fast probabilistic algorithms for hamiltonian circuits and matchings, J. Comput. System Sci. 18 (1979), pp. 155-193.

[Ar] R.G. ARCHIBALD, Introduction to the theory of numbers, Merrill, 1970.

[CoRe] S.A. COOK and R.A. RECKHOW, Time bounded random access machines, J. Comput. System Sci. 7 (1973), pp. 354-375.

[DoGa] W.F. DOWLING and J.H. GALLIER, Linear-time algorithms for testing the satisfiability of propositional Horn formulas, J. Logic Prog. 3 (1984), pp. 267-284.

[Di] E.W. DIJKSTRA, Guarded commands, nondeterminacy and the formal derivation of programs, Comm. A.C.M. 18 (1975), pp. 453-457.

[vEB] P. Van EMDE BOAS, Space measures for storage modification machines, Inf. Proc. Letters, 30 (1989), pp. 103-110.

[Grae] E. GRAEDEL, On the notion of linear time, Proc. 3rd Italian Conf. Theoret. Comput. Sci., Mantova 1989, World Scientific Publ. Co., 323-334 (also to appear in a Special Issue of Internat. Journal of Foundations Comput. Sci.).

[Gr1] E. GRANDJEAN, Universal quantifiers and time complexity of random access machines, Math. Syst. Th. 18, (1985), pp. 171-187.

[Gr2] E. GRANDJEAN, A nontrivial lower bound for an NP problem on automata, SIAM J. Comput. 19 , (1990), pp. 438-451.

[Gr3] E. GRANDJEAN, RAMs with polynomially compact memory are efficiently simulated by RAMs with almost linearly compact memory, Abstracts of A.M.S. 90T-68-33 Issue 67 vol. 11 n. 2 (March 1990) p. 238.

[Gr4] E.GRANDJEAN, RAMs can be simulated in linear time by RAMs with compact memory, Abstracts of A.M.S. 90T-68-146 Issue 70 vol. 11 n. 4 (August 1990) p. 357.

[Gr5] E. GRANDJEAN, Invariance properties of RAMs and linear time, Technical Report L.I.U.C. Univ. Caen FRANCE 90-11 (december 1990).

[GrRo] E. GRANDJEAN and J.M. ROBSON, RAM with compact memory, Abstracts of A.M.S. 90T-68-34 Issue 67 vol. 11 n. 2 (March 1990) p. 238.

[Gu] Y. GUREVICH, Kolmogorov machines and related issues: The column on logic in computer science, Bull. EATCS 35 (1988), pp. 71-82.

[GuSh] Y. GUREVICH and S. SHELAH, Nearly linear time, in Meyer Taitslin (Eds.), Springer-Verlag Berlin, 1989, LNCS 363, pp. 108-118

[HoUl] J.E. HOPCROFT and J. D. ULLMAN, Introduction to automata theory, languages and computation, Addison-Wesley, Reading, MA, 1979.

[KvLP] J. KATAJAINEN, J. van LEUWEN and M. PENTTONEN, Fast simulation of Turing machines by random access machines, SIAM J. Comput. 17 (1988), pp. 77-88.

[KoUs] A.N. KOLMOGOROV and V.A. USPENSKY, On the definition of an algorithm, Uspekhi Mat. Nauk 13:4 (1958), pp. 3-28 (Russian) or AMS translation, ser. 2, vol 21 (1963), pp. 217-245.

[Mi] M. MINOUX, LTUR: a simplified linear-time resolution algorithm for Horn formulae and computer implementation, Inf. Proc. Letters 29 (1988) pp. 1-12.

[Ro] J.M. ROBSON, Random access machines with multi-dimensional memories, Inf. Proc. Letters 34 (1990) pp. 265-266.

[St] A. SCHMITT, On the computational power of the floor function, Inf. Proc. Letters 14 (1982), pp 1-3.

[Sr] C.P. SCHNORR, Satisfiability is quasilinear complete in NQL, J. Assoc. Comput. Mach., 25 (1978), pp. 136-145.

[Se1] A. SCHÖNHAGE, Storage modification machines, SIAM J. Comput. 9 (1980), pp. 490-508.

[Se2] A. SCHÖNHAGE, A nonlinear lower bound for random access machines under logarithmic cost, J. Assoc. Comput. Mach. 35 (1988), pp. 748-754.

[SlvEB] C. SLOT and P. van EMDE BOAS, The problem of space invariance for sequential machines, Inform. and Comput. 77 (1988), pp.93-122 .

[Wi] J. WIEDERMANN, Deterministic and nondeterministic simulation of the RAM by the Turing machine, in Proc. IFIP Congress 83, R.E.A. Mason ed., North Holland, Amsterdam 1983, pp. 163-168.

[Wi2] J. WIEDERMANN, Normalizing and accelerating RAM computations and the problem of reasonable space measures, TR OPS-3 / 1990 (June 1990), Dpt of Programming Systems, Bratislava, Czechoslovakia.

Randomness and Turing Reducibility Restraints

KAROL HABART

ABSTRACT. A definition of random sequences equivalent to the one of Martin-Löf and Schnorr motivated by the hierarchy of Turing reducibility restraints is introduced and compared with different similarily obtained notions.

Randomness is a phenomenon that has been found fascinating and often also obscure for ages. Nowadays it is described by means of a probabilistic theory based on measure theory. Though, this approach, which is due to Kolmogorov, replaced an earlier attempt to formalize randomness — Richard von Mises introduced a theory, where *random sequence* was a fundamental concept (see [**9**]). It was in the 30th when the mathematical community rejected this approach and voted for Kolmogorov's formalization.

However, there are some problems concerning randomness which Kolmogorov's theory do not cope with properly, e.g. just the question, how to characterize random sequences. For such reasons, and also because Richard von Mises' approach is of constructivistic flavour, there are contemporarily attempts to develop probability theory in his spirit (see [**9**]). Besides, Kolmogorov succeeded later in introducing a measure for the inherent quantity of randomness (of information) in (finite) strings. This notion, referred to as Kolmogorov complexity, became a widely and successfully used tool in various branches of computer science, especially in complexity theory at proving lower bounds (see [**4**]).

The intuition behind the notion "random sequence" is quite obvious — e.g. the sequence of 0's and 1's put down as the result of repeated coin tossing is random. Everybody feels, that a sequence beginning with 0110100101011 ... might be random, but a recursive sequence 0101010101010101 ... hardly. But according to classical probability theory one cannot distinguish between recursive and nonrecursive sequences! What should the requirements on a random sequence be?

The first naive attempt might be: x is a random sequence iff $P(x)$ holds for every property P such that

$$\mu\{y : P(y)\} = 1. \tag{1}$$

But no such sequence exists! (Take e.g. the properties $y \neq c$ for fixed sequences c.) The way how to overcome this obstacle is to restrict something in (1). E.g. Lutz (see [**7**]) affects the measure in defining his notion "rec-random". Though, the most widely accepted definition of random sequences is due to Martin-Löf (see [**8**]), who restricts the class of properties in (1) to the so called "recursive sequential tests".

Later Schnorr (see[**11**]) gave a definition of random sequences equivalent to the one of Martin-Löf using a variant of Kolmogorov complexity. One great advantage of this definition is the absence of any notion from classical theory of probability.

The main result of this paper is another equivalent definition of random sequences. It uses "incompressibility of sets", a notion induced by the hierarchy of Turing reducibility restraints: In [**5**] a hierarchy of recursive functions was introduced with respect to their role as oracle-restraints (the restraints are called "bounds" in that paper) in the wtt-reduction of sets (f is a restraint in a reduction if it majorizes the use function in that reduction) as follows: f is *below* g in the hierarchy, denoted $f \ll g$, iff each subset of ω (the natural numbers) that is weak truth-table reducible to an oracle via restraint f must be weak truth-table reducible to an oracle via restraint g, too. A function f is maximal in this hierarchy, if every subset of ω can be wtt-reduced via f. Now $A \subseteq \omega$ is incompressible with respect to this hierarchy if it can be reduced only via a maximal function. Subsets of ω correspond in a natural way to infinite sequences of 0's and 1's. Thus a variant of the above mentioned incompressibility is proved to be equivalent to randomness as defined by Martin-Löf and Schnorr (Theorem 2) and it is shown, that the original notion of incompressibility is (in a sense) much weaker (Theorem 4).

What is this further definition of randomness good for?

Firstly, the advantage of the framework proposed here lies in his uniformity. In the Martin Löf-Schnorr framework there is non-uniformity due to the effect that for each length there exists a distinct descriptive program for the initial segment of this length and these programs do not need to have any kind of coherence at all. In the system presented here the oracles are defined for the entire sequence once for all, and the oracle-restraints become some sort of complexity measure for the Turing reduction implemented by the oracle machine.

Secondly, the concept of defining classes of random sequences by presenting a reducibility along with a class of oracle-restraints provides a uniform way to define several types of randomness. Theorem 2 is an argument for calling them "randomness" types. This is of particular interest when we seek for recursive sequences which are random in some sense (the random sequences according to Martin-Löf and Schnorr are nonrecursive). Such sequences are referred to as "pseudorandom". E.g. instead of Turing reducibility some resource bounded reducibility could be investigated. Pseudorandom sequences play an important role in connection with probabilistic algorithms (see [7]).

Thirdly, the hierarchy of reducibility restraints naturally induces a hierarchy of randomness, i.e. we can not only distinguish between random and nonrandom sequences, but we can also express things like "more or less random". Moreover, according to Theorem 2 this is in some sense a natural refinement of a widely accepted notion of randomness. This hierarchy motivates the development of a nonclassical probability theory, where e.g. things like incomparable probabilities may occur.

However, the aim of this paper is quite modest: to analyse the correlation between random sequences and Turing reducibility restraints.

⋆ ⋆ ⋆

Throughout of this paper ω^ω denotes the set of functions from ω to ω, $\mathcal{R}$ denotes the recursive functions; $f \lceil A$ means the restriction of f to the domain A; $f[A]$ denotes the image of A under f , $f^{-1}[A] = \{x : f(x) \in A\}$; $|A|$ denotes cardinality of A, 2^A the power set of the set A; $2^{<\omega}$ denotes finite sequences of 0's and 1's; if σ is a finite string (i.e. ranging over $2^{<\omega}$) we use the length function $\mathrm{lh}(\sigma) = \mu x[x \notin \mathrm{dom}\,\sigma] = |\,\mathrm{dom}\,\sigma|$. We use $\sigma\hat{}\rho$ for the concatenation of the strings σ and ρ.

We shall identify sets with their characteristic functions. Let $[0, n) = \{x \in \omega : x < n\}$ and for $A \in 2^\omega$ and natural numbers x, y with $x \leq y$ we define $A[x, y)$ to be the string σ of length $y - x$ where $\sigma(i) = A(x + i)$. Let $< e >^B$ denote the partial recursive function with index e relative to the oracle B. The value $u_e^B(x)$ of the use function in the reduction of $< e >^B$ to B at argument x equals the maximum number used in the computation if $< e >^B (x)$ converges. Let $< e >_f^B$ denote $< e >^B$ if the function f majorizes the use function in the reduction of $< e >^B$ to B, $\emptyset$ otherwise. Let

$$\mathcal{S}(f) = \{S \subseteq \omega : (\exists B \subseteq \omega)(\exists e \in \omega)[S =< e >_f^B]\},$$

i.e. $\mathcal{S}(f)$ denotes the set of all subsets of ω which are Turing reducible to an oracle via restraint f. We want to point out here, that $A \in \mathcal{S}(f)$ is equivalent to "A is f-compressible" as defined in [**6**].

We can put up our main notion now:

Def.1. Let $\mathcal{C}$ be a subset of ω^ω. A subset A of ω is called *$\mathcal{C}$-incompressible* iff

$$(\forall f \in \mathcal{C})[A \in \mathcal{S}(f) \longrightarrow \mathcal{S}(f) = 2^\omega].$$

Note. The notion "incompressibility" in earlier paragraphs was just $\mathcal{R}$-incompressibility according to Def.1. We shall use "incompressibility" for "ω^ω-incompressibility" and "$\mathcal{C}$-compressible" instead of "not $\mathcal{C}$-incompressible" in the sequel.

The following two lemmata are, with minor changes, consequences of the results in [**5**]:

Lemma 1. *For every $f \in \omega^\omega$ holds:*

$$\mathcal{S}(f) = 2^\omega \longrightarrow \sup_{x \in \omega}(|f^{-1}[[0,x)]| - x) < \infty.$$

Proof. The proof is based on a counting argument using the obvious fact that

$$(2) \qquad |\{< e >_f^B \lceil f^{-1}[[0,x)] : B \in 2^\omega\}| \leq 2^x \quad (x, e \in \omega).$$

Let the supremum be infinite. Then there are strictly growing sequences $\{x_n\}_{n=0}^\infty$ and $\{y_n\}_{n=0}^\infty$ of natural numbers so that

$$(3) \qquad |f^{-1}[[0,x_0)] \cap [0,y_0)| > x_0$$

and

$$(4) \qquad |f^{-1}[[0,x_{n+1})] \cap ([0,y_{n+1}) \setminus [0,y_n))| > x_{n+1} \quad (n \in \omega).$$

We shall construct a sequence $C \in 2^\omega$ so that for each $B \in 2^\omega$ C does not coincide with $< 0 >_f^B$ on $f^{-1}[[0,x_0)] \cap [0,y_0)$ and it does not coincide with $< n+1 >_f^B$ on $f^{-1}[[0,x_{n+1})] \cap ([0,y_{n+1}) \setminus [0,y_n))$. Let

$$\begin{aligned} A_0 &= \{\sigma \in 2^{<\omega} : \operatorname{lh} \sigma = y_0 \wedge \\ &(\exists B \in 2^\omega)(\forall z \in f^{-1}[[0,x_0)] \cap [0,y_0))[\sigma(z) = < 0 >_f^B (z)]\}. \end{aligned}$$

Because of (2) and (3) $|A_0| < 2^{y_0}$ so that there is a string $\rho \notin A_0$ of length y_0. Put $C\lceil[0, y_0) = \rho$.

Now let $C\lceil[0, y_n)$ be defined. Let for each $n \in \omega$

$$\begin{aligned} A_{n+1} = \{\sigma \in 2^{<\omega} : \operatorname{lh}\sigma = y_{n+1} - y_n \wedge \\ (\exists B \in 2^\omega)(\forall z \in f^{-1}[[0, x_{n+1})] \cap ([0, y_{n+1}) \setminus [0, y_n))) \\ [\sigma(z - y_n) = < n+1 >_f^B (z)]\}. \end{aligned}$$

Because of (2) and (4) $|A_{n+1}| < 2^{y_{n+1}-y_n}$ so that there is a string $\rho \notin A_{n+1}$ of length $y_{n+1} - y_n$. Put $C(y_n + i) = \rho(i)$ $(i < \operatorname{lh}\rho)$. We thus defined $C\lceil[0, y_{n+1})$.

It is obvious that $C \notin \mathcal{S}(f)$. This contradicts $\mathcal{S}(f) = 2^\omega$. **q.e.d.**

Lemma 2. *If $f \in \mathcal{R}$ then also*

$$\sup_{x\in\omega}(|f^{-1}[[0, x)]| - x) < \infty \longrightarrow \mathcal{S}(f) = 2^\omega.$$

Proof. Let $\sup_{x\in\omega}(|f^{-1}[[0, x)]| - x) = d - 1$. We define a recursive function g by induction as follows:

(5) $\quad g(0) = d + f(0)$

(6) $\quad g(x+1) = \max([0, 1 + d + f(x+1)) \setminus g[[0, x+1)]) \quad (x \in \omega).$

In order to have g well defined we need to show for every $x \in \omega$:

$$[0, 1 + d + f(x)) \setminus g[[0, x)] \neq \emptyset.$$

Assume $[0, 1 + d + f(x)) \setminus g[[0, x)] = \emptyset$ for some x and let it be the least such x. Then $g[[0, x)] \cap [0, 1 + d + f(x)) = [0, 1 + d + f(x))$. Choose y maximal with $g[[0, x)] \cap [0, y) = [0, y)$. Obviously $y \geq 1 + d + f(x)$.

Claim. *For each $z < x$ if $g(z) < y$ then $d + f(z) < y$.*

Proof. Assume $g(z) < y$ and $d + f(z) \geq y$. Owing to (5) $z \neq 0$. It is $y \in [0, 1 + d + f(z))$ and because by (6) $g(z) = \max([0, 1 + d + f(z)) \setminus g[[0, z)])$ we have $y \in g[[0, z)]$. Then $y \in g[[0, x)]$, too. Because $g[[0, x)] \cap [0, y) = [0, y)$ is assumed, it is $g[[0, x)] \cap [0, 1 + y) = [0, 1 + y)$. This contradicts the choice of y.

Because for each $z < x$ by (6) $g(z) \notin g[[0,z)]$, g is injective on x and from $[0,y) \subseteq g[[0,x)]$ follows that $|g^{-1}[[0,y)]| = y$. Hence, because by our claim $g^{-1}[[0,y)] \subseteq f^{-1}[[0,y-d)]$, $|f^{-1}[[0,y-d)]| \geq y$. Then $|f^{-1}[[0,y-d)]| - (y-d) \geq d$. This contradicts the fact that $d-1$ is the supremum in $\{|f^{-1}[[0,x)]| - x : x \in \omega\}$. Thus g is well defined and injective. Moreover,

$$(7) \qquad g(x) \leq f(x) + d \qquad (x \in \omega)$$

is immediate from the definition of g.

Let $A \subseteq \omega$ be arbitrary. Choose $\sigma \in 2^{<\omega}$ so that $\mathrm{lh}\,\sigma = d$ and $\sigma(g(x)) = 1$ iff $g(x) < d$ and $x \in A$. Choose $B \subseteq \omega$ so that $g(x) - d \in B$ iff $g(x) \geq d$ and $x \in A$. This is all possible because of the injectivity of g.

Finally choose $e \in \omega$ so that

$$< e >^S (x) = \begin{cases} \sigma(g(x)) & \text{if} \quad g(x) < d \\ S(g(x) - d) & \text{if} \quad g(x) \geq d. \end{cases}$$

Then by (7) $A = < e >^B_f$. **q.e.d.**

We shall use, as already mentioned, Schnorr's standard characterization of random sequences, as given e.g. in [**1**],[**2**],[**3**] and [**9**]. This characterization uses Chaitin's modification of Kolmogorov complexity. We shall introduce this notion now:

Def.2. A *prefix algorithm* is a partial recursive function[1] $P : 2^{<\omega} \to 2^{<\omega}$ which has a prefix-free domain, i.e. for every $\sigma, \rho \in \mathrm{dom}\,P$ neither $\sigma \subset \rho$, nor $\rho \subset \sigma$ ($\subset$ means strict inclusion).

There exists a recursive enumeration of the set of prefix algorithms. Thus we may define a universal prefix algorithm U specified by the requirement that on inputs of the form $\rho = 0^e\hat{\ }1\hat{\ }\sigma$ (i.e. a sequence of e zeroes followed by a one, followed by a string σ), U simulates the action of the e-th prefix algorithm in a standard enumeration on input σ.

Def.3. Let U be a universal prefix algorithm. The *complexity* (also called *information*) $\mathrm{I}(\sigma)$ of $\sigma \in 2^{<\omega}$ is defined to be

$$\mathrm{I}(\sigma) = lh(\rho) \quad \text{if } \rho \text{ is the shortest input such that } U(\rho) = \sigma.$$

The following easy lemma will be needed later on:

---

[1] Fix a coding of the elements of $2^{<\omega}$ by natural numbers. Then a function on $2^{<\omega}$ is partial recursive iff so is the corresponding function on the codes.

Lemma 3. *There is a constant $c \in \omega$ so that for arbitrary $A \in 2^\omega$ and $x, y \in \omega$, $x \leq y$ holds:*

$$\mathrm{I}(A[x,y)) \leq \mathrm{I}(A[0,y)) + 2\log x + c.$$

Proof sketch. One needs only $2\log x + 1$ bits to code the number x in a prefix-free coding ($0^{\log x}\hat{\ }1\hat{\ }\rho_x$ where ρ_x is the dyadic code of x) and from a program that computes $A[0,y)$ and from the number x one gets $A[x,y)$ by an algorithm of constant length. **q.e.d.**

The definition of random sequences according to Martin-Löf uses the notion of recursive sequential test. The equivalent definition given by Schnorr in [**11**] (Theorem 3) is:

Theorem 1. $A \in 2^\omega$ *is random (with respect to the Lebesgue measure λ) iff*

$$(8) \qquad (\exists m)(\forall n)[\mathrm{I}(A[0,n)) > n - m].$$

The main result of this paper is the following theorem:

Theorem 2. *Let M be the set of all non-descending functions in ω^ω. Then for every $A \in 2^\omega$ holds:*

$$A \text{ is } M\text{-incompressible} \quad \longleftrightarrow \quad A \text{ is random.}$$

Proof. "$\longrightarrow$" Assume that A is not random. Then according to Theorem 1

$$(\forall m)(\exists n)[\mathrm{I}(A[0,n)) \leq n - m,$$

i.e. there is a strictly growing sequence $\{n_k\}_{k=0}^{\infty}$ so that $n_0 = 0$ and

$$\mathrm{I}(A[0,n_{k+1})) < n_{k+1} - 2\log n_k - c - n_k - 1 \qquad (k \in \omega)$$

where c is taken from Lemma 3. According to Lemma 3 then

$$(9) \qquad \mathrm{I}(A[n_k, n_{k+1})) \leq n_{k+1} - n_k - 1.$$

Now let σ_k be the shortest input for U computing $A[n_k, n_{k+1})$. Let B be the infinite string gained by the concatenation of strings $\sigma_0\hat{\ }\sigma_1\hat{\ }\sigma_2\ldots$. Because the σ_k are in the prefix-free domain of U, there is a reduction of

A to B where at computing the values in $A[0, n_k)$ only $\sigma_0\hat{}\ldots\hat{}\sigma_k$ from B is used. At argument x the algorithm works as follows: It applies U to the oracle (an infinite string). After a termination (the inputs of U are prefix-free) a string ρ_0 was computed. Now if $\mathrm{lh}\,\rho_0 \leq x$, U is applied to the rest of the oracle (i.e. the string $\sigma_1\hat{}\sigma_2\hat{}\ldots$). Thus a sequence $\{\rho_k\}_{k=0}^{n}$ of strings is produced until $\sum_{k=0}^{n} \mathrm{lh}\,\rho_k > x$ and $\rho_n(x - \sum_{k=0}^{n-1} \mathrm{lh}\,\rho_k)$ is given as output. If at some intermediate stage U does not terminate, the oracle function at argument x is not defined.

If this oracle machine is applied to the oracle B, the produced ρ_k are just the $A[n_k, n_{k+1})$ and the use function f of this reduction at argument $n_k - 1$ is obviously $(\sum_{i \leq k} \mathrm{lh}\,\sigma_i) - 1$ whence by (9) $f(n_k - 1) \leq n_k - k - 1$. According to the monotonicity of f and Lemma 1 then $\mathcal{S}(f) \neq 2^\omega$. This contradicts the M-incompressibility of A.

"$\longleftarrow$" Assume A is M-compressible. Then $A = < e >_f^B$ for some $B \in 2^\omega$, $e \in \omega$ and $f \in M$ with $\mathcal{S}(f) \neq 2^\omega$.

Claim. $\sup_{n \in \omega}(n - f(n)) = \infty$.

Proof. Assume $n \leq f(n) + d$ $(n \in \omega)$ for some d. Then for arbitrary $C \in 2^\omega$ we can define an oracle D by $D(i) = C(i + d)$ $(i \in \omega)$ and an obvious algorithm with index e' satisfying

$$< e' >^S (x) = \begin{cases} C(x) & \text{if } x < d \\ S(x) & \text{elsewhere,} \end{cases}$$

so that $C = < e' >^D_{\lambda n\,(n-d)}$, i.e. $\mathcal{S}(f) = 2^\omega$. Contradiction.

By this claim then

(10) $$(\forall m)(\exists n)[n \geq f(n) + 2m].$$

Given a natural number m, let e'' be the index of an oracle machine that subsequently computes the values $A(i) = < e >^B (i)$ for $i = 0, \ldots, n$ until the smallest n is found so that the use function of the reduction of A to B at value n is less or equal to $n - 2m$. Thus according to (10), for each $m \in \omega$ there is $n_m \in \omega$ so that the string $A[0, n_m + 1)$ is produced from the number m and the string $B[0, n_m - 2m + 1)$ by a uniform algorithm. Because m has a code shorter than $m + 1$ in a prefix-free coding (e.g. $0^m\hat{}1$), $\mathrm{I}(A[0, n_m + 1)) \leq n_m - m + d$ for a suitable constant d. This contradicts the randomness of A. **q.e.d.**

Corollary 1. *There is $S \in 2^\omega$ so that*

$$\mathrm{I}(S[0,n)) \geq n \qquad (n \in 2^\omega). \tag{11}$$

Proof. Assume there is no such S. Let A be random (such strings exist, see e.g. [9]). Define a sequence $\{n_k\}_{k=0}^{\infty}$ as follows: $n_0 = 0$ and $n_{k+1} = \min\{n : \mathrm{I}(A[n_k, n)) < n - n_k\}$. This sequence is well defined, because if for some k $\{n : \mathrm{I}(A[n_k, n)) < n - n_k\} = \emptyset$, then S defined by $S(x) = A(x + n_k)$ $(x \in \omega)$ would satisfy (11). Now according to the first part of the proof of Theorem 2 if there is a sequence $\{n_k\}_{k=0}^{\infty}$ with $\mathrm{I}(A[n_k, n_{k+1})) \leq n_{k+1} - n_k - 1$ (see (9)), then A is M-compressible. This contradicts the randomness of A. **q.e.d.**

We want to compare M-incompressibility with R-incompressibility now. First we shall give a characterization of R-incompressibility by means of the complexity I similar to the characterization of random sequences. It will be an "effective" version of (8). Then we shall make a comparison.

Theorem 3. *For every $A \in 2^\omega$ holds: A is R-compressible iff there is a recursive function g so that*

$$(\forall n \in \omega)[\mathrm{I}(A[0, g(n))) \leq g(n) - n]. \tag{12}$$

Proof. "$\longrightarrow$" If A is R-compressible, then for some recursive function f and $B \in 2^\omega$, $e \in \omega$ we have $A = \langle e \rangle_f^B$ and $\mathcal{S}(f) \neq 2^\omega$. According to Lemma 2 for each $y \in \omega$ there is $x \in \omega$ so that

$$|f^{-1}[[0, f(x)+1)] \cap [0, x+1)| \geq f(x) + y$$

and, moreover, we can effectively find the least such x. There is thus a recursive function h such that

$$(\forall n \in \omega)[|[0, h(n)+1) \cap f^{-1}[[0, f(h(n))+1)]| \geq f(h(n)) + n + 1]. \tag{13}$$

Now, having a number n, using the oracle algorithm with index e we can compute $A \lceil [0, h(2^n)+1)$ from $B \lceil [0, f(h(2^n))+1)$ and a string σ coding the values of A at $x \leq h(2^n)$ with $f(x) > f(h(2^n))$. Because n has a code shorter than $n+1$ in a prefix-free coding, for a suitable constant d holds then

$$\mathrm{I}(A \lceil [0, h(2^n)+1)) \quad \leq \quad f(h(2^n)) + 1 + n + \mathrm{lh}\,\sigma + d.$$

Because by (13) $\mathrm{lh}\,\sigma \le h(2^n) - f(h(2^n)) - 2^n$ we infer that

$$\mathrm{I}(A \lceil [0, h(2^n)+1)) \quad \le \quad h(2^n) - (2^n - n - 1 - d)$$

whence, if we put $g(n) = h(2^{n+k}) + 1$ $(n \in \omega)$, where k is suitably large, we get (12).

"$\longleftarrow$" Let $A \in 2^\omega$ and g be a recursive function satisfying (12). Then there is a strictly growing recursive function h so that

$$(14) \qquad \begin{array}{rcl} \mathrm{I}(A[0, h(0))) & \le & h(0) \\ \mathrm{I}(A[0, h(n+1))) & \le & h(n+1) - h(n) - 1 \qquad (n \in \omega). \end{array}$$

Now define the functions f and k by:

$$\begin{array}{rcll} k(x) & = & \min\{m : h(m) > x\} & (x \in \omega), \\ f(x) & = & h(k(x)) - k(x) - 1 & (x \in \omega). \end{array}$$

It is immediate that f and k are recursive and that f is non-descending. Let σ_n be the shortest input of U computing $A[0, h(n))$ $(n \in \omega)$. Let B be the infinite string obtained by concatenating the strings $\sigma_0\,\hat{}\,\sigma_1\,\hat{}\,\sigma_2 \ldots$. Because the σ_m are elements of the prefix-free domain of U, there is a reduction of A to B, where at computing $A(x)$ only $\sigma_0\,\hat{}\ldots\hat{}\,\sigma_{k(x)}$ is needed. The algorithm works at argument x as follows: It applies U to the oracle. After a termination a string ρ_0 was produced. Now if $\mathrm{lh}\,\rho_0 \le x$, U is applied to the rest of the oracle. Thus a sequence $\{\rho_m\}_{m=0}^n$ of strings is produced until $\mathrm{lh}\,\rho_n > x$. Then $\rho_n(x)$ is given as output. If at some intermediate stage U does not terminate, the oracle function at argument x is not defined.

If this oracle function is applied to the oracle B, the produced strings ρ_m are just $A[0, n_m)$ and A is reduced to B so that the use function f' at argument x is obviously

$$\left(\sum_{m=0}^{k(x)} \mathrm{lh}\,\sigma_m\right) \quad - \quad 1.$$

Hence by (14) $f'(x) \le f(x)$ $(x \in \omega)$, i.e. $A \in \mathcal{S}(f)$. Because by the definition of f for every $n \in \omega$ we have $[0, h(n)) \subseteq f^{-1}[[0, h(n) - n)]$, according to Lemma 1 $\mathcal{S}(f) \ne 2^\omega$. **q.e.d.**

Corollary 2. *Let N be the set of all non-descending recursive functions. Then for every $A \in 2^\omega$ holds:*

$$A \text{ is } R\text{-incompressible} \quad \longleftrightarrow \quad A \text{ is } N\text{-incompressible.}$$

Proof. According to the first part of the proof of Theorem 3 for each R-compressible set A (12) holds and according to the second part of the proof each such set is N-compressible. **q.e.d.**

In the proof of the next theorem we shall use the notion "hyperimmune set". We shall introduce here an equivalent definition of it according to Kuznecov, Medvedev and Uspenskij (see [**13**]): If $A = \{a_0 < a_1 < a_2 < \ldots\}$ is an infinite set, the *principal function* of A is p_A, where $p_A(n) = a_n$. An infinite set A is *hyperimmune* iff no recursive function f majorizes p_A.

The last theorem shows that randomness is stronger than R-incompressibility:

Theorem 4. *There are $2^{\aleph_0}$ M-compressible sets which are R-incompressible.*

Proof. Let $H \subseteq \omega$ be a hyperimmune set (see [**10**],[**13**]). For every sequence $B \in 2^\omega$ we shall define a sequence $\{\sigma_n^B\}_{n=0}^\infty$ of elements of $2^{<\omega}$ as follows: Let $m = \min H$. Take S from Corollary 1 and put $\sigma_0^B = S[m]$. Now let $\sigma_0^B, \ldots, \sigma_n^B$ be already defined, $\sigma_i^B \subset \sigma_j^B$ for $i < j$. We consider the set C_n^B of all "0-prolongations" of σ_n^B, i.e. let

$$C_n^B \;=\; \{\rho \in 2^{<\omega} : \rho \supseteq \sigma_n^B \wedge (\forall x \in [0, \mathrm{lh}\,\rho) \setminus [0, \mathrm{lh}\,\sigma_n^B))[\rho(x) = 0]\}.$$

Claim. *Let $\mu \in C_n^B$ and $\mathrm{lh}\,\mu = 3lh\sigma_n^B$. Then*

$$\mathrm{I}(\mu) \;\leq\; \mathrm{lh}\,\mu - n. \tag{15}$$

Proof. Given a string $\rho \in 2^{<\omega}$, we can uniformly compute its "0-prolongation" ν of length $3\,\mathrm{lh}\,\rho$. Because there is a code of ρ of length $2\,\mathrm{lh}\,\rho$ in a prefix-free coding (e.g. $\rho(0)\hat{\ }0\hat{\ }\rho(1)\hat{\ }0\hat{\ }\ldots\hat{\ }\rho(\mathrm{lh}\,\rho - 1)\hat{\ }1$), it is

$$\mathrm{I}(\nu) \;\leq\; 2\,\mathrm{lh}\,\rho + d$$

for suitable d. Because $\sigma_i^B \subset \sigma_j^B$ for $i < j$, $\mathrm{lh}\,\sigma_n^B \geq \mathrm{lh}\,\sigma_0^B + n$. If $\mathrm{lh}\,\sigma_0^B$ is sufficiently large (and this can be done by omitting a number of smallest elements in H), then (15) follows.

Choose the shortest $\tau \in C_n^B$ so that $\mathrm{I}(\tau) \leq \mathrm{lh}\,\tau - n$. Let $p = \min\{x \in H : x > \mathrm{lh}\,\tau + 1\}$. Put $\rho_{n+1}^B = \tau\hat{\ }B(n)\hat{\ }S[0, p - \mathrm{lh}\,\tau - 1)$. Obviously $\sigma_{n+1}^B \supset \sigma_n^B$ and $\mathrm{lh}\,\sigma_{n+1}^B = p$, i.e.

$$\mathrm{lh}\,\sigma_{n+1}^B \quad \in \quad H. \tag{16}$$

We define $A_B \in 2^\omega$ by $A_B = \bigcup_{n\in\omega} \sigma_n^B$. It is immediate that if $B' \neq B''$, then $A_{B'} \neq A_{B''}$. According to the construction of A_B, each $\sigma_n^B \subset A_B$ has a "0-prolongation" $\tau \subset A_B$ with $\mathrm{I}(\tau) \leq \mathrm{lh}\,\tau - n$, i.e. for every $n \in \omega$ there is $m \in \omega$ so that $\mathrm{I}(A_B[0, m)) \leq m - n$. Hence A_B is M-compressible.

We want to show that A_B is R-incompressible. Assume the contrary. According to Theorem 3 there is then a strictly growing recursive function g such that

$$(17)\quad \begin{array}{lcll} \mathrm{I}(A_B[0, g(0))) & \leq & g(0) - 1 & \\ \mathrm{I}(A_B[0, g(n+1))) & \leq & g(n+1) - 6g(n) - c - 3 & (n \in \omega), \end{array}$$

where c is taken from Lemma 3. We shall show, that g majorizes H, i.e. that

$$|[0, g(n) + 1) \cap H| \quad \geq \quad n + 1 \qquad (n \in \omega).$$

This is a contradiction to the hyperimmunity of H.

Assume $[0, g(0) + 1) \cap H = \emptyset$. Then $\min H > g(0)$ and $A_B[0, g(0)) = S[0, g(0))$ whence $\mathrm{I}(A_B[0, g(0))) \geq g(0)$. This contradicts (17).

Now assume $|[0, g(n)+1) \cap H| \geq n+1$ and $|(g(n+1)+1) \cap H| < n+2$. Hence

$$([0, g(n+1) + 1) \setminus [0, g(n) + 1)) \cap H \;=\; \emptyset. \tag{18}$$

Let $m = \max\{k : \mathrm{lh}\,\sigma_k^B \leq g(n)\}$. Then by (16) and (18) $\mathrm{lh}\,\sigma_{m+1}^B > g(n+1)$ and by the construction of A_B we have $\sigma_{m+1}^B = \rho\hat{\ }B(m)\hat{\ }S[0, \mathrm{lh}\,\sigma_{m+1}^B - \mathrm{lh}\,\rho - 1)$ where according to our claim

$$\mathrm{lh}\,\rho \;\leq\; 3\,\mathrm{lh}\,\sigma_m^B \;\leq\; 3g(n). \tag{19}$$

Because $\mathrm{I}(\nu) > 0$ for every string ν, by (17) $g(n+1) > 6g(n) + 3$, so that according to (19) $\mathrm{lh}\,\rho + 2 < g(n+1)$. Thus

$$\sigma_{m+1}^B \restriction [0, g(n+1)) = A_B[0, g(n+1)) = \rho\hat{\ }B(m)\hat{\ }S[0, g(n+1) - \mathrm{lh}\,\rho - 1).$$

Hence by Lemma 3

$$\mathrm{I}(S[0, g(n+1) - \mathrm{lh}\,\rho - 1)) \quad \leq \quad \mathrm{I}(A_B[0, g(n+1))) + 2\log(\mathrm{lh}\,\rho + 1) + c$$

and because we may assume that $\mathrm{lh}\,\rho+1 > 3$, i.e. $2\log(\mathrm{lh}\,\rho+1) \leq \mathrm{lh}\,\rho+1$,

$$\mathrm{I}(S[0, g(n+1) - \mathrm{lh}\,\rho - 1)) \quad \leq \quad \mathrm{I}(A_B[0, g(n+1))) + \mathrm{lh}\,\rho + 1 + c$$

whence by (17)

$$\mathrm{I}(S[0, g(n+1) - \mathrm{lh}\,\rho - 1)) \quad \leq \quad \mathrm{lh}\,\rho + g(n+1) - 6g(n) - 2.$$

Hence by (19)

$$\mathrm{I}(S[0, g(n+1) - \mathrm{lh}\,\rho - 1)) \quad \leq \quad g(n+1) - \mathrm{lh}\,\rho - 2.$$

This contradicts the choice of S. **q.e.d.**

It is known, that the set of random sequences has measure 1 (with respect to the Lebesgue measure λ). Thus the set of R-incompressible and M-compressible sets has measure 0 and cardinality $2^{\aleph_0}$.

Theorem 4 shows that making our notion "M-incompressible" more uniform by requiring the restraint function to be recursive (and thus obtaining R-incompressibility) leads to a considerable weakening of the original notion. That this weakening was caused by uniformization can be seen by comparing theorems 1 and 3. Similarily, if we add a certain uniformity feature to recursive sequential tests in Martin-Löf's definition of randomness, we obtain total recursive sequential tests (see e.g. [**9**], [**12**]) and randomness will be weakened to so called *weak randomness*. Thus uniformization seems generally cause strict weakening of notions of randomness.

As for the relationship between weak randomness and R-incompressibility, it can be shown that the former implies the latter (a weaker form of Theorem 4 follows then immediately from a result of Schnorr in [**12**]), but the converse seems not to hold (compare Theorem 5.4.1.3 with definitions 3.2.1.2 and 3.2.1.3 in [**9**]).

It is an open question, whether an analogy of Corollary 2 holds, i.e. whether incompressibility is equivalent to M-incompressibility. The author conjectures, that incompressibility is a stronger property than randomness.

REFERENCES

1. Chaitin G.J.: A theory of program size formally identical to information theory, J.Ass. Comp. Mach. 22 (1975), 329–340.
2. Chaitin G.J.: Algorithmic information theory, J.Res. Dev. 21 (1977), 350–359; 496.
3. Dies J.E.: Information et complexité, Ann.Inst.Henri Poincaré B 12 (1976), 365–390; ibidem 14 (1978), 113–118.
4. Gacs P.: Lecture notes on descriptional complexity and randomness, BUCS Tech Report #87-013 (1987), Boston University.
5. Habart K.: Bounds in weak truth-table reducibility, to appear in: Notre Dame Jour. of Form. Logic.
6. Kobayashi K.: On compressibility of infinite sequences, Res Report (1981), Tokyo Inst.of Technology.
7. Lutz J.H.: Pseudorandom sources for BPP, Tech Report #87-22 (1987), Iowa State University.
8. Martin-Löf P.: On the definition of random sequences, Information and Control 9 (1966), 602–619.
9. Lambalgen M.van: Random sequences, PhD Thesis (1987), Univ.Amsterdam.
10. Rogers H.jr.: "Theory of recursive fnnctions and effective computability," McGraw-Hill, London, 1967.
11. Schnorr C.P.: Process complexity and effective random tests, J.C.S.S. 7 (1973), 376–388.
12. Schnorr C.P.: "Zufälligkeit und Wahrscheinlichkeit," Lecture Notes in Mathematics, Springer-Verlag, 1971.
13. Soare R.I.: "Recursively enumerable sets and degrees," Springer Verlag, Berlin and Heidelberg, 1987.

Comenius University
Philosophical Faculty
Dept. of Logic
Gondova 2
818 01 Bratislava
Czechoslovakia

Towards an efficient Tableau Proof Procedure for Multiple-Valued Logics*

Reiner Hähnle

Institute for Logic, Complexity and Deduction Systems
University of Karlsruhe, Am Fasanengarten 5
7500 Karlsruhe
Federal Republic of Germany
haehnle@ira.uka.de

Abstract

One of the obstacles against the use of tableau-based theorem provers for non-standard logics is the inefficiency of tableau systems in practical applications, though they are highly intuitive and extremely flexible from a proof theoretical point of view. We present a method for increasing the efficiency of tableau systems in the case of multiple-valued logics by introducing a generalized notion of signed formulas and give sound and complete tableau systems for arbitrary propositional finite-valued logics.

Introduction

One of the main advantages of the method of semantic tableaux [Smu68, Bet86] is that it yields analytic proof theories for a wide variety of standard and non-standard logics within a single framework. With relatively minor modifications tableau proof systems can be designed for such different logics as temporal, intuitionistic and multiple-valued logics [Wol81, Fit83, Sch89]. In addition, one could easily obtain tableau proof systems, which combine several non-standard concepts, a feature which seems to be interesting e.g. in circuit validation [KW90], natural language processing [FHLvB85] or semantics of logic programs [She89]. Also, avoidance of normal forms is necessary for the potential application of high-level heuristics.

But there are two major obstacles against the use of tableau systems in automated theorem proving without further modifications. First, the search process tends to be much more inefficient than in, say, resolution provers, if no extra care is taken. But recent research showed that it is possible to reach a similar performance as with resolution-based provers [OS86, OS88]. And [Fit90] shows that completeness proofs for tableau systems that have been tuned for automated theorem proving are still much more transparent than their resolution counterparts. Second, the modifications of standard tableau proof systems to adapt them to non-standard logics are, though highly intuitive, usually not very efficient when one asks for performance. In this paper we concentrate on the second problem and on propositional multiple-valued logics. Our work is part of the TCG Project involving the construction of a tableau-based automated theorem prover for multiple-valued logics, a prototype of which is currently being implemented [Hä90].

It should be mentioned that there exists at least one other approach to automated theorem proving in multiple-valued logics. In a series of papers (see e.g. [Sta90]) Stachniak developed resolution style systems for logics with finitely many truth values. While in his systems the underlying

*This work is supported by IBM Germany and is a collaboration of the University of Karlsruhe and the IWBS at IBM Germany in Heidelberg.

logics are specified by consequence relations, we will assume that our logics are given by a tabular semantics (cf. [Wój88]).

The paper is organized as follows: In section 1 we introduce some mathematical concepts and specify syntax and semantics of the class of languages under consideration. In section 2 we present our variant of a tableau-based calculus, in section 3 we give proofs of soundness and completeness for our system and we conclude with section 4, summarizing what has been gained.

1 Preliminaries, Syntax, Semantics

We recall some concepts from universal algebra, e.g. to be found in [BS81].

Definition 1.1 (Abstract Algebra of finite Type, Homomorphism)
A **finite type** $\mathcal{F} = \{f_1, \ldots, f_r\}$ *is an indexed set of symbols, each of them having assigned an arity by a mapping* $m : \mathcal{F} \rightarrow Nat$. *Let* $\mathcal{F}_n$ *denote the operators with arity* n. *Constants are treated as 0-ary functions.*

An **abstract algebra of type** $\mathcal{F}$ *or* **Ω-algebra** *is a non-empty universe A together with a family of mappings such that for all n and each member f in* $\mathcal{F}_n$ *there is a corresponding* **fundamental operation** $f^A : A^n \rightarrow A$. *If convenient, the abstract algebra* $< A, \{f_i^A \mid 1 \leq i \leq r\} >$ *and its universe A are denoted with the same symbol.*

Let A, B *be abstract algebras of the same type and* $h : A \rightarrow B$ *any mapping. If for all* $f \in \mathcal{F}_n, n \in Nat$ *and* $a_1, \ldots, a_n \in A$

$$h(f^A(a_1, \ldots, a_n)) = f^B(h(a_1), \ldots, h(a_n))$$

holds, then h is called **homomorphism from A to B**.

Let $\mathcal{F} = \{F_1, \ldots, F_r\}$ be a set of logical connectives and $L_0 := \{p_i \mid i \in Nat\}$ the set of propositional variables or **atomic formulas**, which has to be disjoint with $\mathcal{F}$. With L we denote the abstract algebra that is freely generated over L_0 in the class of algebras with type $\mathcal{F}$. Thus we have

$$\begin{array}{l} L_{i+1} = L_i \cup \{F_j(X_1, \ldots, X_{m(j)}) \mid X_1, \ldots, X_{m(j)} \in L_i, F_j \in \mathcal{F}\} \\ L = \bigcup\{L_i \mid i \in Nat\} \end{array}$$

as the universe of L.

L_i denotes the **formulas of depth i**. We call L **(propositional) language**, the members of L are called **(propositional) (L-)formulas**.

Let $N = \{0, 1, \ldots, (n-1)\}$ be the finite set of **truth values** and $D \subseteq N$ the set of **designated truth values**. Furthermore let us denote with $n = |N|$ and $d = |D|$ the number of elements in N and D resp. Though all nonnegative values are possible for n and d, we are only interested in the nontrivial cases where $n \geq 2$ and $d \geq 1$.

Let $A =< N, \{f_i \mid 1 \leq i \leq r\} >$ be an algebra of the same type as L. Then we call the pair $\mathcal{A} =< A, D >$ **a structure** for L and the f_i **interpretations** of the F_i. A defines the semantics of the logical operators. We say that $\mathcal{L} =< L, \mathcal{A} >$ is an **n-valued propositional logic with d designated truth-values.**

A **propositional (A-)valuation** of L is a homomorphism v from L to $\mathcal{A}$. A set M of L-formulas is called **(A-)satisfiable**, if there is a valuation v from L to $\mathcal{A}$ such that for any $X \in M$ $v(X) \in D$ holds. In this case v is called **($\mathcal{A}$-)model** for M. If $\{X\}$ is satisfiable for any $\mathcal{A}$-valuation, X is called **tautology**. Due to the universal mapping property (since L was freely generated) it is sufficient to define v on L_0 and then extend it uniquely to L.

Example 1.1 *As the set of logical operators we take* $\mathcal{F} = \{\wedge, \vee, \supset, \neg, \nabla, \sim\}$ *with arities* $m(\wedge) = 2, m(\vee) = 2, m(\supset) = 2, m(\neg) = 1, m(\sim) = 1, m(\nabla) = 1$ *and as truth values* $N = \{0, 1, 2\}, D = \{2\}$. *Their meaning (the abstract algebra A) is given by the following truth tables:*

$\wedge$	0	1	2
0	0	0	0
1	0	1	1
2	0	1	2

$\vee$	0	1	2
0	0	1	2
1	1	1	2
2	2	2	2

$\supset$	0	1	2
0	2	2	2
1	2	2	2
2	0	1	2

$\neg$	
0	2
1	1
2	0

$\sim$	
0	2
1	2
2	0

∇	
0	0
1	2
2	2

Note that we could have defined disjunction and conjunction alternatively as

$$v(X_1 \vee X_2) = max(v(X_1), v(X_2))$$

$$v(X_1 \wedge X_2) = min(v(X_1), v(X_2))$$

resp. There are many alternatives to our definition of implication, but this is not the issue that interests us here. Let us refer to the logic as defined above with the symbol $\mathcal{L}_3$.

2 Semantic Tableaux

Our goal is to give a tableau proof system for propositional multiple-valued logics with the following features:

- We want a generic proof system, i.e. it should yield a sound and complete set of tableau rules for any logic given to it.
- We do not want to have redundancy in proofs due to the formulation of the tableau rules alone.

The first task was begun by Surma [Sur84] and completed by Carnielli [Car87], who provided a generic tableau proof system as proposed for multiple-valued first-order logics with arbitrary logical connectives and generalized quantifiers. Unfortunately, Carnielli's system does not fulfill the second requirement. To explain this further, let us consider the signed version (see [Smu68, Fit90]) of a tableau proof system for standard logic: A tableau branch may be considered as a set of formulas together with a certain assignment of truth-values. The sign attached to each formula in the branch says that the truth-value of the formula should be the one associated with its sign. The tableau rules provide all significant possibilities to extend a set M of signed formulas preserving consistency. If we can arrive after a number of rule applications at a tableau branch that contains instances of all atomic formulas occurring in M at least once, arbitrarily signed, but non-contradictory, then we are able to construct a model for the formulas on the branch. If this is the case, we say that the branch is *open*. Let us call a tableau *closed* if it is fully expanded and contains no open branches. For the moment, assume that M is a singleton, say $M = \{FX\}$ (where F stands for *false*). Then a closed tableau for $\{FX\}$ represents the fact that there is no way to construct a model where X is false, so X must be a tautology.

Turning to three-valued logics we only need to introduce a third sign, corresponding to the third truth-value (say *undefined*) and define the appropriate rules, but the last step above is no longer valid, since *not false* may be *true* as well as *undefined*. To get a proof of the validity of X we have in fact to construct *two* closed tableaux, namely one with root FX and another one with root UX for the refutation of both non-designated truth-values. In the case of a logic with $(n-d)$ non-designated truth-values this amounts to the construction of $(n-d)$ closed tableaux for the proof of one single theorem. Also, the additional rules tend to be more complicated than the classical ones, as the following example shows:

Example 2.1 *3-valued tableau rules for* $\vee$*:*

$$\frac{F\ X_1 \vee X_2}{\begin{array}{c} F\ X_1 \\ F\ X_2 \end{array}} \qquad \frac{U\ X_1 \vee X_2}{\begin{array}{c|c|c} F\ X_1 & U\ X_1 & U\ X_1 \\ U\ X_2 & U\ X_2 & F\ X_2 \end{array}}$$

On the other hand, inspection of sample proofs shows that there is much redundancy in the proof trees, e.g. in the three-valued case most of the structure and formulas of the tableau for $F\ X$ are also part of the $U\ X$-tableau, even if they contribute nothing to the refutation of UX, and vice versa. We present a systematic way to get rid of this kind of redundancy, resulting in a proof system, where only *one* closed tableau has to be generated to prove the validity of a formula in an arbitrary multiple-valued propositional logic.

One approach to increase efficiency would of course be to perform the steps that are identical in all or in some of the proof trees at the same time (possibly using structure sharing), i.e. to search for the refutation of all non-designated truth-values in parallel. But, as always when one is making algorithms and representations trickier, this leads to a fairly complex proof procedure involving much bookkeeping and hence a cryptical completeness proof. A far more satisfying solution can be achieved on a logical level.

To be specific, consider the signed $\mathcal{L}_3$-formula $T \sim A$. Application of the corresponding tableau rule from [Car87] or [Sur84] yields two new branches with extensions FA and UA, resp. But encountering this formula during a proof does not give rise to any logical reason to split the proof in two cases "$v(A) = 0$" and "$v(A) = 1$" resp. So our idea is to increase the expressivity of the signs in order to be able to state conditions like "$v(A) = 0$ *or* $v(A) = 1$" or equivalently "$v(A) \neq 2$" within a single signed formula and thus to decrease the number of new branches per rule application significantly. It is noteworthy that neither the idea of enriching the syntax of signs nor of interpreting them semantically in a different way is new. The first has been used in tableau systems for modal logics for a long time (see e.g. in [Fit83]); on the other hand, in [Fit89] Fitting denoted upper and lower bounds in a lattice of truth values with single signs. What we will do is to systematically exploit both ideas at the same time.

Definition 2.1 (Sign, Signed Formula)
Let L be any language and D and N be defined as above. Then we define the **set of signs** *as $\mathcal{S} = \{S_i \mid i \in 2^N\}$. For any logic $\mathcal{L}$ we fix a certain set of signs $\mathcal{S}_\mathcal{L} \subseteq \mathcal{S}$ which satisfies $\{S_{\{0\}}, \ldots, S_{\{n-1\}}\} \subseteq \mathcal{S}_\mathcal{L}$[1]. From now on a logic will be a triple $\mathcal{L} = < L, \mathcal{A}, \mathcal{S}_\mathcal{L} >$. With $I_\mathcal{L} = \{i \mid S_i \in \mathcal{S}_\mathcal{L}\}$ we denote the set of allowed indices of signs. With the same symbol we identify the abstract algebra generated by $I_\mathcal{L}$ that has the same type as A and whose fundamental operations are defined by $f^{I_\mathcal{L}}(i_1, \ldots, i_m) = \bigcup\{f^A(j_1, \ldots, j_m) \mid j_k \in i_k, 1 \leq k \leq m\}$. From the context it will always be clear which is meant.*

If X is an L-formula and $S_i = S_{\{i_0,\ldots,i_r\}}$ a sign, then we call the string $S_i(X)$ **signed** *(L-)***formula**. *L^* is the set of signed formulas in a logic $\mathcal{L}$, i.e. all signed L-formulas with signs from $\mathcal{S}_\mathcal{L}$. The members of L^* will be called* $I_\mathcal{L}$-**signed formulas.**

In the above definition we have deliberately admitted $S_\emptyset$ and S_N as signs. While the following definitions and theorems exclude the former implicitely, the latter would be perfectly right, though it is hard to imagine any meaningful application for it.

[1]Otherwise it is not guaranteed that all rules can be properly stated.

Example 2.2 *We define for $\mathcal{L}_3$ the set of signs $\{S_{\{0\}}, S_{\{1\}}, S_{\{2\}}, S_{\{0,1\}}\}$ which for convenience we rewrite as $\{F, U, T, (F|U)\}$.*

The intended interpretation of a signed formula $(F|U)(X)$ then is "$v(X) = 0$ or $v(X) = 1$".

Now we are ready to define the tableau rules. We assume familiarity with trees, a formal treatment of proof trees can be found in [Smu68].

Definition 2.2 (Tableau Rule)
*Let $X = F(X_1, \ldots, X_m)$ be an L-formula in the logic $\mathcal{L} =< L, \mathcal{A}, S_{\mathcal{L}} >$. An ($\mathcal{L}$-)***tableau rule** *is a function $\pi_{i,F}$ which assigns to a signed formula $S_i(X) \in L^*$ a tree with root $S_i(F(X_1, \ldots, X_m))$, called* **premise**, *and the linear subtrees*

$$\{S_{j_1}(X_{i_1}) \circ \ldots \circ S_{j_t}(X_{i_t}) \mid j_1, \ldots, j_t \in I_{\mathcal{L}}, t \leq m \text{ and } H_i(F; j_1, \ldots, j_t) \text{ holds}\},$$

called **extensions**[2], *such that*

(T0) *for any $(z_1, \ldots, z_m) \in f^{-1}(i)$ there is an extension $S_{j_1}(X_{i_1}) \circ \ldots \circ S_{j_t}(X_{i_t})$ with $z_{i_k} \in j_k$ for $1 \leq k \leq t$ and the set of extensions is minimal with respect to this condition*[3].

The set of all extensions of a tableau rule is called **conclusion.**

The condition $H_i(F; j_1, \ldots, j_t)$ means, there exists a homomorphism $h : L \to I_{\mathcal{L}}$, satisfying (T1)–(T4) below:

(T1) *$h(X_{i_k}) = j_k$ for $1 \leq k \leq t$.*

(T2) *If f is the interpretation of F, then $f(v_1, \ldots, v_m) \in i$ must hold, where $v_{i_k} \in h(X_{i_k})$ for $1 \leq k \leq t$ and all other arguments are arbitrary.*

(T3) *There is no j'_k with $|j'_k| > |j_k|$ for $1 \leq k \leq t$ that satisfies (T1) and (T2).*

(T4) *There is no t' with $t' < t$ that satisfies (T1) and (T2).*

If no such homomorphism exists, no rule for the specific combination of formula and sign is defined.

Though this definition seems to be fairly abstract, for any given logic it essentially boils down to the usual tableau rules plus the extra feature of more general signs. To provide a better understanding of how the tableau rules are generated, we give an informal description of the process:

Remember that the extensions are thought to be disjunctively connected while the formulas within an extension are conjunctively connected.

The conclusion of a tableau rule for a sign i and connective F can be thought of as a minimal generalized sum-of-products representation of the two-valued function that holds the entry *true* in its truth table on each place where the truth table of F holds a member of i and holds *false* otherwise.

Each extension corresponds to a product term in this representation. A geometrical interpretation would associate a partial cover of entries in the hypercube that constitutes the truth table of F with an extension. All extensions taken together are a total cover.

- Condition (T0) ensures that all entries from i are covered in some extension and minimizes the number of extensions.

[2]Extensions are treated like sets and thus of all subtrees that differ only in the ordering of their signed formulas only one appears as an extension of the rule.

[3]Already in the two-valued case there may be more than one minimal (in our sense) set of extensions for a signed formula, so we need this minimality condition; see [Due88, p. 12f] for an example.

- Condition (T1) defines the interesting part of h.
- Condition (T2) guarantees soundness.
- Condition (T3) represents the strategy to split the proof tree as late as possible, in other words, to keep the signs as general as possible.
- (T4) minimizes the number of subformulas within the extensions and prevents redundant extensions.

Example 2.3 *Consider the truth table of disjunction in $\mathcal{L}_3$ as defined above. Find the tableau rule for $S_{\{1\}}(X_1 \vee X_2)$. We have to find a minimal set of homomorphisms $h : L \rightarrow I_{\mathcal{L}}$ covering all entries equal to 1. Hereby choose the sets $h(X_i)$ maximal.*

First, $X_1 \mapsto \{1\}, X_2 \mapsto \{0,1\}$ defines the partial cover...

x_1/x_2	0	1	2
0	0	1	2
1	1	1	2
2	2	2	2

...adding the partial cover that corresponds to $X_1 \mapsto \{0,1\}, X_2 \mapsto \{1\}$ yields

x_1/x_2	0	1	2
0	0	1	2
1	1	1	2
2	2	2	2

Obviously all of the conditions (T1)–(T4) are satisfied. And since both partial covers are essential and together represent a total cover, condition (T0) also holds.

Example 2.4 *From the homomorphisms that define the cover of the entries equal to 1 we can immediately extract the tableau rule:*

$$\frac{U\ (X_1 \vee X_2)}{\begin{array}{c|c} (F|U)\ X_1 & U\ X_1 \\ U\ X_2 & (F|U)\ X_2 \end{array}}$$

Note that the entry for $X_1 = X_2 = 1$ in the truth table of disjunction is covered by both extensions. The rule is considerably simpler than the one from Example 2.1.

In the Appendix a sound and complete tableau system for $\mathcal{L}_3$ can be found.

Tableaux are by the tableau rules finitely generated trees, their nodes being labeled with signed formulas. A **branch** is a path through a proof tree, beginning with the root and ending with a leaf. Usually we identify a branch with the set of signed formulas that is equal to its label set.

Definition 2.3 (Propositional Tableaux)
Let M be a nonempty finite set of $I_{\mathcal{L}}$-signed formulas. Then a **(propositional) tableau** *for M can be constructed in one of the following ways:*

- *A linear tree, where each formula of M occurs exactly once as a label is a tableau for M.*
- *Let T be a tableau for M and B a branch of T, containing a signed formula $S_i(F(X_1, \ldots, X_m))$. If $\pi_{i,F}$ is defined and has extensions $E_1, \ldots, E_n$, append to T at the end of B n linear subtrees containing the signed formulas in $E_1, \ldots, E_n$, resp. in an arbitrary sequence. The resulting tree is again a tableau for M.*

Definition 2.4 (Open, Closed)
A tableau branch is called **closed** *if one of the following conditions is satisfied:*

- *It contains a* **complementary atom set**, *i.e. signed atomic formulas* $S_{i_1}(p), \ldots, S_{i_n}(p)$ *with* $\bigcap_{j=1}^{n} i_j = \emptyset$
- *It contains a non-atomic signed formula for which no rule is defined*[4].

A **branch** *that is not closed is called* **open**. *An open branch, for which any rule application yields formulas, that are already on the branch, is called* **exhausted**. *A* **tableau** *is called* **closed** *if each of its branches is closed, and* **open** *otherwise. A tableau is called* **complete** *if each of its branches is either closed or exhausted.*

Example 2.5 *We prove that the formula* $\neg A \supset (\sim A \wedge \neg A)$ *is a* $\mathcal{L}_3$*-tautology by constructing a closed tableau with root* $(F|U)\neg A \supset (\sim A \wedge \neg A)$. *The existence of such a closed tableau tells us that in any possible valuation the truth value of the formula in question can neither be 0 nor 1, so we can conclude that indeed it must be a tautology. In the following tableau the numbers of the formulas are marked with right brackets, whereas the numbers of the parent formulas are indicated by full bracketed numbers. At the end of each branch the numbers of the complementary formulas are stated. The tableau rules used here refer to the Appendix.*

(-) 1) $(F|U)\neg A \supset (\sim A \wedge \neg A)$

(1) 2) $T\neg A$
(1) 3) $(F|U) \sim A \wedge \neg A$

(2) 4) FA

(3) 5) $(F|U) \sim A$ — (5) 6) TA — closed(4,6)

(3) 7) $(F|U)\neg A$ — (7) 8) TA — closed(4,8)

(3) 7) $(F|U)\neg A$ — (7) 9) UA — closed(4,9)

The proof of this theorem in Carnielli's system requires the construction of two trees, one of which is considerably more complex than the one above.

We close this section with an appropriate definition of satisfiability of branches and tableaux, which we shall need for the formulation of the main lemma in the soundness proof.

Definition 2.5 (Satisfiability of Branches)
A set B of **signed formulas** *is called* **satisfiable**, *if there is a valuation v such that for all* $S_i(X) \in B$ $v(X) \in i$ *holds. In this case we say that v is a* **model** *for B. A* **branch** *is satisfiable iff its label set is. A* **tableau** *is satisfiable iff it contains at least one satisfiable branch.*

3 Soundness and Completeness

3.1 Soundness

Lemma 3.1 (Satisfiability Preservation of Tableau Rules) *Let T be a satisfiable tableau and suppose T′ was created by rule application to an arbitrary formula in T. Then T′*

[4]This case corresponds to closure of branches that contain e.g. $T \perp$ in classical logic.

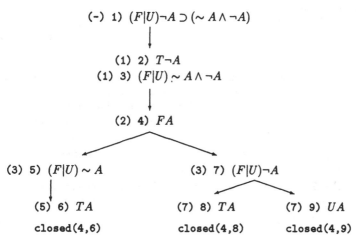

is also satisfiable.

Proof: T contains at least one satisfiable branch B. If the formula in the rule application was not in B, B is unchanged and hence still satisfiable.

On the other hand, let $S_i(F(X_1, \ldots, X_m)) \in B$ be the formula that supplied the premise for rule application and let v be a valuation that satisfies B. For such a valuation by definition we always have $v(F(X_1, \ldots, X_m)) \in i$. Since v is a homomorphism,

$$v(F(X_1, \ldots, X_m)) = f(v(X_1), \ldots, v(X_m)) \in i$$

holds.

Let $S_{j_1}(X_{i_1}) \circ \ldots \circ S_{j_t}(X_{i_t})$ be an extension obtained by applying (T0) to $(v(X_1), \ldots, v(X_m)) \in f^{-1}(i)$. Take any i_k: By (T0) we have $v(X_{i_k}) \in j_k$. Together with the assumption that v was a model for B we have the satisfiability of $B \cup \{S_{j_1}(X_{i_1}), \ldots, S_{j_t}(X_{i_t})\}$, which concludes the proof. ∎

Now soundness follows easily:

Theorem 3.1 (Soundness) *Let A be any L-formula. If there is a closed tableau with root $S_{N-D}(A)$, then A is a tautology.*

Proof: Let T be such a tableau. T cannot be satisfiable. For assume B is an arbitrary branch in T. Since T is closed, B either contains a complementary atom set or a signed formula with no corresponding rule definition. Obviously, no valuation v that satisfies B can exist, since in the first case, it would be no mapping, in the second case it would be only partially defined. Since this holds for arbitrary branches, T is not satisfiable.

The next step is to show by a straightforward induction, using the above lemma, that any tableau with satisfiable root also must be satisfiable.

Together we have that T is not satisfiable, so the root $S_{N-D}(A)$ is not satisfiable, which means by definition for all valuations v that $v(A) \notin N - D$ iff for all valuations v $v(A) \in D$ iff A is a tautology. ∎

3.2 Completeness

The completeness proof for our system will be quite straightforward and will closely follow the lines of standard tableau completeness proofs (see e.g. [Fit90]), but in order to be able to deal with generalized signs we will have to make appropriate modifications of the definitions of Hintikka Set and Analytic Consistency Property. Then we proceed as usual, proving first Hintikka's Lemma, and second a model existence theorem, which in turn yields completeness. For the sake of modularity and flexibility we prefer the formulation with analytic consistency properties over a more direct one. Then it is easy to extend the proofs to first-order formulas or infinite sets of formulas. Also other standard results like strong completeness and compactness may easily be obtained, though we do not include them here.

Definition 3.1 (Hintikka Set)
A set H of $I_{\mathcal{L}}$-signed formulas is called a **Hintikka set** *iff it is atomically consistent and downward saturated, or more precisely, if the following two conditions hold:*

(H1) *For all propositional variables $p \in L_0$: If $S_{i_1}(p), \ldots, S_{i_n}(p) \in H$ then $\bigcap_{j=1}^{n} i_j \neq \emptyset$.*

(H2) *If $S_i(F(X_1,\ldots,X_m)) \in H$ then $\pi_{i,F}$ is defined and at least one of the hereby determined extensions*[5] $\{S_{j_1}(X_{i_1}),\ldots,S_{j_t}(X_{i_t})\}$ *is also in H.*

A Hintikka set H' is called **saturated Hintikka set** *or* **model set** *iff in addition to the above stated conditions it is atomically complete and upward saturated, i.e.*

(H3) *For all propositional variables $p \in L_0$ there exists an $i \in I_{\mathcal{L}}$ such that $S_i(p) \in H'$.*

(H4) *If $i \in I_{\mathcal{L}}$ then $S_i(F(X_1,\ldots,X_m)) \in H'$, whenever at least one of the extensions $\{S_{j_1}(X_{i_1}),\ldots,S_{j_t}(X_{i_t})\}$ determined by $\pi_{i,F}$ is in H'.*

Note that by (H1) and (H2) it is impossible that $S_{\emptyset}(X)$ for any $X \in L$ ever occurs in a Hintikka set.

Theorem 3.2 (Hintikka's Lemma) *Every Hintikka set H can be extended to a saturated Hintikka set H'.*

Proof: Let H be a Hintikka set and $L_0 := \{p_i \mid i \in Nat\}$ an enumeration of the propositional variables. We extend H to a saturated Hintikka set H' in the following way:

$$H_0 = H \cup \{S_{\{0\}}(p_i) \mid i \in Nat \text{ and } S_j(p_i) \in H \text{ for no } j \in I_{\mathcal{L}}\}$$
$$H_{i+1} = H_i \cup \{S_j(F(X_1,\ldots,X_m)) \mid j \in I_{\mathcal{L}}, \pi_{j,F} \text{ defined and at least one of the extensions } \{S_{j_1}(X_{i_1}),\ldots,S_{j_t}(X_{i_t})\} \text{ determined by } \pi_{j,F} \text{ is in } H_i\}$$
$$H' = \bigcup\{H_i \mid i \in Nat\}$$

First we extend H such that it assigns a definite truth value (we took 0, but it is arbitrary) to each variable not already occurring in H, then we inductively take all L-formulas into account.

For H' (H1) holds, because let $p \in L_0$, then either there exists a $j \in I_{\mathcal{L}}$ such that $S_j(p) \in H$ and nothing is changed by the construction, so (H1) still holds, or $S_j(p) \in H$ for no $j \in I_{\mathcal{L}}$, then $S_{\{0\}}(p)$ is added and since this is the only occurrence of p (H1) holds trivially. (H3) and (H4) hold by construction of H'. To see that (H2) holds, let $S_j(F(X_1,\ldots,X_m)) \in H'$ and $j \in I_{\mathcal{L}}$. Then either already $S_j(F(X_1,\ldots,X_m)) \in H$ and (H2) is inherited from H, or $S_j(F(X_1,\ldots,X_m))$ was generated during the construction in some $H_i, i > 0$. Then, by definition, at least one of the extensions $\{S_{j_1}(X_{i_1}),\ldots,S_{j_t}(X_{i_t})\}$ determined by $\pi_{j,F}$ is in H_{i-1} and (H2) is inherited from H_{i-1}. ∎

Definition 3.2 (Analytic Consistency Property)
A family Γ ranging over sets of $I_{\mathcal{L}}$-signed formulas is called an **Analytic Consistency Property (ACP)** *iff for all $K \in \Gamma$ the following conditions hold:*

(F) *Γ is of* finite character, *i.e. K belongs to Γ iff all finite subsets of K belong to Γ.*

(ACP1) *For all propositional variables $p \in L_0$ holds: If $S_{i_1}(p),\ldots,S_{i_n}(p) \in K$ then $\bigcap_{j=1}^{n} i_j \neq \emptyset$.*

(ACP2) *If $S_i(F(X_1,\ldots,X_m)) \in K$ then $\pi_{i,F}$ is defined and for at least one of the extensions $C = \{S_{j_1}(X_{i_1}),\ldots,S_{j_t}(X_{i_t})\}$ $K \cup C \in \Gamma$.*

If $K \in \Gamma$ then K is called **Γ-consistent**. *While Γ has finite character, from $K' \subseteq K$ and $K \in \Gamma$ we always have $K' \in \Gamma$.*

[5] Here and in the following we view extensions as sets.

Theorem 3.3 (Model Existence) *Let Γ be an ACP and K a set of $I_{\mathcal{L}}$-signed formulas. If K is Γ-consistent then there exists a valuation v, such that $v(X) \in j$ holds, whenever $S_j(X) \in K$, in other words, v is a model for K.*

Proof: In a first step we will carry out a Lindenbaum-type construction restricted to ACP's in order to find a L^*-maximal element M in Γ (this corresponds to Tukey's lemma in the denumerable case), then we show that M is a Hintikka set, so we can use it to define an appropriate valuation.

Let $\{Z_1, Z_2, \ldots\}$ be an enumeration of all signed formulas in L^* and define C_n for $n \geq 0$ as follows:

$$C_0 = K$$
$$C_{n+1} = \begin{cases} C_n \cup \{Z_i\} & \text{if } C_n \cup \{Z_n\}\Gamma\text{-konsistent} \\ C_n & \text{otherwise} \end{cases}$$

Clearly, all C_n are members of Γ and, ordered by inclusion, are building a chain in Γ. We define

$$M = \bigcup_{n \geq 0} C_i$$

and thus have:

1. M is L^*-maximal in Γ, since
   (a) Let $K \subseteq M$ be arbitrary, but finite. Hence we have some C_n with $K \subseteq C_n$ and while $C_n \in \Gamma$, we have also that $K \in \Gamma$ because of the finite character of Γ. Thus we have $K \in \Gamma$ for all finite $K \subseteq M$ and so $M \in \Gamma$, again because of the finite character of Γ.
   (b) Assume there were $M' \subseteq L^*$ with $M \subset M' \in \Gamma$, $M \neq M'$. So we must have some $Z_n \in M'$ with $Z_n \notin M$. By definition, we have $C_n \subseteq M \subset M'$, hence $C_n \cup \{Z_n\} \subseteq M'$. By the finite character of Γ we know that $C_n \cup \{Z_n\} \in \Gamma$. But then, by definition, $C_{n+1} = C_n \cup \{Z_n\}$ thus yielding $Z_n \in M$, which is a contradiction.
2. (a) and (ACP1) imply (H1), (b) and (ACP2) imply (H2) for M, so M is indeed a Hintikka set. According to Hintikka's lemma we can extend M to a saturated Hintikka set $\bar{M}$.

It remains to show that $\bar{M}$ determines a model for K. For this purpose we fix an arbitrary v for $p \in L_0$ such that:

$$v(p) \in i \text{ iff } S_i(p) \in \bar{M}$$

Since $\bar{M}$ is a saturated Hintikka set, (H1) guarantees that v is welldefined, (H3) that it is totally defined on L_0. We extend v to a homomorphism from L to $\mathcal{A}$ and show by induction on the depth of X that $X \in L$ and $S_j(X) \in \bar{M}$ imply $v(X) \in j$[6].

The case when X is atomic is settled by definition of v.

Suppose that $S_j(X) = S_j(F(X_1, \ldots, X_m)) \in \bar{M}$. According to (H2) there is at least one extension determined by $\pi_{j,F}$ with $\{S_{j_1}(X_{i_1}), \ldots, S_{j_t}(X_{i_t})\} \subseteq \bar{M}$. The induction hypothesis yields $v(X_{i_k}) \in j_k$ for $1 \leq k \leq t$. With this we can conclude, using the homomorphism h that defines the extension:

[6]Note that in the proof we don't make use of (H4). In fact, using (H4) we could show the other direction as well, namely that for any $X \in L$ there exists a $j \in I_{\mathcal{L}}$ such that $v(X) \in j$ implies $S_j(X) \in \bar{M}$.

$v(F(X_1,\ldots,X_m))$
$= f(v(X_1),\ldots,v(X_{i_1}),\ldots,v(X_{i_t}),\ldots,v(X_m))$ (v homomorphism)
$\in j$ (by induction hypothesis, (T1), (T2))

So we have indeed constructed a model for $\bar{M}$ and the theorem follows from the fact that $K \subseteq \bar{M}$.

■

Theorem 3.4 (Completeness) *If A is a tautology then there exists a closed tableau with root $S_{N-D}(A)$.*

Proof: Since A is a tautology, for all valuations $v(A) \in D$ must hold. Now suppose no closed tableau with root $S_{N-D}(A)$ exists. It follows that there exists at least one exhausted tableau with root $S_{N-D}(A)$, containing an exhausted, open branch M. Define $\mathcal{B}$ as the set of all finite tableau branches that cannot be closed. For all $B \in \mathcal{B}$ we have:

- For all propositional variables $p \in L_0$ holds: If $S_{i_1}(p),\ldots,S_{i_n}(p) \in B$ then $\bigcap_{j=1}^{n} i_j \neq \emptyset$, otherwise B would be closed.
- If $S_i(F(X_1,\ldots,X_m)) \in B$ then $\pi_{i,F}$ is defined and for at least one of the hereby determined extensions $C = \{S_{j_1}(X_{i_1}),\ldots,S_{j_t}(X_{i_t})\}$ $B \cup C \in \mathcal{B}$. For assume $\pi_{i,F}$ were not defined, then B were closed and if for no C $B \cup C \in \mathcal{B}$, then B could be closed later on.
- Clearly, $\mathcal{B}$ has finite character.

Putting the facts together, we have that $\mathcal{B}$ is an ACP and $\{S_{N-D}(A)\}$ is $\mathcal{B}$-consistent, since $\{S_{N-D}(A)\} \subseteq M \in \mathcal{B}$. Now, from the model existence theorem we know that there exists a valuation v with $v(A) \in N - D$ and this is the contradiction we have been looking for.

■

4 Conclusion

We presented a generic tableau proof system for propositional multiple-valued logics that is more efficient and elegant than its predecessors. For support of the efficiency claim, consider the $\mathcal{L}_3$-tautology $((\ldots(p_1 \vee p_2) \vee \ldots \vee p_n) \vee \sim p_1)$. It is easy to see that there is a proof of linear size wrt n in our system, while the shortest proof in Carnielli's system is of exponential size wrt n.

The achievement was gained by generalizing signs from truth values to sets of truth values. We emphasize that the improvements were made on a logical rather than on an algorithmical level by enriching the language, so we can use our tableau system for standard tableau provers with minor modifications. Another advantage of this approach is the compatibility with techniques that are set on the bookkeeping level e.g. indexing schemes or weighting.

The extension of the technique to first-order multiple-valued logics is possible if some restrictions on allowed signs, connectives and quantifiers are imposed. A follow-up to this paper concerned with multiple-valued predicate logic is available [Hä91]. To keep the paper short we have excluded the notion of *systematic tableaux* which is needed for mechanizing tableau proofs and which requires no further modifications for the use in our framework.

Acknowledgements

I would like to thank Peter H. Schmitt for many helpful discussions and suggestions during the composition of this paper. I took advantage from the comments of Alan Shepherd and the remarks of an anonymous referee.

Appendix: A Tableau System for $\mathcal{L}_3$

Rules for $\vee$:

$$\frac{T\ X_1 \vee X_2}{T\ X_1 \mid T\ X_2} \qquad \frac{U\ X_1 \vee X_2}{\begin{array}{c|c} U\ X_1 & (F|U)\ X_1 \\ (F|U)\ X_2 & U\ X_2 \end{array}} \qquad \frac{F\ X_1 \vee X_2}{\begin{array}{c} F\ X_1 \\ F\ X_2 \end{array}} \qquad \frac{(F|U)\ X_1 \vee X_2}{\begin{array}{c} (F|U)\ X_1 \\ (F|U)\ X_2 \end{array}}$$

Rules for $\wedge$:

$$\frac{T\ X_1 \wedge X_2}{\begin{array}{c} T\ X_1 \\ T\ X_2 \end{array}} \qquad \frac{U\ X_1 \wedge X_2}{\begin{array}{c|c|c} U\ X_1 & U\ X_1 & T\ X_1 \\ T\ X_2 & U\ X_2 & U\ X_2 \end{array}} \qquad \frac{F\ X_1 \wedge X_2}{F\ X_1 \mid F\ X_2} \qquad \frac{(F|U)\ X_1 \wedge X_2}{(F|U)\ X_1 \mid (F|U)\ X_2}$$

Rules for $\supset$:

$$\frac{T\ X_1 \supset X_2}{(F|U)\ X_1 \mid T\ X_2} \qquad \frac{U\ X_1 \supset X_2}{\begin{array}{c} T\ X_1 \\ U\ X_2 \end{array}} \qquad \frac{F\ X_1 \supset X_2}{\begin{array}{c} T\ X_1 \\ F\ X_2 \end{array}} \qquad \frac{(F|U)\ X_1 \supset X_2}{\begin{array}{c} T\ X_1 \\ (F|U)\ X_2 \end{array}}$$

Rules for $\neg$:

$$\frac{T\ \neg X}{F\ X} \qquad \frac{U\ \neg X}{U\ X} \qquad \frac{F\ \neg X}{T\ X} \qquad \frac{(F|U)\ \neg X}{U\ X \mid T\ X}$$

Rules for $\sim$:

$$\frac{T\ \sim X}{(F|U)\ X} \quad \text{(no rule defined for } U\ \sim X\text{)} \quad \frac{F\ \sim X}{T\ X} \qquad \frac{(F|U)\ \sim X}{T\ X}$$

Rules for ∇:

$$\frac{T\ \neg X}{U\ X \mid T\ X} \quad \text{(no rule defined for } U\ \nabla X\text{)} \quad \frac{F\ \nabla X}{F\ X} \qquad \frac{(F|U)\ \nabla X}{F\ X}$$

References

[Bet86] E. W. Beth. Semantic entailment and formal derivability. In Karel Berka and Lothar Kreiser, editors, *Logik-Texte. Kommentierte Auswahl zur Geschichte der modernen Logik*, pages 262–266. Akademie–Verlag, Berlin, 1986.

[BS81] Stanley Burris and H.P. Sankappanavar. *A Course in Universal Algebra*, volume 78 of *Graduate Texts in Mathematics*. Springer, New York, 1981.

[Car87] Walter A. Carnielli. Systematization of finite many-valued logics through the method of tableaux. *Journal of Symbolic Logic*, 52(2):473–493, June 1987.

[Due88] Gerhard W. Dueck. *Algorithms for the Minimization of Binary and Multiple–Valued Logic Functions*. PhD thesis, University of Manitoba, Winnipeg, 1988.

[FHLvB85] Jens Erik Fenstad, Per-Kristian Halvorsen, Tore Langholm, and Johan von Benthem. Equations, schemata and situations: A framework for linguistic semantics. Technical Report CSLI–85–29, Center for the Studies of Language and Information Stanford, 1985.

[Fit83] Melvin C. Fitting. *Proof Methods for Modal and Intutionistic Logics*. Reidel, Dordrecht, 1983.

[Fit89] Melvin C. Fitting. Negation as refutation. In *LICS 1989 Proceedings*, 1989.

[Fit90] Melvin C. Fitting. *First–Order Logic and Automated Theorem Proving*. Springer, New York, 1990.

[Hä90] Reiner Hähnle. Spezifikation eines Theorembeweisers für dreiwertige First–Order Logik. IWBS Report 136, Wissenschaftliches Zentrum, IWBS, IBM Deutschland, September 1990.

[Hä91] Reiner Hähnle. Uniform notation of tableaux rules for multiple–valued logics. In *To appear in Proceedings International Symposium on Multiple–Valued Logic, Victoria*, 1991.

[KW90] T. Kropf and H.-J. Wunderlich. Hierarchische Testmustergenerierung für sequentielle Schaltungen mit Hilfe von Temporaler Logik. II. ITG/GI Workshop Testmethoden und Zuverlässigkeit von Schaltungen und Systemen, 1990.

[OS86] F. Oppacher and E. Suen. Controlling deduction with proof condensation and heuristics. In Jörg H. Siekmann, editor, *Proc. 8th International Conference on Automated Deduction*, pages 384–393, 1986.

[OS88] F. Oppacher and E. Suen. HARP: A tableau-based theorem prover. *Journal of Automated Reasoning*, 4:69 – 100, 1988.

[Sch89] Peter H. Schmitt. Perspectives in multi–valued logic. Proceedings International Scientific Symposium on Natural Language and Logic, Hamburg, 1989.

[She89] John C. Sheperdson. A sound and complete semantics for a version of negation as failure. *Theoretical Computer Science*, 65:343–371, 1989.

[Smu68] Raymond Smullyan. *First-Order Logic*. Springer, New York, second edition, 1968.

[Sta90] Z. Stachniak. Note on resolution approximation of many-valued logics. In *20th International Symposium on Multiple-Valued Logic, Charlotte*, pages 204–209, May 1990.

[Sur84] Stanisław J. Surma. An algorithm for axiomatizing every finite logic. In David C. Rine, editor, *Computer Science and Multiple–Valued Logics*, pages 143–149. North–Holland, Amsterdam, 1984.

[Wój88] Ryszard Wójcicki. *Theory of Logical Calculi*. Reidel, Dordrecht, 1988.

[Wol81] Pierre Wolper. Temporal logic can be more expressive. In *Proceedings 22nd Annual Symposium on Foundations of Computer Science*, pages 340 – 348, 1981.

Interactive Proof Systems: Provers, Rounds, and Error Bounds

Ulrich Hertrampf *   *Klaus Wagner* †

Institut für Informatik
Universität Würzburg
D-8700 Würzburg, Germany

Abstract

We introduce generalized multi-prover interactive proof systems and the associated polynomial time complexity classes $IP(m,r,\frac{1}{h})$, which depend on the number m of provers, number r of rounds and the value $\frac{1}{h}$ by which the error is bounded away from one half. In this denotation the class $IP(m,r)$ of languages accepted by ordinary IP-systems with m provers and r rounds appears as $IP(m,r,\frac{1}{6})$, whereas we define $IP'(m,r)$ to be the union of all $IP(m,r,\frac{1}{h})$ with an arbitrary polynomial h. We prove several simulation theorems that enable us to prove most of the known relations between different IP-classes and a collapse of the IP′ hierarchy to essentially only four classes, namely

$$\begin{aligned} IP'(1,1) = IP(1,1) &\subsetneq IP'(1,poly) = IP(1,poly) = \mathrm{PSPACE} \\ &\subseteq IP'(2,1) = IP(poly,1) \\ &\subseteq IP'(2,2) = IP(poly,poly) = \mathrm{NEXPTIME} \end{aligned}$$

Finally we show how to reduce the space needed by an interactive proof system introducing one additional prover.

1 Introduction

Interactive proof systems have been introduced in 1985 independently (and in different versions) by GOLDWASSER, MICALI and RACKOFF [GMR85] and by BABAI [Ba85]. The idea of interactive proof systems extends the idea of *efficient proofs* which has led to the class IP. The *verifier* of an interactive proof system is a probabilistic polynomial time machine which can communicate with a *prover* (that is a deterministic machine whose complexity is not bounded) who tries to convince the verifier to accept. A language has an interactive proof system if for instances belonging to the language there is a prover who can convince the verifier to accept with probability at least $\frac{2}{3}$ and if for instances not

*e-mail: hertramp@informatik.uni-wuerzburg.dbp.de

†e-mail: wagner@informatik.uni-wuerzburg.dbp.de

belonging to the language the prover can convince the verifier to accept with probability less than $\frac{1}{3}$. The class of languages which can be accepted by interactive proof systems is denoted by IP.

Thus IP seems to be a moderate BPP-like extension of NP. Indeed, it is proved in [Ba85] that interactive proof systems with a constant number of rounds (a round consists of a query of the verifier and the answer of the prover) accept exactly the languages of the class BP·NP which is the nondeterministic version of BPP. The class BP·NP includes obviously NP and it is included in the class Π_2^p of the polynomial hierarchy [Ba85]. However BOPPANA, HÅSTAD and ZACHOS [BHZ87] proved that BP·NP does not include co-NP unless the polynomial hierarchy collapses to $\Sigma_2^p = \Pi_2^p$. But also an unbounded number of rounds did not seem to increase the power of interactive proof systems considerably, since FORTNOW and SIPSER [FS88] could show that there is an oracle B such that $co - NP^B \not\subseteq IP^B$.

Knowing these results it was a great surprise when LUND, FORTNOW, KARLOFF and NISAN [LFKN89] proved that not only co-NP but also the whole polynomial hierarchy (and even $P^{\#P}$) is included in IP. This research was crowned eventually by SHAMIR'S [Sh89] result IP = PSPACE.

A natural extension of interactive proof systems are multiple interactive proof systems introduced by BEN-OR, GOLDWASSER, KILIAN and WIGDERSON [BGKW88]. Here the verifier can communicate with an unbounded number of provers in such a manner that no prover has access to the informations exchanged between the verifier and the other provers. In [BGKW88] it is shown that two provers are as powerful as an arbitrary (i.e. polynomial) number of provers. Let MIP be the class of all languages accepted by multi prover interactive proof systems. These systems seem to be more powerful than proof systems with one prover. CAI, CONDON and LIPTON [CCL90] showed that every IP language can be accepted by an interactive proof system with two provers using only one round (a round in a multi prover interactive proof system consists of at most one query per prover and their answers, where first the queries to all provers are sent and then all answers are given). To complete the picture BABAI, FORTNOW and LUND [BFL90] extended the method of [LFKN89] and [Sh89] to prove MIP = NEXPTIME. For a very detailed overview on interactive proof systems see [Ba90].

Regarding the number of rounds the results for one prover and multi-prover interactive proof systems differ considerably. Whereas for one-prover interactive proof systems only a linear round speed up is known ([Ba85], [GS86], [BM88]) we have some more dramatic collapses for multi-prover systems. FORTNOW, ROMPEL and SIPSER [FRS88] proved that two rounds are as powerful as arbitrarily (i.e. polynomially) many rounds. (In fact, they claimed even that three provers and two rounds are sufficient but their proof turned out to be wrong [FRS90].) Denoting for $f, g : \mathbb{N} \mapsto \mathbb{N}$ by $IP(f,g)$ the class of all languages accepted by an interactive proof system with at most $f(n)$ provers and using at most $g(n)$ rounds for inputs of length n we can express the results mentioned above as follows (*poly* is the set of polynomials): PSPACE = IP = $IP(1, poly) \subseteq IP(2,1) \subseteq IP(2, poly)$ $= IP(poly, poly) =$ NEXPTIME.

It is also possible to reduce both the number of provers and the number of rounds, but this causes a larger error bound, i.e. the error bound shifts near to $\frac{1}{2}$. Unfortunately no amplification lemma is known which would allow to shift the error bound back. Amplification can only be made on the cost of more provers or more rounds. To study this interplay between the number of provers, the number of rounds and the error bound we introduce in this paper the three-component class $IP(f, g, \frac{1}{h})$ of all languages acceptable by interactive proof systems with at most $f(n)$ provers, using at most $g(n)$ rounds and having an error bounded away from $\frac{1}{2}$ by $\frac{1}{h(n)}$ for inputs of length n. We prove four simulation theorems which have besides others the following corollaries (where any $a \in (0, \frac{1}{2})$ could replace $\frac{1}{6}$):

$IP(poly, 1, \frac{1}{poly}) = IP(2, 1, \frac{1}{poly}) = IP(poly, 1, \frac{1}{6})$ and

$IP(poly, poly, \frac{1}{poly}) = IP(2, 2, \frac{1}{poly}) = IP(2, poly, \frac{1}{6}) = IP(poly, 2, \frac{1}{6})$

Defining $IP'(f, g) = IP(f, g, \frac{1}{poly})$ (which seems to be very natural because of the missing amplification lemma) we obtain

$IP'(poly, 1) = IP(poly, 1) = IP'(2, 1)$

$IP'(poly, poly) = IP(poly, poly) = IP'(2, 2)$

We conjecture that even $IP'(2, 1) = IP(2, 1)$ and $IP'(2, 2) = IP(2, 2)$ hold true.

Finally we give a simulation theorem showing that the space used by the verifier of a multi-prover interactive proof system can be reduced logarithmically on the cost of using one more prover.

2 Generalized Multi-Prover Interactive Proof Systems

In this section we give the basic definitions about generalized multi-prover interactive proof systems.

A multi-prover interactive proof system is a probabilistic Turing machine V (which is usually complexity-bounded) called the verifier. Besides the input tape and the working tapes the verifier has two special tapes, the communication tape and the address tape. Via these tapes the verifier can communicate with an unbounded number of deterministic machines $P_1, P_2, P_3, \ldots$ called the provers. This communication between the verifier and the provers is made in several rounds. One round consists of a first stage in which V gives queries to several provers (one query per prover) and a second stage where these provers give their answers to V. Queries and answers are written on the communication tape. The address tape is a read-write tape for the verifier and a read-only tape for the provers. Its contents determines which prover is allowed to read or write on the communication tape. This is organized in such a manner that no prover can read any query to or answer from another prover.

Thus we can think of the provers as recursive functions as follows. If the prover P_j is queried in the rounds $i_1, i_2, \ldots, i_r$ and the queries are $z_1, z_2, \ldots, z_r$ then P_j's answer in round i_m, $(m \leq r)$ is $P_j(z_1, z_2, \ldots, z_m)$. Note that the difference between a prover and an oracle is the following: if the same query is repeated an oracle gives the same answer while a prover can give a different answer depending on the whole "history" of queries.

Definition 1 Let $f, g, h : \mathbb{N} \mapsto \mathbb{N}$. A language L has a polynomial time interactive proof system with $f(n)$ provers, $g(n)$ rounds and error bound $\frac{1}{h(n)}$ if and only if there exists a polynomial time verifier V such that

(i) for every x the verifier V activates at most $f(|x|)$ provers in at most $g(|x|)$ rounds.

(ii) $x \in L \longrightarrow$ there exist provers $P_1, P_2, \ldots, P_{f(|x|)}$ such that prob(V accepts x with $P_1, P_2, \ldots, P_{f(|x|)}) \geq \frac{1}{2} + \frac{1}{h(|x|)}$.

(iii) $x \notin L \longrightarrow$ for all provers $P_1, P_2, \ldots, P_{f(|x|)}$ holds prob(V accepts x with $P_1, P_2, \ldots, P_{f(|x|)}) \leq \frac{1}{2} - \frac{1}{h(|x|)}$.

Definition 2 For bounding functions f, g, h we define

$L \in IP(f, g, \frac{1}{h}) \iff L$ has a polynomial time interactive proof system with $f(n)$ provers, $g(n)$ rounds, and error bound $\frac{1}{h(n)}$.

For classes $\mathcal{F}$, $\mathcal{G}$, and $\mathcal{H}$ of bounding functions we define

$$IP(\mathcal{F}, \mathcal{G}, \tfrac{1}{\mathcal{H}}) = \bigcup_{f \in \mathcal{F}} \bigcup_{g \in \mathcal{G}} \bigcup_{h \in \mathcal{H}} IP(f, g, \tfrac{1}{h}).$$

Particularly we consider the class *const* of all positive constants and the class *poly* of all polynomials. In this notation the classes IP and MIP appear as the classes $IP(1, poly, \frac{1}{6})$ and $IP(poly, poly, \frac{1}{6})$, resp., and for $f, g : \mathbb{N} \longmapsto \mathbb{N}$, the class $IP(f, g)$ of [FRS88] appears as $IP(f, g, \frac{1}{6})$.

3 The Simulation Theorems

In this section we prove three simulation theorems that enable us to reduce for a class $IP(f, g, \frac{1}{h})$ each of the functions f, g, h paying by an increase of another one. First we show in Subsection 3.1 how to reduce the number of rounds by increasing the number of provers, in Subsection 3.2 a reduction of provers is proved which causes a lower error bound. Subsection 3.3 shows how to increase the error bound by an increase of rounds.

3.1 Rounds vs. Provers

Our technique for reducing the number of rounds is the following: We use a prover to give us the whole history of the simulated protocol. But then we check whether the history was a correct history. Thus an optimal round reduction with this technique can not reduce the number of rounds below 2. We give a reduction to 2 rounds:

Theorem 3 $IP(f, g, \frac{1}{h}) \subseteq IP(f \cdot g + 1, 2, \frac{1}{h})$

Proof: Let $L \in IP(f, g, \frac{1}{h})$ via the verifier V. We define a new verifier V' as follows:

On input x generate t random bits $r_1, \ldots, r_t$, where t is the time that V computes (at most) on input x.

Round 1: $V' \longrightarrow$ Prover 1: $r_1, \ldots, r_t$
Prover 1 $\longrightarrow V'$: a list of pairs $(\alpha_{jk}, \beta_{jk})$ where $1 \leq j \leq f(|x|)$, $1 \leq k \leq g(|x|)$.

Now V' checks whether V with random sequence $r_1, \ldots, r_t$ would in round k really send α_{jk} as a question to prover j, under the assumption that prover i had answered the question α_{il} with β_{il} ($1 \leq i \leq f(|x|)$, $l < k$). If this is not true then V' rejects. If V with the given history would reject, then V' also rejects.

For each pair (j, k), where $1 \leq j \leq f(|x|)$ and $1 \leq k \leq g(|x|)$:

Round 2: $V' \to$ Prover $(j-1) \cdot g(|x|) + k + 1$: $\quad \alpha_{j1}, \ldots, \alpha_{jk}$
Prover $(j-1) \cdot g(|x|) + k + 1 \to V'$: $\quad \gamma_{jk}$

Now V' checks, whether $\beta_{jk} = \gamma_{jk}$ for all j and k. If so, then V' accepts, otherwise it rejects.

Claim: $\quad \exists P_1, P_2, \ldots$ s.t. prob$(V, P_1, P_2, \ldots$ accepts $x) \geq \delta \iff$
$\exists P_1', P_2'$ s.t. prob$(V', P_1', P_2'$ accepts $x) \geq \delta$

This claim follows from direct simulation in both directions. Note that for the backward direction the j-th prover of the new protocol can give in round k the answer of the $((j-1) \cdot g(|x|) + k + 1)$-th prover of the old protocol.

From the claim we know that V' accepts the same language as V with the same error bound. The number of rounds is 2, and the number of provers used by V' is $f \cdot g + 1$. That proves the theorem. ∎

3.2 Provers vs. Error Bound

Similar to our rounds reduction technique we want to use one prover to play the parts of all provers in the reduction of provers. But again we have to check the validity of that prover, in order to keep him from using his global information to give answers that the independent provers of the original system would not be able to give. Thus we need a second prover controlling the first one. Note that it is very unlikely that one can prove a reduction of the number of provers to one, because this would imply NEXPTIME = $IP(poly, poly, \frac{1}{6}) \subseteq IP(1, poly, \frac{1}{poly}) =$ PSPACE (see Theorem 11 below).

Theorem 4 $IP(f, g, \frac{1}{h}) \subseteq IP(2, g, \frac{1}{2 \cdot f \cdot h})$

Proof: Let $L \in IP(f, g, \frac{1}{h})$ via verifier V. We define a new verifier V' as follows:

On input x generate t random bits $r_1, \ldots, r_t$, where t is the time that V computes (at most) on input x. Generate two random numbers $s_1, s_2 \in \{1, \ldots, f(|x|)\}$.

Round i ($1 \leq i \leq g(|x|)$):

$V' \to$ Prover 1: $\quad (\alpha_{1i}, \alpha_{2i}, \ldots, \alpha_{f(|x|),i})$
$V' \to$ Prover 2: $\quad (s_1, \alpha_{s_1,i})$
Prover 1 $\to V'$: $\quad (\beta_{1i}, \beta_{2i}, \ldots, \beta_{f(|x|),i})$
Prover 2 $\to V'$: $\quad \gamma_i$

Here α_{ji} is the i-th question V would send to prover j if on the previous questions α_{mk} he received answers β_{mk} ($1 \leq m \leq f(|x|)$, $k < i$).

If there is an i with $\gamma_i \neq \beta_{s_1,i}$ then V' rejects, otherwise V' accepts if and only if $s_2 \neq 1$ or V with answers β_{ji} from prover j to question α_{ji} would accept.

<u>Claim:</u> $x \in L \Rightarrow \exists P_1', P_2'$: prob$(V', P_1', P_2'$ accepts $x) \geq 1 - \frac{1}{2 \cdot f(|x|)} + \frac{1}{f(|x|) \cdot h(|x|)}$
$x \notin L \Rightarrow \forall P_1', P_2'$: prob$(V', P_1', P_2'$ accepts $x) \leq 1 - \frac{1}{2 \cdot f(|x|)} - \frac{1}{f(|x|) \cdot h(|x|)}$

Once the claim is proved we are ready, because an easy translation by introducing an a priori rejection with appropriate probability shifts the probabilities down to around one half.

To prove the claim we first assume $x \in L$. Then there exist provers $P_1, P_2, \ldots$ such that prob$(V, P_1, P_2, \ldots$ accepts $x) \geq \frac{1}{2} + \frac{1}{h(|x|)}$. Now prover P_1' can act in his k-th component like P_k, and P_2' can answer question (j, α) like P_j would answer question α. Then, with $\delta :=$ prob$(V, P_1, P_2, \ldots$ accepts $x)$ we have

$$Pr(x) := \text{prob}(V', P_1', P_2' \text{ accepts } x) \geq \delta + (1-\delta) \cdot (1 - \tfrac{1}{f(|x|)})$$

the latter part stemming from the case that $s_2 \neq 1$. So

$$Pr(x) \geq \delta + (1-\delta) \cdot (1 - \tfrac{1}{f(|x|)}) = 1 - \tfrac{1}{f(|x|)} + \tfrac{\delta}{f(|x|)} \geq 1 - \tfrac{1}{2 \cdot f(|x|)} + \tfrac{1}{f(|x|) \cdot h(|x|)}$$

This proves the claim for $x \in L$.

Now let $x \notin L$, i.e for all $P_1, P_2, \ldots$, prob$(V, P_1, P_2, \ldots$ accepts $x) \leq \frac{1}{2} - \frac{1}{h(|x|)}$. Assume arbitrary provers P_1', P_2' for V'. Define the following partition $R = R_1 \cup R_2 \cup R_3$ for the set R of all random sequences $r_1, \ldots, r_t$:

$$R_1 := \left\{ r : \begin{array}{l} \text{a) } \forall s_1 \forall i : \ \gamma_i(r, s_1) = \beta_{i,s_1}(r) \\ \text{b) } V \text{ with the history simulated by } V' \text{ with } P_1' \text{ on} \\ \quad \text{random sequence } r \text{ would accept} \end{array} \right\}$$

(here $\gamma_i(r, s_1)$ means the value of γ_i that would appear in the protocol of V', if the random sequence was r and the first random number was s_1, and $\beta_{i,s_1}(r)$ means the value of β_{i,s_1} if the random sequence was r.)

$$R_2 := \left\{ r : \begin{array}{l} \text{a) } \forall s_1 \forall i : \ \gamma_i(r, s_1) = \beta_{i,s_1}(r) \\ \text{b) } V \text{ with the history simulated by } V' \text{ with } P_1' \text{ on} \\ \quad \text{random sequence } r \text{ would not accept} \end{array} \right\}$$

$$R_3 := \{ r : \ \exists s_1 \exists i : \ \gamma_i(r, s_1) \neq \beta_{i,s_1}(r) \}$$

Obviously $\frac{|R_1|}{|R|} \leq \frac{1}{2} - \frac{1}{h(|x|)}$, because otherwise we could construct provers accepting x with verifier V and probability greater than $\frac{1}{2} - \frac{1}{h(|x|)}$, in contradiction to $x \notin L$.

Now

$$Pr(x) := \text{prob}(V', P_1', P_2' \text{ accepts } x) \leq \tfrac{|R_1|}{|R|} + \tfrac{|R_2|}{|R|} \cdot (1 - \tfrac{1}{f(|x|)}) + \tfrac{|R_3|}{|R|} \cdot (1 - \tfrac{1}{f(|x|)}),$$

the first term reflecting the cases that V' has to accept anyway, the second term the case that $s_2 \neq 1$ (which happens with probability $1 - \frac{1}{f(|x|)}$), and the third term the case that although $\gamma_i(r, s_1) \neq \beta_{i,s_1}(r)$ for at least one pair (i, s_1), the verifier V' does not recognize that, because V' did not randomly choose this s_1, which happens at most with probability $1 - \frac{1}{f(|x|)}$. So we conclude that

$$\begin{aligned} Pr(x) &\leq \frac{|R_1|+|R_2|+|R_3|}{|R|} \cdot \left(1 - \frac{1}{f(|x|)}\right) + \frac{|R_1|}{|R|} \cdot \frac{1}{f(|x|)} \\ &\leq 1 - \frac{1}{f(|x|)} + \left(\frac{1}{2} - \frac{1}{h(|x|)}\right) \cdot \frac{1}{f(|x|)} = 1 - \frac{1}{2 \cdot f(|x|)} - \frac{1}{f(|x|) \cdot h(|x|)} \end{aligned}$$

That completes the proof of Theorem 4. ■

3.3 Error Bound vs. Rounds (or Provers)

We want to show now that a small error bound can be increased to the constant value $\frac{1}{6}$ by increasing the number of rounds. We do this by sequentially repeating the original computation of the verifier. We certainly can make the new verifier act in each repetition exactly like the original verifier, but we have to prove that the provers cannot do any better than also act in each repetition the same way they would act in the original system. Once this is shown, we can view the repetitions as independent events and apply the technique that SCHÖNING [Sc86] used to amplify BPP-systems, especially the following proposition:

Proposition 5 (SCHÖNING [Sc86], Lemma 3.4)
Let E be some event that occurs with probability at least $\frac{1}{2} + d$, $0 < d < \frac{1}{2}$. Then E occurs within t independent trials (t odd) more than $\frac{t}{2}$ times with probability at least

$$1 - \frac{1}{2} \cdot (1 - 4d^2)^{t/2}.$$

Theorem 6 $IP(f, g, \frac{1}{h}) \subseteq IP(f, (h+1)^2 \cdot g, \frac{1}{6})$

Proof: We first have to show that the n-th execution of the same protocol (with new random bits) does not depend on the first $n-1$ executions, as long as the verifier acts always in the same way. But assume to the contrary that the provers could achieve a higher acceptance probability exploiting their knowledge (depending on the particular random sequences) about the first $n-1$ executions. Then we can construct new provers for the original system making the provers act as if there had been $n-1$ previous executions with suitable random sequences. Note that the provers can easily compute optimal random sequences for the hypothetical first $n-1$ executions! So the n-th execution can not be better than the execution of the original system, but it certainly does not have to be worse, since the provers can always act like the ones of the original system. This proves the independence of serial executions.

Now let V be a verifier accepting L with $f(n)$ provers, $g(n)$ rounds and $\frac{1}{h(n)}$ error bound. On input x let V' be the verifier that $(h(|x|)+1)^2$ times iterates the protocol of V (we assume $h(|x|)$ is even, otherwise even $(h(|x|))^2$ iterations suffice), and accepts if more than half of the iterations simulated accepting protocols. Since we can view the iterations as independent, we can apply Proposition 5 and conclude that V' accepts with probability not less than

$$1-\frac{1}{2}\cdot\left(1-4\cdot\left(\frac{1}{h(|x|)+1}\right)^2\right)^{(h(|x|)+1)^2/2} \geq 1-\frac{1}{2}\cdot\left[\left(1-\frac{4}{(h(|x|)+1)^2}\right)^{(h(|x|)+1)^2/4}\right]^2$$

$$\geq 1-\frac{1}{2}\cdot\left(\frac{1}{e}\right)^2 \geq \frac{2}{3}$$

(recall that the function $(1-\frac{1}{x})^x$ approximates $\frac{1}{e}$ from below!), and so the error bound is greater than $\frac{1}{6}$. Note that V' still uses at most $f(|x|)$ provers, and and the number of rounds is at most $(h(|x|)+1)^2\cdot g(|x|)$, as required. ∎

Corollary 7 $IP(f,g,\frac{1}{h}) \subseteq IP((h+1)^2\cdot f,g,\frac{1}{6})$

Proof: The parallel execution of MIP computations, where every execution has its own set of provers, is trivially independent. So we can apply the same amplification technique as in Theorem 6. ∎

4 Consequences for Ordinary IP-systems

In the following we will as usual write $IP(f,g)$ instead of $IP(f,g,\frac{1}{6})$. First we investigate the case where at least one of the parameters is *poly*. We show how one can derive from our simulation theorems in Section 3 the already known result $IP(2,poly) = IP(poly,2) = IP(poly,poly) =$ MIP ([BGKW88], [FRS88]).

Theorem 8
$IP = IP(1,poly) \subseteq IP(poly,1) \subseteq IP(poly,2) = IP(2,poly) =$ MIP.

Proof: It suffices to prove

a) $IP(1,poly) \subseteq IP(poly,1)$

b) $IP(poly,poly) \subseteq IP(2,poly)$

c) $IP(2,poly) \subseteq IP(poly,2)$

Part a) follows from [CCL90], where they show that even $IP(2,1)$ contains IP = PSPACE. Part b) can be shown by first reducing the number of provers (Theorem 4), which results in a lower error bound, and then increasing the error bound again (Theorem 6). Finally, part c) follows directly from Theorem 3. ∎

Now we give the relationships for classes, where at least one parameter is *const*, but none is *poly*:

Theorem 9
$IP(1,const) \subseteq IP(const,1) \subseteq IP(const,2) = IP(2,const) = IP(const,const)$.

Proof: The proof is essentially the same as for Corollary 8. ∎

The relationships between different IP-classes are shown in Figure 4.1.

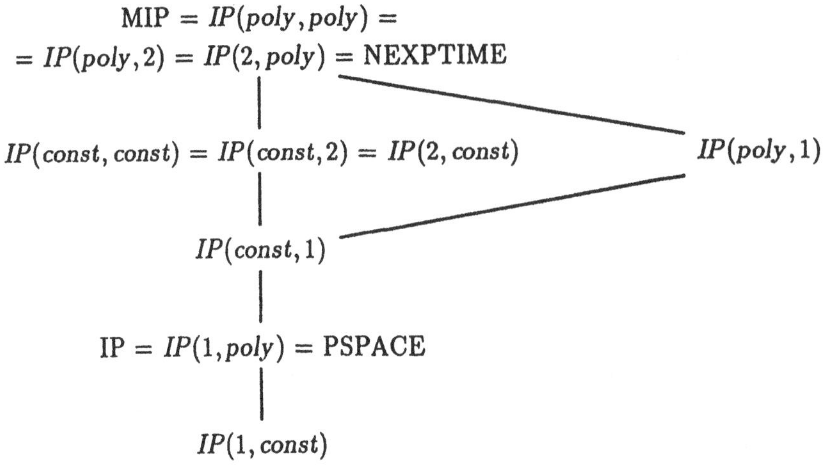

Figure 4.1: The most important IP-classes

5 Consequences for Generalized IP-systems

In this section we use the notation $IP'(f,g)$ to denote $IP(f,g,\frac{1}{poly})$. Obviously we have

Proposition 10 For all bounding functions f and g holds $IP(f,g) \subseteq IP'(f,g)$.

We show the collapse of the IP$'$-hierarchy to essentially only four classes and give some relations between IP$'$-classes and IP-classes.

Theorem 11

a) $IP'(2,2) = IP'(poly,poly) = IP(poly,poly)$

b) $IP'(2,1) = IP'(poly,1) = IP(poly,1)$

c) $IP'(1,poly) = IP(1,poly)$

d) $IP'(1,1) = IP'(1,const) = IP(1,const) = IP(1,1)$

Proof:

a) By Theorem 3, $IP'(poly,poly) \subseteq IP'(poly,2)$, and by Theorem 4, $IP'(poly,2) \subseteq IP'(2,2)$. Finally Theorem 6 yields $IP'(2,2) \subseteq IP(2,poly)$.

b) By Theorem 4 we obtain $IP(poly,1) \subseteq IP'(2,1)$, and then using Corollary 7 we get $IP'(poly,1) \subseteq IP(poly,1)$.

c) Use Theorem 6.

d) In the case of only one prover there is a general amplification scheme without increase of provers or rounds (cf. [BM88]). Thus $IP'(1, const) = IP(1, const)$. Finally, $IP(1, const) = IP(1,1)$ ([Ba85], [GS86]) completes the proof.

∎

Thus the combined hierarchy of IP and IP′ classes looks the way it is shown in Figure 5.1.

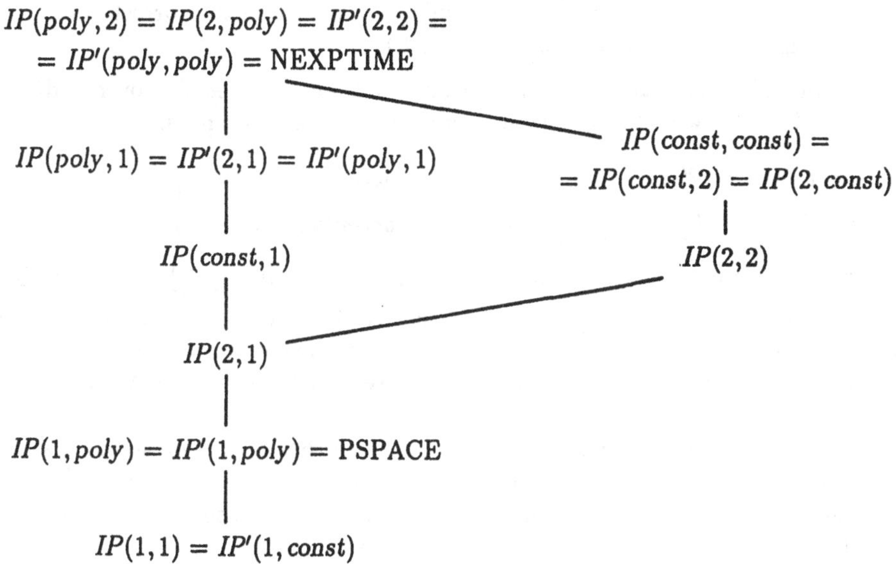

Figure 5.1: The most important IP and IP′ classes

6 Space bounded MIP-Systems

In this section we will show how one can reduce the space used by the verifier of a polynomial time multi-prover interactive proof system (for short: MIP system) on the cost of using more provers. The fundamental theorem for that space reduction is the following:

Theorem 12 Let L be accepted by an MIP system with an $s(n)$ space bounded verifier and k provers (k a constant). Then L can be accepted by an MIP system with a $\log s(n)$ space bounded verifier and $k+1$ provers.

Proof: Let V be an $s(n)$ space bounded verifier for L using k provers. Define a new verifier V' as follows:

On input x generate $\log s(|x|)$ random bits, interpreted as an address *adr* on the worktape of V. Generate another $\log s(|x|)$ random bits, interpreted as a number $z \in \{1, \dots, s(n)\}$.

Use one special prover that has the task to simulate the worktape of V. Now simulate step by step the work of V, only all accesses to the worktape are done by a query to the special prover. On logarithmic space (in s), V' is able to remember the head position on the worktape of V, and every time the head position is *adr*, V' can control whether the special prover gives the right content of that cell (from the beginning, V' can store the correct value of tape cell *adr* using only one extra tape cell).

V' accepts, if there was never a wrong answer from the special prover and either $z \neq 1$ or V with the simulated history would have accepted.

Claim: $x \in L \Rightarrow \exists P'_1, P'_2, \dots$: prob($V', P'_1, P'_2, \dots$ accepts x) $\geq 1 - \frac{1}{3 \cdot s(|x|)}$

$x \notin L \Rightarrow \forall P'_1, P'_2, \dots$: prob($V', P'_1, P'_2, \dots$ accepts x) $\leq 1 - \frac{2}{3 \cdot s(|x|)}$

Once the claim is proved we obtain easily an $IP(k+1, poly, \frac{1}{12 \cdot s(n)})$ system for L by shifting the probability from $1 - \frac{1}{2 \cdot s(n)}$ to $\frac{1}{2}$. Theorem 6 then yields an $IP(k+1, poly)$ system for L. Note that the additional space needed by the construction in the proof of Theorem 6 is only logarithmically in $s(n)$, because we only have to count the number of repetitions.

To prove the claim we proceed very similar to the proof of the analogous claim in Section 3.2. In case $x \in L$, we get the acceptance probability $\delta + (1-\delta) \cdot (1 - \frac{1}{s(|x|)})$, the latter part stemming from the case that $z \neq 1$. Here δ is the acceptance probability of V on x. Since $\delta \geq \frac{2}{3}$, a simple computation gives the above estimation.

Now let $x \notin L$. Then we give a partition of the set R of possible random sequences (of the original verifier):

$R_1 := \{r$: the special prover never lies in a protocol with this random sequence, and V with this random sequence would accept$\}$,
$R_2 := \{r$: the special prover never lies in a protocol with this random sequence, and V with this random sequence would reject$\}$,
$R_3 := \{r$: the special prover lies at least once with this random sequence$\}$.

We then get an acceptance probability of at most

$$\frac{|R_1|+|R_2|+|R_3|}{|R|} \cdot \left(1 - \frac{1}{s(|x|)}\right) + \frac{|R_1|}{|R|} \cdot \frac{1}{s(|x|)} \leq 1 - \frac{2}{3 \cdot s(|x|)}$$

because at most one third of all r can be in R_1, if $x \notin L$. ■

Corollary 13 Every language in NEXPTIME = MIP can be accepted by an MIP-system with an $O(\log^{(k)} n)$ space bounded verifier and $k+2$ provers. (By $\log^{(k)}$ we denote $\log \log \dots \log$ (k times).)

Proof: By Theorem 8 every set from MIP can be accepted by an MIP system with a polynomially space bounded verifier and two provers. Now applying k times Theorem 12 gives the result. ■

7 Concluding remarks

Our main attention in this paper was on interactive proof systems with more than one prover. So far no amplification lemma for such systems is known which would justify to fix the error bound arbitrarily without any influence to the power of the interactive proof systems. Thus we introduced error bound dependent IP classes the largest of which allow an error bounded away from $\frac{1}{2}$ by $\frac{1}{h}$ where h is an arbitrary polynomial. For these classes $IP'(f,g)$ we could show that there are at most two different classes for systems with at least two provers, namely $IP'(2,1) = IP'(poly,1)$ and $IP'(2,2) = IP'(poly,poly)$. For the classes with fixed error bound we do not know such a strong result. Particularly, we do not know whether $IP(2,1) = IP(poly,1)$ and $IP(2,2) = IP(poly,poly)$. This is equivalent to $IP(2,1) = IP'(2,1)$ and $IP(2,2) = IP'(2,2)$, resp. Moreover, this is equivalent to the validity of an amplification lemma for one-round systems or two-round systems, resp. We conjecture that this is true, i.e. that the hierarchy of all IP classes with more than one prover collapses to at most two classes as shown in figure 7.1. In this case the study of the class $IP(2,1)$ which is located between PSPACE and NEXPTIME would be of particular interest.

$$IP(2,2) \stackrel{!}{=} IP'(poly,poly) = \text{NEXPTIME}$$

$$|$$

$$IP(2,1) \stackrel{!}{=} IP'(poly,1)$$

$$|$$

$$IP(1,poly) = \text{PSPACE}$$

Figure 7.1: Conjectures

8 Acknowledgements

This work has been started during a joint seminar week of the structural complexity groups of the Technical University Berlin, the University Jena, and the University Würzburg. This seminar week took place in June 1990 and was the first which became possible after the revolution in East Germany. We thank all the participants for the many fruitful discussions. We are grateful to Uwe Schöning (Ulm) for several helpful hints.

References

[Ba85] L. Babai. Trading group theory for randomness. 17th *Annual ACM Symposium on Theory of Computing (STOC)*, 421–429, 1985.

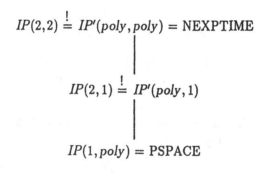

[Ba90] L. Babai. E-mail and the unexpected power of interaction. 5th *Structure in Complexity Theory (IEEE)*, 30–44, 1990.

[BFL90] L. Babai, L. Fortnow, C. Lund. Non-deterministic exponential time has two-prover interactive protocols. University of Chicago Technical Report 90–03, 1990.

[BGKW88] M. Ben-Or, S. Goldwasser, J. Kilian, A. Wigderson. Multi-prover interactive proofs: How to remove the intractability assumptions. 20th *Annual ACM Symposium on Theory of Computing (STOC)*, 113–131, 1988.

[BHZ87] R. Boppana, J. Håstad, S. Zachos. Does co-NP have short interactive proofs? *Information Processing Letters* 25, 127–132, 1987.

[BM88] L. Babai, S. Moran. Arthur-Merlin games: A randomized proof system, and a hierarchy of complexity classes. *Journal of Computer and System Science* 36 2, 254–276, 1988.

[CCL90] J. Cai, A. Condon, R. Lipton. PSPACE is provable by two provers in one round. Manuscript, 1990.

[FRS88] L. Fortnow, J. Rompel, M. Sipser. On the power of multi-prover interactive protocols. 3rd *Structure in Complexity Theory (IEEE)*, 156–161, 1988.

[FRS90] L. Fortnow, J. Rompel, M. Sipser. Errata for on the power of multi-prover interactive protocols. 5th *Structure in Complexity Theory (IEEE)*, 318–319, 1990.

[FS88] L. Fortnow, M. Sipser. Are there interactive protocols for co-NP languages? *Information Processing Letters* 28, 249–251, 1988.

[GMR85] S. Goldwasser, S. Micali, C. Rackoff. The knowledge complexity of interactive proof systems. 17th *Annual ACM Symposium on Theory of Computing (STOC)*, 291–304, 1985.

[GS86] S. Goldwasser, M. Sipser. Private coins versus public coins in interactive proof systems. 18th *Annual ACM Symposium on Theory of Computing (STOC)*, 59–68, 1986.

[LFKN89] C. Lund, L. Fortnow, H. Karloff, N. Nisan. The polynomial time hierarchy has interactive proofs. E-mail announcement, 1989.

[Sc86] U. Schöning. *Complexity and Structure*. Springer-Verlag *Lecture Notes in Computer Science* 211, 1986.

[Sh89] A. Shamir. IP = PSPACE. E-mail announcement, 1989, also in 31st *Annual Symposium on Foundations of Computer Science (FOCS)*, 11–15, 1990.

Logics for Belief Dependence

Zhisheng Huang
Department of Mathematics and Computer Science
University of Amsterdam, The Netherlands

Abstract

In this paper, we investigate the theoretical foundations of belief dependence in multiple agent environment, where agents may rely on someone else about their beliefs or knowledge. Several logics for belief dependence are introduced and studied. First of all, we try to formalize the problem of belief dependence in the framework of general epistemic logics, by which we will argue that general epistemic logic is not appropriate to formalize the problem of belief dependence. Then, based on an approach which is similar to Fagin and Halpern's general awareness logic, we present the second logic for belief dependence, which is called a syntactic approach. The third logic is an adapted possible world logic, where sub-beliefs are directly introduced in the models.

1 Introduction

One of the important topics of research in logics of computer science and artificial intelligence is to study the problem of reasoning about knowledge, especially, in multiple agent environment. Recently reasoning about knowledge in multiple agent environment has found many applications such as distributed knowledge-bases, communication and cooperation for multi agents planning[1, 3, 4, 10, 12].

However, little attention has been paid to study the problem of belief dependence in multiple agent environment, where agents may rely on someone else about their beliefs and knowledge. As is well known, in multiple agent environment, it is frequently beneficial to enable agents to communicate their knowledge or beliefs among agents, because these agents generally may have limited resources, or may lack computation capability for some specified problems or facts.

Although there have been attempts to study the problem of the communication of belief and knowledge among agents [4, 12], the existing formalisms generally focus on the problem of communication, in which some main features about belief dependence, such as suspicion, indirect dependence, are rarely formalized. In this paper, we would like to develop a formal theory of belief dependence which serves as a foundation for understanding rational behaviour of artificial agents in multiple agent environment. Moreover, we expect that the proposed formalism would be expressive and natural enough to specify knowledge and belief passing and dependence among artificial agents, which can be found some applications in those relevant fields such as knowledge acquisition, machine learning, human-computer interaction, distributed artificial intelligence and distributed network systems.

2 The Problem of Belief Dependence

2.1 Compartmentalized Information and Incorporated Information

Just like human beings, artificial agents (computers, knowledge bases, robots, and processes) get information from someone else, and then assimilate the information. In the existing approaches to formalize the procedure of artificial agents' information assimilation, others' knowledge and beliefs are simply accepted or refused, which are handled with by different strategies. Because others' knowledge and beliefs are often contradict each other, many parts of the information may be refused. In order to solve the problem, a natural approach is to introduce the notion of probability-based beliefs, by which an agent may have contradict beliefs, because they can be indexed by different probabilities. However, as far as I know, there exist no strong psychological evidences which shows that it is necessary to use the notion of probability in human cognitive activities. An alternative approach is to introduce the notion of *society of minds*[2]. Formally, the notion of "society of minds" means that each agent possesses its own cluster of beliefs, which may contradict each other. Each cluster of beliefs is connected with each mind frame. However, if accepted information is simply separated in different mind frames, it is hard to say that agents can assimilate others' knowledge efficiently and can enlarge his knowledge and beliefs.

In the studies of incorporating new information into existing world knowledge of human beings, cognitive psychologists make a distinction between compartmentalized information and incorporated information. As Potts et al. point out in[11]:

> ...it is unlikely that subjects in most psychology experiments incorporate the new information they learn into their existing body of world knowledge. Though they certainly use their existing world knowledge to help comprehend the new material, the resulting amalgam of new information, and the existing world knowledge used to understand it, is isolated as a unit unto itself: it is *compartmentalized.*

We also believe that an appropriate procedure to assimilate others' knowledge and beliefs should pass the following two phases: compartmentalized information and incorporated information. Formally, *compartmentalized information* are those fragment of information which are accepted and remembered as isolated beliefs which are somewhat different from those beliefs are completely believed. Whereas *incorporated information* consists of those beliefs are completely believed by the agents.

2.2 Some Syntactic Considerations for Logics of Belief Dependence

There are some important and fundamental notions in logics for belief dependence. First of all, there is the general notion about knowledge and beliefs. Therefore, in our logics for belief dependence, general epistemic and doxastic operators are used to represent agents' knowledge and beliefs. For the sake of convenience, just like those in general epistemic logics, we use $L_i\varphi$ to represent that agent i knows or believes the formula φ. As is well known, L is interpreted as an epistemic operator, if the logic system is a **S5** system, whereas L is a doxastic operator if the system is a weak **S5** system.

In the existing epistemic logics, agents generally make no distinction among sources of those knowledge and beliefs. However, in real life, human beings seem not to be so naive. When peoples get information from outside, they generally keep in minds about the sources of information, They know from whom the information comes at the first phase of information assimilation, although they may finally forget these sources at all. Sometimes they even may make appraisal of agents who send the information to him. We call the phenomenon in which agents track sources of information *source indexing.*

In order to formalize the compartmentalized information and source indexing, in the logics of belief dependence, a natural strategy is to introduce a *compartment modal operator* $L_{i,j}$. Intuitively, we can give $L_{i,j}\varphi$ an interpretation: "agent i believes φ due to agent j". From the point of view of minds society, $L_{i,j}\varphi$ can be more intuitively interpreated as "agent i believes φ on the mind frame indexed j". Sometimes we call $L_{i,j}\varphi$ agent i's *sub-belief,* and $L_{i,j}$ is called *sub-belief operator.* $L_{i,i}\varphi$ naturally means that agent i believes φ, which semantically corresponds to the modal operator for knowledge and beliefs in general epistemic logics. Sometimes we use $L_i\varphi$ as an abbreviation of $L_{i,i}\varphi$.

Both sub-beliefs and general beliefs have close relationships with the truth and falsity of beliefs. Sometimes we need a neutral[1] modal operator $D_{i,j}$ for belief dependence logics. $D_{i,j}$ is called *dependent operator,* or alternatively *rely-on operator.* Intuitively, we can give $D_{i,j}\varphi$ a number of interpretations:"agent i relies on agent j about the formula φ", "agent i depends on agent j about believing φ", "agent j is the credible advisor of agent i about φ", even specially in distributed process networks, "processor i can obtain the knowledge about φ from processor j". There are two kinds of interpretations about dependent operator $D_{i,j}$. One is *explicit dependence,* which says that belief dependence is explicitly known by believers. In other words, that means the axiom $D_{i,j}\varphi \rightarrow L_iD_{i,j}\varphi$ holds. The other one is implicit dependence, in which believers do not necessarily know their dependence.

However, it should be noted that $L_{i,i}\varphi$ is not necessarily equal to $L_i\varphi$. As is well known, general epistemic logics suffer from the problem of logical omniscience. The so-called *logical omniscience* means that agents are assumed to be intelligent that they must know all valid formulas, and that their knowledge is closed under implication, so that if an agent knows p, and that p implies q, then the agent must also know q. However, in computer science, even in real life, agents are not such ideal reasoners. In order to provide a more realistic representation of human reasoning, there are various attempts to deal with the problem of logical omniscience[3, 6, 7, 10]. In [10], Levesque first presents the notions of *explicit belief* and *implicit belief.* Explicit beliefs are those beliefs an agent actually has, whereas implicit beliefs consist of all of the logical consequences of an agent's explicit beliefs. In [3], Fagin and Halpern point out that 'lack of awareness" is one of sources of logical omniscience. They argue that one cannot say he knows p or does not know p if p is a concept he is completely unaware of. In order to solve the problem of awareness, Fagin and Halpern offer a solution in which one can decide on a metalevel what formulas an agent is supposed to be aware of. In their general awareness logic, implicit beliefs are represented as $L_i\varphi$, whereas explicit beliefs are defined as $L_i\varphi \wedge A_i\varphi$, where $A_i\varphi$ means that agent i is aware of φ. In [6], we argue that the notion of belief dependence can be viewed

[1]Because we consider the axiom $D_{i,j}\varphi \equiv D_{i,j}\neg\varphi$ as a fundamental axiom about $D_{i,j}$

as an intuitive extension to the notion of awareness, since we can define $A_i\varphi \equiv \exists j D_{i,j}\varphi$. This means that agent i is aware of φ if and only if agent i believes in himself about φ or agent i could get the truth of the formula φ by consulting his adviser about φ. Moreover, at least, we can define $A_i\varphi \equiv D_{i,i}\varphi$, therefore, $L_{i,i}\varphi$ is not necessarily equal to $L_i\varphi$. From the point of view of explicit beliefs and implicit beliefs, $L_i\varphi$ can be interpret as implicit belief, whereas $L_{i,i}\varphi$ can be interpret as explicit beliefs if we define $L_{i,i}\varphi \equiv D_{i,i}\varphi \wedge L_i\varphi$.

Supposed we have a set $\mathbf{A_n}$ of n agents, and a set Ψ_0 of primitive propositions, the language $\mathbf{L}$ for belief dependence logics is the minimal set of formulas closed the following syntactic rules:

(i) $\mathbf{true} \in \mathbf{L}$
(ii) $p \in \Psi_0 \Rightarrow p \in \mathbf{L}$
(iii) $\varphi \in \mathbf{L}, \psi \in \mathbf{L} \Rightarrow \varphi \wedge \psi \in \mathbf{L}$,
(iv) $\varphi \in \mathbf{L} \Rightarrow \neg\varphi \in \mathbf{L}$,
(v) $\varphi \in \mathbf{L}, i \in \mathbf{A_n} \Rightarrow L_i\varphi \in \mathbf{L}$
(vi) $\varphi \in \mathbf{L}, i, j \in \mathbf{A_n} \Rightarrow L_{i,j}\varphi \in \mathbf{L}$
(vii) $\varphi \in \mathbf{L}, i, j \in \mathbf{A_n} \Rightarrow D_{i,j}\varphi \in \mathbf{L}$

Logical connectives such as $\rightarrow$ and $\vee$ are defined in terms of $\neg$ and $\wedge$ as usual, and **false** is an abbreviation of $\neg$**true**.

In some special belief dependence logics, among the three belief dependence modal operators, some may be defined by others. For example, the sub-belief modal operator can be defined by the general epistemic operator and the dependent operator, i.e. $L_{i,j}\varphi \stackrel{\text{def}}{=} D_{i,j}\varphi \wedge L_j\varphi$, if we suppose that the communications between agents are reliable, and every teller is honest. Moreover, sometimes we may view the general epistemic operator as a kind of special sub-epistemic operator, i.e. $L_i\varphi \stackrel{\text{def}}{=} L_{i,i}\varphi$. Therefore, sometimes we need some sub-language for belief dependence logics. We define the language $\mathbf{L_D}$ as the minimal set of formulas closed by the syntactic rules (i),(ii),(iii),(iv),(v), and (vii). Moreover, the language $\mathbf{L_L}$ is defined by the rules (i),(ii),(iii),(iv),(v) and the language $\mathbf{L_{Lij}}$ is defined by the rules (i),(ii),(iii),(iv), and (vi).

2.3 General Scenario

We have argued that an appropriate procedure for formalizing information assimilation should pass two phases: compartmentalized and incorporated information. In the logics for belief dependence, compartmentalized information corresponds to sub-beliefs $L_{i,j}\varphi$ for agent i. Whereas incorporated information corresponds to general beliefs of agent i, namely, $L_i\varphi$.

For multiple agent environment, we assume that some primitive rely-on relations about some propositions among those agents can be decided on the metalevel. We call the assumption *initial role-knowledge assumption.* We believe that the assumption is appropriate and intuitive. That is because, in multiple agent environment, some agents have to possess some minimal knowledge about someone else, in order to guarantee their communications. In many application situations, primitive rely-on relations are easy to be captured, because primitive rely-on relation can be generally viewed those which have no relationship with the problem how the agents solve the conflict of their own beliefs with new information. In other words, in a reliable communication network, if we assume that agents are honest, no-doubt and something else, primitive rely-on relations often collapse into primitive communication relations.

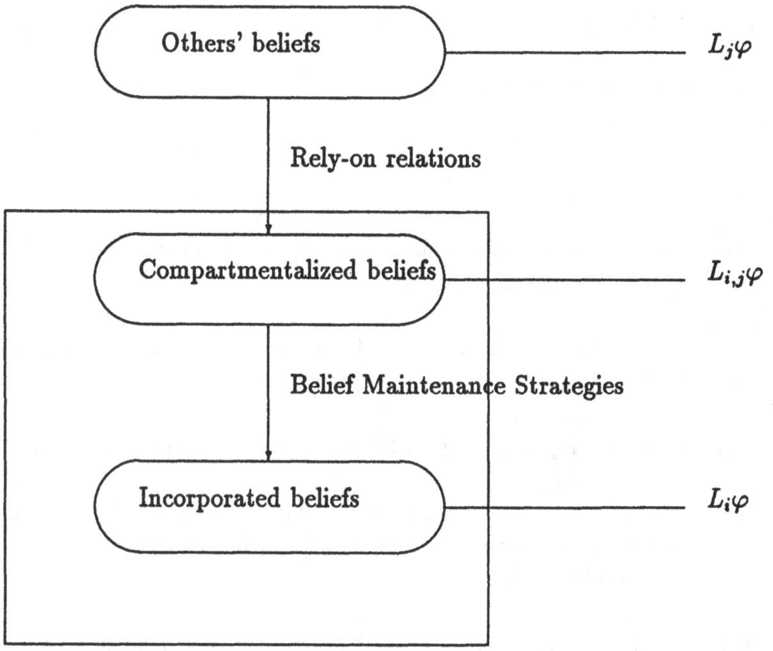

Figure 1: General Scenario

Therefore, based on the primitive rely-on relations, we can capture a complete knowledge about agents' sub-beliefs by using the logics for belief dependence. Furthermore, based on the complete sets concerning agents' sub-beliefs, we can figure out some agents' appraisal information about others. In the next section we will propose some role-appraisal axioms such as "fool believer", and "stubborn believer". Based on these role-appraisal information, it is possible to determine some rational belief maintenance strategies, by which we can figure out whether and how compartmentalized beliefs can be assimilated into the incorporated beliefs for some agents. However, in this paper, we would like to focus on the formalism concerning the first phase of information assimilation. That is, we will focus on the problem how the complete sub-belief and the complete rely-on relations can be captured, basing on the primitive rely-on relations. As far as the second phase is concerned, we will discuss the problem in the further papers[8]. The general scenario about the formalism of belief dependence is shown in the figure.

3 Formalizing Belief Dependence

3.1 Belief Dependence Systems Based on Epistemic Operator and Dependent Operator

In this subsection, first of all, we would like to present a belief dependence axiom system, basing on general epistemic operator and dependent operator. Naturally, weak-S5 system remains to be the subsystem of belief dependence system. Here is a logic system for belief dependence, which is called **L5⁻+D4** system:

Axioms:

(L1) All instances of propositional tautologies.

(L2) $L_i\varphi \wedge L_i(\varphi \to \psi) \to L_i\psi$.

(L3) $\neg L_i$**false**.

(L4) $L_i\varphi \to L_iL_i\varphi$.

(L5) $\neg L_i\varphi \to L_i\neg L_i\varphi$.

The axioms above consist of a weak-S5 modal logic system. Moreover, we select the following axioms as axioms about dependent operator:

(D1) $D_{i,j}\varphi \equiv D_{i,j}\neg\varphi$.

(Neutral axiom. Rely on someone else about φ iff rely on about the negation of φ. It seems to be the most fundamental axiom for dependent operator.)

(D2) $D_{i,j}\varphi \wedge D_{i,j}(\varphi \to \psi) \to D_{i,j}\psi$.

(Closure under implication, for dependent operator, closing under implication is intuitive.)

(D3) $D_{i,j}\varphi \wedge D_{i,j}\psi \to D_{i,j}(\varphi \wedge \psi)$.

(Closure under conjunction. Because we index sub-beliefs simply by agent name, this requires that beliefs which come from the same agent should be consistent. Therefore, we consider the axiom a reasonable one.)

(D4) $D_{i,j}\varphi \to L_iD_{i,j}\varphi$.

(Positive explicit dependent axiom. As it is argued above, the axiom means that dependency is explicitly known by believer.)

Rules of Inference:

(R1) $\vdash \varphi, \vdash \varphi \to \psi \Rightarrow \vdash \psi$.

(R2) $\vdash \varphi \Rightarrow \vdash L_i\varphi$.

So far we have not present any axiom concerning sub-belief operator $L_{i,j}$ in the logic system **L5⁻+D4**. If we suppose that the communications in the system are reliable and every agent is honest, then a plausible definition about the sub-belief operator can be represented as follows:

Definitions:

(Lijdf) $L_{i,j}\varphi \equiv D_{i,j}\varphi \wedge L_j\varphi$.

3.2 Belief Dependence System Based on Sub-belief Operator

Based on the sub-belief operator, we also can present logic systems for belief dependence. The following axiom system is called to be a **Lij5⁻+D** belief dependence logic system:

Axioms

(L1) All instances of propositional tautologies.

(Lij2) $L_{i,j}\varphi \wedge L_{i,j}(\varphi \to \psi) \to L_{i,j}\psi$.

(Just like those in general epistemic logics, sub-beliefs are closed under logical implication.)

(Lij3) $\neg L_{i,j}$**false**.

(This axiom means that an agent never believe the false fact from someone else, including himself.)

(Lij4) $L_{i,j}\varphi \to L_iL_{i,j}\varphi$.

(Positive introspective axiom for sub-beliefs.)

(Lij5) $\neg L_{i,j}\varphi \to L_i\neg L_{i,j}\varphi$.

(Negative introspective axiom for sub-beliefs.)

Rules

(R1) $\vdash \varphi, \vdash \varphi \rightarrow \psi \Rightarrow \vdash \psi$.
(RLij) $\vdash \varphi \rightarrow \vdash L_{i,j}\varphi$.
Definitions
(Ddf) $D_{i,j}\varphi \stackrel{\text{def}}{=} L_{i,j}\varphi \vee L_{i,j}\neg\varphi$.
(If agent i believes φ or believes $\neg\varphi$ from agent j, then this means that agent i rely on agent j about φ.)
(Ldf) $L_i\varphi \stackrel{\text{def}}{=} L_{i,i}\varphi$.
(This definition means that we make no distinction between implicit beliefs and explicit beliefs.)

Claim 3.1 *The logic system* **L5⁻+D4** *is a subsystem of the logic system* **Lij5⁻+D**. *That is,* $\mathbf{L5^-{+}D4} \subset \mathbf{Lij5^-{+}D}$.

3.3 Formalizing Suspicion and Other Features

Based on the three modal operators concerning belief dependence, namely, the general epistemic operator L_i, the sub-belief operator $L_{i,j}$, and the dependent operator $D_{i,j}$, we can formalize many important and interesting features about belief dependence. The following axioms can be some candidates for formalizing belief dependence.
(a) No-doubt Axiom
$L_{i,j}\varphi \rightarrow L_iL_j\varphi$. (*Whatever come from someone else is believed to be true.*)
We know that $L_{i,j}\varphi$ is not necessarily equal to $L_iL_j\varphi$. However, in the no-doubt belief dependence system, the sub-belief $L_{i,j}\varphi$ implies $L_iL_j\varphi$.
(b) Honesty Axiom
$L_{i,j}\varphi \rightarrow L_j\varphi$. (*Sub-beliefs are actually teller's beliefs.*)
Therefore, if we select the definition $L_{i,j}\varphi \stackrel{def}{=} D_{i,j}\varphi \wedge L_j\varphi$, then this means that in the system every agent is honest.
(c) Negative Explicit Dependent Axiom
$\neg D_{i,j}\varphi \rightarrow L_i\neg D_{i,j}\varphi$.
(*If agent i does not rely on agent j about φ, then agent i will know the fact.*)
(d) Consultation Axiom
$D_{i,j}\varphi \rightarrow L_jD_{i,j}\varphi$.
(*Agent i asks for the information about φ from agent j, and believes what is told. Therefore, agent j knows his relied on.*)
(e) Confidence Axiom
$L_i\varphi \wedge D_{i,j}\varphi \rightarrow L_iL_j\varphi$.
(*Agent i believes his dependent beliefs are actually true.*)
(f) Fool Believer Axiom
$L_i\varphi \rightarrow \exists j L_{i,j}\varphi \quad (j \neq i)^2$. (*All of his beliefs come from someone else.*)
(g) Stubborn Believer Axiom
$L_{i,j}\varphi \rightarrow L_i\varphi$. (*He never believes those coming from someone else.*)
(h) Communicative Agent Axiom
$L_i\varphi \rightarrow \exists j L_{j,i}\varphi \quad (j \neq i)$. (*All of his beliefs are believed by someone else.*)
(i) Cautious Believer Axiom

[2] Although we do not introduce any quantifer and equality in the language **L**, however, because we generally consider a finite agent set, say $\mathbf{A_n} = \{i_1, ..., i_n\}$, the formula $\exists j L_{i_l,j}\varphi \ (j \neq i_l)$ can be viewed to be an abbreviation for the formula $L_{i_l,i_1}\varphi \vee ... \vee L_{i_l,i_{l-1}}\varphi \vee L_{i_l,i_{l+1}}\varphi \vee ... \vee L_{i_l,i_n}\varphi$

$L_{i,j}\varphi \rightarrow \exists k L_{i,k}\varphi (k \neq j)$. *(He believes those which is believed by more than two agents.)* Moreover, based on those operators, we can formalize the notion of suspicion as follows: $\text{Suspect}_i\varphi \stackrel{def}{=} (\exists j)(L_{i,j}\varphi \wedge \neg L_i L_j \varphi)$.
(Agent i suspects φ if and only if there exists some agent j such that agent i believes φ from j, but agent i does not believe that agent j believes φ.)

Propositions 3.1 *For the system* **Lij5$^-$+D***:*
(a)$\mathit{Suspect}_i\varphi \rightarrow L_i \mathit{Suspect}_i\varphi$.
(If agent i suspects φ, then he can know his suspicion.)
(b)$\neg \mathit{Suspect}_i\varphi \rightarrow L_i \neg \mathit{Suspect}_i\varphi$.
(If agent i does not suspect φ, then he knows that fact.)

Proof:(a)$\text{Suspect}_i\varphi \equiv (\exists j)(L_{i,j}\varphi \wedge \neg L_i L_j\varphi)$
$\Rightarrow L_{i,j}\varphi \wedge \neg L_i L_j \varphi \Rightarrow L_i L_{i,j}\varphi \wedge L_i \neg L_i L_j \varphi$
$\Rightarrow L_i(L_{i,j}\varphi \wedge \neg L_i L_j\varphi) \Rightarrow L_i \text{Suspect}_i\varphi$.
(b)$\neg\text{Suspect}_i\varphi \equiv (\forall j)(\neg L_{i,j}\varphi \vee L_i L_j\varphi)$
$\Rightarrow (\forall j)(L_i \neg L_{i,j}\varphi \vee L_i L_i L_j \varphi) \Rightarrow L_i((\forall j)(\neg L_{i,j}\varphi \vee L_i L_j\varphi)) \Rightarrow L_i \neg \text{Suspect}_i\varphi$.

3.4 Formalizing Indirect Dependence

In multiple agent environment, knowledge and beliefs may be transitive among agents. Therefore, we would like to extend the definition of dependent beliefs into indirect dependent beliefs as follows:
We define that $D^+_{i,j}\varphi \stackrel{\text{def}}{=} D_{i,j_1}\varphi \wedge D_{j_1,j_2}\varphi \wedge ... \wedge D_{j_m,j}\varphi$, $(i \neq j_1)$, and
$D^*_{i,j}\varphi \stackrel{\text{def}}{=} D^+_{i,j}\varphi \vee D_{i,j}\varphi$.

Propositions 3.2 *(Transitivity of Indirect Dependence)*
(a) $D^*_{i,j}\varphi \wedge D^*_{j,k}\varphi \rightarrow D^*_{i,k}\varphi$.
(b) $D^+_{i,j}\varphi \wedge D^+_{j,k}\varphi \rightarrow D^+_{i,k}\varphi$.
More generally, we have:
(c) for any $x, y, z \in \{*, +\}$ $(i \neq j), (j \neq k)$,
$D^x_{i,j}\varphi \wedge D^y_{j,k}\varphi \rightarrow D^z_{i,k}\varphi$.

We also would like to define indirect sub-beliefs for the agent set $\mathbf{A}_n$ as follows:
$L^1_{i,j}\varphi \stackrel{\text{def}}{=} D_{i,j}\varphi \wedge L_j\varphi$
$L^m_{i,j}\varphi \stackrel{\text{def}}{=} D_{i,j'}\varphi \wedge L^{m-1}_{j',j}\varphi$.
$L^*_{i,j}\varphi \stackrel{\text{def}}{=} [\vee^n_{k=1}]L^k_{i,j}\varphi$.

Propositions 3.3
(a) Coincidence
$L^*_{i,j}\varphi \equiv D^*_{i,j}\varphi \wedge L_j\varphi$.
(b) Consistence
$L^*_{i,j}\varphi \rightarrow \neg L^*_{i,j}\neg\varphi$.
Proof $L^*_{i,j}\varphi \Rightarrow D^*_{i,j}\varphi \wedge L_j\varphi \Rightarrow L_j\varphi \Rightarrow \neg L_j\neg\varphi$
$\Rightarrow \neg L_j \neg\varphi \vee \neg D^*_{i,j}\neg\varphi \Rightarrow \neg L^*_{i,j}\neg\varphi$.
(c) Same-source-propagation
$D^*_{i,k}\varphi \wedge L^*_{j,k}\varphi \rightarrow L^*_{i,k}\varphi$.
(d) Strong-consistence

$L^*_{i,j}\neg\varphi \rightarrow (\forall k)(\neg L^*_{k,j}\varphi)$.
(e) No-same-source-assertion
$L^*_{i,j}\varphi \wedge \neg L^*_{k,j}\varphi \rightarrow \neg D^*_{k,j}\varphi$.
Proof $L^*_{i,j}\varphi \wedge \neg L^*_{k,j}\varphi \Rightarrow D^*_{i,j}\varphi \wedge L_j\varphi \wedge (\neg D^*_{k,j}\varphi \vee \neg L_j\varphi)$
$\Rightarrow D^*_{i,j}\varphi \wedge L_j\varphi \wedge \neg D^*_{k,j}\varphi \Rightarrow \neg D^*_{k,j}\varphi$.

4 Semantics Models of Belief Dependence

4.1 L-Model of Belief Dependence: An Approach Based on General Epistemic Logic

In this section, we try to define the dependent operator by general doxastic and epistemic operator, by which we can study the problem of belief dependence in the general epistemic logics framework. $D_{i,j}\varphi$ means that agent i relies on agent j about believing φ . Formally, there might exist many different interpretations about the dependent operator. In other words, there are many semantically interpretations about the meaning of "rely on". Here are some of definitions:
(Ddf1) $D_{i,j}\varphi \stackrel{\text{def}}{=} (L_j\varphi \rightarrow L_i\varphi) \wedge (L_j\neg\varphi \rightarrow L_i\neg\varphi)$.
(If agent j believes φ, so does agent i; if agent j believes φ is false, agent i believes φ is false as well.)
(Ddf1') $D_{i,j}\varphi \stackrel{\text{def}}{=} (L_j\varphi \equiv L_i\varphi)$.
(If agent j believes φ, so does agent i; if agent j does not believe φ, neither does agent i)
(Ddf2) $D_{i,j}\varphi \stackrel{\text{def}}{=} L_i(L_j\varphi \rightarrow L_i\varphi) \wedge L_i(L_j\neg\varphi \rightarrow L_i\neg\varphi)$.
(Agent i believes that if agent j believes φ, then so does agent i, agent j believes its false, so does agent i.)
(Ddf2') $D_{i,j}\varphi \stackrel{\text{def}}{=} L_i(L_j\varphi \equiv L_i\varphi)$.
(Agent i believes that agent j believes φ iff agent i believes φ).
(Ddf3) $D_{i,j}\varphi \stackrel{\text{def}}{=} (L_iL_j\varphi \rightarrow L_i\varphi) \wedge (L_iL_j\neg\varphi \rightarrow L_i\neg\varphi)$.
(If agent i believes that agent j believes φ, then agent i will believe it; if agent i believes agent j believes φ is false, then agent i will also believe that φ is false.)
Of those definitions, (Ddf2) and (Ddf2') are the definitions of explicit dependence, because they say that agent i believes the dependent relation. Whereas other definitions are implicit. Moreover, (Ddf1) seems to be a simple one, but it is completely implicit. (Ddf3) can be viewed as a semi-implicit one since agent i's dependent beliefs depend on parts of its own beliefs. (Ddf1') is a symmetric definition. However, dependent relations are not intuitively symmetric. Although (Ddf2') is not symmetric, "≡" still make the definition too strong. Therefore, we view the definitions (Ddf1), (Ddf2), and (Ddf3) are more reasonable and acceptable.

For those three definitions (Ddf1), (Ddf2), and (Ddf3), we know that the neutral axiom (D1), namely, $D_{i,j}\varphi \equiv D_{i,j}\neg\varphi$, holds in any epistemic logics systems. Moreover, we naturally expect that the closure under conjunction axiom will hold for those definitions. Unfortunely, we have the following result.

Claim 4.1 *In any possible world semantic model for the epistemic operator L_i, $D_{i,j}\varphi \wedge D_{i,j}\psi \wedge \neg D_{i,j}(\varphi \wedge \psi)$ is satisfiable if $D_{i,j}\varphi$ is defined by (Ddf1), (Ddf2), or (Ddf3).*

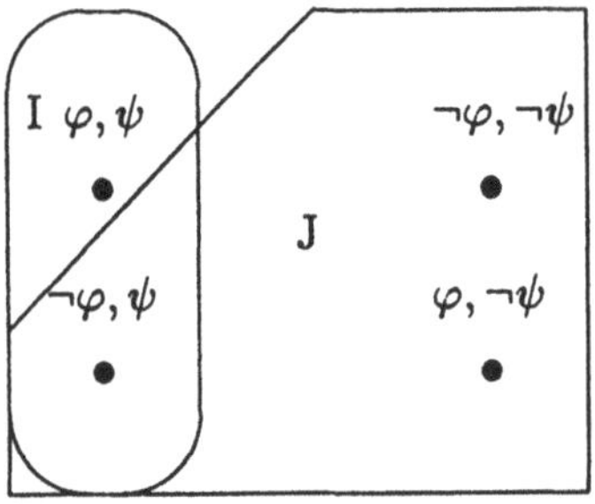

Figure 2: Satisfibility for (Formula 2)

Proof: For the definition (Ddf1),
$D_{i,j}\varphi \wedge D_{i,j}\psi \wedge \neg D_{i,j}(\varphi \wedge \psi)$
$\equiv (L_j\varphi \rightarrow L_i\varphi) \wedge (L_j\neg\varphi \rightarrow L_i\neg\varphi) \wedge (L_j\psi \rightarrow L_i\psi) \wedge (L_j\neg\psi \rightarrow L_i\neg\psi)$
$\wedge \neg((L_j(\varphi \wedge \psi) \rightarrow L_i(\varphi \wedge \psi)) \wedge (L_j\neg(\varphi \wedge \psi) \rightarrow L_i\neg(\varphi \wedge \psi)))$
$\equiv (L_j\varphi \rightarrow L_i\varphi) \wedge (L_j\neg\varphi \rightarrow L_i\neg\varphi) \wedge (L_j\psi \rightarrow L_i\psi) \wedge (L_j\neg\varphi \rightarrow L_i\neg\psi)$
$\wedge ((L_j(\varphi \wedge \psi) \wedge \neg L_i(\varphi \wedge \psi) \vee L_j(\neg\varphi \vee \neg\psi) \wedge \neg L_i(\neg\varphi \vee \neg\psi))$ (Formula 1)
Moreover, let (Formula 2) be the formula $\neg L_j\varphi \wedge \neg L_j\neg\varphi \wedge \neg L_j\psi \wedge \neg L_j\neg\psi \wedge \neg L_i(\neg\varphi \vee \neg\psi) \wedge L_j(\neg\varphi \vee \neg\psi)$.
We know that if (Formula 2) is satisfiable, then so is (Formula 1), because we have:
$\neg L_j\varphi \Rightarrow (L_j\varphi \rightarrow L_i\varphi)$
$\neg L_j\neg\varphi \Rightarrow (L_j\neg\varphi \rightarrow L_i\neg\varphi)$
$\neg L_j\psi \Rightarrow L_j\psi \rightarrow L_i\psi$
$\neg L_j\neg\psi \Rightarrow L_j\neg\psi \rightarrow L_i\psi$
$L_j(\neg\varphi \vee \neg\psi) \wedge \neg L_i(\neg\varphi \vee \neg\psi) \Rightarrow L_j(\neg\varphi \vee \neg\psi) \wedge \neg L_i(\neg\varphi \vee \neg\psi)$.
It is easy to show that (Formula 2) is satisfiable. One of the cases is shown in the figure. The cases of (Ddf2) and (Ddf3) can be similarly shown. □

From the above argument, we know that general epistemic logics are not an appropriate means to formalize the problem of belief dependence, since some intuitive properties such closure under conjunction cannot be formalized efficiently. However, in order to make a comparison with the other semantic models which would be studies in the next subsections, we would like to include the semantic model of epistemic logic as a kind of model for belief dependence, although it is a weak one, which does not explicitly represent the belief dependence at all. For the sake of notation consistency, we therefore have the following definition:

Definition 4.1 *(***Belief Dependence L-model***)*
A belief dependence L-model is a tuple $M = (S, \pi, \mathcal{L})$
where S *is a set of states,* $\pi(s, .)$ *is a truth assignment for each state* $s \in S$, *and* $\mathcal{L} : \mathbf{A_n} \rightarrow 2^{S\times S}$, *which consists of* n *binary accessibility relations on* S.

4.2 D-Model of Belief Dependence: A Syntactic Approach

We have known that sub-beliefs can be defined directly from the dependent operator and the general epistemic operator, namely, $L_{i,j}\varphi \equiv D_{i,j}\varphi \wedge L_j\varphi$. Therefore, to formulate belief dependence, naturally, an approach is to add dependent structure to general Kripke model of epistemic logics. The approach is similar to Fagin and Halpern's general awareness logic[3]. The general idea is that one can decide on a metalevel what formulas each agent is supposed to rely on others. By this approach, what we can do is to introduce dependent formula sets for each agent and each state, namely, formula sets $\mathcal{D}(i,j,s)$. The formula $\varphi \in \mathcal{D}(i,j,s)$ means that agent i relies on agent j about the formula φ. Therefore, we call it a syntactic approach.

In [6], we have presented a syntactic approach about modelling of belief dependence. However, in [6], the added dependent structure is a dependent function $\mathcal{D}_i : \mathbf{L_D} \times S \rightarrow \mathbf{An} \cup \{\lambda\}$. $\mathcal{D}_i(\varphi, s) = j$ means that in the state s agent i relies on agent j about φ, where λ means nobody. The dependent function requires each agent has only one credible advisor for each formula in each state, which seems not to be a flexible formalism for modelling belief dependence. In this paper, we would like to extend the dependent structure to formula sets, which allows that each agent has more than one credible advisor. Formally, we have the following definition:

Definition 4.2 *(***Belief dependence D-model***)*
A belief dependence D-model is a tuple $M = (S, \pi, \mathcal{L}, \mathcal{D})$
where S *is a set of states,* $\pi(s,.)$ *is a truth assignment for each state* $s \in S$, *and* $\mathcal{L} : \mathbf{A_n} \rightarrow 2^{S\times S}$, *which consists of* n *binary accessibility relations on* S, $\mathcal{D} : \mathbf{A_n} \times \mathbf{A_n} \times S \rightarrow 2^{\mathbf{L_D}}$.

The truth relation $\models$ is defined inductively as follows:

$M, s \models p$,	where p is a primitive proposition, iff $\pi(s,p)$ =true,
$M, s \models \neg\varphi$	iff $M, s \not\models \varphi$
$M, s \models \varphi_1 \wedge \varphi_2$	iff $M, s \models \varphi_1 \wedge M, s \models \varphi_2$,
$M, s \models L_i\varphi$	iff $M, t \models \varphi$ for all t such $(s,t) \in \mathcal{L}(i)$
$M, s \models D_{i,j}\varphi$	iff $\varphi \in \mathcal{D}(i,j,s)$.

We say a formula φ is *valid in structure* M if $M, s \models \varphi$ for all possible worlds s in M; φ is *satisfiable in* M if $M, s \models \varphi$ for some possible worlds in M. We say φ is *valid* if it is valid in all structures; φ is *satisfiable* if it is satisfiable in some structure.

For D-models, we define sub-beliefs as $L_{i,j}\varphi \overset{\text{def}}{=} D_{i,j}\varphi \wedge L_j\varphi$, which means that system is honest because the honesty axiom $L_{i,j}\varphi \rightarrow L_j\varphi$ holds.

In the definition about D-model of belief dependence, we have placed no restriction on the dependent formula sets. To capture certain properties for belief dependence, we may well add some restrictions on the dependent formula sets. Some typical restrictions we may want to add to $\mathcal{D}(i,j,s)$ can be expressed by some closure properties under the logical connectives and modal operators.

Definition 4.3 *A dependent formula set* $\mathcal{D}(i,j,s)$ *is said to be:*
(a) closed under negation, iff $\varphi \in \mathcal{D}(i,j,s) \leftrightarrow \neg\varphi \in \mathcal{D}(i,j,s)$.
(b) closed under conjunction, iff $\varphi \in \mathcal{D}(i,j,s) \wedge \psi \in \mathcal{D}(i,j,s) \rightarrow (\varphi \wedge \psi) \in \mathcal{D}(i,j,s)$.
(c) decomposable under conjunction, iff $\varphi \wedge \psi \in \mathcal{D}(i,j,s) \rightarrow \varphi, \psi \in \mathcal{D}(i,j,s)$.
(d) closed under implication, iff $\varphi \in \mathcal{D}(i,j,s) \wedge (\varphi \rightarrow \psi) \in \mathcal{D}(i,j,s) \rightarrow \psi \in \mathcal{D}(i,j,s)$.

Definition 4.4 *A D-model for belief dependence* $M = (S, \pi, \mathcal{L}, \mathcal{D})$ *is an* **L5⁻+D4** *D-model, if it satisfies the following conditions:*
(a) Each accessibility relation $\mathcal{L}(i,s)$ *is serial, transitive, and Euclidean,*
(b) Each dependent formula set $\mathcal{D}(i,j,s)$ *is closed under negation, implication, and conjunction,*

(c) For any formula φ, if $\varphi \in \mathcal{D}(i,j,s)$, then $\varphi \in \mathcal{D}(i,j,t)$ for all of states t such that $< s,t > \in \mathcal{L}_i$.

In order to show that soundness and completeness of the system **L5⁻+D4** for **L5⁻+D4** D-models, we can use the standard techniques, namely, techniques of canonical structures[3, 6, 9]. First, we need the following definitions: A formula p is *consistent* (with respect to an axiom system) if $\neg p$ is not provable. A finite set $\{p_1, ..., p_k\}$ is consistent exactly if the formula $p_1 \wedge ... \wedge p_k$ is consistent. An infinite set of formulae is consistent if every finite subset of it is consistent. A set F of formulae is a *maximal consistent set* if it is consistent and any strict superset is inconsistent. As it is pointed out in [3], using standard techniques of propositional reasoning we can show:

Lemma 4.1 *In any axiom system that includes (L1) and (R1):*
(1) Any consistent set can be extended to a maximal consistent set.
(2) If F is a maximal consistent set, then for all formulas φ and ψ:
(2.a) either $\varphi \in F$ or $\neg\varphi \in F$,
(2.b) $\varphi \wedge \psi \in F$ iff $\varphi \in F$ and $\psi \in F$,
(2.c) if $\varphi \in F$ and $\varphi \rightarrow \psi$, then $\psi \in F$,
(2.d) if φ is provable, then $\varphi \in F$.

Theorem 4.1 **L5⁻+D4** *belief dependence systems are sound and complete for any* **L5⁻+D4** *D-model.*

Proof:. Soundness is evident. For the completeness, we must show every valid formula is provable. Equivalently, we should show that every consistent formula is satisfiable. A canonical structure Mc is constructed as follows:
$Mc = (S, \pi, \mathcal{L}c, \mathcal{D}c)$
where
$S = \{s_v | V \text{ is a maximal consistent set}\}$,
$\pi(s_v, p) =$true, if $p \in V$; false, if $p \notin V$,
$\mathcal{L}c(i) = \{< s_v, s_w > | L_i^-(V) \subseteq W\}$, for any $i \in \mathbf{A_n}$
where $L_i^-(s_v) \stackrel{\text{def}}{=} \{\varphi | L_i\varphi \in V\}$
$\mathcal{D}c(i,j,s_v) = \{\varphi | D_{i,j}\varphi \in V\}$.

First, we show that Mc is an **L5⁻+D4** D-model. Axioms (L3), (L4), and (L5) guarantee that $\mathcal{L}c(i)$ is serial, transitive, and Euclidean. As for as the dependence formula set $\mathcal{D}c$ is concerned, we have:
$\varphi \in \mathcal{D}c(i,j,s_v) \Rightarrow D_{i,j}\varphi \in V$ (By the definition of Mc)
$\Rightarrow D_{i,j}\neg\varphi \in V$ (By axiom (D1) and lemma (2.c))
$\Rightarrow \neg\varphi \in \mathcal{D}c(i,j,s_v)$ (by the definition of Mc)
Therefore, the dependence formula sets are closed under negation. The cases concerning closure under conjunction and implication can be similarly shown. Moreover, for any formula φ,
$\varphi \in \mathcal{D}c(i,j,s_v) \Rightarrow D_{i,j}\varphi \in V$ (By the definition of Mc)
$\Rightarrow L_iD_{i,j}\varphi \in V$ (By axiom (D4) and lemma (2.c))
$\Rightarrow D_{i,j}\varphi \in W$ for all W such that $< s_v, s_w > \in \mathcal{L}c(i)$ (By the definition of Mc)
$\varphi \in \mathcal{D}c(i,j,s_w)$ for all s_w such that $< s_v, s_w > \in \mathcal{L}c(i)$.
Therefore, Mc is an **L5⁻+D4** D-model.

In order to show every formula φ is satisfiable, we should show $\varphi \in V \Leftrightarrow Mc, s_v \models \varphi$. we can show that by induction on the structure of formulas. □

4.3 Lij-Model: An Adapted Possible World approach

D-models of belief dependence is a syntactic approach, which does not somewhat coincide with possible world semantics for epistemic logics. Moreover, L-models of belief dependence, which is based on general epistemic logics, suffer from the problem in which dependent operator can not be intuitively handled with. Therefore, in this section, we would like to present a third logic for belief dependence. The ideas behind the third logic are to adapt possible world semantics for modelling belief dependence by directly introduction of sub-belief structures. Formally, we have the following definition:

Definition 4.5 *(***Belief dependence Lij-model***)*
A belief dependence Lij-model is a tuple $M = (S, \pi, \mathcal{L})$ where S is a set of states, $\pi(s,.)$ is a truth assignment for each state $s \in S$, and $\mathcal{L}: \mathbf{A_n} \times \mathbf{A_n} \to 2^{S\times S}$, which consists of $n \times n$ binary accessibility relations on S.

The relation $\models$ is similarly defined inductively as follows:
$M, s \models p$, where p is a primitive proposition, iff $\pi(s,p)$ =true,
$M, s \models \neg\varphi$ iff $M, s \not\models \varphi$
$M, s \models \varphi_1 \wedge \varphi_2$ iff $M, s \models \varphi_1 \wedge M, s \models \varphi_2$,
$M, s \models L_{i,j}\varphi$ iff $M, t \models \varphi$ for all t such $(s,t) \in \mathcal{L}(i,j)$.
$L_{i,j}\varphi$ means that due to agent j, agent i believes the formula φ. In Lij-models, we intuitively consider $L_{i,i}\varphi$ as its general epistemic interpretation, namely, $L_i\varphi$. Just like the cases in epistemic logics, sometimes we hope that the axiom $L_i\varphi \to L_iL_i\varphi$ holds. Similarly, for sub-beliefs, we generally hope that the axiom $L_{i,j}\varphi \to L_iL_{i,j}\varphi$ holds. In order to formulate those properties, first of all, we need the following definitions:

Definition 4.6 *(***Left-closed accessibility relations***)*
For any Lij model $M = (S, \pi, \mathcal{L})$, an accessibility relation $\mathcal{L}(i,j)$ is a left-closed relation, if $\mathcal{L}(i,i) \circ \mathcal{L}(i,j) \subseteq \mathcal{L}(i,j)$ holds.

Propositions 4.1 *For any Lij model in which every accessibility relation is left-closed, the axiom $L_{i,j}\varphi \to L_iL_{i,j}\varphi$ holds.*

Definition 4.7 *(***Almost-Euclidean accessibility relations***)*
For any Lij-model $M = (s, \pi, \mathcal{L})$, an accessibility relation $\mathcal{L}(i,j)$ is an almost-Euclidean relation, if $(t,u) \in \mathcal{L}(i,j)$ whenever $(s,u) \in \mathcal{L}(i,j)$ and $(s,t) \in \mathcal{L}(i,i)$.

Propositions 4.2 *For any Lij-model in which every accessibility relation is almost-Euclidean the axiom $\neg L_{i,j}\varphi \to L_{i,i}\neg L_{i,j}\varphi$ holds.*

Definition 4.8 *For any accessibility relation $R \subseteq S \times S$, R is said to be:*
(a) a **D***-relation, if it is serial, namely, for each $s \in S$ there is some $t \in S$ such that $(s,t) \in R$.*
(b) a **4***-relation, if it is transitive, namely, if $(s,u) \in R$ whenever $(s,t) \in R$ and $(t,u) \in R$.*
(c) a **5***-relation, if it is Euclidean, namely, if $(t,u) \in R$ whenever $(s,t) \in R$ and $(s,u) \in R$.*
(d) a **4****-relation, if it is a left-closed relation.*
(e) a **5****-relation, if it is an almost-Euclidean relation.*

Definition 4.9 *(D4*5* Lij-model)*
An Lij-model $M = (S, \pi, \mathcal{L})$ is a **D4*5*** *Lij-model, if every accessibility relation on S is serial, left-closed, and almost-Euclidean.*

Theorem 4.2 Lij5⁻+D *belief dependence logics are sound and complete for any* **D4*5*** *Lij-model.*

Proof. Soundness is evident, and completeness can be proved in an analogous fashion to the theorem about the **L5⁻+D4** systems. We define the canonical structure $Mc = (S, \pi, \mathcal{L}c)$as follows:

$S = \{s_v | V \text{ is a maximal consistent set}\}$,
$\pi(s_v, p) =$true, if $p \in V$; false, if $p \notin V$,
$\mathcal{L}c(i,j) = \{(s_v, s_w) | L^-_{i,j}(V) \subseteq W\}$.
where $L^-_{i,j}(V) \stackrel{\text{def}}{=} \{\varphi | L_{i,j}\varphi \in V\}$.

First, we show that Mc is a **D4*5*** Lij-model. Axiom (Lij3) guarantees every $\mathcal{L}c(i,j)$ is serial. For any $(s_v, s_w) \in \mathcal{L}c(i,i)$, and $(s_w, s_{w'}) \in \mathcal{L}c(i,j)$,
We have $L^-_{i,i}(V) \subseteq W$ and $L^-_{i,j} \subseteq W'$.
$L_{i,j}\varphi \in V \Rightarrow L_i L_{i,j} \in V$ (Axiom(Lij4) and lemma(2.c))
$\Rightarrow L_{i,j}\varphi \in W \quad (L^-_{i,i}(V) \subseteq W)$
$\Rightarrow \varphi \in W' \quad (L^-_{i,j}(W) \subseteq W')$

Therefore, every accessibility relation is a **4***-relation. Furthermore, for any $(s_v, s_w) \in \mathcal{L}c(i,j)$, and $(s_v, s_{w'}) \in \mathcal{L}c(i,i)$,
We have $L^-_{i,j}(V) \subseteq W$ and $L^-_{i,i}(V) \subseteq W'$.
$L_{i,j}\varphi \in W' \Rightarrow L_{i,i}L_{i,j}\varphi \in V \quad (L^-_{i,i}(V) \subseteq W')$
$\Rightarrow \neg L_{i,i}\neg L_{i,j}\varphi \in V$ (Axiom $L_{i,i}\psi \to \neg L_{i,i}\neg\psi$)
$\Rightarrow L_{i,j}\varphi \in V$ (Axiom ($Lij5$))
$\Rightarrow \varphi \in W \quad (L^-_{i,j}(V) \subseteq W)$
Therefore, every accessibility relation is a **5***-relation, i.e., Mc is a **D4*5*** Lij-model. Moreover, we can show that $Mc, s_v \models \varphi$ iff $\varphi \in V$.□

5 Conclusions

We have proposed several approaches for belief dependence logics. All of those approaches can capture certain properties concerning belief dependence. There might exist many different criteria to appraise those approaches. We suggest some main criteria as follows:
i) *Efficiency Adequacy*: Approaches can efficiently formalize the fundamental features such as closure, suspicion, indirect dependence, role-appraisal.
ii) *Coherence Adequacy*: Approaches can be captured intuitively, in which semantics models should be naturally connected.
iii) *Avoidance of Logical Omniscience*: Approaches do not suffer from the problem of logical omniscience.

The approach about D-model is based on a syntactic strategy, which is amalgamated with general possible world approach. The approach concerning L-model actually is a general epistemic logics approach, which fail to capture some important features of dependent operator. The approach of Lij-model seems to be a more reasonable and acceptable one, since which can capture many intuitive properties concerning the dependent operator, although the approach suffers from the problem of logical omniscience, just like those in

Approaches	*Efficiency*	*Coherence*	*Avoidance of Logical Omniscience*
D-model	Yes	No	Yes
L-model	No	Yes	No
Lij-model	Yes	Yes	No

Figure 3: Summaries about Approaches

general epistemic logics. The comparison is shown in the figure.

Acknowledgements
The author would like to thank Peter van Emde Boas, Karen Kwast, and Sieger van Denneheuvel for stimulating discussions, and John-Jules Meyer for useful suggestions.

References

[1] Bond, A. H. and Gasser, L.,(eds.), *Readings in Distributed Artificial Intelligence,* Morgan-Kaufmann, San Mateo, CA, 1988.

[2] Doyle, J., A society of mind, in: *Proceedings IJCAI-83*, Karlsruhe, F. R. G. (1983) 309-314.

[3] Fagin, R. F. and Halpern, J. Y., Belief, Awareness, and Limited Reasoning, in: *Artificial Intelligence,* 34 (1988) 39-76.

[4] Halpern, J. K. and Fagin, R. F., Modelling knowledge and action in distributed systems, *Distributed Computing,* 3(1989) 159-177.

[5] J. Hintikka, *Knowledge and Belief,* Cornell University Press, 1962.

[6] Zhisheng Huang, Dependency of Belief in Distributed Systems, in: M. Stokhof and L. Torenvliet (eds.) *Proceedings of the Seventh Amsterdam Colloquium*, 637-662, 1990.

[7] Zhisheng Huang and Karen Kwast, Awareness, Negation, and Logical Omniscience, to appear in: *Proceedings of European Workshop on Logics in Artificial Intelligence*, (JELIA'90), 1990.

[8] Zhisheng Huang and Peter van Emde Boas, Belief Dependence, Revision and Persistence, in preparation.

[9] G. E. Hughes and M. J. Cresswell, *A Companion to Modal Logic*, Methuen, 1984.

[10] Levesque, H. J., A logic of implicit and explicit belief, in: *Proceedings AAAI-84* Austin, TX (1984) 198-202.

[11] G. R. Potts, M. F. ST. John, and D. Kirson, Incorporating New Information into Existing World Knowledge, *Cognitive Psychology*, 21(1989) 303-333.

[12] Eric Werner, Toward a Theory of Communication and Cooperation for Multiagent Planning, in: M. Y. Vardi (ed.) *Proceedings of TARK 1988,* Morgan-Kaufmann, (1988), 129-143.

A Generalization of Stability and its Application to Circumscription of Positive Introspective Knowledge [1]

Jan Jaspars

Institute for Language Technology & AI
Tilburg University
PO Box 90153
5000 LE TILBURG
THE NETHERLANDS
EMAIL: jaspars@kub.nl

1 Introduction

Stability was introduced by Stalnaker [Mooa] in order to describe the content of a logically omniscient agent's knowledge. This definition has been used extensively [Mooa] [Moob] [HaM] [Kon] for formalization of circumscription of knowledge and belief; 'what are the stable sets that correspond to *only* knowing/believing a certain formula φ' and 'what are its (non-monotonic) consequences'. Halpern and Moses [HaM] gave an S5-formalization of consistent stable sets, and nice syntactic (disjunction rule) and semantic (minimal models) classification of formulae that can be circumscribed in this way. These *honest* formulae are those that can be known *only*. Should an agent give a dishonest message, it would violate the *conversational maxim of quantity* [Gri]; the agent must know *more* than just this information. If we take an answer of a conversational partner, to be the *only* knowledge he has, we have to exclude more epistemic messages than only the inconsistent ones. If a partner says "I know p or I know q", we cannot both obey the maxim of quantity and have a meaningful interpretation of this message. If this message was the *only* knowledge of our partner (on p and q) we would end up with an inconsistency. We may infer that our partner neither knows p nor q, which contradicts his original message. Such a message is therefore *dishonest*, but not inconsistent. Technically speaking, there is no unique minimal interpretation (stable set) for this message.

Stalnaker's definition of stability assumes full introspective capacities of cognitive agents. In philosophical debates the negative aspect of this full introspection – an agent knows/believes all his ignorances/disbeliefs – has generally been rejected [Hin] [Len]. The question which arises is, whether circumscription techniques, using full introspection, could be re-implemented into weaker modal systems for knowledge (and belief). In this paper we offer a generalization of Stalnaker's definition and show how it can be used for knowledge circumscription in S4: *positive introspective knowledge.*

Circumscription for S4 was already defined in [Var], by means of an alternative to standard Kripke-semantics (possible world semantics). We define circumscription of knowledge in terms of Kripke-

[1] This paper contains a somewhat revised summary of the second chapter of my master's thesis [Jasa], which was supervised by Johan van Benthem at the University of Amsterdam. I would like to thank him for his stimulating advice and his introduction to most of the concepts that are used in this paper. Furthermore, thanks to Wietske Sijtsma and Arthur van Horck for checking earlier drafts of this paper.

semantics. This enables us to characterize honest formulae in S4 both syntactically and semantically. This comes very close to the classifications of honesty in S5 [HaM]. This is made possible by our generalization of stability. Furthermore, the proposed generalization could reveal circumscription techniques for other modal logics [2].

2 Some formal definitions

Here we give the formal language that we will be dealing with:

Definition 1

> $\mathcal{L}^{\Box}$ *is the smallest set, such that*
>
> $I\!P \subset \mathcal{L}^{\Box}$, *where* $I\!P$ *is a finite set of primitive propositions.*
>
> $\varphi \in \mathcal{L}^{\Box} \Longrightarrow \neg\varphi \in \mathcal{L}^{\Box}$
>
> $\varphi \in \mathcal{L}^{\Box}, \psi \in \mathcal{L}^{\Box} \Longrightarrow (\varphi \wedge \psi) \in \mathcal{L}^{\Box}$
>
> $\varphi \in \mathcal{L}^{\Box} \Longrightarrow \Box\varphi \in \mathcal{L}^{\Box}$
>
> *Furthermore we use well-known abbreviations such as* $\top, \bot, \vee, \rightarrow$ *and* $\Diamond$ ($\top := p \vee \neg p, \bot := \neg\top, \varphi \vee \psi := \neg(\neg\varphi \wedge \neg\psi), \varphi \rightarrow \psi := (\neg\varphi \vee \psi), \Diamond\varphi := \neg\Box\neg\varphi$). $\Box\varphi$ *should be interpreted as 'the agent knows that* φ*'.*

The following axioms will be of importance in this paper:

1 All propositional tautologies are theorems.

2 $\vdash \Box(\varphi \rightarrow \psi) \rightarrow (\Box\varphi \rightarrow \Box\psi)$

3 $\vdash \Box\varphi \rightarrow \varphi$

4 $\vdash \Box\varphi \rightarrow \Box\Box\varphi$

5 $\vdash \neg\Box\varphi \rightarrow \Box\neg\Box\varphi$

R1 $\vdash \varphi \Rightarrow \vdash \Box\varphi$

R2 $\varphi, \varphi \rightarrow \psi \vdash \psi$

Logics consisting of at least **1** + **2** + **R1** + **R2** are *normal modal logics*. The modal logic only consisting of these axioms and inference-rules is called the *minimal normal modal logic* and is generally denoted by **K**. **4** and **5** are the axioms of positive and negative introspection respectively. **K** + **4** + **5** is known as **K45**. **K** + **3** + **4** is known as **S4**. The latter logic is Hintikka's axiomatization of the knowledge of an idealized rational agent [Hin]. Besides positive introspection, the content of such an agent's knowledge is taken to be true (3). **S5** denotes **K** + **3** + **5**. Because $\vdash_{\mathbf{S5}}$ **4** this logic is the full introspective logic of knowledge.

In this paper we will extensively use the in many respects useful notion of maximal consistent sets [HuC] with respect to a normal modal logic L.

Definition 2

> *Let* L *be a normal modal logic. A set of formulae* $\Gamma \subseteq \mathcal{L}^{\Box}$ *is said to be L-consistent iff for any* $\{\alpha_1, .., \alpha_m\} \subseteq \Gamma \ : \not\vdash_L \neg(\alpha_1 \wedge .. \wedge \alpha_m)$*. A set* $\Gamma \subseteq \mathcal{L}^{\Box}$ *is maximal L-consistent iff there is no proper L-consistent extension of* Γ*:* $\forall\Delta \supseteq \Gamma \ : \ \Delta$ *is L-consistent only if* $\Delta = \Gamma$.

[2] For example Lenzen's axiomatization of knowledge S4.2 [Len].

This notion of maximal L-consistent sets have wide popularity because they are the main ingredient in Henkin style completeness proofs for modal logics [HuC]. In this paper it will show its use once again.

3 Stable sets

Stalnaker's original definition of stable sets for $\mathcal{L}^{\Box}$ is as follows [Mooa]:

Definition 3

A set $S \subseteq \mathcal{L}^{\Box}$ is stable if S obeys the following four criteria

- *All propositional tautologies are contained by S.*
- *If $\varphi \in S$ and $\varphi \rightarrow \psi \in S$, then also $\psi \in S$.*
- $\varphi \in S \Longleftrightarrow \Box\varphi \in S$
- $\varphi \notin S \Longleftrightarrow \neg\Box\varphi \in S$

Moore used this definition for founding his autoepistemic logic [Mooa]. Halpern and Moses [HaM] added a fifth criterion: propositional consistency of S. They used these sets for localizing minimal knowledge: "What does someone's knowledge look like when he *only* knows a certain formula φ". It turns out that Stalnaker's original definition is identical to the knowledge that is contained in maximal **K45**-consistent sets. **K45** is the modal logic that Moore used (for belief). Adding Halpern and Moses' extra criterion gives us precisely the maximal S5-consistent sets, and not very surprisingly, this was the modal system that they used for knowledge. These observations entail the following alternative general definition of stability for arbitrary normal modal logics.

Definition 4

A set S of epistemic formulae is said to be stable with respect to a certain normal modal (epistemic) logic L (L-stable) iff there exists a maximal L-consistent set Γ such that $S = \Box^{-}\Gamma = \{\varphi \mid \Box\varphi \in \Gamma\}$.

Observation 1

For simple modal systems such as **S4**, **S5** and **K45** we inherit from their Henkin-style completeness [HuC] and the generalized definition of stability the following soundness/completeness observation:

$\vdash_L \Box\varphi \Leftrightarrow \varphi \in S$ for all L-stable sets S, and

$\vdash_L \neg\Box\varphi \Leftrightarrow \varphi \notin S$ for all L-stable sets S, with $L \in \{\mathbf{S4},\mathbf{S5},\mathbf{K45}\}$.

The following theorem transforms our theoretical generalization into a 'Stalnaker-like' format for all normal modal logics, which looks epistemically more plausible.

Theorem 1

A set $S \subseteq \mathcal{L}^{\Box}$ is L-stable iff the following three criteria hold for S:

- $\vdash_L \varphi \Longrightarrow \varphi \in S$
- $\varphi \in S \;\&\; \varphi \rightarrow \psi \in S \Longrightarrow \psi \in S$
- $\Box S \cup \neg\Box\overline{S} := \{\Box\psi \mid \psi \in S\} \cup \{\neg\Box\psi \mid \psi \notin S\}$ *is L-consistent.*

proof Let Γ be a maximal L-consistent set. $\vdash_L \varphi \Rightarrow \varphi \in \Box^-\Gamma$ and $\varphi \in \Box^-\Gamma, \varphi \to \psi \in \Box^-\Gamma \Longrightarrow \psi \in \Box^-\Gamma$ are simple consequences of the normality of L. Furthermore, we check easily $\{\Box\varphi \mid \varphi \in \Box^-\Gamma\} \cup \{\neg\Box\varphi \mid \varphi \notin \Box^-\Gamma\} \subseteq \Gamma$, and therefore the first set must at least be L-consistent.

Let S be a set that satisfies the given three criteria. By taking a maximal L-consistent extension Γ of $\{\Box\psi|\psi \in S\} \cup \{\neg\Box\psi|\psi \notin S\}$, we always have $\Box^-\Gamma = S$.

The first two clauses of these three criteria for stability indicate the logically omniscient capacities of a rational agent. The third requirement says that the knowledge together with the ignorance determined by S may not be contradictory with respect to L: the epistemic logic that we equip our cognitive agent with.

4 Circumscription of knowledge: honesty

The remainder of the paper is concerned with an application of this generalization to circumscription of positive introspective knowledge, that is specification of the minimal knowledge states in **S4**. In [HaM] this idea has been developed for **S5**. With the help of our generalization of stability, we come to a classification of formulae φ that can be circumscribed in **S4**, quite similar to what Halpern and Moses found for **S5**. A formal idea which first comes to mind, of *only* knowing a formula φ is *only* knowing its logical L-consequences. This set is denoted by S^L_φ.

$$S^L_\varphi = \{\psi \in \mathcal{L}^\Box \mid \vdash_L \Box\varphi \to \psi\}$$

In full introspective systems like **K45** and **S5** knowledge (or belief) interferes with ignorance (or disbelief). So, knowing only the consequences of a formula φ, leads to knowledge of ignorance of the non-consequences: $\neg\Box\overline{S^L_\varphi}$. The interplay of ignorance and knowledge in modal logics containing the axiom of negative introspection (5) causes us to be careful in defining minimal knowledge states. If one only knows a certain proposition, this causes that one is ignorant about 'the rest', and henceforth, one knows this ignorance and should therefore be a part of the knowledge state. Formally, if S is an L-stable set and L is a normal modal logic containing **5**, then for every $\varphi \notin S$ we conclude $\neg\Box\varphi \in S$. For **K45** and **S5** this means that stable sets for these two systems can never properly contain one another [Mooa] [HaM].

Observation 2

Let L be a full introspective modal logic (**4** and **5** containing). If S and S' are two L-stable sets and $S \subseteq S'$ then $S = S'$.

Clearly, this means that the normal subset relation cannot serve as an order with which we can minimize. In the case of **S5** a subset relation on the propositional (objective or $\Box$-free) part is defined. Such an order is justified by the following observation [Mooa] [HaM].

Observation 3

Every **S5**-stable set S is fully determined by its propositional part.

$$\{\varphi \in S \mid \varphi\ \Box\text{-free}\} \subseteq S' \ \&\ S' \text{ is S5-stable} \Longrightarrow S = S'$$

Definition 5

Let S and S' be two **S5**-*stable sets.*

$$S \sqsubseteq_{prop} S' \text{ iff } \{\varphi \in S \mid \varphi\ \Box\text{-free}\} \subseteq \{\varphi \in S' \mid \varphi\ \Box\text{-free}\}.$$

An **S5**-*stable set S is said to be minimal for a formula φ iff $\varphi \in S$, and for every S' that is* **S5**-*stable: $\varphi \in S' \Leftrightarrow S \sqsubseteq_{prop} S'$. Formulae that have such a minimal* **S5**-*stable set are said to be* **S5**-*honest [HaM].*

This definition is justified by the observation that **S5**-stable sets are totally determined by their propositional ($\Box$-free) part. For **S4** the initial intuition to look at S^L_φ as minimal, suffices. The ordinary subset-relation is suitable here.

Definition 6

An **S4**-*stable S set is said to be minimal for a formula φ iff $\varphi \in S$, and for every S' that is* **S4**-*stable: $\varphi \in S' \Leftrightarrow S \subseteq S'$. Formulae that have such a minimal* **S4**-*stable set are said to be* **S4**-*honest.*

Theorem 2

If φ is **S4**-*honest, it has only one minimal* **S4**-*stable set: $S^{\mathbf{S4}}_\varphi$.*

proof By theorem 1 we learn that for every stable set which contains φ, must include $S^{\mathbf{S4}}_\varphi$. So, if $S^{\mathbf{S4}}_\varphi$ is **S4**-stable, it must be minimal.

Suppose that φ has a minimal **S4**-stable set S, but let $S^{\mathbf{S4}}_\varphi$ not be stable. Because $S^{\mathbf{S4}}_\varphi \subseteq S$ (theorem 1), there exists $\alpha \in S$ and $\alpha \notin S^{\mathbf{S4}}_\varphi$. From this last conclusion we learn that $\Box\varphi \wedge \neg\alpha$ is **S4**-consistent. Let Γ be a maximal **S4**-consistent extension of $\{\Box\varphi \wedge \neg\alpha\}$. By $\Box\alpha \notin \Gamma$ ($\vdash_{\mathbf{S4}} \neg\alpha \rightarrow \neg\Box\alpha$), we see that $\alpha \notin \Box^-\Gamma$, but $\varphi \in \Box^-\Gamma$. Because $\Box^-\Gamma$ is a **S4**-stable set, and $S \not\subseteq \Box^-\Gamma$, it violates the minimality of S.

Example 1

- $\Box p \vee \Box q$ is both **S5**- and **S4**-dishonest. Its **S4**-dishonesty is instantly clear:

  $$\{\Box(\Box p \vee \Box q), \neg\Box p, \neg\Box q\} \subseteq \Box S^{\mathbf{S4}}_{\Box p \vee \Box q} \cup \neg\Box\overline{S}^{\mathbf{S4}}_{\Box p \vee \Box q}$$

  Evidently, $\Box S^{\mathbf{S4}}_{\Box p \vee \Box q} \cup \neg\Box\overline{S}^{\mathbf{S4}}_{\Box p \vee \Box q}$ is **S4**-inconsistent.

- $\Box p \vee \Box\neg\Box p$ is **S5**-honest. It is even an **S5**-theorem. Knowing only this formula is the same as knowing nothing in **S5**. It is therefore not only an honest formula , but also innocent. $\Box p \vee \Box\neg\Box p$ turns out to be **S4**-dishonest.

- $\Box p \vee q$ is **S5**-dishonest, but **S4**-honest (see also sec. 7).

Observation 4

Because $S^{\mathbf{S4}}_\varphi$ always obeys the first two criteria listed in theorem 1, a formula φ is **S4**-honest iff $\Box S^{\mathbf{S4}}_\varphi \cup \neg\Box\overline{S}^{\mathbf{S4}}_\varphi$ is **S4**-consistent.

5 Syntactic classification of honest formulae

Halpern and Moses gave a syntactic classification of honest formulae by a modification of the so-called *rule of disjunction* [HuC].

Definition 7

A formula φ obeys the rule of disjunction (RD) with respect to a normal modal logic L iff for every finite sequence $\alpha_1, .., \alpha_m$ of formulae :

$$\vdash_L \Box\varphi \rightarrow (\Box\alpha_1 \vee .. \vee \Box\alpha_m) \Longrightarrow \exists i \in \{1, .., m\} : \vdash_L \Box\varphi \rightarrow \alpha_i \textit{ holds}$$ [3]

The modification of Halpern and Moses for $L = \mathbf{S5}$ of this rule of disjunction is to consider the sequences $\{\alpha_i\}_i$ as consisting only of propositional formulae. This modified rule is called the propositional rule of disjunction. The formulae which obey this modified rule with respect to S5 are precisely the S5-honest formulae. In case of S4, this modification is not necessary.

Theorem 3

A formula φ is S4-honest iff it obeys the rule of disjunction with respect to S4 and $\Box\varphi$ is S4-consistent.

proof Let φ be S4-honest. $\Box\varphi$ must be S4-consistent. Suppose furthermore that φ does not satisfy RD, then there exists a finite sequence of formulae $\{\alpha_1, .., \alpha_m\}$ such that $\vdash_{\mathbf{S4}} \Box\varphi \rightarrow (\Box\alpha_1 \vee .. \vee \Box\alpha_m)$ and $\not\vdash_{\mathbf{S4}} \Box\varphi \rightarrow \alpha_i$ for all $i \in \{1, .., m\}$. This means that $\neg\Box\alpha_i \in \neg\Box\overline{S}_\varphi^{\mathbf{S4}}$ for all $i \in \{1, .., m\}$, but on the other hand $\Box(\Box\alpha_1 \vee .. \vee \Box\alpha_m) \in \Box S_\varphi^{\mathbf{S4}}$, and so $\neg\Box\overline{S}_\varphi^{\mathbf{S4}} \cup \Box S_\varphi^{\mathbf{S4}}$ is S4-inconsistent. From theorem 1 we learn that $S_\varphi^{\mathbf{S4}}$ is not stable, and this contradicts φ's S4-honesty.

Suppose φ is S4-dishonest, then $\Box S_\varphi^{\mathbf{S4}} \cup \neg\Box\overline{S}_\varphi^{\mathbf{S4}}$ is S4-inconsistent (observation 4).
Let $\neg\Box\overline{S}_\varphi^{\mathbf{S4}} \neq \emptyset$, then there is a sequence of formulae $\varphi_1, .., \varphi_n, \psi_1, .., \psi_m$ where $\varphi_i \in S_\varphi^{\mathbf{S4}}$ and $\psi_j \notin S_\varphi^{\mathbf{S4}}$, such that

$$\vdash_{\mathbf{S4}} \neg(\Box\varphi_1 \wedge .. \wedge \Box\varphi_n \wedge \neg\Box\psi_1 \wedge .. \wedge \neg\Box\psi_m)$$

We can reformulate this as follows:

$$\Leftrightarrow \vdash_{\mathbf{S4}} \neg(\Box(\varphi_1 \wedge .. \wedge \varphi_n) \wedge \neg\Box\psi_1 \wedge .. \wedge \neg\Box\psi_m)$$

$$\Leftrightarrow \vdash_{\mathbf{S4}} \neg\Box(\varphi_1 \wedge .. \wedge \varphi_n) \vee \Box\psi_1 \vee .. \vee \Box\psi_m$$

Because $\varphi_i \in S_\varphi^{\mathbf{S4}}$ for all $i \in \{1, .., n\}$ we have $\vdash_{\mathbf{S4}} \Box\varphi \rightarrow \Box(\varphi_1 \wedge .. \wedge \varphi_n)$ and so $\vdash_{\mathbf{S4}} \Box\varphi \rightarrow \Box\psi_1 \vee .. \vee \Box\psi_m$. But $\not\vdash_{\mathbf{S4}} \Box\varphi \rightarrow \psi_j$ for all $j \in \{1, .., m\}$ because $\psi_j \notin S_\varphi^{\mathbf{S4}}$ and so φ does not satisfy RD.
If $\neg\Box\overline{S}_\varphi^{\mathbf{S4}} = \emptyset$, then $\vdash_{\mathbf{S4}} \Box\varphi \rightarrow \alpha$ for all $\alpha \in \mathcal{L}^\Box$, and so $\vdash_{\mathbf{S4}} \Box\varphi \rightarrow \bot$, or shorter $\vdash_{\mathbf{S4}} \neg\Box\varphi$, and so $\Box\varphi$ is S4-inconsistent.

Example 2 From RD we immediately see that $\Box p \vee \Box q$ is S4-dishonest.

$$\vdash_{\mathbf{S4}} \Box(\Box p \vee \Box q) \rightarrow \Box p \vee \Box q, \text{ but } \not\vdash_{\mathbf{S4}} \Box(\Box p \vee \Box q) \rightarrow \Box p \text{ and } \not\vdash_{\mathbf{S4}} \Box(\Box p \vee \Box q) \rightarrow \Box q.$$

S4-dishonesty of $\Box p \vee \Box\neg\Box p$ can be shown similarly. S4-honesty of $\Box p \vee q$ will be shown in section 7, by a model-theoretic technique.

---

[3]Maybe the reader expects $\vdash_L \Box\varphi \rightarrow \Box\alpha_i$ here. We have orthodox reasons for leaving the $\Box$ out. In standard literature such as [HuC] we find this definition of RD for modal systems (not for individual formulae). Note that in the case of S4 and S5 : $\vdash_L \Box\varphi \rightarrow \Box\alpha_i \Longleftrightarrow \vdash_L \Box\varphi \rightarrow \alpha_i$. Maybe, if one would be interested in defining a rule of disjunction for weaker modal system in order to have syntactic correspondence with honesty for such a system, the $\Box$-definition might be the appropriate one.

6 Semantic classification: minimal models

Semantic classification of both S5- and S4-honesty can be given by Kripke- or possible world semantics.

Definition 8

A Kripke-frame is a pair $\langle W, R\rangle$ with W a non-empty set of worlds and R a binary accessibility-relation on $W \times W$. A Kripke-model is a triple $\langle W, R, V\rangle$ such that $\langle W, R\rangle$ is a Kripke-frame and V is a valuation on worlds: $V : W \times \mathbb{P} \longrightarrow \{0,1\}$.

Truth-assignment to formula by a world w in a model $M = \langle W, R, V\rangle$ is recursively determined by the following composition:

$$M, w \models p \Leftrightarrow V(w,p) = 1$$

$$M, w \models \neg\varphi \Leftrightarrow M, w \not\models \varphi$$

$$M, w \models \varphi \wedge \psi \Leftrightarrow M, w \models \varphi \;\&\; M, w \models \psi$$

$$M, w \models \Box\varphi \Leftrightarrow (\forall v \in W : wRv \Rightarrow M, v \models \varphi)$$

As is known from standard modal logic [HuC] S5 is sound and complete with respect to Kripke-models with an equivalent accessibility-relation. S4 is sound and complete with respect to the class of transitive reflexive models. A further restriction is to consider only the *generated* cases of these classes, that is: there exists a world (*generator*) from which all other worlds in the model are accessible in a certain number of steps. Because the two classes mentioned above are both reflexive and transitive this amounts to strongly generated equivalent and strongly generated transitive reflexive models. This means that there exists a strong generator (SG) in these models, that is a world g from which all worlds (g itself included) are accessible. In the case of S5 we end up with universal models (any pair of worlds is mutually related). Halpern and Moses considered this relatively small and uniform class for defining *minimal S5-models*. We consider the class of strongly generated transitive reflexive models in order to define minimality for models in the case of S4. We will abbreviate this class as SG-S4-models.

Because maximal S4- and S5-consistent sets correspond with the local theories of worlds (the set of formulae that they assign *true*), and because the knowledge in the theories of strong generators correspond to the global theories of models, we obtain the minimization of models freely from our definition of stable sets.

Definition 9

The theory of a model M, $Th(M)$, is the set of formulae that are verified by all its worlds:

$$Th(M) := \{\varphi \mid M, w \models \varphi \text{ for all } w \text{ in } M\}$$

Observation 5

For strongly generated models we obtain:

$$Th(M) = \{\varphi \mid M, g \models \Box\varphi \text{ and } g \text{ SG in } M\}$$

By the fact that all generated S5- and S4-models have such a SG, S5- and S4-stable sets can be characterized as total theories of models.

Definition 10

A universal model is said to be minimal for a formula φ iff $M \models \varphi$ (for all $w \in M$) and for every universal model M': $M' \models \varphi \Longleftrightarrow Th(M) \sqsubseteq_{prop} Th(M')$.

Theorem 4 *[HaM]*

A formula has a minimal universal model iff it is S5-honest.

proof Simply by the observations made above.

By observation 5 the order on S4-stable sets can be adopted for SG-S4-models as well.

Definition 11

An SG-S4-model M is said to be minimal for a formula φ iff $M \models \varphi$ (for all $w \in M$) and for every universal model M': $M' \models \varphi \Longleftrightarrow Th(M) \subseteq Th(M')$.

Theorem 5

A formula has a minimal S4- model iff it is S4-honest.

7 A construction of minimal models: amalgamation

In the case of S5, minimal models are constructed fairly easily. One takes all (distinguishable) models that verify a formula φ and unites them, in the sense that one gathers all worlds from these models, copies the local valuations in this 'union', and stretches the universal accessibility-relation over it. If this model still verifies φ, this construction must be a minimal model for φ, and so it must be S5-honest. Conversely this construction succeeds for every S5-honest formula.

The construction of S4-minimal models is less straightforward. Here we use a technique which is known as *amalgamation*[4].

Definition 12

Let $\{M_i\}_{i\in I} = \{\langle W_i, R_i, V_i\rangle\}_{i\in I}$ be a family of SG S4-models. An amalgamation of $\{M_i\}_{i\in I}$ is a model $M^* = \langle W^*, R^*, V^*\rangle$ such that

$$W^* = \bigsqcup_{i\in I} W_i \sqcup \{w^*\} := \bigcup_{i\in I}\{\langle w, i\rangle | w \in W_i\} \cup \{w^*\}\ ;\ w^* \notin \bigsqcup W_i$$

$$xR^*y \Leftrightarrow x = w^* \text{ or } x = \langle w, i\rangle \text{ and } y = \langle v, i\rangle \text{ and } wR_iv \text{ for certain } i \in I$$

$$V^*(\langle w, j\rangle, p) = V_j(w, p) \text{ for all } j \in I, p \in I\!P.$$

Observation 6

- An amalgamation is always a SG -S4-model itself, and w^* is the only SG $(w^*R^*w^*)$.
- V^* can be chosen freely in w^*. So there are $2^{|I\!P|}$ amalgamations of $\{M_i\}_{i\in I}$.
- $M_i, w \models \varphi \Leftrightarrow M^*, \langle w, i\rangle \models \varphi$ for all $w \in M_i, \varphi \in \mathcal{L}^{\Box}$.

How do we construct a minimal model for a given formula φ with this notion of amalgamation? We take a family $\{M_\alpha\}_{\alpha\in \overline{S}_\varphi^{S4}}$, where for every formula α with $\not\vdash_{S4} \Box\varphi \to \alpha$ a model M_α is chosen such that it falsifies $\Box\varphi \to \alpha$ in a SG s $(M_\alpha, s \models \Box\varphi \wedge \neg\alpha)$. Next we amalgamate the family $\{M_\alpha\}_{\alpha\in \overline{S}_\varphi^{S4}}$ into $\langle W^*, R^*, V^*\rangle$ with root w^* and choose V^* such that $M^*, w^* \models \varphi$. Henceforth, we infer $\varphi \in Th(M^*)$,

[4] For an extensive survey on amalgamations, in a more general setting, see [HuC]. The definition of amalgamations can be traced back to [LeS].

and for any $\alpha \in \overline{S}_{\varphi}^{\mathbf{S4}}$ we have $M^*, w^* \models \neg\Box\alpha$ ($\alpha \notin Th(M^*)$). From the former section we can conclude that M^* is a minimal model for φ. The illustration below depicts the structure of this minimizing amalgamation.

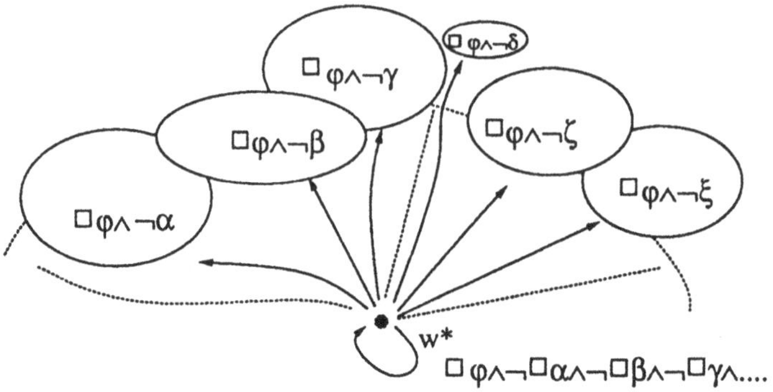

Of course, it is not that simple. We cannot always find a valuation V^* such that $M^*, w^* \models \varphi$. It could very well be the case, that for all amalgamations φ will be falsified in the root. In the following definition we characterize the class for which this will not be the case for any amalgamation.

Definition 13

A formula $\varphi \in \mathcal{L}^{\Box}$ is 'rootable' if for every non-empty family of SG *S4-models $\{M_i\}_{i \in I}$ where $\varphi \in Th(M_i)$ for every $i \in I$, there exists an amalgamation M^* of $\{M_i\}_{i \in I}$ such that $\varphi \in Th(M^*)$.*

Theorem 6

If a formula $\varphi \in \mathcal{L}^{\Box}$ is rootable, then it is S4-honest.

proof $\Rightarrow$: Suppose φ is S4-dishonest. By the rule of disjunction we conclude that there exists a finite sequence of formulae $\{\alpha_1, .., \alpha_m\}$ such that $\vdash_{\mathbf{S4}} \Box\varphi \rightarrow \Box\alpha_1 \vee .. \vee \Box\alpha_m$ (*) and for all $i \in \{1, .., m\}$: $\not\vdash_{\mathbf{S4}} \Box\varphi \rightarrow \alpha_i$. Take for all $i \in \{1, .., m\}$ a model M_i such that for some SG g_i in M_i: $M_i, g_i \models \Box\varphi \wedge \neg\alpha_i$. For every amalgamation M^* of $\{M_i\}_{i=1}^{m}$ we obtain $\forall i \in \{1, .., m\} : M^*, w^* \models \neg\Box\alpha_i$ for the root w^* in M^*, and so by (*) $\Box\varphi \notin Th(M^*)$.

Example 3 Now S4-honesty for $\Box p \vee q$ is immediately clear. We can always choose q to hold in a root of an amalgamation. Consistent propositional formulae can always be forced in a root, and therefore, are all S4-honest. Formulae of the form $\neg\Box\varphi$ are also rootable and therefore S4-honest, whenever $\Box\neg\Box\varphi$ is S4-consistent. Whatever the valuation is, knowledge about ignorance in the family of models that are amalgamated is always inherited by the root.

Example 4 In [Jasa] rootability was proposed as a sufficient condition for S4-honesty. From the following example we learn that this condition is not necessary.

Consider $\varphi = \Box p \vee \Box q \vee (\Diamond(p \wedge q) \wedge \Diamond(\neg p \wedge \neg q))$, and the following simple pair of SG-S4-models.

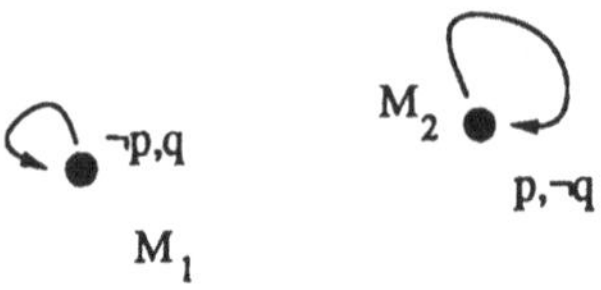

Both models contain φ: $\varphi \in Th(M_i), i = 1,2$. Amalgamating these two models into a model M^*, with a root w^*, we would end up with $M^*, w^* \models \neg\Box p \wedge \neg\Box q$, whatever the valuation in the root, w^*, would be. However, no valuation forces φ to hold, and so φ is not rootable, but .. *poly-rootable.*

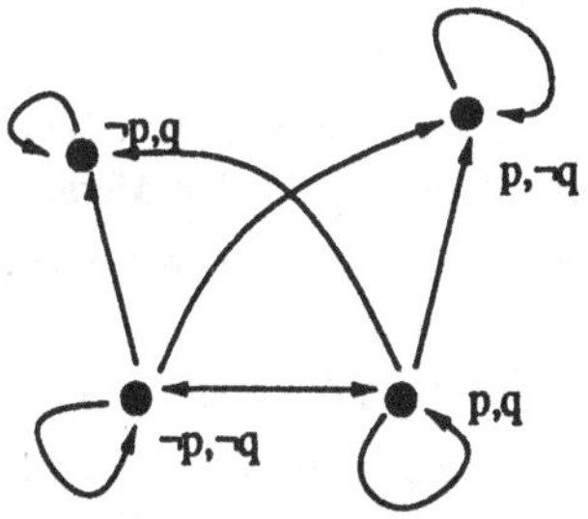

By poly-rootability we mean that we choose more than one root. This set of roots must form a SG-S4-frame itself, and furthermore, we take the accessibility relation to hold between any root and any world in the family of models that is to be poly-amalgamated. Such a poly-amalgamation of M_1 and M_2 is depicted in the last figure. It may be clear that the third disjunct of φ holds in both roots. Because this validity of φ does not depend on the family that we amalgamate, this construction would always give us a model for φ, and it is therefore poly-rootable.

Such poly-rootability would give us an equivalent of S4-honesty. The proof of theorem 6 would still hold, and vice versa, it turns out that we can always choose the minimal model of the formula, presupposing its S4-honesty, as root-model. This can hardly be called a construction, because we use a minimal model to construct a minimal model.

Concluding remarks

We have shown that the notion of stability, that was introduced by Stalnaker, can be generalized to any normal modal system of one's own choice. This generalization allowed us to define circumscription of S4-knowledge, and furthermore, we managed to obtain clear syntactic and semantic results, similar to Halpern and Moses' [HaM] for S5. The only complication left is the construction of minimal models, which looks troublesome. The question arises as to whether there exists some friendlier way of constructing minimal models somewhere between amalgamation and poly-amalgamation.

Further research

One point of interest that first springs to mind is: 'could this generalization of stability give us uniform minimization techniques for variable modal systems?'. The order of S5-stable sets ($\sqsubseteq_{prop}$) and S4-stable sets ($\subseteq$) have individual character. What is the formal relation between these orders and the modal logics, and can we predict how minimization orders would look for other modal systems?

The strange thing is that circumscription in epistemic logics leads to enormous Kripke-models. The more alternatives an agent considers as uncertainties, the less knowledge he has. This disagrees with initial motivations of circumscription [McC]. In partial modal logics, which I study at the moment, this nasty property of possible worlds semantics disappears [Jasb]. One of the concerns in this research is how the techniques that were mentioned in this paper can be transferred to partial modal systems.

References

[Gri] H.P. Grice, "Logic and conversation," in *Speech Acts, Syntax and Semantics*, P.Cole & J. Morga eds. #III: Speech Acts, Academic Press, New York, 1975.

[HaM] J.Y. Halpern & Y.O. Moses, "Towards a theory of knowledge and ignorance: a preliminary report," *Proceedings of the AAAI Workshop on Non-Monotonic Reasoning*, New Paltz, NY, 1984, 125–14

[Hin] J. Hintikka, *Knowledge and Belief: An Introduction to the Logic of the Two Notions*, Corn University Press, Ithaca N.Y., 1962.

[HuC] G.E. Hughes & M.J. Cresswell, *A Companion to Modal Logic*, Methuen, New York, 1984.

[Jasa] J.O.M. Jaspars, "Kennis en Eerlijk in S4 Kripke Modellen (Knowledge and Honesty in S4 Krip Models)," University of Amsterdam, M.-thesis, supervised by J.F.A.K. van Benthem, Amsterda 1988.

[Jasb] J.O.M. Jaspars, "Theoretical Circumscription in Partial Modal Logic," in *Logics in AI / JELIA'* J. van Eijck, ed., Lecture Notes in Artificial Intelligence #478, Springer Verlag, Heidelberg, 19 303–318.

[Kon] K. Konolige, "On the relation between default and autoepistemic logic," *Artificial Intelligence* (1988), 343–382.

[LeS] E.J. Lemmon & D.S. Scott, *The 'Lemmon Notes': An Introduction to Modal Logic*, Blackwe Oxford, 1977.

[Len] W. Lenzen, "Recent work in epistemic logic," *Acta Philosophica Fennica* 30 (1978), 1–219.

[McC] J. McCarthy, "Circumscription – A form of non-monotonic reasoning," *Artificial Intelligence Jour* 13 (1980), 27–39.

[Mooa] R.C. Moore, "Semantical considerations on non-monotonic reasoning," in *Proceedings of the Eig International Joint Conference on Artificial Intelligence*, Morgan Kaufmann, Los Altos, 1983, 27 279.

[Moob] R.C. Moore, "Possible world semantics for autoepistemic logic," in *Proceedings of the AAAI Wo shop on Non-Monotonic Reasoning*, New Paltz, NY, 1984, 344–354.

[Var] M.Y. Vardi, "Model-theoretic analysis of monotone knowledge," in *Proceedings of the Ninth Int national Joint Conference on Artificial Intelligence*, Morgan Kaufmann, Los Altos, 1985, 509–51

THE COMPLEXITY OF ADAPTIVE ERROR-CORRECTING CODES

DANIELE MUNDICI
Department of Computer Science, University of Milan,
via Moretto da Brescia 9,
20133 Milan, Italy

Someone thinks of a number between one and one million (which is just less than 2^{20}*).*
Another person is allowed to ask up to twenty questions,
to each of which the first person is supposed to answer only yes or no.
Obviously the number can be guessed by asking first:
Is the number in the first half million ?
then again reduce the reservoir of numbers in the next question by one-half, and so on.
Finally the number is obtained in less than $\log_2(1000000)$.
Now suppose one were allowed to lie once or twice,
then how many questions would one need to get the right answer ?

S.M. ULAM, *Adventures of a Mathematician,* Scribner's, New York, 1976, page 281.

Introduction The questions and answers interchanged between Questioner and Responder (the latter being conveniently identified with Pinocchio) in Ulam's game with k lies do not behave as propositions in classical logic: two *opposite answers* to the same repeated question do not lead to inconsistency; the conjunction of two *equal answers* is more informative than a single answer. In [6] the author showed that the Lukasiewicz (k+2)-valued sentential calculus [10] provides a natural logic for these propositions. In this paper, we apply Lukasiewicz logic to measure the complexity of Ulam's game.

Suppose the number x to be guessed belongs to a certain search space $\{0,1,\ldots,2^n-1\} \cong \{0,1\}^n$, and Pinocchio is allowed to lie at most k times. We call n the dimension of the search space. Let q be the smallest number obeying Hamming's inequality [11, p. 59]

(1) $$2^q \geq 2^n\left(\binom{q}{0}+\binom{q}{1}+\ldots+\binom{q}{k}\right).$$

Then any strategy to guess x will require at least q questions. Conversely, as proved in [8], [2], [3], q questions are sufficient for the case of one and (with the only exception of $n=2$, where one needs 8 questions) for two lies. The corresponding optimal searching strategies are described in these papers.

In the 0-lie case, an optimal strategy [1, p. 6 and p. 62] to guess a number $x \in \{0,1\}^n$ is a sequence of questions, i.e., Boolean functions $f_1,\ldots,f_n$ of n variables, such that for each choice of the answers $b_1,\ldots,b_n \in \{\text{yes, no}\} \cong \{1, 0\}$, the function $f_1^{b_1} \wedge\ldots\wedge f_n^{b_n}$ differs from zero in exactly one point of $\{0,1\}^n$. Here, $f^1 = f$, and $f^0 = 1-f$.

Turning to $k \geq 1$ lies, a particular case of Ulam's game occurs when all questions precede (or, are independent of) the answers. In the terminology of [1, p. 9] this is the *nonadap-*

tive case of the game. Optimal nonadaptive searching strategies immediately yield optimal k-error correcting codes. However [11], very few such codes are known when $k \geq 2$. By contrast, already in the case of two lies, our optimal *adaptive* strategies are more efficient than their nonadaptive counterparts. For instance, we have

PROPOSITION. 90 *questions are sufficient to find a number having* 78 *bits with two lies, while* 90 *questions are not sufficient if all questions precede all answers.*

Proof. The first statement is a particular case of the main theorem of [3]. As for the second statement, if 90 questions preceding all answers were sufficient to find a number with 78 bits, then the set $\{0,1\}^{78}$ would admit a 2-error correcting code with codewords of length 90. This code would be perfect and nontrivial, thus contradicting a well-known negative result [11, 7.5.18, 7.7.5]. QED

The complexity of coding and decoding Adaptive strategies may find concrete applications in the natural variant of Ulam's game where Pinocchio need not know which of his answers are false. For instance, in the theory of communication with feedback [4], [9], one may consider the case of a low-power transmitter P sending us binary numbers of length q coding binary numbers of length n. P may be thought of as a sincere Pinocchio answering our questions from a very distant place: during travel, distortion may transform any bit b into $1-b$. At most k bits may be so distorted. Using a powerful transmitter, we can however inform P (with no distortion) whether b or $1-b$ was actually received. If the coding algorithm is easy and available to P, then from our single-bit feedback, P can immediately deduce what the next question would be; accordingly, his next bit is the answer to this question. Using the decoding algorithm, we recover Pinocchio's number from the q bits received from him. An *optimal* strategy is one that minimizes the number q. It is then of some interest to investigate the complexity of coding and decoding algorithms of optimal adaptive strategies. For the sake of simplicity, in this paper we only consider the case of two lies.

The *coding algorithm* of an adaptive strategy is the following algorithm

INPUT: the dimension n of the search space, together with bits $b_1,...,b_t$ $(0 \leq t < q)$, recording a sequence of answers (yes = 1, no = 0).

OUTPUT: the $(t+1)$th question, a *question* being represented by an n-ary Boolean formula.

Similarly, the *decoding* algorithm is the following algorithm

INPUT: a string of exactly q bits giving a complete sequence of answers.

OUTPUT: the coded number (if any), and the number of lies told by Pinocchio.

The *complexity* of either algorithm is defined as in [5]. A strategy is *regular* iff for each $i = 1, 2,..., n,$ the ith question is given by "is the ith binary digit of x equal to 1 ?"

1.1 THEOREM. *For any regular optimal strategy with two lies, the decoding algorithm is computable in deterministic polynomial time iff so is the coding algorithm.*

1.2 LEMMA (Lies have short legs). *For any optimal strategy to search an n-bit number with two lies, the number q of questions obeys the inequality*

$$q \leq n + 4 + 2\log_2 n. \tag{2}$$

Proof. By [3, Theorem 1], q is the smallest number obeying Berlekamp's inequality (1), except for $n = 2$, in which case $q = 8$. Direct inspection then shows that (2) holds for all $n <$ 30. On the other hand, for $n \geq 30$, letting $s = q - n$, from the first part of the proof of [3, Lemma 7], together with [3, Theorem 1] again, we get

$$s \leq n/3. \tag{3}$$

From the assumed minimality of q in (1), it follows that

$$2^{s-1} < (q^2 - q + 2)/2. \tag{4}$$

A straightforward computation using (3) and (4) now yields $2^s < 2 + s^2 + n^2 + 2sn - s - n < s^2 + n^2 + 2sn \leq 4n^2$, which implies (2). This proves the lemma.

1.3 *Proof of Theorem 1.1.* We shall now interpret Ulam's game with two lies in the 4-valued Lukasiewicz calculus [10]. We assume familiarity with [6]. We let the map

$$x \in \{0,1\}^n \rightarrow x^\mu \in \{1/3, 2/3\}^n$$

transform each coordinate x_i of x into 1/3 or 2/3 according as $x_i = 0$ or $x_i = 1$, respectively. For each question $Q \subseteq \{0,1\}^n$ we let $Q^\mu = \{x^\mu \mid x \in Q\}$. The positive answer Q^{yes} to Q is canonically transformed by μ into the function $f: \{1/3, 2/3\}^n \rightarrow \{2/3, 1\}$ such that $f(y) = 1$ if $y \in Q^\mu$, and $f(y) = 2/3$, otherwise. The negative answer to Q coincides with the positive answer to the opposite question $\{0,1\}^n \setminus Q$. The initial state of knowledge is the constant function 1. In this way, Pinocchio's answers to questions $Q_1,\ldots,Q_t$ determine t functions $f_1,\ldots,f_t: \{1/3, 2/3\}^n \rightarrow \{2/3, 1\}$, and the state of knowledge after these answers is recorded by the Lukasiewicz conjunction $f = f_1 \bullet \ldots \bullet f_t$, where $a \bullet b = \max(0, a + b - 1)$. A number $y \in \{1/3, 2/3\}^n$ falsifies zero, one, two, or more than two of the answers iff $f(y)$ equals 1, 2/3, 1/3, 0, respectively. Thus, states of knowledge are represented by sentences in the Lukasiewicz calculus. Conversely, every sentence ϕ represents the state of knowledge given by the restriction to $\{1/3, 2/3\}^n$ of the McNaughton function of ϕ. Following tradition [5], we use such expressions as "easy" and "short" as abbreviations of "computable in deterministic polynomial time", and "having polynomially bounded length" (with respect to input length).

Claim 1. If coding is easy, then so is decoding.

Indeed, given a sequence of bits $b_1,...,b_q$, we can quickly determine Pinocchio's number, and the number of falsified answers, as follows: First, using (1), the number n is easily determined from q. By the assumed regularity of our strategy, from the first n answers $b_1,...,b_n$ we obtain sets $X, Y, Z \subseteq \{0,1\}^n$, X with 1 element, Y with n elements, Z with $n(n-1)/2$ elements, respectively containing the numbers that falsify zero, one, and two of the first n answers. Let $X^\mu, Y^\mu, Z^\mu \subseteq \{1/3, 2/3\}^n$ be the image of X, Y, Z, respectively. We can easily write down a sentence ψ whose McNaughton function f takes value 1 over X^μ, value 2/3 over Y^μ, value 1/3 over Z^μ, and value 0 for the remaining points of $\{1/3, 2/3\}^n$. For every $t = n+1,..., q$, let the n-ary Boolean formula β_t represent the tth question after answers $b_1,...,b_n,...,b_{t-1}$. Let the Boolean formula α_t be given by $\alpha_t = \beta_t$ if b_t = 1, and α_t = not(β_t) if b_t = 0. Since by assumption each β_t is easily obtainable from $b_1,...,b_n,...,b_{t-1}$, we can quickly write down sentences $\phi_{n+1},..., \phi_q$ in the Lukasiewicz calculus such that for each t the McNaughton function f_t of ϕ_t satisfies the conditions $f_t(x^\mu) = 1$ if $\alpha_t(x) = 1$, and $f_t(x^\mu) = 2/3$ if $\alpha_t(x) = 0$, $x \in \{0,1\}^n$. The McNaughton function $g = f \bullet f_{n+1} \bullet ... \bullet f_q$ is the state of knowledge over $\{1/3, 2/3\}^n$ after the q answers [6]. By the assumed optimality of our strategy (and assuming $n \neq 2$, without loss of generality), at most one $y \in \{1/3, 2/3\}^n$ satisfies $g(y) \neq 0$. It is easy to decide whether any such y exists, and, if this is the case, to find it. As a matter of fact, our search is restricted to the small set $W = X^\mu \cup Y^\mu \cup Z^\mu$, and g is the McNaughton function of a short conjunction of easily obtainable sentences. If for all $y \in W$ we have $g(y) = 0$, the sequence of answers $b_1,...,b_q$ codes no number (however, this can only happen if Pinocchio has told three or more lies, against the rules of the game). Otherwise, if y is the only element of W such that $g(y) \neq 0$, then the number of answers falsified by y is given by $3 - 3g(y)$. The proof of Claim 1 is complete.

Claim 2. If decoding is easy, then so is coding.

Indeed, given n together with bits $b_1,...,b_t$ $(0 \leq t < q)$, a fast procedure yielding the $(t+1)$th question is as follows: Without loss of generality, $n \neq 2$. We can safely concentrate on the case $t \geq n$; for otherwise, by the assumed regularity of our strategy, the $(t+1)$th question is trivial. Since our strategy is optimal, every sequence of bits $\mathbf{b} = b_1,...,b_t, b_{t+1},...,b_q$ of length q extending $b_1,...,b_t$ (in the sense that the initial segment of $\mathbf{b}$ of length t equals $b_1,...,b_t$) determines a set $S_\mathbf{b} \subseteq \{0,1\}^n$ containing at most one element $x_\mathbf{b}$ (this is Pinocchio's number), and also determines the number $k_{\mathbf{b}} \leq 2$ of answers falsified by $x_{\mathbf{b}}$. Let $\phi_{\mathbf{b}}$ be a sentence in the Lukasiewicz calculus whose McNaughton function $f_{\mathbf{b}}$ satisfies $f_{\mathbf{b}}(x_{\mathbf{b}}^{\mu}) = 1 - k_{\mathbf{b}}/3$, and $f_{\mathbf{b}}(y) = 0$ for the remaining $y \in \{1/3, 2/3\}^n$. By Lemma 1.2, **b** is not much longer than $b_1,...,b_t$, and by hypothesis, $x_{\mathbf{b}}$ and $k_{\mathbf{b}}$ are easily computed from **b**; thus, $\phi_{\mathbf{b}}$ can be quickly written down. Define the map $f^+ : \{1/3, 2/3\}^n \to \{0, 1/3, 2/3, 1\}$ by

(5) $\quad f^+(y) = \max\{f_{\mathbf{b}}(y) \mid \mathbf{b}$ ranging over all extensions of $b_1,...,b_t, 1$ of length $q\}$.

Similarly, let f^- be defined by

(6) $\quad f^-(y) = \max\{f_{\mathbf{b}}(y) \mid \mathbf{b}$ ranging over all extensions of $b_1,...,b_t, 0$ of length $q\}$.

Then f^+ (resp., f^-) is the state of knowledge over $\{1/3, 2/3\}^n$ after answers $b_1,...,b_t, 1$ (resp., after answers $b_1,...,b_t, 0$). Further, the function $h = \max(f^+, f^-)$ is the state of knowledge over $\{1/3, 2/3\}^n$ after answers $b_1,...,b_t$. Let $\mathrm{supp}(h) = \{y \in \{1/3, 2/3\}^n \mid h(y) > 0\}$, and $Q_{b_1,...,b_t} = \{y \in \{1/3, 2/3\}^n \mid f^+(y) > f^-(y)\}$. Then $Q_{b_1,...,b_t}$ is the set of points in $\mathrm{supp}(h)$ satisfying the positive answer to the $(t+1)$th question. Let $Q \subseteq \{0,1\}^n$ be defined by $Q^{\mu} = Q_{b_1,...,b_t}$. Then Q may be safely identified with the $(t+1)$th question. Let ξ be an n-ary Boolean formula representing the characteristic function of Q. By Lemma 1.2, the number of extensions **b** of $b_1,...,b_t$ is small, whence so is the number of elements of $Q_{b_1,...,b_t}$. Moreover, since for every **b** the sentence $\phi_{\mathbf{b}}$ can be easily written down, ξ too, can be obtained in polynomial time, as required. Having completed the proof of Claim 2, the proof of Theorem 1.1 is complete. QED

Remark. Note the role of the "weak disjunction" connective (connective **A**, in the traditional Polish notation of [10]) in the above reconstruction (5), (6) of states of knowledge starting from final states of knowledge.

1.4 THEOREM. *The coding and decoding algorithms of the adaptive and optimal strategies described in* [3] *are computable in deterministic polynomial time.*

Proof. Since such strategies are regular, by Theorem 1.1 it is sufficient to prove that their decoding algorithms are easy. We assume familiarity with [2] and [3]. In the framework of [2], a state of knowledge over $\{0,1\}^n$ is a triplet (X, Y, Z) where X contains the points falsifying no answer, Y contains those falsifying one answer, and Z those falsifying two answers. We let x, y, z denote the number of elements of X, Y, Z respectively. By definition, the weight of (x, y, z) before j questions is given by

$$w_j(x, y, z) = x(j(j-1)/2 + j + 1) + y(j + 1) + z.$$

We say that state (X, Y, Z) is *j-complete* iff $w_j(x, y, z) = 2^j$. Given bits $b_1,...,b_q$, a fast procedure yielding Pinocchio's number is as follows: Using (1) we immediately compute n. We can safely assume $n \geq 30$. Let $s = q-n$. By regularity, after the first n answers $b_1,...,b_n$ the state of knowledge is given by the triplet X_n, Y_n, U_n of subsets of $\{0,1\}^n$, where X_n only contains the element falsifying none of the first n answers (namely, the binary number $b_1...b_n$), Y_n contains the elements falsifying one answer, U_n those elements falsifying two answers. Note that $x_n = 1$, $y_n = n$, $u_n = n(n-1)/2$; s questions are still to be asked. Let $d = 2^s - w_s(x_n, y_n, u_n)$. Then by (1), $d \geq 0$. We then complete the triplet (X_n, Y_n, U_n) by adding new elements to U_n, until the weight becomes equal to 2^s. More precisely, let D be the set of the first d elements of $\{0,1\}^n \setminus (X_n \cup Y_n \cup U_n)$. By Lemma 1.2, the set $Z_n = U_n \cup D$ still has a small number of elements, which can be easily listed. Further, the s-complete triplet (X_n, Y_n, Z_n) is still s-solvable (in the sense of [2]) using (a trivial modification of) the strategy in [3].

Starting from (X_n, Y_n, Z_n), we shall inductively produce the $(t+1)$th question, for each $t = n, n+1,...,q-1$ and, using answer b_{t+1}, the next state $(X_{t+1}, Y_{t+1}, Z_{t+1})$. Assuming (X_t, Y_t, Z_t) to be $(q-t)$-complete and $(q-t)$-solvable, $(X_{t+1}, Y_{t+1}, Z_{t+1})$ will turn out to be $(q-t-1)$-complete and $(q-t-1)$-solvable. From the final state (X_q, Y_q, Z_q) we shall easily obtain Pinocchio's number and the number of falsified answers. Suppose (X_t, Y_t, Z_t) has already been obtained. We proceed by cases:

Case 1: $x_t = 1$ and y_t is nonzero and even.
Then, proceeding as in [3, p. 73], we let the $(t+1)$th question Q be given by
$Q \cap X_t = X_t$,
$Q \cap Y_t =$ the first element of Y_t,
$Q \cap Z_t =$ the first r elements of Z_t, where $r = 2^{q-t-1} - y_t - ((q-t)^2 + (q-t))/2 - y_t$.

Case 2: $x_t = 1$ and y_t is odd.
Then as in [3, p. 74], we let the $(t+1)$th question Q be given by
$Q \cap X_t = X_t$,
$Q \cap Y_t = \emptyset$
$Q \cap Z_t =$ the first p elements of Z_t, where $p = 2^{q-t-1} - y_t - ((q-t)^2 - (q-t))/2 - 1$.

Case 3: $x_t = 1$ and $y_t = 0$.
Then, as in [3, Lemma 3], we let the $(t+1)$th question Q be given by
$Q \cap X_t = X_t$,
$Q \cap Y_t = \emptyset$
$Q \cap Z_t =$ the first m elements of Z_t, where $m = 2^{q-t-1} - ((q-t)^2-(q-t))/2 - 1$.

Case 4: $x_t = 0$.

Then, as in [3, Lemma 4], we let the $(t+1)$ th question Q be given by

$Q \cap X_t = \emptyset$

$Q \cap Y_t$ = the first $\lfloor y_t/2 \rfloor$ elements of Y_t, where $\lfloor y_t/2 \rfloor$ = greatest integer $\leq y_t/2$

$Q \cap Z_t$ = the first i elements of Z_t, where, $i = 2^{q-t-1} - y_t + \lfloor y_t/2 \rfloor (1-q+t)$.

According as $b_{t+1} = 1$ or $b_{t+1} = 0$, we can immediately compute the state of knowledge $(X_{t+1}, Y_{t+1}, Z_{t+1})$ resulting from the answer to Q. The analysis in [3] ensures that if (X_t, Y_t, Z_t) is $(q-t)$-complete and $(q-t)$-solvable, then $(X_{t+1}, Y_{t+1}, Z_{t+1})$ is $(q-t-1)$-complete and $(q-t-1)$-solvable. The above discussion shows that $(X_{t+1}, Y_{t+1}, Z_{t+1})$ can be quickly obtained from (X_t, Y_t, Z_t) and b_{t+1}.

Proceeding in this way, when $t = q$, we are left with a 0-complete and 0-solvable triplet (X_q, Y_q, Z_q); thus in particular, $X_q \cup Y_q \cup Z_q$ contains exactly one element y. Since our search of y is restricted to the set $X_n \cup Y_n \cup Z_n$, y can be easily computed. Now, if y belongs to the spurious set D, we declare that the input sequence of answers $b_1,...,b_q$ codes no number. Otherwise, y is Pinocchio's number, and the number of falsified answers is zero, one, or two according as y is a member of X_q, Y_q, or Z_q, respectively. Since the sequence of triplets (X_n, Y_n, Z_n), $(X_{n+1}, Y_{n+1}, Z_{n+1})$, ..., (X_q, Y_q, Z_q) is so easily obtainable from $b_1,...,b_q$, we conclude that Pinocchio's number and the number of lies can be computed in deterministic polynomial time. QED

Remarks. (i) The author has written an efficient program for the above searching strategies with two lies.

(ii) Since the many-valued calculus of Lukasiewicz is deeply related to approximately finite-dimensional (AF) C^*-algebras [7], and since AF C^*-algebras are useful in the description of quantum spin systems, one can use a game-theoretic reformulation of the notions of *state*, *yes-no observable*, *spectrum*, to investigate the combinatorial complexity of searching strategies over such systems.

References

[1] M. AIGNER, "Combinatorial Search", Wiley-Teubner, New York-Stuttgart, 1988.

[2] J. CZYZOWICZ, D. MUNDICI, A. PELC, Solution of Ulam's problem on binary search with two lies, *J. Combinatorial Theory, Series A*, **49** (1988) 384-388.

[3] J. CZYZOWICZ, D. MUNDICI, A. PELC, Ulam's searching game with lies, *J. Combinatorial Theory, Series A*, **52** (1989) 62-76.

[4] R.L. DOBRUSHIN, Information transmission in a channel with feedback, *Theory of Probability and Applications,* **34** (1958) 367-383.

[5] M.R. GAREY, D.S. JOHNSON, Computers and Intractability, W.H. Freeman, San Francisco 1979.

[6] D. MUNDICI, The logic of Ulam's game with lies, *Cambridge Studies in Probability, Induction, and Decision Theory,* to appear.

[7] D. MUNDICI, Interpretation of AF C^*-algebras in Lukasiewicz sentential calculus, *J. Functional Analysis,* **65** (1986) 15-63; see also: *J.Algebra* **98** (1986) 76-81; *ibid.*, **105** (1987) 236-241; *ibid.*, **113** (1988) 89-109; *Lecture Notes in Computer Science* **270** (1987) 256-264; *Theoretical Computer Science* **52** (1987) 145-153; *Advances in Mathematics* **68** (1988) 23-39; *Contemporary Mathematics, AMS,* **69** (1988) 209-227; *Proceedings Logic Colloquium '88, Studies in Logic and the Foundations of Mathematics,* North-Holland, Amsterdam (1989) pp. 61-77.

[8] A.PELC, Solution of Ulam's problem on searching with a lie, *J. Combinatorial Theory, Series A,* **44** (1987) 129-140.

[9] D. SLEPIAN (Editor), "Key Papers in the Development of Information Theory", IEEE Press, New York, 1974 (contains [4]).

[10] A.TARSKI, J.LUKASIEWICZ, Investigations into the Sentential Calculus, Chapter IV in: Logic, Semantics, Metamathematics, Oxford University Press, 1956, pp. 38-59.

[11] J.H.VAN LINT, Introduction to Coding Theory, Springer, Berlin, 1982.

RAMSEY'S THEOREM IN BOUNDED ARITHMETIC

Pavel Pudlák
Mathematical Institute
Praha 1, Žitná 25
Czechoslovakia

Abstract: *We shall show that the finite Ramsey theorem as a Δ_0 schema is provable in $I\Delta_0+\Omega_1$. As a consequence we get that propositional formulas expressing the finite Ramsey theorem have polynomial-size bounded-depth Frege proofs.*

1 Introduction

In fragments of arithmetic such as $I\Delta_0$ and $I\Delta_0+\Omega_1$ the exponentiation is only a partial function. Many combinatorial and number theoretical statements, though they are Π_1, need for their proofs the Π_2 axiom Exp saying that exponentiation is a total function. This leads to the question what Π_1 principles are needed to derive Π_1 consequences of exponentiation over $I\Delta_0$ or $I\Delta_0+\Omega_1$ as a base theory. Let us consider Δ_0-PHP, the Δ_0 Pigeon Hole Principle, as an example. This is the schema, for every Δ_0 formula φ,

$\forall u\ [\forall x\leq u\ \exists y<u\ \varphi(x,y) \rightarrow \exists x_0,x_1\leq u\ \exists y<u\ \varphi(x_0,y)\ \&\ \varphi(x_1,y)]$.

The meaning of the schema is that no mapping from $[0,u]$ to $[0,u-1]$ is one-to-one. This is a Π_1 formula derivable in $I\Delta_0$+Exp but very likely not derivable in $I\Delta_0$ itself (maybe even not in $I\Delta_0+\Omega_1$), see [A]. A weaker version of PHP which says only that there is no one-to-one mapping from $[0,2u]$ to $[0,u]$, i.e.

$\forall u\ [\forall x<2u\ \exists y<u\ \varphi(x,y) \rightarrow \exists x_0,x_1<2u\ \exists y<u\ \varphi(x_0,y)\ \&\ \varphi(x_1,y)]$,

is derivable in $I\Delta_0+\Omega_1$, see [PWW]. We shall denote it by Δ_0-WPHP.

The difficulty of proving Δ_0-PHP is caused by the fact that there is no Δ_0 definition for counting the number of elements that satisfy a Δ_0 formula. (This follows from a result of Toda [T], provided that Polynomial Hierarchy does not collapse). Thus there is another interesting question: what are all arithmetical consequences of "having counting functions". This could be formalized by adding a new function symbol for each Δ_0 formula with a corresponding axiom and by extending the schema of induction to bounded formulas containing also the new symbols. PHP is a typical statement derivable using counting.

A natural generalization of the Pigeon Hole Principle is the finite Ramsey theorem. This is usually stated as follows. Let k,m,r be positive numbers, k<m, then there exists n such that if we color the k element subsets of an n element set X by r colors, then there exists Y, an m element

number n must be exponential in m, thus it cannot be presented as a Π_1 sentence.

We shall consider a Π_1 form of the finite Ramsey's theorem. In order to simplify the matter let us take the special case of k=2 and r=2. Then for each coloring of pairs on an n element set X by two colors there is a set Y of size $\lceil 1/2 \lfloor \log_2 n \rfloor \rceil$ such that all pairs of elements of Y have the same color. (The bound to the size $\lceil 1/2 \lfloor \log_2 u \rfloor \rceil$ is not the best one, we shall discuss this question later). We formalize this statement as follows: For every bounded formula $\varphi(x,y,z)$, possibly with other parameters,

$$\forall u[\forall x,y<u\ \exists z<2\ \varphi(x,y,z) \rightarrow \exists z<2\ \exists Y\subseteq u\ (|Y|\geq \lceil 1/2 \lfloor \log_2 u \rfloor \rceil\ \&$$
$$\&\ \forall x,y\in Y(x<y \rightarrow \varphi(x,y,z)))].$$

This is not a Π_1 formula, however if we extend the language by a function symbol for $x^{\lceil \log_2 x \rceil}$, we can talk about subsets of u of size $\log_2 u$. This is quite justified in $I\Delta_0+\Omega_1$, since the axiom Ω_1 just says that $x^{\lceil \log_2 x \rceil}$ is a total function. We can work also in the equivalent theory S_2 which has the smash function x#y with the same growth rate as $x^{\lceil \log_2 x \rceil}$, see [B1]. For larger k and r the set Y is even smaller, thus also the general finite Ramsey theorem can be formalized in this way.

The purpose of this paper is to show that such formalizations of finite Ramsey's theorem are derivable from the Pigeon Hole Principle. In fact they are derivable from the weaker version Δ_0-WPHP, hence they are provable in $I\Delta_0+\Omega_1$. Using a translation of bounded formulas into propositional formulas we conclude that propositional formulas expressing the finite Ramsey theorem have polynomial-size bounded-depth Frege proofs. For a discussion of related questions on propositional proof systems see section 4. We shall also discuss some complexity theoretical questions concerning Ramsey's theorem and present principles, which are candidates for natural counting principles not derivable from the Pigeon Hole Principle.

2 The proof of the Ramsey schema

We shall prove the Ramsey schema for pairs and then, in the next section, sketch the proof in the general case. We shall use the language of arithmetic extended by $x^{\lceil \log_2 x \rceil}$, thus in bounded (i.e. Δ_0) formulas we can quantify sets of logarithmic size.

Theorem 1

The following sentence is provable in $I\Delta_0+\Omega_1$ for every Δ_0 formula $\varphi(x,y,z)$:

$$\forall u\ \forall r>1[\forall x,y<u\ \exists z<r\ \varphi(x,y,z) \rightarrow \exists z<r\ \exists Y\subseteq u\ (|Y|\geq \lceil r^{-1} \lfloor \log_r u \rfloor \rceil\ \&$$

We shall talk about a *function* $G(x,y)=z$, instead of the formula $\varphi(x,y,z)$. In the schema above we do not require that φ defines a function, but because we work in $I\Delta_0$, we can always replace φ by a formula which defines a function.

The proof is based on the following lemma.

Reduction Lemma 1 (in $I\Delta_0+\Omega_1$)
Let a Δ_0 definable function $G: u \times u \to r$ be given, $u=r^s$. Then there exist $x_0<x_1< \dots <x_{s-1}<u$ and $\alpha_0,\alpha_1,\dots,\alpha_{s-2}<r$ such that for every $i<j<s$, $G(x_i,x_j) = \alpha_i$.

We shall prove the theorem using this lemma. The lemma gives us a coloring of $x_0,\dots,x_{s-2}$ by r colors, namely, x_i is colored by α_i. Since the set $\{x_0,\dots,x_{s-2}\}$ has only logarithmic length, it can be coded by a number. For coded sets we can define the size in $I\Delta_0+\Omega_1$. Hence we can find a subsequence of $x_0,\dots,x_{s-2}$ of length $\lceil s-1/r \rceil$ whose elements have the same color. Now we can add to this sequence also x_{s-1} and we have the required subset.

It remains to prove the lemma. We shall use the following notation. We identify u with the set of numbers less than u; $[u]^{\le k}$ is the set of all subsets of u having at most k elements.

Let G be given. Define a relation $R \subseteq [u]^{\le s} \times [r]^{\le s}$ by

$$R((x_0,\dots,x_j),(\alpha_0,\dots,\alpha_h)) \equiv_{df} x_0=0,\ x_0< \dots <x_j,\ h+1=j,$$
$$\forall i<t\le j\ G(x_i,x_t)=\alpha_i\ ,\ \text{and}$$
$$\forall i<j\ \ x_{i+1} \text{ is minimal such that } \forall t\le i\ G(x_t,x_{i+1})=\alpha_t.$$

There is a Δ_0 formula (in the extended language) which defines R. To prove the reduction lemma we need to show

$$(*)\quad \exists x_0,\dots,x_{s-1}\exists\alpha_0,\dots,\alpha_{s-2}\ R((x_0,\dots,x_{s-1}),(\alpha_0,\dots,\alpha_{s-2})).$$

Arguing in $I\Delta_0+\Omega_1$, assume that it is not so. Then we shall define, using a bounded formula, a one-to-one function $F:r^s \to r^{s-1}$. By the result of Wilkie mentioned above such a function cannot exist, hence the lemma will be proved. The function F is defined as follows.

$F(0) =_{df} 0;$

if $x>0$ then

$F(x) =_{df} y$, such that y is minimal satisfying the following formula

$$\exists j<s-1\ \exists x_0<\dots<x_j\ \exists\alpha_0,\dots,\alpha_{j-1}<r$$
$$[R((x_0,\dots,x_j),(\alpha_0,\dots,\alpha_{j-1}))\ \&\ x=x_j\ \&\ y=\sum_{i=0}^{j-1}(\alpha_i+1)r^{i-1}].$$

The sum in the formula can be equivalently defined by a bounded formula, hence F is defined by a bounded formula. We have to prove that F is defined for all $x<u$, F is one-to-one and maps r^s into r^{s-1}.

$\& \ x_j<x \ \& \ \forall i\leq j \ G(x_i,x)=\alpha_i \ ].$

So $Q((x_0,\ldots,x_j),x)$ is like $R((x_0,\ldots,x_j,x),(\alpha_0,\ldots,\alpha_j))$, except that x need not be minimal.

Claim 1. For every $j<s$, and $x<u$, either

(1) $\exists x_0,\ldots x_j \ Q((x_0,\ldots,x_j),x)$, or

(2) $\exists i<j \ \exists x_0,\ldots,x_i \ \exists\alpha_0,\ldots,\alpha_{i-1} \ R((x_0,\ldots,x_i,x),(\alpha_0,\ldots,\alpha_{i-1}))$.

Proof

We shall prove the claim by Δ_0 induction. For $x=0$, we have trivially (2). Let x, $0<x<u$ be given.

(i) Let $j=0$. Then put $x_0=0$ and $\alpha_0=G(0,x)$. Then we have (1).

(ii) Suppose the claim holds for j. If we have (2) for j, then we have it also for j+1. Thus suppose we have (1) for j, i.e.

$R((x_0,\ldots,x_j),(\alpha_0,\ldots,\alpha_{j-1})) \ \& \ \forall i\leq j \ G(x_i,x)=\alpha_i$

for some $x_0,\ldots,x_j,\alpha_0,\ldots,\alpha_{j-1},\alpha_j$. If x is minimal such that it satisfies this formula, then we have

$R((x_0,\ldots,x_j,x),(\alpha_0,\ldots,\alpha_{j-1},\alpha_j))$,

hence we get (2) for j+1. If x is not minimal, then, by Δ_0 induction, there exists some minimal element satisfying this formula and less than x. Denote this element by x_{j+1} and then we have (1) for j+1. Thus the claim is proved.

Consider $j=s-2$ in the claim. Suppose (1) holds for some $x<u$. Take the minimal such x. Then, clearly, we have

$R((x_0,\ldots,x_{s-2},x),(\alpha_0,\ldots,\alpha_{s-2}))$,

for some $x_0,\ldots,x_{s-2},\alpha_0,\ldots,\alpha_{s-2}$. But this is a contradiction with our assumption (*). Hence for $j=s-2$ and every $x<u$ we must have (2). This proves that F(x) is always defined for $x<u$. Also we have the following bound

$$F(x) = \sum_{t=0}^{i-1}(\alpha_t+1)r^{t-1} \leq \sum_{t=0}^{s-3} r.r^{t-1} = r(r^{s-2}-1) < r^{s-1},$$

since $i<j=s-2$. Hence it remains to prove that F is one-to-one.

Claim 2. $\sum_{i=0}^{j-1}(\alpha_i+1)r^{i-1} = \sum_{i=0}^{j'-1}(\alpha'_i+1)r^{i-1}$ implies $j=j'$ and $\alpha_0=\alpha'_0,\ldots,\alpha_{j-1}=\alpha'_{j-1}$.

Proof

This is just the r-adic representation of numbers.

Claim 3. For every $j\leq s$,

$$R((x_0,\ldots,x_j),(\alpha_0,\ldots,\alpha_{j-1})) \ \& \ R((x'_0,\ldots,x'_j),(\alpha_0,\ldots,\alpha_{j-1})) \rightarrow$$
$$\rightarrow x_0=x'_0,\ldots,x_j=x'_j.$$

Proof

determined by x_i and $\alpha_0,\ldots,\alpha_i$.

Now we can prove that F is one-to-one. Suppose $F(x)=F(x')=y$. We have the sequences $(x_0,\ldots,x_j,x)$, $(\alpha_0,\ldots,\alpha_j)$ and $(x'_0,\ldots,x'_j,x')$, $(\alpha'_0,\ldots,\alpha'_j)$ which witness $F(x)=y$ and $F(x')=y$ respectively. By Claim 2 we know that $(\alpha_0,\ldots,\alpha_j) = (\alpha'_0,\ldots,\alpha'_j)$. Then by Claim 3 we get $(x_0,\ldots,x_j,x) = (x'_0,\ldots,x'_j,x')$, hence, in particular, $x=x'$. This finishes the proof of the lemma, hence also of the theorem. □

3 General Ramsey schema

We can state the Ramsey schema in the same way as above for every fixed k. The schema can be stated and proved also for k as a parameter. Then we have to talk about the codes of sequences of length k and assume that such sequences exist for u.

Theorem 2
The following is provable in $I\Delta_0+\Omega_1$ for every Δ_0 definable function G:
$\exists\varepsilon>0\ \forall u,r,k\ (G:[u]^k\to r\ \Rightarrow\ \exists\alpha<r\ \exists Y\subseteq u\ (Y\subseteq G^{-1}(\alpha)\ \&\ |Y|\geq\varepsilon\log_r^{(k)}u))$.

Here $\log_r^{(k)}$ denotes k-times iterated logarithm and ε is a rational. By writing $G:[u]^k\to r$ we state implicitly that u^k exists. We shall state the corresponding reduction lemma.

Reduction Lemma 2 (in $I\Delta_0+\Omega_1$)
Let a Δ_0 definable function $G:[u]^k\to r$ be given, $u=r^s$. Then there exist $X\subseteq u$, $\binom{|X|}{k-1}\geq s$, and $A:X^{[k-1]}\to r$ such that for every $(x_0,\ldots,x_{k-1})\in[X]^k$, $G((x_0,\ldots,x_{k-1})) = A((x_0,\ldots,x_{k-2}))$.

In this statement it is sufficient to consider only sets X of size logarithmic in u, thus this statement can be presented as a Δ_0 formula too. The proof of this lemma is essentially the same as the proof of Lemma 1. The difference is only that we have the function A instead of the sequence $\alpha_0,\ldots,\alpha_{k-2}$. We shall leave the details to the reader.

Clearly, the reduction lemma reduces the theorem from k to k-1. Note that in order to prove the theorem we do not need to apply the reduction lemma above more than once. After the first reduction we have already a set which can be coded, thus the rest of the proof can just follow the classical proof. The classical proof is based also on successive reductions, but once the sets in question are coded, everything is easy to formalize.

Finally let us consider also the case when k=1, i.e. a direct

Δ_0-PHP, hence we are not able to prove it in $I\Delta_0+\Omega_1$ alone.

Proposition 1

The following is provable in $I\Delta_0$ + Δ_0-PHP for every Δ_0 definable function G: $\forall u,r(G: u \to r \ \&\ \exists y(y \geq u^{u/r}) \ \Rightarrow\ \exists\alpha\ |G^{-1}(\alpha)| \geq u/r\ )$.
(The last inequality can be expressed by a bounded formula with a parameter of size $u^{u/r}$.)

Proof

Let G, u and r be given. Suppose $|G^{-1}(\alpha)|<u/r$, for every $\alpha<r$. Denote by $k=\lceil u/r\rceil-1$. Thus $|G^{-1}(\alpha)|\leq k$, for every $\alpha<r$. Define a one-to-one Δ_0 mapping $F: u \to kr$ by

$$F(x) =_{df} k.G(x)+m-1, \quad \text{where } x \text{ is the m-th element in } G^{-1}(G(x)).$$

We can express the condition "x is the m-th element" using a bounded formula with parameter $y\geq u^{u/r}$. Since $k<u/r$, we have $kr<u$. Thus this is not possible by Δ_0-PHP. □

Let us note that if we required only say $|G^{-1}(\alpha)| \geq u/r-1$, then the statement would be provable using the Δ_0-WPHP, hence already in $I\Delta_0+\Omega_1$.

4 Ramsey's theorem in propositional calculus

Let $P(x_1,\ldots,x_k)$ be a new predicate symbol. Let $\psi(x,P)$ be a bounded formula in the language augmented by P with x as the only free variable. Let $n\geq 0$ be given. We define a propositional translation $\{\psi(x,P)\}_n$ of $\psi(x,P)$. First evaluate all terms in $\psi(n,P)$. Let m-1 be their maximal value. Introduce m^k propositional variables $p_{i_1,\ldots,i_k}$, $i_1,\ldots,i_k<m$. Then translate atomic formulas of the form $\tau(n)=\sigma(n)$, τ,σ terms, as *truth* or *falsehood* according to the truth of the equation and translate $P(\tau_1(n),\ldots,\tau_k(n))$ as $p_{i_1,\ldots,i_k}$, where $i_1,\ldots,i_k$ are the values of the terms $\tau_1(n),\ldots,\tau_k(n)$. The propositional connectives remain as they stand. The quantifiers are replaced by conjunctions and disjunctions. Since m is bounded by a polynomial in n, the size of $\{\psi(x,P)\}_n$ is polynomial in n too. Let us assume that we have only $\&,\vee$ and $\neg$ as connectives. Define *the depth* of a propositional formula to be the number of alternations of connectives. Then the depth of $\{\psi(x,P)\}_n$ depends only on ψ and does not depend on n.

Consider $I\Delta_0(P)$, the theory obtained from $I\Delta_0$ by extending the schema of induction to all bounded formulas in the extended language. Suppose we have a proof p of $\forall x\psi(x,P)$ in $I\Delta_0(P)$. By cut-elimination we can assume that all formulas in p but the last one are bounded. For a given n we can replace induction inferences by polynomially many propositional inferences and thus we obtain an induction-free proof of $\psi(n,P)$ whose size is polynomial in n.

polynomial size proof of $\{\psi(x,P)\}_n$. This proof has also bounded depth and it is a proof in a Frege system. (For the definition of Frege proof systems see [CR].) This was a sketch of a proof of the following theorem.

Theorem 3 (Paris, Wilkie [PW1]).
Suppose $I\Delta_0(P)$ proves $\forall x\psi(x,P)$. Then $\{\psi(x,P)\}_n$ have polynomial-size bounded-depth proofs in a Frege system. □

Let m and k be numbers. Let R(m,k) be the following proposition

$$\bigvee_{\substack{X\subseteq m\\ |X|=k}} \Big(\bigwedge_{\substack{i,j\in X\\ i<j}} p_{ij} \vee \bigwedge_{\substack{i,j\in X\\ i<j}} \neg p_{ij} \Big).$$

Clearly $R(m,\lceil 1/2\lfloor \log_2 m\rfloor\rceil)$ expresses the same form of Ramsey's theorem as we have considered above.

Theorem 4.
$R(m,\lceil 1/2\lfloor \log_2 m\rfloor\rceil)$ have polynomial-size bounded-depth Frege proofs.

Proof-sketch

We shall derive the theorem from (the proof of) Theorems 1 and 3. First it is easy to check that Theorem 1 can be strengthened by taking $I\Delta_0(P)+\Omega_1$ instead of $I\Delta_0+\Omega_1$ and by taking the schema with the predicate symbol P instead of a bounded formula φ. Further we need to express Ramsey's theorem by a bounded formula which does not use $x^{\lceil \log_2 x\rceil}$ as a function symbol. This can be done by writing a formula in which the parameter is an upper bound to the codes of subsets of u of size $\lceil 1/2\lfloor \log_2 u\rfloor\rceil$. Thus we obtain a formula $\psi(x,P)$ which expresses Ramsey's property of partitions of $[u]^2$ for u such that $u^{\lceil \log_2 u\rceil}\le x$. The translation $\{\psi(x,P)\}_n$ is not exactly $R(m,\lceil 1/2\lfloor \log_2 m\rfloor\rceil)$, but these two formulas are equivalent and there is a polynomial-size bounded-depth proof of this equivalence. It remains to show that $\forall x\psi(x,P)$ is provable in $I\Delta_0(P)$. The proof of Δ_0-WPHP actually shows that for every $\Delta_0(P)$ definable function F

$$I\Delta_0(P) \vdash \forall x,u(\ u^{\lceil \log_2 u\rceil}\le x\ \&\ F:2u \to u \Rightarrow F \text{ is not one-to-one}).$$

This is just what is needed in the proof of $\forall x\psi(x,P)$, hence we can apply Theorem 3 to construct polynomial-size bounded-depth proofs of $R(m,\lceil 1/2\lfloor \log_2 m\rfloor\rceil)$. □

Krishnamurthy [K] was the first to ask about the length of proofs of formulas $R(m,j)$. He considered true formulas $R(m,j)$ with a more precise bound j, but still quite close to ours ($\lceil 1/2\lfloor \log_2 m\rfloor\rceil - c \le j \le \lceil 1/2\lfloor \log_2 m\rfloor\rceil$ for a constant c). Thus Theorem 4 gives a partial answer to his question. It is possible that the more precise bound considered by Krishnamurthy can be obtained using PHP instead of WPHP. Using the result of Buss [B2] that the

Frege systems (without the additional condition of bounded depth).

Let us note that much more is known about the propositional version of the PHP, which can be thought of as the simplest case of the Ramsey Theorem. Haken [H] proved that proofs of PHP in the resolution system are exponential; Ajtai [A] showed a superpolynomial lower bound for bounded depth Frege systems.

5 Related questions in complexity theory and open problems

First we shall discuss how good are our estimates for the size of a homogeneous subset. We shall confine ourselves to the simplest case k=2 and r=2. The following estimate is due to P. Erdös.

Theorem 5

For every n, there exists $G:[n]^2 \to 2$ such that every homogeneous subset for G has size at most $2.\log_2 n + o(\log_2 n)$. □

The current best lower bound is only $1/2.\log_2 n+\Omega(\log_2\log_2 n)$, which is slightly better than we have in $I\Delta_0+\Omega_1$. Note that the proof of Theorem 5 is nonconstructive. It is an open problem, whether such functions can be constructed in some sense effectively. Note that if we could define such functions by Δ_0 formulas then they would be also computable in space O(log n).

The best bound to the running time needed to construct such graphs is $O(n^{O(\log n)})$ and it is also due to Erdös. Let us briefly sketch this proof. Let x_{ij}, $i,j<n$ be variables. Consider the following polynomial

$$p_{n,k}(x) =_{df} \sum_{\substack{Y\subseteq n \\ |Y|=k}} \left[\prod_{i,j\in Y} x_{ij} + \prod_{i,j\in Y}(1-x_{ij}) \right].$$

One can compute that for some $k=2.\log_2 n + o(\log_2 n)$ we have

$p_{n,k}(1/2,\dots,1/2) <1$,

where we have substituted 1/2 for all variables. Now we shall replace successively the entries 1/2 by 1 or 0. Since the polynomial is linear in each variable, we can always choose the possibility in which the value of the polynomial does not increase. Thus eventually we have $p_{n,k}(a)<1$ for some vector a of 1's and 0's, (in fact $p_{n,k}(a)=0$, since it must be an integer). This means that the mapping defined according to a does not have a homogeneous subset of size k. The only time consuming part of the computation is the computation of the values of the polynomial.

Working on the same problem from a different side, Frankl and Wilson [FW] showed that there are explicitly definable graphs, in particular

It is possible to try to prove larger lower bounds to the homogeneous sets for Δ_0 definable functions using techniques developed in circuit complexity. In particular, if a function $G:[n]^2 \to 2$ is definable by a Δ_0 arithmetical formula, then it is also definable using bounded depth and unbounded fan-in Boolean formula. It is conceivable that such functions have always homogeneous sets larger than arbitrary functions. Similar questions have been considered in [PRS] and [R]. In [PRS] it was shown that graphs with small complexity (for a suitable definition of complexity) have a certain Ramsey type property which is not shared by all graphs. On the other hand, by [R], graphs defined by polynomial size formulas behave very much like random graphs.

* * *

Let us end this discussion by suggesting some Δ_0 schemata which might be not derivable using Δ_0-PHP. These schemata are motivated by the following problem. We have shown in [KPT] that $I\Delta_0+\Omega_1$ is not finitely axiomatizable provided Polynomial Hierarchy does not collapse. This proof uses certain counting argument and therefore we are not able to formalize it in $I\Delta_0+\Omega_1$. If we could do it it may have some interesting consequences.

We shall talk about *directed graphs* instead of mappings $G:n\times n \to 2$. A graph is called *a tournament*, if for each two distinct vertices there is an arrow in exactly one direction. A subset of vertices X is called *a dominating set*, if each vertex v of the graph is either in X, or there is an arrow from X to v. A well known theorem states that in a tournament with n vertices there is a dominating set of size $O(\log n)$.

Problem 1. Is it provable in $I\Delta_0+\Omega_1$ or in $I\Delta_0+\Omega_1+\Delta_0$-PHP that every Δ_0 definable tournament on $[0,u-1]$ has a dominating set of size $O(\log u)$?

An attempt to answer positively this problem led to the following question.

Problem 2. Is it provable in $I\Delta_0+\Omega_1$ or in $I\Delta_0+\Omega_1+\Delta_0$-PHP that for every Δ_0 definable functions F, G_0, G_1 the following situation is not possible

$F: 2u \to 2$, $G_0, G_1: u^2+1 \to 2u$, $\forall x<u(\ F(G_0(x)=0\ \&\ F(G_1(x))=1\ )$,

$\forall x,y<u(\ x\neq y \Rightarrow (G_0(x),G_1(x))\neq(G_0(x),G_1(x))\ )$?

Acknowledgement. I am indebted to Vojtěch Rödl for several discussions about finite Ramsey's theorem.

References

[A] M.Ajtai, *The complexity of the pigeonhole principle*, 29-th Symp. on Foundations of Comp. Sci. (1988), pp.346-355.

[B1] S.Buss, *Bounded Arithmetic*, Bibliopolis, 1986.

[CR] S.A.Cook, R.A.Reckhow, *The relative efficiency of propositional proof systems*, Journ. Symb. Logic 44, (1979), pp.36-50.

[FW] P.Frankl, R.M.Wilson, *Intersection theorems with geometric consequences, Combinatorica* 1(4), (1981), pp.357-368.

[H] A.Haken, *The intractability of resolution*, Theor. Comp. Sci. 39, (1985), pp.297-308.

[KPT] J.Krajíiček, P.Pudlák, G.Takeuti, *Bounded arithmetic and the polynomial hierarchy*, Annals of Pure and Applied Logic, to appear.

[K] B.Krishnamurthy, *Short proofs for tricky formulas*, Acta Informatica 22, (1985), pp.253-275.

[PW1] J.Paris, A.Wilkie, *Δ_0 sets and induction*, in Proc. Jadwisin Logic Conference, Poland, Leeds Univ. Press, 1981, pp.237-248.

[PW2] J.Paris, A.Wilkie, *Counting Δ_0 sets*, Fundamenta Mathematicae 127, (1987), pp. 67-76.

[PWW] J.B.Paris, A,J,Wilkie and A.R.Woods, *Provability of the Pigeon Hole Principle and the existence of infinitely many primes*, JSL 53/4, (1988), pp. 1235-1244.

[PRS] P.Pudlák, V.Rödl, P.Savický, *Graph complexity*, Acta Informatica 25, (1988), pp.515-535.

[R] A.A.Razborov, *Formulas of bounded depth in basis* {&,⊕} *and some combinatorial problems*, in Složnost' algoritmov i prikladnaja matematičeskaja logika, S.I.Adjan editor, 1987.

[T] S.Toda, *On the computational power of* PP *and* ⊕PP, 30-th Symp. on Foundations of Comp. Sci., (1989), pp.514-519.

Nontrivial Lower Bounds for some NP-problems on Directed Graphs

Solomampionona RANAIVOSON,
Laboratoire d'Informatique de l'Université de Caen,
Esplanade de la Paix - 14032 Caen cedex - France. tel. 31 45 56 16

__Abstract.__ NP-complete problems are believed to be not in P. But only a very few NP-complete problems, and none concerning graph theory, are proved to have a nontrivial time lower bound (i.e. not to be solvable in linear time on a DTM (i.e. deterministic Turing machine) . A problem $L \in NP$ is linearly NP-complete if any problem in Ntime (n) can be reduced to it in linear time on a DTM. It follows from the separation result between deterministic and nondeterministic linear-time complexity classes [PPST83], that a linearly NP-complete problem has a nontrivial time lower bound. We present in this paper the first natural problems on graphs which are linearly NP-complete.

__Section 0. Introduction__

Many well known combinatoric problems which have been studied for a long time are NP-complete. It is believed that NP-complete problems can't be solved in deterministic polynomial time. But, strangely, until recentely, none of these longstanding problems, as Cook noticed in [Co83], has been proved to have even a nontrivial time lower bound (i.e. proved not to be solvable in linear time on a Deterministic Turing Machine (on a DTM for short)). By defining each NTIME (n) problem with a second-order existential sentence whose all second-order variables are unary function symbols and which has only a single first order variable, GrandJean [Gr90a] proved that the problem RISA (Reduction of Incompletely Specified Automaton, problem recorded in [GaJo79]) is __linearly NP-complete__, i.e. that each problem in NTIME (n) can be reduced to it in linear time on a DTM. It follows from this and the separation result : $\bigcup_{c \geq 1} \text{Dtime}(c\,n) \subsetneq \text{NTIME}(n)$, of Paull, Pippenger, Szemeredi and Trotter, that the problem RISA is not computable in linear time on a DTM.

Let G be a directed graph. A __condensation__ of G is a directed graph G' obtained from G by identifying some vertices of G and keeping the edges of G. The problems we prove linearly NP-complete consist of checking, for a given graph G which satisfies a given property $\mathcal{P}$ (ex. acyclicity and outdegree bounded by a fixed integer $d \geq 2$) wether yes or no G has a condensation G' which also satisfies $\mathcal{P}$ but which is simpler than G (ex. with less vertices or edges than G).

Here is the structure of the paper. In section 1, we essentially give the proof of the linearly NP-completeness of a problem of contraction of partial functions (problem Contract 1) which is proposed in [Gr90b] to be a generic tool for the proof of other linearly NP-complete problems. Our problems of condensation of graph and the main theorem are given in section 2. The main theorem is proved in section 4, by reducing the problem Contract 2 (which is a variant of problem Contract 1 presented in section 3) in linear time on a DTM to each of our problems of condensation of graph. The proof of some long and combinatorial lemmas will be given in section 5.

Section 1. Preliminaries

Definition A [Gr88]. A problem $L \in NP$ is linearly NP-hard (or linearly NP-complete) if each problem in NTIME (n) is reducible to L in linear time on a deteministic Turing machine (DTM for short).

Let $L \in NP$. Assume L is linearly NP-hard. It is then easy to prove that any problem $L' \in NTime(n^k)$ is reducible to L in time $O(n^k)$ (with the same k) on a DTM : i.e. L is (linearly) NP-complete. Moreover, $L \notin \bigcup_{c \geq 1} Dtime(c\, n)$, by the separation lemma :

Lemma A [PPST83] $\bigcup_{c \geq 1} Dtime(c\, n) \subsetneq \bigcup_{c \geq 1} NTime(c\, n)$

The notion of linear NP-completeness was first introduced by Grandjean in [Gr88, Gr90a] where is proved the lemma B below.

Definition B [Gr90a]. Let Φ be a first order sentence of type $\mathcal{T} = \{S_1, \dots, S_p, U_1, \dots, U_k\}$, where the S_i (called specified symbols) are all unary function symbols and the U_j (called unspecified symbols) are any (relation, constant or function) symbols. The generalized spectrum of Φ, denoted GenSPECTRUM (Φ), is the set of the finite structures $<\{0, \dots, M-1\}, S_1, \dots, S_p>$ that have an expansion of the form $<\{0, \dots, M-1\}, S_1, \dots, S_p, U_1, \dots, U_k>$ that satisfies Φ

Lemma B [Gr90a]. For each $L \in NTime(n)$ there is a first-order sentence $\Phi \equiv \forall y\, \Psi(y)$ such that L is reducible to GenSPECTRUM (Φ) in linear time on a DTM, with

Ψ (y) being a conjunction of equalities σ_i (y) = τ_i (y), each, of the form $f_k \dots f_1$ (y) = $h_l \dots h_1$ (y), where y is the only variable of Ψ, the f_j and h_j are unary function symbols, and $k \geq 1$ and $l \geq 0$.

In [Gr90b], Grandjean proposes the problem Contract 1 below, which is just another formulation of the spectrum lemma above, to be a generic tool for the proof of other linearly NP-complete problems.

Problem Contract 1. Contraction of partial functions (r is a fixed integer which is sufficientely large)

Instance. A set of r unary function symbols $F = \{f_0, \dots, f_{r-1}\}$, a set of constants $A = \{a_0, \dots, a_{k-1}\}$, a set of variables $X = \{x_0, \dots x_{n-1}\}$ and a conjunction Γ of equalities of the form : g (u) = v, with 1) $g \in F$ and $u, v \in A \cup X$, 2) all the terms g (u) are distinct and are in (lexical) order.

Question : is Γ satisfiable on A ? i.e. are there an application *val* from $A \cup X$ to A (with *val* (a) = a if $a \in A$) and (partial) functions $f_0', \dots, f'_{r-1}$ from A to A such that if $s, t \in A \cup X \cup Y$ and $h < r$ and f_h (s) = t is an equality of Γ then f_h' (*val* (s)) = *val* (t) ?

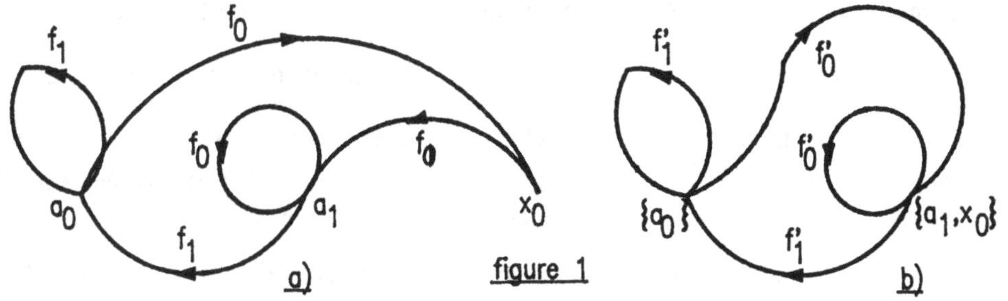

figure 1

The figure 1.a) illustrates an instance $\mathcal{S}$ of problem Contract 1, with $F = \{f_0, f_1\}$, $A = \{a_0, a_1\}$, $X = \{x_0\}$ and $\Gamma \equiv f_0(a_0) = x_0 \wedge f_1(a_0) = a_0 \wedge f_0(a_1) = a_1 \wedge f_1(a_1) = a_0 \wedge f_0(x_0) = a_1$. Figure 1.b) illustrates a contraction of $\mathcal{S}$ on A with *val* $(x_0) = a_1$.

In the lemma C below, the formulas Φ of lemma B are normalized in order to make possible in linear time on a DTM, the reduction of their generalized spectrum to the problem Contract 1 (because of the condition 2) of the problem Contract 1).

Lemma C [Gr90b]. Assume S is the generalized spectrum of a sentence of the form $\forall y\, \Psi(y)$ such that Ψ (y) is a conjunction $\bigwedge_i \sigma_i(y) = \tau_i(y)$ of equalities, each, of the form $f_k \dots f_1$ (y) = $h_l \dots h_1$ (y), where y is the only variable of Ψ, the f_j and h_j are unary function symbols, and $k \geq 1$ and $l \geq 0$. Then S is also the generalized spectrum of a sentence of the same form above but which in addition satisfies the following conditions :

a) no term or subterm of the form g (y) with g specified, appears in Ψ

b) no left-member σ_i (y) of some equality is the subterm of an other left-member σ_j (y) (for $i \neq j$) or of a right-member τ_i (y) of an equality (in our meaning, each term is also its subterm).

Proof. The sentence Ψ is transformed gradually in 3 steps.

1) Suppress subterms g (y) with g specified by replacing $\bigwedge_i \sigma_i (y) = \tau_i (y)$ by $I (y) = y \wedge \bigwedge_i \sigma_i (I (y)) = \tau_i (I (y))$, where I is a new non specified function symbol, which intuitively is the identity function. Hence the condition a) is satisfied, and it will be preserved during the next two steps.

2) Prevent two equalities to have the same left-member. For that, consider the partition of the set of the terms of Ψ into classes of equalities. More precisely, two terms t and t' of Ψ belong to the same class if an equality between t and t' can be derived (syntactically) by reflexive-symmetrical-transitive closure from the equalities in Ψ. The sentence Ψ is replaced by a new sentence Ψ (which remains semantically equivalent to the old Ψ) as follows : for each class $\mathcal{C}$ of the partition, choose a term t among the shortest terms in $\mathcal{C}$ and put in Ψ the conjunction $\Gamma_{\mathcal{C}} = \bigwedge_{j=1}^{c} t_j = t$, where $t_1, \dots , t_c$ is the list of the other terms in $\mathcal{C}$. Φ is the conjunction of the conjunctions $\Gamma_{\mathcal{C}}$, for all the classes $\mathcal{C}$.

3) Prevent each left-member of an equality to be the proper subterm of a left-member, or the subterm of a right-member of some equality. Replace the new conjunction $\bigwedge_i \sigma_i (y) = \tau_i (y)$ obtained in 2) by the new formula Ψ (y) below :

$Id (y) = y \wedge \bigwedge_i Id (\sigma_i (y)) = \tau_i (y)$, where Id is a new unspecified function symbol (which intuitively is also the identity function).

To check that this new formula Ψ satisfies the condition b), it may be noticed that, each left-member (of an equality) is of the form id (t) with t a term in which doesn't appear the symbol id. No left-member is then the proper subterm of an other left-member. As the symbol id also doesn't appear in any right-member, no left-member is a subterm of some right-member. And last, two distinct equalities can't have the same left-member (because of 2) and the fact that a left-member of an equality is never of the form y. That concludes the proof of lemma C □

Let us now reduce the generalized spectrum of a sentence $\Phi \equiv \forall y \, \Psi (y)$ of lemma C to the problem Contract 1. To each finite structure $\mathcal{S} = \langle [0, M-1], S_1, \dots , Sp \rangle$, where $S_1, \dots , Sp$ are the specified symbols of Φ, we associate an instance $\mathcal{D} = (A \cup X , f_0, \dots , f_{r-1}, \Gamma)$ of problem Contract 1 as follow : the set of function symbols

$\{f_0, \dots, f_{r-1}\}$ is the type of Ψ, $A = \{\overline{e} \mid e \in [0, M-1]\}$, $X = \{\overline{t(e)} \mid e \in [0, M-1]$ and $t(y)$ is a term $\neq y$ which is either a proper subterm of a left-member $\sigma_j(y)$ or a subterm of a right-member $\tau_j(y)$ of some equality of $\Psi\}$ and Γ is the conjunction of :

a) the specified equalities $\bigwedge_{i=1}^{p} \bigwedge_{e=0}^{M-1} S_i(\overline{e}) = \overline{e'}$, where e' is the specified value of $S_i(e)$, and

of the equalities in b) and c) below which are derived from each equality of the form $f_k f_{k-1} \dots f_1(e) = h_m h_{m-1} \dots h_1(e)$ of Ψ :

b) definition of subterms :

$$\bigwedge_{i=1}^{k-1} \bigwedge_{e=0}^{M-1} f_i\,(\overline{f_{i-1} \dots f_1(e)}) = \overline{f_i f_{i-1} \dots f_1(e)} \text{ and } \bigwedge_{i=1}^{m} \bigwedge_{e=0}^{M-1} h_i\,(\overline{h_{i-1} \dots h_1(e)}) = \overline{h_i h_{i-1} \dots h_1(e)}$$

c) explicit equalities : $\bigwedge_{e=0}^{M-1} f_k\,(\overline{f_{k-1} \dots f_1(e)}) = \overline{h_m h_{m-1} \dots h_1(e)}$.

The list of all the equalities given above is sorted in lexical order in Γ. Since we started from a sentence $\Phi \equiv \forall y\, \Psi(y)$ of lemma C, it is easy to check that Γ satisfies the condition 2) in the instance of problem Contract 1. Moreover $\mathcal{D}$ can be computed from $\mathcal{S} = \langle [0, M[, S_1, \dots, S_p \rangle$ in $O(M \log M)$ on a DTM, i.e. in linear time . Note that the size of the finite structure $\mathcal{S}$ is $\theta(M \log M)$.

It is clear that $\mathcal{S}$ has an expansion satisfying the sentence $\forall y\, \Psi(y)$ (i.e. $\mathcal{S} \in S$) iff its associated conjunction (of equalities) Γ is satisfiable. This proves that problem S (which is any problem of NTIME (n)) is reducible in linear time to problem Contract 1.

In conclusion, the problem Contract 1 is linearly NP-complete

Section 2. The main results

All the graphs we consider are directed and may countain edges of the form (x, x). The set of the points (resp. edges) of a graph G is denoted *Vertices* (G) (resp. *Edges* (G)). If e = (x, y) is an edge of G then x is a father of y and y is a son of x , x (resp. y) is the initial (resp. final) point of e. The outdegree of a point x of G, denoted $d_G^+(x)$, is the number of the sons of x . G is of outdegree d if the outdegree of each point of G is at most d.

Definition 1 [Ha65] Let G=(V, E) be a directed graph. Let V' be a partition of V. The condensation of G with respect to the partition V' of V is the directed graph G' = (V', E') such that there is an edge of G' from any $S \in V'$ to any $T \in V'$ iff there is an edge of G from some vertex $s \in S$ to some vertex $t \in T$. If π denotes the canonical projection associated to the partition V' of V then the couple (π, G') is called a <u>condensation of</u> G.

<u>Example :</u> figure 2 below ilustrates the notion of condensation of graphs

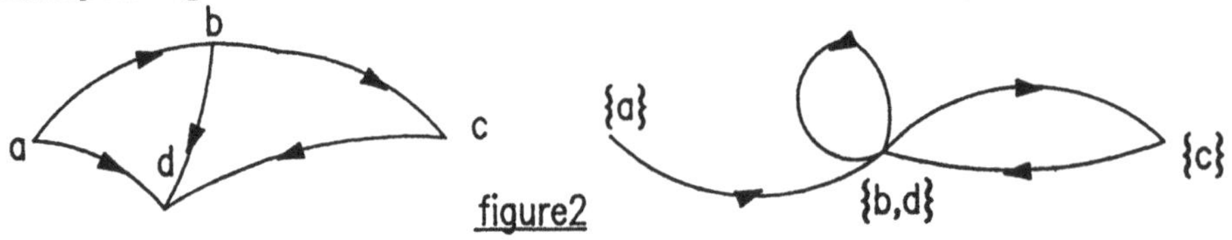

<u>figure2</u>

<u>Main theorem</u> : the following two problems are linearly NP-complete (d being a fixed integer ≥ 2) :

<u>Problem 1.</u> Vertex-bounded condensation of an acyclic graph of outdegree d.

<u>Instance :</u> an acyclic directed graph G = (V, E) of outdegree d, an integer $K \leq |V|$. <u>Question</u> : is there a condensation (π, G') of G such that G' is acyclic, of outdegree d and has at most K vertices ?

<u>Problem 2 :</u> edge-bounded condensation of an acyclic graph.

<u>Instance :</u> an acyclic directed graph G = (V, E), and an integer $H \leq |E|$. <u>Question :</u> is there a condensation (π , G') of G such that G' is acyclic and has at most H edges ?

<u>Remarks.</u> For problem 1, the theorem is proved for d = 2 but the proof can easily be extended for any fixed $d \geq 2$. Variant of problem 1, in which the notion of <u>outdegree</u> is replaced by <u>indegree</u>, is also linearly NP-complete. The results of the main theorem above is optimal as the diagram below shows (LNPC means linearly NP-complete) :

Properties $\mathcal{P}$ of G and of the condensation G'	Vertex-bounded condensation	Edge-bounded condensation
Acyclicity and outdegree not bounded	P	LNPC
Acyclicity and outdegree $\leq$ d (d fixed $\geq$ 2)	LNPC	LNPC
Acyclicity and outdegree d = 1	P	P

For lack of place we omit the proofs of the "Polynomial" results, which are much easier than the proof of the main theorem.

The same method as in [Gr88] can be applied to prove that problems 1 and 2 are solvable in linear time, with a bounded number of alternation, on an alternating Turing machine.

The variants of the problems 1 and 2, in which the acyclicity constraint is missing, have trivial solution. So we only will consider acyclic graphs in the sequel. <u>Convention</u> : a condensation (π, G') of G sucht that G' is acyclic is called <u>an acyclic condensation</u> of G.

Section 3. A generic problem for the proof of the main theorem

It is the variant of problem Contract 1 below, in which auxiliary variables have been introduced to make each function f_h totally defined on the domain $A \dot{\cup} X$, that we will reduce directely to each of the problems 1 and 2.

Problem Contract 2 (r is a fixed integer which is sufficientely large).

Instance : a set of r unary function symbols $F = \{f_0, \dots, f_{r-1}\}$. A set of constants $A = \{a_0, \dots, a_{k-1}\}$. A set of (main) variables $X = \{x_0, \dots x_{n-1}\}$. A set of (auxiliary) variables $Y = \{y_0, \dots y_{m-1}\}$. A conjunction Γ of equalities of the form : $g(s) = t$, with 1) $g \in F$, $s \in A \cup X$, $t \in A \cup X \cup Y$, 2) all the terms $g(s)$ are distincts and are in (lexical) order, 3) for each $g \in F$ and each $s \in A \cup X$, the term $g(s)$ occurs in Γ, 4) if $s \in A$ then $t \in A \cup X$, 5) each auxiliary variable occurs in Γ (in a second member).

Question is Γ satisfiable on A ? i.e. are there an application *val* from $A \cup X \cup Y$ to A (with *val* $(s) = s$ if $s \in A$) and r functions $f_0', \dots, f'_{r-1}$ from A to A such that for all $f_h \in F$ and $s, t \in A \cup X \cup Y$, if $f_h(s) = t$ is an equality in Γ then $f_h'(val(s)) = val(t)$, ?

Notation 1. We denote an instance of problem Contract 2 by $\mathcal{D} = (A \cup X \cup Y, f_0, \dots, f_{r-1}, \Gamma)$. We denote a solution (also called a contraction on A) of $\mathcal{D}$ by $\mathcal{S} = (val, f_0', \dots, f'_{r-1})$

Note that no term of the form $f_h(y)$ appears in $\mathcal{D}$ for any auxiliary variable $y_i \in Y$. Here is a simplified algorithm which transforms an instance $(A \cup X, f_0, \dots, f_{r-1}, \Gamma)$ of the problem Contract 1 into an instance $(A \cup X \cup Y, f_0, \dots, f_{r-1}, \Gamma)$ of the problem Contract 2

```
For every term fl (ai) missing in Γ :
    add a new variable xj in X ;
    add the equality fl (ai) = xj to Γ;
Y ← Ø
For every term h = fl (xj) missing in Γ :
    add a new variable yh in Y ;
    add the equality fl (xj) = yh to Γ
```

It is clear that the equalities added to Γ by this algorithm don't modify Γ semantically. Since r is a fixed integer and the terms $f_l(a_i)$ et $f_l(x_j)$ are given in (lexical) order, this is a linear-time reduction (on a DTM) of the problem **Contract 1** to the problem Contract 2 , which therefore is linearly NP-complete.

Section 4. The proof of the main theorem

We reduce, in linear time on a DTM, the problem Contract 2 to the problems 1 and 2. Let us fix once and for all an instance $\mathcal{D} = (A \cup X \cup Y, f_0, \dots, f_{r-1}, \Gamma)$ of problem Contract 2 . We associate to $\mathcal{D}$ an acyclic directed graph of outdegree 2, denoted $G = (V, E)$ and two integers $K \leq |V|$ and $H \leq |E|$. We then prove that the following 3 statements are equivalent : 1) $\mathcal{D}$ has a contraction on A, 2) G has a an acyclic condensation (π, G') with G' of outdegree 2 and with at most K vertices, 3) G has a an acyclic condensation (π, G') with G' having at most H edges.

The idea in the construction of G from $\mathcal{D}$, is to represent each element $s \in A \cup X$ by a subgraph S of G, called gadget-graph, in which are distinguished r "output" levels (levels 0 to r−1, r being the number of function symbols of $\mathcal{D}$) and one "input" level (level r). Each equality $f_h(s) = t$ in Γ is then represented by a special edge, called function-edge, from the "output" level number h of S to the "input" level of T, S and T being the gadget-graphs associated respectively to s and t.

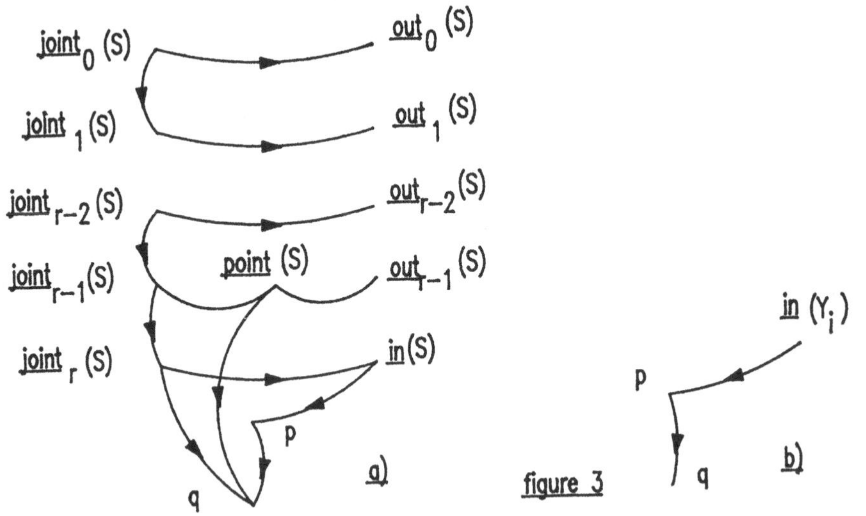

figure 3

Definition 1 (Gadget-graphs). Let $\mathcal{D} = (A \cup X \cup Y, f_0, \dots, f_{r-1}, \Gamma)$ an instance of problem Contract 2. We associate to each $s \in A \cup X$ (resp. $y_i \in Y$) the graph denoted S (resp. Y_i) of figure 3,a) (resp. 3,b). S (resp. Y_i) is called the gadget-graph associated to s (resp. y_i). If $s \in A$ then S is said constant. Y_i and S such that $s \in X$ are said variable.

Formally : $\mathit{Vertices}(S) = \bigcup_{h=0}^{r-1} \{\text{joint}_h(S), \text{out}_h(S)\} \cup \{\text{point}(S), \text{joint}_r(S), \text{in}(S), p, q\}$

$$\mathcal{E}dges\ (S) = \bigcup_{h=0}^{r-2} \{(\underline{joint}_h\ (S),\ \underline{out}_h\ (S)),\ (\underline{joint}_h\ (S),\ \underline{joint}_{h+1}\ (S))\}\ \cup$$
$$\{(\underline{joint}_{r-1}(S),\ \underline{point}\ (S)),\ (\underline{joint}_{r-1}(S),\ \underline{joint}_r\ (S)),\ (\underline{point}\ (S),\ \underline{out}_{r-1}\ (S)),\ (\underline{point}\ (S),\ q)\}$$
$$\cup\ \{(\underline{joint}_r\ (S),\ \underline{in}\ (S)),\ (\underline{joint}_r\ (S),\ q),\ (\underline{in}\ (S),\ p),\ (p,\ q)\}$$
$$\mathcal{V}ertices\ (Y_j) = \{\underline{in}\ (Y_j),\ p,\ q\}\ .\ \mathcal{E}dges\ (Y_j) = \{(\underline{in}\ (Y_j),\ p),\ (p,\ q)\}$$

Notations. Lower-case letters with or without subscripts denote elements of $A \cup X \cup Y$ (ex. a_j, x_j, y_h, s ...). The corresponding upper-case letters denote the corresponding gadget-graphs (ex. A_j, X_j, Y_h, S ...)

Definition 2 ($G = (V, E)$). To any instance $\mathcal{D} = (A \cup X \cup Y, f_0, \ldots, f_{r-1}, \Gamma)$ of problem Contract 2, is associated some directed graph $G = (V, E)$ which is the union of :

- the gadget-graphs S associated to the elements s of $A \cup X \cup Y$
- the edges, called function-edges, of the form $(\underline{out}_h\ (S),\ \underline{in}\ (T))$ such that $f_h\ (s) = t$ is an equality in Γ
- the edge, called special-edge, $(\underline{in}\ (A_{k-1}),\ q)$ (a_k is the last element of A)
- edges, called bridge-edges, added between the constant-gadget graphs, such that there is in G a path, denoted ConstPath, wich contains each of the vertices of the constant gadget-graphs. ConstPath is :

$$(\underline{joint}_0\ (A_0),\ \underline{out}_0\ (A_0),\ \underline{joint}_0\ (A_1), \underline{out}_0\ (A_1),\ \ldots\ ,\ \underline{joint}_0\ (A_{k-1}),\ \underline{out}_0\ (A_{k-1}),$$
$$\underline{joint}_1\ (A_0),\ \underline{out}_1\ (A_0),\ \underline{joint}_1\ (A_1),\ \underline{out}_1\ (A_1),\ \ldots\ ,\ \underline{joint}_1\ (A_{k-1}),\ \underline{out}_1\ (A_{k-1}),$$
$$\ldots ,$$
$$\underline{joint}_{r-1}\ (A_0),\ \underline{point}\ (A_0),\ \underline{out}_{r-1}(A_0),\ \underline{joint}_{r-1}\ (A_1),\ \ldots\ ,\ \underline{point}\ (A_0),\ out_{r-1}\ (A_{k-1}),$$
$$\underline{joint}_r\ (A_0),\ \underline{in}\ (A_0),\ \underline{joint}_r\ (A_1),\ \ldots\ ,\ \underline{joint}_r\ (A_{k-1}),\ \underline{in}\ (A_{k-1}),$$
$$p\ ,\ q)$$

Notations. The vertices of the constant gadget-graphs are said constant. Their set is denoted by V_{const}. The set of the edges departing from the constant vertices is denoted E_{const}. The vertices of G which are not constant are said variable.

Definition 3 (Successor function on the constant points). A successor function, denoted Succ, is defined on ConstPath : $\underline{Succ}\ (\underline{joint}_0\ (A_0)) = \underline{out}_0\ (A_0)\ ,\ \ldots\ ,\ \underline{Succ}\ (p) = q$. Only $\underline{Succ}\ (q)$ is not defined. A predecessor function, denoted Pred, which is the inverse of Succ, is also defied on the contant points (except on $\underline{joint}_0\ (A_0)$) .

Definition 4. (Level and nature of a point). Two partitions, denoted respectively *Levels* and *Natures*, are defined on V. The (r + 3) elements of *Levels* are :

$\underline{level}_0 = \{\underline{joint}_0 (S), \underline{out}_0 (S)\}_S$, ... , $\underline{level}_{r-2} = \{\underline{joint}_{r-2} (S), \underline{point} (S), \underline{out}_{r-2} (S)\}_S$, $\underline{level}_{r-1} = \{\underline{joint}_{r-1} (S), \underline{point} (S), \underline{out}_{r-1} (S)\}_S$, $\underline{level}_r = \{\underline{joint}_r (S), \underline{in} (S)\}_S$, $\underline{level}_{r+1} = \{p\}$ and $\underline{level}_{r+2} = \{q\}$. The elements of *Natures* are $\underline{Joints} = \{\underline{joint}_h (S)\}_{h, S}$, $\underline{Outs} = \{\underline{Out}_h (S)\}_{h, S}$, $\underline{Ins} = \{\underline{in} (S)\}_S$, $\underline{Point} = \{\underline{point} (S)\}_S$, $\{p\}$ and $\{q\}$. Two points of V are of the same level (resp. nature) if they belong both to the same element of *Levels* (resp. *Natures*).

Example : figure 4 a) represents an instance $\mathcal{D}$ of problem Contract 2, figure 5 a) represents the graph G = (V, E) associated to $\mathcal{D}$.

figure 4

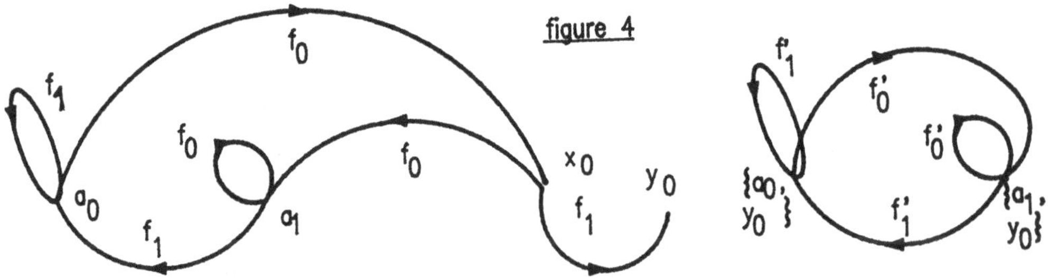

a)

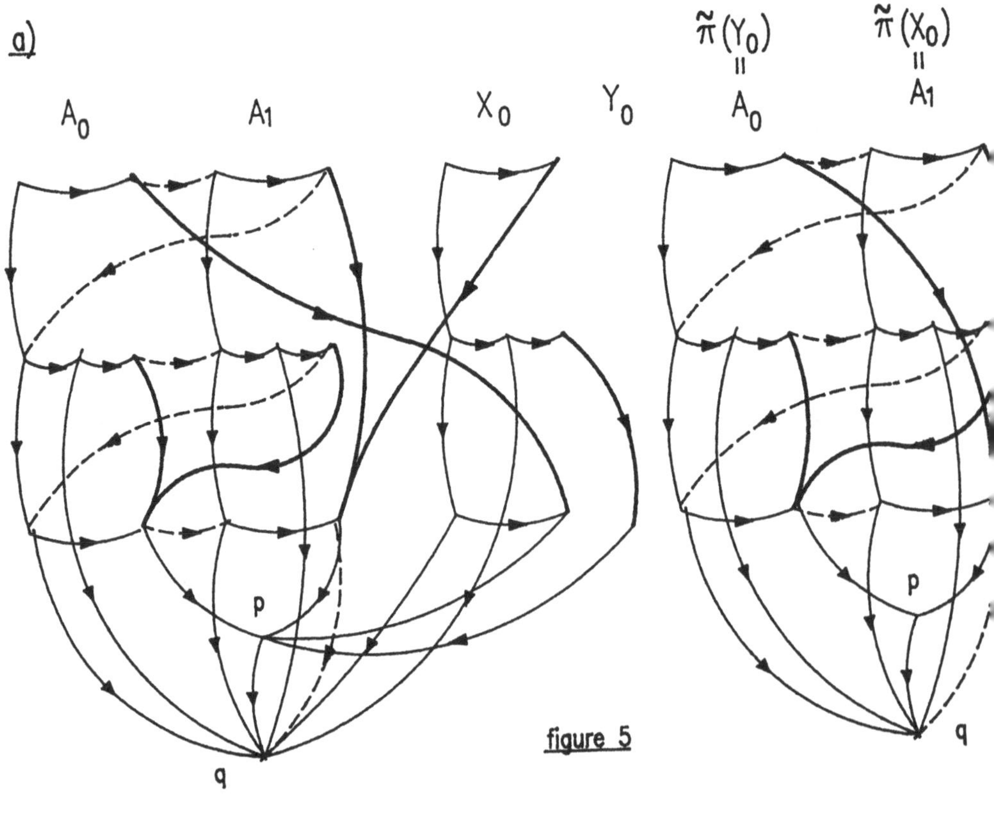

figure 5

Lemma 1 a) G is acyclic and of outdegree 2

b) $d^+_G(p) = 1$, $d^+_G(q) = 0$

c) Each point of G is the origin of a path which arrives at q. Each point of G, except q, is the origin of a path which arrives at p.

d) Each constant point of G is exactly of outdegree 2, except p and q.

e) The outdegree of each constant point of G can't be augmented (i.e. no other edge departing from any constant point can be added to G) else G is no longer acyclic or no longer of outdegree 2

Proof a) and b) follows from the definition of G. c) and d) result from the definition of G and the condition 3) of the problem Contract 2. e).If we add a new edge from a constant point $s \neq p, q$, then $d^+_G(s) = 3$ by d). If we add in G a new edge (q, s) then this edge and the path from s to q (by c) will form a cycle of G. Do the same with p □

Lemma 2 If a constant point s has a variable son, denoted s_2, then s is of the form $\underline{out}_h(A_i)$, s_2 is of the form $\underline{in}(X_j)$, the other son of s is constant : it is $s_1 = \underline{Succ}(s)$

Proof Immediate consequence of the definition of G.

Definition 5 (singular points) a constant point s of G which has a son which is variable is called a singular point of G (see lemma 2).

Here are the instances of the problems 1, 2 which we associate to $\mathcal{D}$:

Problem 1 : the directed graph G = (V, E) and the integer $K = |V_{const}|$.

Problem 2 : the directed graph G = (V, E) and the integer $H = |E_{const}|$.

Since r is a fixed integer, it is obvious that G, $|V_{const}|$ and $|E_{const}|$ are computable in linear time from $\mathcal{D}$ on a DTM.

The following definition is usefull to handle gadget-graphs.

Definition 4 Let $\mathcal{D} = (A \cup X \cup Y, f_0, \dots, f_{r-1}, \Gamma)$ be an instance of problem Contract 2 and G = (V, E) the graph associated to $\mathcal{D}$. Let π be any application from V to V_{const}. A gadget-graph T is mapped by π respectfully to some constant gadget-graph A_i, which we denote $\tilde{\pi}(T) = A_i$, if π maps each point s of T to the point of A_i of the same nature and same level as s.

Remark : if T is of the form Y_h then a necessary and sufficient condition for $\tilde{\pi}$ (T) = A_j is that q, p and in (T) are mapped by π respectively to p, q and in (A_j).

Intuitively, it remains, for the proof of the main theorem, to show that $\mathcal{D}$ has a contraction on A iff G has a "good" condensation G' : i.e. that G' has the same form as G but without the variable gadget-graphs and the edges departing from the variable points (G' is obtained from G by confounding "respectfully" each variable gadget-graph with some constant gadget-graph). Deducing a "good" condensation of G from a contraction on A of $\mathcal{D}$ is easy (lemma 3) ; the converse of it (lemma 4) is difficult, it needs a long preparation.

Lemma 3 If there are r functions $f_0', \ldots, f'_{r-1}$ from A to A and an application *val* from $A \cup X \cup Y$ to A (with *val* (a) = a if $a \in A$), such that $\mathcal{S} = (val, f_0', \ldots, f'_{r-1})$ satisfies Γ, i.e. : if $f_h(s) = t$ is an equality of Γ then $f_h'(val(s)) = val(t)$

then there is an acyclic condensation (π, G') of G such that G' = (V', E') satisfies

(r1) G' is of outdegree 2 and $|V'| \le |V_{const}|$

(r2) $|E'| \le |E_{const}|$

Proof. Assume there exists a contraction $\mathcal{S} = (val, f_0', \ldots, f'_{r-1})$ on A of $\mathcal{D}$. Without loss of generality we can suppose that $\mathcal{S}$ is a "minimal" contraction on A of $\mathcal{D}$, i.e. it satisfies : $\forall\, a_i, a_j \in A\ (f_h'(a_i) = a_j) \Leftrightarrow (\exists\, s, t \in A \cup X \cup Y\ (f_h(s) = t$ is an equality of $\Gamma \wedge val(s) = a_i \wedge val(t) = a_j))$.

From $\mathcal{S}$ we define the following condensation (π, G') of G : 1) $\pi : V \rightarrow V_{const}$ such that $\tilde{\pi}(S) = A_i$ iff $val(s) = a_i$, 2) G' = (V', E') is the union of the constant gadget-graphs, the bridge-edges of G, the special edge (in (A_{k-1}), q), and the function-edges of the form ($\underline{out}_h$ (A_i), in (A_j)) such that $f_h'(a_i) = a_j$. It is then easy to check that G' is acyclic and satisfies (r1) and (r2) □

Lemma 4 Let (π, G') an acyclic condensation of G such that G' = (V', E') satisfies one of the following constraints (r1) or (r2) :

(r1) G' is of outdegree 2 and $|V'| \le |V_{const}|$

(r2) $|E'| \le |E_{const}|$

Then V' = V_{const} (i.e. each element of V' contains a single element of V_{const} and for this reason, is identified to it) and $\mathcal{S} = (val, f_0', \ldots, f'_{r-1})$ defined below is a solution of $\mathcal{D}$:

1) *val* : $A \cup X \cup Y \rightarrow A$ such that $val(s) = a_i$ iff $\tilde{\pi}(S) = A_i$

2) $f_l' : A \rightarrow A$ such that $f_l'(a_i) = a_j$ iff ($\underline{out}_l$ (A_i), in (A_j)) $\in$ E'

To prove lemma 4, we first show that *val* is a total function on $A \cup X \cup Y$ and that each f_l' is a (single-valued) function on A ; then we have to check that they satisfy the conjunction Γ of $\mathcal{B}$. We begin by showing that, for an acyclic condensation (π, G') of G, the constraints (r1) and (r2) are each equivalent to the following constraint (r1') : G' is of outdegree 2 and $V' = V_{const}$ (i. e. each element of V' contains a single element of V_{const}).

Lemma 4.1. Let (π, G') be an acyclic condensation of G. Then, for two distinct constant points a, b of G, $\pi(a) \neq \pi(b)$. Thus each constant point s of G can be identified with its image $\pi(s)$ (for each $s \in V_{const}$, $\pi(s) = s$) and $V_{const} \subseteq V'$.

Proof. It results from the acyclicity of G' and the path ConstPath in G : no two distinct constant points of G can have the same image by π otherwise the portion of ConstPath between them becomes a cycle in G' □

Lemma 4.2 For an acyclic condensation (π, G') of G, the statement (r1) is equivalent to the statement (r1') : G' is of outdegree 2, $\pi(s) = s$ if $s \in V_{const}$ and $V' = V_{const}$.

Proof. It reults from lemma 4.1 and the fact that $|V'| = |V_{const}|$ □

Convention : in the sequel, G' is acyclic, therefore, by lemma 4.1, we will identify each constant point s of G with $\pi(s)$ without mentionning it.

Lemma 4.3 (the particular points p and q). Let (π, G') be an acyclic condensation of G. Then a) for all $s \in V$, $\pi(s) = \pi(q) \leftrightarrow s = q$, and $\pi(s) = \pi(p) \leftrightarrow s = p$

b) $d^+_G(p) = d^+_{G'}(p) = 1$ and $d^+_G(q) = d^+_{G'}(q) = 0$

Proof It uses the acyclicity of G' and the lemma 1 c). a) It is clear that $\pi(q) = q$. Conversely, suppose $s \neq q$. Then $\pi(s) \neq q$ else the image by π of the path from s to q becomes a cycle which contains q in G'. Do the same with p.

b) $d^+_{G'}(q) = 0$ else there is some $s \in V$ such that $(q, \pi(s)) \in E'$ would form with the image by π of the path from s to q a cycle in G'. Do the same with p. □

Lemma 4.4 For an acyclic condensation (π, G') of G, (r1') implies (r2).

Proof assume (r1') is satisfied. Let $s \in V_{const}$. In case s = p or q, $d^+_{G'}(s) = d^+_{G'}(s)$ (lemma 4.3). In case $s \neq p$ and $s \neq q$, $d^+_{G'}(s) \leq d^+_G(s)$ since $d^+_G(s) = 2$ (lemma 2) and $d^+_{G'}(s) \leq 2$ by (r2). The summation of these equalities, for all constant points gives then (since $V' = V_{const}$ by (r1')) : $|E'| = \sum_{u \in V'} d^+_{G'}(u) \leq \sum_{u \in V'} d^+_G(u) = |E_{const}|$ □

<u>Notations 4</u> : $E_{const-const}$ is the set of the edges of G whose <u>initial</u> and <u>final</u> points are <u>both constant</u>. $E_{const-var}$ is the set of the edges of G whose <u>initial</u> point is <u>constant</u> and whose <u>final</u> point is <u>variable</u>

The following lemma 4.5 showes that (r2) implies (r1') for an acyclic condensation (π, G') of G. Moreover it also states that to obtain E', we only need to consider E_{const} and then move to some constant point each (variable) final point of the function-edges issued from the singular points of G.

<u>Lemma 4.5</u> Let (π, G') be an acyclic condensation of G with $|E'| \leq |E_{const}|$. Let $s \in V_{const} - \{p, q\}$ and s_1 = <u>Succ</u> (s) and s_2 the two sons of s in G .Then

a) $V' = V_{const}$.

b) if s is not a singular point, the sons of s in G' are s_1 and s_2.

c) if s is a singular point, the sons of s in G' are s_1 and $\pi(s_2)$, with $s_1 \neq \pi(s_2)$.

d) for all constant point $s \neq p$ and $s \neq q$, $d^+_{G'}(s) = d^+_G(s) = 2$.

e) E' is the disjoint union of $E_{const-const}$ and $\pi(E_{const-var})$.

<u>Proof (outline)</u> The complete proof is given in section 4. By the fact that the outdegree of each constant point is saturated at its maximum, E_{const} will be sufficient to obtain G' from. We only will have to show that **no** variable son, of the form s_2 = <u>in</u> (X_j), of each singular point s is counfounded by π with its "brother" s_1 = <u>Succ</u> (<u>out</u>$_1$ (A_i)). This implies that $|\pi(E_{const})| = |E_{const}| \geq |E'|$ because constant points are **unvarying** by π (lemma 4.1). And thus $E' = \pi(E_{const})$ and it is obtained from E_{const} , only by moving (by π) the variable son of each singular point to some constant point, without identifying an edge of $E_{const-var}$ with an edge of $E_{const-const}$ and without adding, from the constant points, new edges $e \notin \pi(E_{const})$ (thus $d^+_{G'}(s) = d^+_G(s) \leq 2$) □

<u>Lemma 4.6.</u> The statements (r1), (r1') and (r2) are equivalent for an acyclic condensation (π , G') of G.

<u>Proof</u> it follows from lemmas 4.2, 4.4 and 4.5.

<u>Lemma 4.7.</u> Let (π, G') be an acyclic condensation of G, satisfying (r1'). Then :

a) $(t, p) \in E$ iff t is of the form $\underline{in}$ (S)
b) π ($\underline{in}$ (S)) is of the form $\underline{in}$ (A_j) for any gadget-graph S

Proof. a) By definition of G.

b) Let $t = \underline{in}$ (S). Then $e = (t, p) \in E$ by a). Then $\pi(e) = (\pi(t), p) \in E'$. Thus $\pi(e) \in E_{const-const}$ or $\pi(e) \in \pi(E_{const-var})$ by lemma 4.5 . But the final point of $\pi(e)$ is equal to p. And that is impossible for an element of $\pi(E_{const-var})$ (lemma 4.3). Hence $\pi(e) \in E_{const-const}$, and therefore $\pi(e)$ is of the form $\underline{in}$ (A_j) by a) □

The lemma 4.8 below is an immediate consequence of the lemmas 4.5 and 4.7. Intuitively it asserts that G' is a "good" condensation of G.

Lemma 4.8. Let (π, G') be an acyclic condensation of G, satisfying (r1'). Then :

a) G' is the union of the constant gadget-graphs : $A_0, \dots, A_{k-1}$, the bridge-edges and the special edge ($\underline{in}$ (A_{k-1}), q), and of edges of the form ($\underline{out}_h$ (A_i), $\underline{in}$ (A_j))

b) for each constant gadget-graph A_i and each integer $h < r$, there exists exactly one edge of G' from $\underline{out}_h$ (A_i) to a point of the form $\underline{in}$ (A_j).

Lemma 4.9 : let (π, G') be an acyclic condensation of G, satisfying (r1'). Then :

a) for any gadget-graph T, there exists some constant gadget-graph A_i such that

$$\tilde{\pi}(T) = A_j , \ (\tilde{\pi}(T) = T \text{ if } T \text{ is constant})$$

b) for any function-edge $e = (\underline{out}_l\ (T), \underline{in}\ (S))$, if $\tilde{\pi}(T) = A_i$ and $\tilde{\pi}(S) = A_j$ then

$$\pi(e) = (\underline{out}_l\ (A_i), \underline{in}\ (A_j))$$

Proof. b) is an immediate consequence of a). The proof of a), which essentially uses combinatorics, is developped in the next section □

We now can prove lemma 4. By lemma 4.9 b), each $f_0', \dots, f'_{r-1}$ is a single-valued function on A. Lemma 4.9 a) shows that val is an application from $A \cup X$ to A. Thus it remains to check that $\mathcal{S} = (\mathit{val}, f_0', \dots, f_{r-1}')$ satisfies Γ. That results immediately from the definitions of G and $\mathcal{S}$ and from the lemma 4.9 b). That concludes the proof of lemma 4 □

Lemmas 3 and 4 concludes the proof of the main theorem. In the next section we carry out the proof of lemmas 4.5 and 4.9.

Section 5. Combinatorial lemmas for the proof of the main theorem

The proof of lemma 4.5 uses the claims 1 to 6 below. Each of these claims are under the hypothesis : (π, G') is an acyclic condensation of G such that $|E'| \le |E_{const}|$.

Claim 1 Let $s \in V_{const} - \{p, q\}$ and $s_1 = \underline{Succ}(s)$ and s_2 the sons of s in G. Then :

a) if s is not a singular point, s_1 and s_2 are two distinct sons of s in G': $d^+_{G'}(s) \geq 2$

b) if $d^+_{G'}(s) = 1$ then s_2 is variable, s is a singular point and $\pi(s_2) = s_1$

c) if $\pi(s_2) = s_1$ and if t_1 and t_2 are the sons of s_1 in G, then t_1, t_2 and p are three distinct sons of s_1 in G', thus $d^+_{G'}(s_1) \geq 3$)

Proof. a) if s is not a singular point, s_1 and s_2 are both constant and thus invariant by π (lemma 4.1). b) Obviously s is a singular point by a) and $\pi(s_2) = s_1$, else $d^+_{G'}(s) \geq 2$.

c) Let $\pi(s_2) = s_1$. Then s is a singular point. Let then $(s, s_2) = (\underline{out_l}(A_i), \underline{in}(X_j))$. Then $(s_1, p) \in E'$ since (s_2, p) is an edge of G. Then, by lemmas 4.1 and 4.4 , s_1 has in G' three distinct sons : p, t_1 and t_2, where t_1 and t_2 are the sons (both constant and different from p) of s_1 in G □

Claim 2 Let $s \in V_{const} - \{p, q\}$. Then either $d^+_{G'}(s) = 1$ and then $d^+_{G'}(Succ(s)) \geq 3$ and $d^+_{G'}(s) + d^+_{G'}(Succ(s)) \geqslant 4 = d^+_G(s) + d^+_G(Succ(s))$, or $d^+_{G'}(s) \geq 2 = d^+_G(s)$

Proof By the fact that $d^+_G(s) \geq 1$ (lemma 1) and by claim 1 □

Claim 3 : $\sum_{s \in V_{const}} d^+_{G'}(s) = \sum_{s \in V_{const}} d^+_G(s) = |E_{const}| = |E'|$

Proof $|E_{const}| \geq |E'| \geq \sum_{s \in V_{const}} d^+_{G'}(s) \geq \sum_{s \in V_{const}} d^+_G(s) = |E_{const}|$.

the third inequality is obtained by the summation for all $s \in V_{const} - \{p, q\}$ of the inequalities in claim 2 and by the fact that $d^+_{G'}(s) = d^+_G(s)$ if $s = p, q$ (lemma 4.4) □

Claim 4 $V' = V_{const}$

Proof From each point t of G is at least issued an edge. Thus $d^+_{G'}(\pi(t)) \geq 1$. So, if there was $t \in V$ such that $\pi(t) \notin V_{const}$ then we would have the contradiction below :

$$|E_{const}| = |E'| = \sum_{s \in V'} d^+_{G'}(s) \geq d^+_{G'}(\pi(t)) + \sum_{s \in V_{const}} d^+_{G'}(s) \geq 1 + |E_{const}|$$ □

Claim 5 Let $t_1, \ldots, t_\alpha$ be the list of the singular points t such that $d^+_{G'}(t) = 1$. Let $s \in V_{const} - \{p, q\}$ and $s_1 = \underline{Succ}(s)$ and s_2 the distinct sons of s in G). Then

a) $d^+_{G'}(s) = 3$ iff s is of the form $Succ(t_i)$ $(i = 1, \ldots, \alpha)$

b) if s is not of the form t_i nor of the form Succ (t_i) $(i = 1, \dots, \alpha)$, $d^+_{G'}(s) = 2$

c) if $\pi(s_2) = s_1$, $d^+_{G'}(s) = 1$ (it is the converse of the point c) of claim 1)

Proof : Let $t_1, \dots, t_\alpha$ be the list (perhaps empty) of the singular points such that $d^+_{G'}(t) = 1$. By claim 1, $d^+_{G'}(\underline{\text{Succ}}(t_i)) \geq 3$ for each $i = 1, \dots, \alpha$. Let $u_1, u_2, \dots, u_\beta$ the list of points of $V_{const} - \{p, q\}$ which are not of the form t_i nor of the form Succ (t_i). Then $d^+_{G'}(u_j) \geq 2$ $(j = 1, 2, \dots, \beta)$, by claim 2. Denote by H' (resp. H) the number of edges of G (resp. G') issued from the constant points. Then $|E'| \geq H'$ and

$$H' = \sum_{i=0}^{\alpha} (d^+_{G'}(t_i) + d^+_{G'}(\underline{\text{Succ}}(t_i))) + \sum_{j=0}^{\beta} d^+_{G'}(u_j) + d^+_{G'}(p) + d^+_{G'}(q).$$

We also have $|E_{const}| = \sum_{i=0}^{\alpha} (d^+_G(t_i) + d^+_G(\underline{\text{Succ}}(t_i))) + \sum_{j=0}^{\beta} d^+_G(u_j) + d^+_G(p) + d^+_G(q)$. So,

If there was some t_i such that $d^+_{G'}(\underline{\text{Succ}}(t_i)) > 3$ (resp. some u_j such that $d^+_{G'}(u_j) > 2$)

then : $d^+_{G'}(t_i) + d^+_{G'}(\underline{\text{Succ}}(t_i)) > 4 = d^+_G(t_i) + d^+_G(\underline{\text{Succ}}(s))$ (resp. $d^+_{G'}(u_j) > 2 = d^+_G(u_j)$).

Then we should have $|E'| \geq H' > |E_{const}|$. This contradicts claim 3. Thus $d^+_{G'}(\underline{\text{Succ}}(t_i)) = 3$ $(i = 1, \dots, \alpha)$ and $d^+_{G'}(u_j) = 2$ $(j = 1, \dots, \beta)$. Hence for any $s \in V_{const} - \{p, q\}$, $\pi(s) = 1$ or 2 or 3. If $d^+_{G'}(s) = 3$, then s is of the form Succ (t_i) unless $d^+_{G'}(s) = 2$ or $d^+_{G'}(s) = 1$. We have then proved a) and b).

If $\pi(s_2) = s_1$ then $d^+_{G'}(s_1) \geq 3$ (claim 1). So $d^+_{G'}(s_1) = 3$ and then s_1 is of the form Succ (t_j), i.e. s is of the form t_j □

Claim 6. a) The list $t_1, \dots, t_\alpha$ of singular point such that $d^+_{G'}(t) = 1$ is empty.

b) $d^+_{G'}(s) = 2 = d^+_{G'}(s)$ for each $s \in V_{const} - \{p, q\}$

<u>Proof</u> a) Suppose there exists a point s of the form t_i. Let us get a contradiction from it. Let s_1= <u>Succ</u> (s) ∈ <u>Joints</u> and s_2 = <u>in</u> (X_j) be the two distinct sons of s in G. Then π (s_2) = s_1 (claim 1). Since (<u>joint</u>$_r$ (X_j), s_2) and (<u>joint</u>$_r$ (X_j), q) are edges of G, (π (<u>joint</u>$_r$ (X_j)), s_1) and (π (<u>joint</u>$_r$ (X_j)), q) are edges of G'. As V' = V_{const} (claim 3), π (<u>joint</u>$_r$ (X_j)) is equal to some constant point c ∈ V_{const} with c ≠ p, q (lemma 4.3). Then q and s_1 ∈ <u>Joints</u> are two distinct sons of c in G' : $d^+_{G'}$ (c) ≥ 2 . Let c_1 = <u>Succ</u> (c) and c_2 the sons of c in G . Consider then all the possible cases for c :

- either c belongs to the list t_1, ..., t_α, and then 1= $d^+_{G'}$ (c) ≥ 2 (that is absurd).

- or c is of the form <u>Succ</u> (t_i) and then $d^+_{G'}$ (c) = 3, by claim 4. And since c_1, c_2 and p are three distinct sons of c in G' (claim 1), c_1, c_2 and p are exactly the sons of c in G'. Thus {s_1, q} ⊂ {c_1, c_2, p}, and then necessarily {c_1, c_2} = {q, s_1} by lemma 4.4. That is absurd since, in G, no point c has as sons both q and a point s_1 ∈ <u>Joint</u>.

- or c is not of the form t_i nor of the form <u>Succ</u> (t_j). Then $d^+_{G'}$ (c) = 2 (claim 5.b). Thus π (c_2) ≠ c_1 (claim 5.c). So the sons of c in G' are c_1 and π (c_2). Then {c_1, π (c_2)} = {q, s_1}. This is absurd since nor s_1 ∈ <u>Joint</u> nor π (c_2) is equal to q by lemma 4.3. In conclusion, the existence of a point s ∈ V_{const} − {p, q} of the form t_1,... , t_α leads only to contradictions. Hence no point of G' is of the form t_i or <u>Succ</u> (t_i). In conclusion $d^+_{G'}$ (s) = 2 = d^+_G (s) for every s ∈ V_{const} − {p, q}. □

<u>Proof of the lemma 4.5</u> a) V' = V_{const} by claim 4. b) see claim 1,a).
c) if s is a singular point, π (s_2) ≠ s_1 else $d^+_{G'}$ (s) = 1 (claim 5) and $d^+_{G'}$ (s) = 2 (claim 6) : that is absurd. d) see claim 6.b). e) since π (s_2) ≠ s_1 for each singular point s, $E_{const-const}$ ∩ π ($E_{const-var}$) =∅. On the other hand, E'= $E_{const-const}$ ∪ π ($E_{const-var}$) by b), c) and lemma 4.3. This concludes the proof of the lemma 4.5 □

Let us now prove the point a) of the <u>lemma 4.9</u> of the preceeding section.

The proof uses the lemmas 4.9.1, 2 and 3 below and the claims 7 to , which are all assumed to be under the following <u>hypothesis</u> : (π, G') is an acyclic condensation of G such that (r1') : G' is of outdegree 2 , π (s) = s if s ∈ V_{const} and V' = V_{const}.

Lemmas 4.9.1 and 4.9.2 below are immediat consequences of lemma 4.8.

Lemma 4.9.1 Let s, t $\in V_{const}$. Then :

a) If s $\notin$ Outs and t is a son of s in G', t is also a son of s in G
b) If s $\in$ Outs and t is a son of s in G', t = Succ (s) or t $\in$ $\{in (A_h)\}_h$.

Lemma 4.9.2 Let s, t $\in$ V_{const}. Then :

a) If s $\notin$ Ins and t is a father of s in G', t is also a father of s in G
b) If s $\in$ Ins and t is a father of s in G', t = Pred (s) or t $\in$ $\{out_h (A_i)\}_{h,\, i}$.

Hence, according to the two lemmas above, there is no need to precise in which graph (G or G') a constant point t is the son (resp. the father) of a constant point s, except if s $\in$ Outs (resp. s $\in$ Ins). The following technical lemma will be useful to get contradicions.

Lemma 4.9.3 If s is of the form out_h (T) or point (T) then π(s) is not of the form in (A_i).

Proof Suppose first that s = out_h (T) and π (s) = in (A_i). Since s has, in G, a son of the form in (U), π (s) = in (A_i) has a son of the form π (in (U))=in (A_h) (lemma 4.7). That is impossible by definition of G. Hence π (s) $\neq$ in (A_i).

Suppose next that s = point (T) and π (s) = in (A_i). Since q is a son of s in G, q is a son of π (s) = in (A_i) . That is impossible by definition of G unless i = k−1 (because of the special edge (in A_{k-1}, q). But in case i = k−1, since out_{r-1} (T) is also a son of point (T), π (out_{r-1} (T)) is a son of in (A_{k-1}). That contradicts lemma 4.3, since the sons of in (A_{k-1}) are p and q. Hence π (s) $\neq$ in (A_i) □

Proof of lemma 4.9. We have already proved that π (q) = q, π (p) = p (lemma 4.4), and that π (in (T)) $\in$ Ins (lemma 4.7). From now on we fix : π (in (T)) = in (A_i). Obviously if T is of the form Y_j then proof is finished. Lets us assume hence that T is of the form X_j. First, we directely prove that each point s of T of level h = r, r−1 is mapped by π to the point of A_i of the same level and nature as s. Next we will prove by (descending) induction on h $\in$ [0, r−2] that the image by π of the two sons of $joint_h$ (T) are distincts and that π($joint_h$ (T)) = $joint_h$ (A_i). Last we prove that for any h $\in$ [0, r−2[, π (out_h (T)) = in (A_i)

Claim 7. π ($joint_r$ (T)) = π ($joint_r$ (A_i)).

Proof. Since $joint_r$ (T) is, in G, a father of in (T), π ($joint_r$ (T)) is a father of in (A_i) in G'. Thus π ($joint_r$ (T)) is equal to $joint_r$ (A_i) or to a point of the form out_h (A_j) (lemma 4.9.2 b). But π ($joint_r$ (T)) = out_h (A_j) implies that out_h (A_j) is a father of q = π (q) (since $joint_r$ (T) is a father of q in G) : that is impossible by definition of G. Hence π ($joint_r$ (T)) = $joint_r$ (A_i) □

Claim 8. The images by π of the two sons of $\underline{joint}_{r-1}$ (T) are distincts i.e. π ($\underline{point}$ (T)) $\neq$ π ($\underline{joint}_r$ (T)).

Proof. (by contradiction) Suppose that π ($\underline{point}$ (T)) = π ($\underline{joint}_r$ (T)). Since $\underline{out}_{r-1}$ (T) is a son of $\underline{point}$ (T) in G, π ($\underline{out}_{r-1}$ (T)) is a son of π ($\underline{point}$ (T)) = $\underline{joint}_r$ (A_i) (claim 7). So π ($\underline{out}_{r-1}$ (T)) is equal either to q or to $\underline{in}$ (A_i) . That is impossible, by lemma 4.3 and lemma 4.9.3 □

Claim 9. π ($\underline{joint}_{r-1}$ (T)) = $\underline{joint}_{r-1}$ (A_i)

Proof. Since $\underline{joint}_{r-1}$ (T) is a father of $\underline{joint}_r$ (T) in G, π ($\underline{joint}_{r-1}$ (T)) is a father of $\underline{joint}_r$ (A_i). Thus π ($\underline{joint}_{r-1}$ (T)) is equal to $\underline{joint}_{r-1}$(A_i) (1) or to $\underline{in}$ (A_{i-1}) if i $\neq$ 0 (2) or to $\underline{out}_{r-1}$ (A_{k-1}) if i = 0 (3). Let us show that the cases (2) and (3) are absurd. First , suppose the case (2) true. As $\underline{point}$ (T) is a son of $\underline{joint}_{r-1}$ (T) in G, π ($\underline{point}$ (T)) is a son of $\underline{in}$ (A_{i-1}) and is thus equal to p or $\underline{joint}_r$ (A_i). That is impossible, by lemma 4.3 and claim 8. Next, suppose the case (3) true. Then π ($\underline{point}$ (T)) is a son of $\underline{out}_{r-1}$ (A_{k-1}) in G' since $\underline{point}$ (T) is a son of $\underline{joint}_{r-1}$ (T) in G. Thus π ($\underline{point}$ (T)) is equal to $\underline{joint}_r$ (A_0) or to a point of the form $\underline{in}$ (A_h) (lemma 4.9.1). That is impossible, by claim 8 and lemma 4.9.3. In conclusion, π ($\underline{joint}_{r-1}$ (T)) = $\underline{joint}_{r-1}$(A_i) (1). □

Claim 10. π ($\underline{point}$ (T)) = $\underline{point}$ (A_i)

Proof. Since $\underline{point}$ (T) is a son of $\underline{joint}_{r-1}$ (T) in G, π ($\underline{joint}_{r-1}$ (T)) is a son of $\underline{joint}_{r-1}$ (A_i) . Thus π ($\underline{joint}_{r-1}$ (T)) is equal to $\underline{point}$ (A_i) or $\underline{joint}_r$ (A_i). Hence π ($\underline{point}$ (T)) = $\underline{point}$ (A_i) (claim 8). □

Claim 11. π ($\underline{out}_{r-1}$ (T)) = $\underline{out}_{r-1}$ (A_i)

Proof. As $\underline{out}_{r-1}$(T) is a son of $\underline{point}$ (T) in G ,π ($\underline{out}_{r-1}$ (T)) is a son of $\underline{point}$ (A_i) . Thus π ($\underline{out}_{r-1}$ (T)) is equal to $\underline{out}_{r-1}$ (A_i) or q. Hence π ($\underline{out}_{r-1}$ (T)) = $\underline{out}_{r-1}$ (A_i) since π ($\underline{out}_{r-1}$ (T)) $\neq$ q by lemma 4.3 □

Claim 12 Let h $\in$ [1, r−1]. Assume that for all j $\in$ [h, r−1], a1) the images by π of the two sons in G of $\underline{joint}_j$ (T) are distinct and a2) π ($\underline{joint}_j$ (T)) = $\underline{joint}_j$ (A_i). Then :

b1) the images by π of the two sons in G of $\underline{joint}_{h-1}$ (T) are distinct and

b2) π ($\underline{joint}_{h-1}$ (T)) = $\underline{joint}_{h-1}$ (A_i).

Proof. Suppose that b1) is false. Then π ($\underline{out}_{h-1}$ (T)) = π ($\underline{joint}_h$ (T)). Then π ($\underline{out}_{h-1}$ (T)) = $\underline{joint}_h$ (A_i) by a2). But $\underline{joint}_l$ (A_i) has a son of the form π ($\underline{in}$ (S)), where $\underline{in}$ (S) is a son of $\underline{out}_{h-1}$ (T) in G. That is impossible in G since l $\leq$ r−2 and π ($\underline{in}$ (T)) $\in$ $\underline{Ins}$ (lemma 4.7). Hence b1) is true.

Since $\underline{joint}_{h-1}$ (T) is a father of $\underline{joint}_h$ (T) in G , π ($\underline{joint}_{h-1}$ (T)) is a father of $\underline{joint}_h(A_j)$ by a2). So π ($\underline{joint}_{h-1}$(T)) is equal to $\underline{joint}_{h-1}(A_j)$ (1) or to $\underline{Pred}$ ($\underline{joint}_h(A_j)$) (2). But let us show that (2) is impossible. Suppose that (2) is true. Then π ($\underline{out}_{h-1}$ (T)) is a son of $\underline{Pred}$ ($\underline{joint}_h$ (A_j)) in G', since $\underline{out}_{h-1}$ (T) is a son of $\underline{joint}_{h-1}$ (T) in G. Then π ($\underline{out}_{h-1}$(T)) is equal to $\underline{joint}_h$ (A_j) or to a point of the form $\underline{in}$ (A_f) (lemma 4.9.1). That is impossible, by b1) and by lemma 4.9.3. Hence we have (1) : π ($\underline{out}_{h-1}$(T)) = $\underline{joint}_{h-1}$ (A_j) □

The claims 3, 4 and 6 constitute the proof by induction of the following claim :

Claim 13. for any $h \in [0, r-1]$, the images by π of the two sons in G of $\underline{joint}_{h-1}$ (T) are distinct and π ($\underline{joint}_{h-1}$ (T)) = $\underline{joint}_{h-1}$ (A_j).

Claim 14. π ($\underline{out}_{h-1}$ (T)) = $\underline{out}_{h-1}$ (A_j) for any $h \in [0, r-2]$

Proof : As $\underline{out}_{h-1}$ (T) is a son of $\underline{joint}_{h-1}$ (T) in G, π ($\underline{out}_{h-1}$(T)) is a son of $\underline{joint}_{h-1}$ (A_j) (claim 13). Thus π ($\underline{out}_{h-1}$(T)) is equal to $\underline{out}_{h-1}$ (A_j) or $\underline{joint}_h$ (A_j) because $h - 1 < r - 1$. Hence π ($\underline{out}_{h-1}$(T)) = $\underline{out}_{h-1}$ (A_j) by claim 13 □

This concludes the proof of lemma 4.9

Conclusion

We have proved some linearly NP-complete problems on graphs, which therefore have non trivial lower bound. We noticed that the graphs we considered are given by their "list of edges" (sparse graphs).

By a similar method, we have improved [Ra90] the result of [Gr90a] by proving that the problem RISA (reduction of incompletely specified automatons) is linearly NP-complete if the automatons has a fixed alphabet of 2 (or more) symbols. To prove that, a directed graph of outdegree 2, like the one we use in the present paper is considered as an incompleted automaton : its vertices are the states of the automaton and its edges, labeled α or β, where $\Sigma = \{\alpha, \beta\}$ is the alphabet of the automaton, represent the transition function. Intuitively, the condensation of the graph to K-vertices is equivalent to the K-reduction of the incompleted automaton.

We think that all these are problems of contraction of a structure with constraints.

Acknowledgements : warm thanks to Etienne Grandjean, my thesis adviser.

References

[Co87/88] S. A. COOK, Short propositional formulas represent nondeterministic computations, Information Processing Letters 26 (1987/88) 269-270, North-Holland.

[GaJo79] M. S. GAREY and D. S. JOHNSON, Computers and Intractability : a Guide to the Theory of NP-Completeness. W. H. Freeman, San Francisco, 1979.

[Gr88] E. GRANDJEAN, A natural NP-complete problem with a nontrivial lower bound, SIAM J. Comput. , 1988.

[Gr90] E. GRANDJEAN, A nontrivial lower bound for an NP problem on Automata, SIAM J. Comput. , 1990. pp. 438-451.

[Gr90b] E. GRANDJEAN, personal communication, 1990

[Ha65] F. HARARY, An introduction to the theory of directed graphs, John Wiley and Sons, 1965.

[Ly90] J. F. LYNCH, The quantifier structure of sentences that characterize non-deterministic time-complexity, Logic in computer science, Proceeding colloquium, July 1990.

[Pf73] C. P. PFLEEGER, State reduction in incompletely specified sequential switching functions, IRE Trans, Electron. Comput EC-8 (1959), 356-367.

[PPST83] W. J. PAUL, N. PIPPENGER, E. SZEMERDI and W. T. TROTTER, On determinism versus nondeterminism and related problems, Proc 24 th IEEE Symp. On Foundations of Computer Science. IEE (1983), 428-438.

[Ra90] S. RANAIVOSON, bornes inférieures non triviales pour des problèmes NP en théorie des graphes et des automates, thèse de doctorat, laboratoire d'informatique de l'université de Caen

[Ro87] M. de Rougemont, Second-order and inductive definability on finite structures, Zeitschr. f. math. Logik und Grundalgen d. Math. 33 (1987), 47-63.

Expansions and Models of Autoepistemic Theories

Cecylia M. Rauszer

Institute of Mathematics, University of Warsaw,
PKiN, IXp., 00-901 Warsaw, Poland.

ABSTRACT.

Autoepistemic logic is a logic for modelling the beliefs of an agent who reflects on his own beliefs. In the paper we discuss problems dealing with existence of autoepistemic models and autoepistemic expansions. The necessary and sufficient conditions for existence of autoepistemic models and expansions are formulated.

1. Introduction

Autoepistemic logic, for short *ae-logic* was created by Moore as a theory of an agent reasoning about his own beliefs. The agent's beliefs, called sometimes the state of knowledge of an agent, are meant as a set of sentences of the language $\mathcal{L}_L$ which, in turn, is considered as a language $\mathcal{L}$ of a classical logic augmented by a unary operator L. An intended interpretation of a formula $L\alpha$ is: *α is believed*, or, more precisely: *α occurs on the list of sentences accepted by the agent.* The central role in the investigations of ae-logic is played by stable theories [S]. Namely,

Definition 1.1. A set $S \subset \mathcal{L}_L$ is *stable* if

(1) $S = Cn(S)$,

(2) if $\alpha \in S$, then $L\alpha \in S$,

(3) if $\alpha \notin S$, then $-L\alpha \in S$,

where Cn denotes usual classical consequence and $\alpha \in \mathcal{L}_L$.

Thoroughly the paper, we assume the following conventions:

- p,q,r ... denote ordinary (propositional) formulae, that is, formulae from $\mathcal{L}$. We call them *objective formulae.*
- $a, b, c, \alpha, \beta, \gamma, \ldots, \phi, \psi$, with indices, if necessary, denote formulae from $\mathcal{L}_L$, sometimes called *ae-formulae.*
- f denotes a false formula, that is, the negation $-\alpha$ being defined as the formula $\alpha \rightarrow f$.
- S will always denote a stable set. We assume that all stable sets under consideration are consistent, i.e., $S \neq F_L$, where F_L is the set of all ae-formulae.
- $\perp, \top$ mean logical values, *falsity* and *truth*, respectively.
- $S(A)$ is treated as any stable extension of a set of *initial premises* $A \subset \mathcal{L}_L$. In other words, $S(A)$ is a set containing A and closed under the stability conditions (1)-(3) of Definition 1.1.
- The word *tautology* is reserved for formulae which are valid in classical logic.
- 'α is a tautology' is abbreviated by $\vdash \alpha$.

- The word *model* means a propositional model.
- If v is a valuation, that is, a mapping from F_L into $\{\bot, \top\}$, then $v(X) = \top$ means that for every $\alpha \in X$, $v(\alpha) = \top$, where $X \subseteq F_L$.

Objective formulae play an important role in the investigations of stable sets. The following proposition summarizes the basic results concerning them:

Proposition 1.1. (a) *If S is a stable set then the set $S \cap \mathcal{L}$ is closed under Cn.* (b) *If S_1, S_2 are stable sets and $S_1 \cap \mathcal{L} = S_2 \cap \mathcal{L}$ then $S_1 = S_2$.*

Finally, recall that for every ae-formula α there are $\phi_1, \ldots, \phi_n$ such that for all $i \leq n$ and for some k, l

$$\phi_i = -L\alpha_{i1} \vee \cdots \vee -L\alpha_{ik} \vee L\beta_{i1} \vee \cdots \vee L\beta_{il} \vee p_i$$

and α is tautological equivalent to the conjunction of $\phi_1, \ldots, \phi_n$. This property of ae-formulae was observed by Moore and the conjunction of $\phi_1, \ldots, \phi_n$ is called *conjunction normal form* , abbreviated by CNF, of α.

In the paper we will assume that every set of initial premises possesses the following property: A consists of formulae in CNF such that none of them is a disjunction of $L\beta$'s.

One can easily observe, that the language $\mathcal{L}_L$ of autoepistemic logic may be considered as the language of modal logic with modal operator denoted by L. Let A be a subset of $\mathcal{L}_L$. By *an S4 modal theory based on A* we mean any set $M(A)$ of formulae of $\mathcal{L}_L$ such that $M(A)$ is the smallest set of formulae that follow from A by means of all axioms and rules of S4 modal logic [C]. We say that $M(A)$ is a *maximal* S4 theory if $M(A)$ is consistent and for every $\phi \notin M(A)$ the S4 modal theory $M(A \cup \phi)$ is inconsistent.

Observe that maximal S4 theories obey the principle: $-L\phi$ is a consequence of a set A of initial premises if ϕ is not a consequence of A. Hence, the concept of maximal S4 theories may be used as an alternative interpretation of non-monotonicity of stable sets. In fact, let $M(A)$ be a maximal S4 theory. Then $M(A)$ is consistent, and therefore, there exists a formula ϕ not in $M(A)$. If we add ϕ to A then we have to withdraw some formulae from $M(A)$. If not, then we lose the consistency of $M(A \cup \phi)$ which is a superset of $M(A)$.

Thus, we have

Lemma 1.1.[R] *Let A be a set of formulae of $\mathcal{L}_L$. The following conditions are equivalent:*

(i) *$M(A)$ is closed under the stability conditions.*

(ii) *$M(A)$ is a maximal S4 theory.*

2. Autoepistemic models

In autoepistemic logic a set of ae-formulae is interpreted as a specification of beliefs of an agent reflecting upon his own beliefs. We will call any set of ae-formulae closed under classical consequence an *autoepistemic theory*, abbreviated by ae-theory.

An ae-formula of the form $L\alpha$ will be true with respect to an agent if and only if α is in his set of beliefs. According to Moore [Mo], the truth of an agent beliefs, expressed as an ae-theory is determined by:

(i) propositional formulae which are true in the external world. (The agent world is closed under classical provability).

(ii) which formulae are believed by the agent.

These ideas are formalized by notions of an *autoepistemic interpretation*, for short, ae-interpretation and *autoepistemic models*, for short, ae-models.

Definition 2.1. An *autoepistemic interpretation* v of an ae-theory T is a valuation v of F_L into the truth values set that satisfies the following condititons:

(1) v is the usual truth assignment for objective formulae,

(2) $v(L\alpha) = \top$ if and only if $\alpha \in T$.

Definition 2.2. An ae-interpretation v is an *autoepistemic model of* T provided, for every $\alpha \in T, v(\alpha) = \top$.

One of the main properties of the standard logical calculi states that if a set of formulae is consistent then it possesses a model. It turns out, that for ae-logic, consistency of an ae-theory does not imply the existence of ae-models. To show this, take as a set of initial premises the set $Cn(Lp)$. This set is consistent as $p \notin Cn(Lp)$. Indeed, one can easily check that a formula of the form $Lp \rightarrow p$ is not a tautology. Assume now, that v is an ae-model of $Cn(Lp)$. Then $v(Lp) = \top$ which, by (2) of Definition 2.1, implies $p \in Cn(Lp)$, a contradiction.

Lemma 2.1. *Let T be an ae-theory. The following are equivalent:*

(1) *T has an ae-model,*

(2) *For every ae-formula α, $L\alpha \in T$ implies $\alpha \in T$ and $-L\alpha \in T$ implies $\alpha \notin T$.*

Proof: Suppose (1) and let $L\alpha \in T$. Thus for v being an ae-model for T, $v(L\alpha) = \top$ and therefore $\alpha \in T$. To show the latter implication let $-L\alpha \in T$. Then, for an ae-model v of T, $v(L\alpha) = \bot$ which implies $\alpha \notin T$.

Assume (2). Observe that T is consistent. Otherwise, for some ae-formula α, $L\alpha \wedge -L\alpha \in T$. Hence, by the assumptions $\alpha \in T$ and $\alpha \notin T$, a contradiction. Thus, T is consistent and T has a model. More precisely, there is a valuation v_0 such that for all $\alpha \in T$, $v_0(\alpha) = \top$. Now, we define an interpretation v as follows:

$$v(\alpha) = \begin{cases} v_0(\alpha), & \text{if } \alpha \in T; \\ \top, & \text{if } \alpha \notin T, \alpha = L\beta \text{ and } \beta \in T; \\ \bot, & \text{otherwise.} \end{cases}$$

We show that v is the required ae-model of T. Clearly, v is an ordinary truth assignment for formulae of $\mathcal{L}$. To prove that v is an ae-interpretation observe that if $\alpha \in T$ then $v(L\alpha) = \top$. To show the converse implication, let $\alpha \notin T$. Then, by (2), $L\alpha \notin T$ and, by the definition of v, $v(L\alpha) = \bot$, which completes the proof of Lemma 2.1.

Observe that under our assumption T consists of formulae in CNF such that none of them is a disjunction of $L\beta$'s. Without this property the above lemma is not true. Namely, R.Stark shows that if T is $Cn(Lp \vee Lq)$ then T satisfies condition (2), as for no formula $L\phi$: $L\phi \in T$ or $-L\phi \in T$. However T has no ae-model. Indeed, let v be an ae-interpretation of T. Then $v(Lp) = f$ since $p \notin T$ and also $v(Lq) = f$ since $q \notin T$. Therefore $v(Lp \vee Lq) = f$ and v is not an ae-model of T.

Definition 2.3. An ae-theory T is said to be *semantically complete* if and only if T contains every formula that is true in every ae-model of T.

Moore has observed that the stability conditions precisely characterize the ae-theories which are semantically complete. Namely,

Theorem 2.1. [Mo] **(completeness)** *An ae-theory T is semantically complete if and only if T is stable.*

Definition 2.4. An ae-theory T is *sound with respect to a set A of initial premises* if and only if every ae-interpretation of T that is a model of A is a model of T.

Definition 2.5. An ae-theory T based on a set of initial premises A is *grounded in* A provided T is a subset of the set $Cn(A \cup LT \cup -L\overline{T})$, where $LT = \{L\phi : \phi \in T\}$ and $-L\overline{T} = \{-L\phi : \phi \notin T\}$.

Theorem 2.2. [Mo] **(soundness)** *A consistent ae-theory T is sound with respect to a set of intial premises A if and only if T is grounded in A.*

3. Autoepistemic Expansions

Intuitively, the beliefs of an ideally rational agent ought to be both semantically complete and grounded in his initial premises.

Theorem 2.1 guarantees that if the set of agent's beliefs T is stable then his beliefs are semantically complete. From Theorem 2.2 follows that the set of beliefs of a rational agent consists of no more than all sentences from the set of initial premises and those facts required by the stability conditions, that is to say, T is grounded in the set of his initial premises.

Autoepistemic expansions are meant as sets of beliefs that satisfy the intuitive requirements on the state of knowledge of an agent. A formal definition of the mentioned expansions, for short ae-expansions, is as follows:

Definition 3.1. An ae-theory T is an *ae-expansion of a set of intial premises A* if and only if

$$T = Cn(A \cup LT \cup -L\overline{T}). \tag{1}$$

It is known that

Lemma 3.1.[Mo], [M]. *Every consistent set of objective formulae has a unique ae-expansion.*

As we have mentioned, ae-expansions are considered as sets that reflect all beliefs of an ideally rational agent. However, it is known that if the set A of initial premises contains ae-formulae, then A may have no ae-expansion, several ae-expansions or a unique ae-expansion.

Example 1.

(1) The set $A = \{Lp\}$ has no ae-expansion. Indeed, assuming that T is an ae-expansion of A, we infer that $Lp \in T$ and therefore $p \in T$. Thus

$$\vdash Lp \wedge \bigwedge_i Lb_i \wedge \bigwedge_i -Lc_i \to p,$$

where for all i, $b_i \in T$, $c_i \notin T$. However, the valuation v such that for all i, $v(Lp) = v(Lb_i) = \top$ and $v(Lc_i) = v(p) = \bot$ rejects the above formula.

(2) The set $A = \{-Lp \to q,\ -Lq \to p\}$ has two ae-expansions. Namely, ae-expansions of $Cn(p)$ and $Cn(q)$ are ae-expansions of A.

(3) Let $A = \{-Lp \to q\}$. The ae-expansion of $Cn(q)$ is the only ae-expansion of A.

One can easily check

Proposition 3.1. *Every autoepistemic expansion is a stable set.*

The converse of this proposition is not true, since every consistent set A may be extended to a maximal S4 theory, which however, need not be an ae-expansion of A. An example of a set showing that the converse does not hold is the set$A = \{Lp\}$. Indeed, A may be embedded into maximal S4 theories and none of them is an ae-expansion of A.

For a given set A let

$$T(A) = Cn(A \cup \{\beta :\ \alpha \to \beta \in SUBF(A) \cap S(A) \text{ and } \alpha \in S(A)\}),$$

where $SUBF(A)$ means the set of all subformulas of A.

Proposition 3.2. *If $L\alpha \in A$ and $\alpha \notin T(A)$, then A has no ae-expansion.*

Proof: Let $L\alpha \in A$, $\alpha \notin T(A)$ and let T be an ae-expansion of A. Assume that $\alpha \in T$. Then for some $a_i \in A$, $b_i \in T$ and $c_i \notin T$ a formula γ of the form

$$\bigwedge_i a_i \wedge \bigwedge_i Lb_i \wedge \bigwedge_i -Lc_i \to \alpha$$

is a tautology. Clearly, for all i, $a_i \neq \alpha$. Let us notice that if α is an objective formula p, then for all i, $Lb_i \neq p$. Since the valuation v such that $v(a_i) = v(Lb_i) = \top$ and $v(Lc_i) = v(\alpha) = \bot$ is not a model for γ, we conclude that $p \notin T$.

If α is Lp and for some i_0, $Lb_{i_0} = Lp$, then $p \in T$, a contradiction, By the induction on the complexity of α we can show that for all i, $Lb_i \neq \alpha$. Thus the valuation v defined above rejects γ, which proves that $\alpha \notin T$. Since T is a stable set we infer that $-L\alpha \in T$ and T is inconsistent. Hence, the assumption that A has an ae-expansion leads us to a contradiction.

Lemma 3.2. *If A has an ae-expansion, then $T(A)$ has an ae-model.*

Proof: Let T be an ae-expansion of A. Then T is a stable set containing A. Clearly, $T(A) \subseteq T$. Suppose that $-L\alpha \in T(A)$ and $\alpha \in T(A)$. Then $L\alpha \wedge -L\alpha \in T$, a contradiction.

Thus $-L\alpha \in T(A)$ implies $\alpha \notin T(A)$ which by Lemma 2.1 and the last proposition allows us to infer that A has an ae-model.

The converse of Lemma 3.2 is not true. Namely, let $A = \{-Lp \to p\}$. Since $p \notin Cn(A) = T(A)$, $Cn(A)$ is consistent and it has a model v_0. Let v be a valuation defined as follows:

$$v(\alpha) = \begin{cases} v_0(\alpha), & \text{if } \alpha \in Cn(A); \\ \top, & \text{if } \alpha = p \text{ or } \alpha = L\beta \text{ and } \beta \in A; \\ \bot, & \text{otherwise.} \end{cases}$$

It is easy to observe that v is well defined an ae-interpretation and moreover, v is an ae-model for $Cn(A)$. Notice, that A has no ae-expansion. Indeed, if T is an ae-expansion of A, then T satisfies the equality (1). Observe, that $p \notin T$. Otherwise, for some $b_i \in T$ and $c_i \notin T$,

$$\vdash -Lp \to p \wedge \bigwedge_i Lb_i \wedge \bigwedge_i -Lc_i \to p.$$

Since for all c_i, $c_i \neq -Lp \to p$, the valuation v such that $v(Lp) = v(Lb_i) = \top$ and $v(Lc_i) = v(p) = \bot$ rejects the above formula, which proves that $p \notin T$. Hence, $-Lp \in T$ and $Lp \in T$, and T would be an inconsistent stable set.

Now, we want to present a certain construction that, for some sets of initial premises, guarantees the existence of an ae-expansion.

Recall, that the *L-depth* of an ae-formula α, denoted by $d(\alpha)$ is defined as follows:

- if $\alpha \in \mathcal{L}$ then $d(\alpha)= 0$,
- if $\alpha = L\beta$ then $d(\alpha) = d(\beta) + 1$,
- if $\alpha = -\beta$ then $d(\alpha) = d(\beta)$,
- if $\alpha = \beta_1 \circ \beta_2$ then $d(\alpha) = max\{d(\beta_1), d(\beta_2)\}$, where $\circ \in \{\vee, \wedge, \to, \leftrightarrow\}$.

Let $\mathcal{L}_{L,n}$ be the set of all formulae of $\mathcal{L}_L$ of L-depth at most n. Note, $\mathcal{L} = \mathcal{L}_{L,0}$.

Assume that A is a set of autoepistemic formulae such that for every n, the sets E_n are consistent, where E_n are defined inductively as follows:

$$E_0 = Cn(A).$$

If E_n is constructed (and consistent) let

$$E_{n+1} = Cn(E_n \cup \{L\phi : \phi \in E_n\} \cup \{-L\phi : \phi \in \mathcal{L}_{L,n} - E_n\}).$$

Finally, put

$$E(A) = \bigcup_{n=0}^{\infty} E_n.$$

(Let us mention that the definition of the family $\mathcal{E} = (E_n)_{n\in\omega}$ is a modification of the construction presented by Konolige [K] and Marek [M]).

Example 2.

(1) For the set $A = \{Lq\}$, $q \in \mathcal{L}$, the family $\mathcal{E}$ does not exists. Indeed, for $A = \{Lq\}$ we have

$$E_1 = Cn(Cn(Lq) \cup \{L\phi :\ \phi \in Cn(Lq)\} \cup \{-L\phi :\ \phi \in \mathcal{L}_{L,0} - Cn(Lq)\}).$$

Since $q \notin Cn(Lq)$ ($Lq \rightarrow q$ is not a tautology), we infer that $-Lq \in E_1$ and $Lq \in E_1$, which proves that E_1 is inconsistent. Thus, for the set $A = \{Lq\}$ the family $\mathcal{E}$ cannot be constructed.

(2) If $A = \{-Lp \rightarrow q\}$ then E_0 and E_1 are consistent. However, E_2 is inconsistent. Indeed, one can easily check that $p \vee q \notin E_0$. Hence $-Lp \wedge -Lq \in E_1$ and $q \in E_1$. Thus $-Lq \wedge Lq \in E_2$ which proves that the set $E(\{-Lp \rightarrow q\})$ does not exist.

(3) One can prove that for the set $A = \{-Lp \rightarrow q, q\}$ the set $E(A)$ exists.

From now on, unless stated otherwise, we assume that for all sets under consideration, the family $\mathcal{E} = (E_n)_{n \in \omega}$ is defined.

We show

Lemma 3.3. *E(A) is a consistent stable set.*

Proof: 1. $E(A)$ is consistent. For if it were not, then some finite subset of $E(A)$ would be inconsistent. But clearly every finite subset of $E(A)$ is a subset of some E_n which, in turn, would be inconsistent.

2. $E(A)$ is closed under tautological consequence, i.e., $E(A) = Cn(E(A))$. Clearly, $E(A) \subseteq Cn(E(A))$. Let $\phi \in Cn(E(A))$. Then for some $\alpha_1, \ldots, \alpha_k \in E(A)$, a formula of the form

$$\alpha_1 \wedge \cdots \wedge \alpha_k \rightarrow \phi$$

is a tautology. Hence, there is an n, such that all α_i's, i $\leq$ k, belong to E_n, and, we conclude that $\phi \in E_n \subseteq E(A)$.

3. $E(A)$ satisfies the conditions (2) and (3) of Definition 1.1. In fact, let $\phi \in E(A)$. Then for some n, $\phi \in E_n$ and by the definition, $L\phi \in E_{n+1}$ which is a subset of $E(A)$. Now, let $\phi \notin E(A)$. Then for every n, $\phi \notin E_n$. If the L-depth of ϕ is m, then $\phi \in \mathcal{L}_{L,m} - E_m$ and $-L\phi \in E_{m+1}$, which completes the proof of the lemma.

Lemma 3.4. *If $\phi \in E_n$, $1 \leq n$, then a formula of the form*

$$\bigwedge_i a_i \wedge \bigwedge_i Lb_i \wedge \bigwedge_i -Lc_i \rightarrow \phi$$

is a tautology, where all $a_i \in A$, $b_i \in E_{n-1}$ and $c_i \notin E_{n-1}$.

Proof: By the induction with respect to n.

Theorem 3.1. *The set E(A) is an ae-expansion of A.*

Proof: We have to show that if $E(A) = \bigcup_{n=0}^{\infty} E_n$, where all E_n are defined above, then $E(A)$ obeys the equality (1). The inclusion $\supseteq$ is obvious as $E(A)$ is closed under tautological consequence.

To show the converse inclusion, let allow ϕ be in $E(A)$. Then for some n, $\phi \in E_n$ and by Lemma 3.4 a formula of the form

$$\bigwedge_i a_i \wedge \bigwedge_i Lb_i \wedge \bigwedge_i -Lc_i \rightarrow \phi$$

is a tautology, where all $a_i \in A$, $b_i \in E_{n-1}$ and $c_i \notin E_{n-1}$. Thus, all $Lb_i \in E_n \subseteq E(A)$. To finish the proof, it is enough to show that all c_i are not in $E(A)$. Otherwise, assume that there is an m, such that for some i_0, $c_{i_0} \in E_m$. Then $Lc_{i_0} \in E_{m+1}$.

Case (1). $m < n-1$. Then $Lc_{i_0} \in E_n$ and E_n is inconsistent as $-Lc_{i_0} \in E_n$, a contradiction.

Case (2). $n-1 \leq m$. Then $-Lc_{i_0} \in E_n \subseteq E_{m+1}$ and E_{m+1} is inconsistent as $Lc_{i_0} \in E_{m+1}$, a contradiction.

Hence, for all i, $c_i \notin E(A)$, which proves the required inclusion. Now, we conclude that for the set $E(A)$ the equality (1) holds, that is, $E(A)$ is an ae-expansion of A.

4 Good sets

It follows from Theorem 3.1 that the family $\mathcal{E}$, if exists for a set of initial premises A, then it determines an ae-expansion of A. We have mentioned in the preceding section that for some sets of initial premises the family $\mathcal{E} = (E_n)_{n \in \omega}$ might not be constructed. However, some of them, for instance, the set $A = \{-Lp \rightarrow q\}$, have an ae-expansion. Now, we want to find necessary and sufficient conditions that guarantee the existence of ae-expansions.

For this purpose, we introduce the notion of *good sets*. The concept of these sets has the following motivation:

The existence of an ae-expansion of A is coded in the set of initial premises.

Intuitively, A is a good set provided:

(1) If it is possible to isolate a formula of the form $L\beta \vee q$ from A by means of the stability conditions, then $Cn(A \cup q)$ has an ae-model.

(2) If it is possible to isolate a formula of the form $-L\beta \vee q$ from A by means of the stability conditions, then $Cn(A \cup q)$ has an ae-model.

Definition 4.1. Let $A \subset \mathcal{L}_L$. A is said to be a *good set* if there exists a set $X \subset \mathcal{L}$ such that:

(1) $T = Cn(A \cup X)$ has an ae-model,

(2) If $L\beta \vee q \in SUBF(A) \cap S(T)$ and $A \not\vdash \beta$ then $q \in X$,

(3) If $-L\alpha \vee q \in SUBF(A) \cap S(T)$ and $S(T) \vdash \alpha$ then $q \in X$.

If A is a good set then T is said to be a *better set than A*.

The following proposition is a simple consequence of Definition 4.1:

Proposition 4.1. *Let A be a good set and let $T = Cn(A \cup X)$ be a better set than A. Then $q \in X$ if and only if for some formula β,*

$$-L\beta \rightarrow q \in SUBF(A) \cap S(T) and - L\beta \in S(T)$$

or

$$L\beta \rightarrow q \in SUBF(A) \cap S(T) and L\beta \in S(T),$$

Moreover,

$$X \subseteq \{p_i : -L\alpha_{i1} \vee \cdots \vee -L\alpha_{ik} \vee L\beta_{i1} \vee \cdots \vee L\beta_{il} \vee p_i \in SUBF(A)\}.$$

Example 3

(1) The set $A = \{Lp\}$ is not good. Otherwise, there is T a better set than A. We know that $Lp \nvdash p$. Hence, $f \in S(T)$ and T has no ae-model.

(2) If $A \subset \mathcal{L}$ and A is consistent than A is a good set. Namely, the set $Cn(A)$ is a better set than A.

(3) The set $A = \{-Lp \rightarrow q,\ -Lq \rightarrow p\}$ is good. Namely, it is not difficult to observe that the set $T_1 = Cn(A \cup p)$ as well as $T_2 = Cn(A \cup q)$ are better sets than A.

The idea of good sets may be illustrated in the more conclusive way by the following example:

(4) Let $A = \{p,\ (Lp \wedge -Lq) \rightarrow q\}$. This set is not good. Otherwise, if it is, then there is a better set T for A. Observe, that we can select from A, by means of the stability conditions, a subformula of the form $Lq \vee q$. Clearly, $A \nvdash q$ and by the definition 4.1 (2), $q \in X$ which is a subset of $S(T)$ and $-Lq \in S(T)$, a contradiction.

(5) The set $A = \{-Lp \rightarrow q,\ Lq \rightarrow p\}$ is not good. If yes, then for a better set T than A we have $q \in T$ and $-Lp \in S(T)$. Moreover, $Lq \wedge Lp \in S(T)$. Thus $S(T)$ is inconsistent.

(6) The set $A = \{-Lq \rightarrow p, Lq \rightarrow p \wedge q\}$ is good. Namely, the sets $T_1 = Cn(A \cup p)$ and $T_2 = Cn(A \cup p \wedge q)$ are better sets than A.

The important result of this section is the proof of the theorem that states:

Theorem 4.1. *If A is a good set, then A has an ae-expansion.*

The above theorem follows from the following one:

Theorem 4.2. *If A is a good set, then for the set T being a better set than A there is a family $\mathcal{E}$ of consistent sets E_n.*

Once Theorem 4.2 is proved then the proof of Theorem 4.1 is as follows:

Proof of Theorem 4.1 : If A is a good set, then by Theorem 4.2, for the set T being a better set that A there is a family of consistent sets E_n. On account of Theorem 3.1 the set $E(T) = \bigcup E_n$ is an ae-expansion of T, that is,

$$E(T) = Cn(T \cup LE(T) \cup -L\overline{E(T)}).$$

We prove now that $E(T)$ is an ae-expansion of A. For this purpose we have to show that

$$E(T) = Cn(A \cup LE(T) \cup -L\overline{E(T)}).$$

Clearly

$$Cn(A \cup LE(T) \cup -L\overline{E(T)}) \subseteq Cn(T \cup LE(T) \cup -L\overline{E(T)}).$$

To show the converse inclusion it is enough to prove that

$$X \subseteq Cn(A \cup LE(T) \cup -L\overline{E(T)}),$$

where recall, the set X is a set of objective formulae such that $T = Cn(A \cup X)$ is a better set that A. Without loss of generality we may assume that X is a nonempty set.

For the simplicity denote by Y the set on the right hand side of the considered inclusion. So, we claim that $X \subseteq Y$.

Let $p \in X$. Then there are two possibilities:

Case (1). p is added to X because of a formula of the form $L\beta \vee p$ belongs to $SUBF(A)$ and it is a stable consequence of the set T and β is not a consequence of A. Moreover, $-L\beta$ is a stable consequence of T.

Case (2). p is added to X because of a formula of the form $-L\alpha \vee p$ belongs to $SUBF(A)$ and it is a stable consequence of the set T and α is a stable consequence of T. Hence, we conclude: If p belongs to X then for some m, n a formula of the form

$$L\alpha_1(\rightarrow \cdots \rightarrow (L\alpha_m \rightarrow (-L\beta_1(\rightarrow \cdots \rightarrow (-L\beta_n \rightarrow p)\ldots)))\ldots)$$

occurs as a subformula in A and all $-L\beta_i$ are stable consequences of T, and all $L\alpha_i$ are stable consequences of T. Hence, all $-L\beta_i$ belong to $-L\overline{E(T)}$ and all $L\alpha_i$ belong to $LE(T)$. Because of $-L\overline{E(T)} \cup LE(T) \subseteq Y$, and as Y is closed under provability, we infer that $p \in Y$. Thus the required inclusion holds, which proves that the set $E(T)$ is an ae-expansion of A.

Now, we provide

Proof of Theorem 4.2 : Let A be a good set and T a better set than A, that is, there is a set $X \subset \mathcal{L}$ such that the set $T = Cn(A \cup X)$ satisfies the conditions (1) - (3) of Definition 4.1. We show that for the set T the family $\mathcal{E}$ of consistent sets E_n exists.

Put $E_0 = T$. Clearly, E_0 is consistent as T has an ae-model.

Let

$$E_1 = Cn(E_0 \cup LE_0 \cup \{-Lp : p \in \mathcal{L}_{L,0} - E_0\}).$$

We are going to show that E_1 is consistent. First of all observe, that E_1 has the following property denoted by P(1):

$P_1(1)$ If $L\beta \vee q \in E_1$, then $\beta \in E_0$ or $q \in E_0$,

$P_2(1)$ If $-L\alpha \vee q \in E_1$, then $\alpha \notin E_0$ or $q \in E_0$.

Proof of $P_1(1)$: Suppose that $L\beta \vee q \in E_1$, $\beta \notin E_0$ and $q \notin E_0$. By the definition of E_1 we infer that a formula of the form

$$\bigwedge_i a_i \wedge \bigwedge_i Lb_i \wedge \bigwedge_i - Lc_i \rightarrow L\beta \vee q,$$

where for all i, $a_i, b_i \in E_0 = Cn(A \cup X)$, $c_i \notin E_0$, is a tautology.

If $L\beta \in E_0$, then, because of E_0 has an ae-model, β would be in E_0. Thus $L\beta \notin E_0$. Observe now, that for all i, $a_i \neq L\beta \vee q$. Otherwise, if for some i_0, $a_{i_0} = L\beta \vee q$, then $L\beta \vee q$ belongs to $SUBF(A) \cap S(E_0)$ and, since $\beta \notin E_0$, we infer that $A \nvdash \beta$. However, A is a good set, hence $q \in X$ which is a subset of E_0, a contradiction.

It is obvious, that for all i, $a_i \neq L\beta$ and $b_i \neq L\beta$. Indeed, let if for some i_0, $a_{i_0} = L\beta$, (for $b_{i_0} = L\beta$ the proof in an analogous) then $L\beta \in E_0$, a contradiction. Clearly, for all i, $a_i \neq q$ and $b_i \neq q$.

Now, let v be an ae-model for E_0 such that $v(q) = \bot$. It is obvious that v rejects the above formula, and therefore, we conclude that $P_1(1)$ holds for E_1.

Proof of $P_2(1)$: Similarly, if $-L\alpha \vee q \in E_1$, $\alpha \in E_0$ and $q \notin E_0$, then

$$\vdash \bigwedge_i a_i \wedge \bigwedge_i Lb_i \wedge \bigwedge_i -Lc_i \rightarrow -L\alpha \vee q$$

and for all i, $a_i, b_i \in E_0, c_i \notin E_0$.

Notice, that $-L\alpha \notin E_0$. Indeed, $-L\alpha \in E_0$ contradicts to the existence of ae-model for E_0. Now, if for some i_0, $a_{i_0} = -L\alpha \vee q$, then $-L\alpha \vee q \in SUBF(A)$. Since E_0 is a better set than A and $\alpha \in E_0$, we conclude that $q \in E_0$, a contradiction. Thus, for all i, $a_i \neq -L\alpha \vee q$. Notice, that also for all $i, a_i \neq -L\alpha$ and $b_i \neq -L\alpha$. Moreover, for all i, $a_i \neq q$, $b_i \neq q$ and $c_i \neq \alpha$. Now, observe, that an ae-model for E_0 such that $v(q) = \bot$ is a counter model for the above formula, which completes the proof of P(1) for E_1.

Now, we show that E_1 has an ae-model. By Lemma 2.1 it is enough to prove that

(a) $L\alpha \in E_1$ implies $\alpha \in E_1$,

(b) $-L\alpha \in E_1$ implies $\alpha \notin E_1$.

Proof of (a): Let $L\alpha \in E_1$ and $\alpha \notin E_1$. Then $\alpha \notin E_0$. Clearly, $L\alpha \vee f \in E_1$ and by $P_1(1)$, $f \in E_0$, which implies that E_0 is inconsistent, a contradiction. Hence, we infer that $\alpha \in E_1$.

Proof of (b): Let $-L\alpha \in E_1$ and $\alpha \in E_1$. If $\alpha \in E_0$ then, because of $-L\alpha \vee f \in E_1$ and by $P_2(1)$ we infer that E_0 is inconsistent. So, $\alpha \notin E_0$ and, by the assumption, $\alpha \in E_1$. Thus

$$\vdash \bigwedge_i a_i \wedge \bigwedge_i Lb_i \wedge \bigwedge_i -Lc_i \rightarrow \alpha,$$

where for all $i, a_i, b_i \in E_0$ and $c_i \notin E_0$. Let v be an ae-model for E_0 such that $v(\alpha) = \bot$. Clearly v rejects the above formula that contradicts to the assumption that $\alpha \in E_1$.

Now, we conclude, that if T is a better set than A, than E_1 has an ae-model, and, therefore, E_1 is consistent.

Suppose that the sets $E_0, E_1, \ldots E_k$ have been constructed such that for all $i \leq k, E_i$ has the $P(i)$ property and each of them has an ae-model. Hence, for all $i \leq k$, E_i is consistent.

Put

$$E_{k+1} = Cn(E_k \cup LE_k \cup \{-L\beta : \beta \in \mathcal{L}_{L,k} - E_k\}).$$

By the assumption, E_k has an ae-model. Now, we prove that the set E_{k+1} has the P(k+1) property, that is,

$P_1(k+1)$ if $L\beta \vee q \in E_{k+1}$ then $\beta \in E_k$ or $q \in E_k$,

$P_2(k+1)$ if $-L\alpha \vee q \in E_{k+1}$, then $\alpha \notin E_k$ or $q \in E_k$.

Proof of $P_1(k+1)$: Let $L\beta \vee q \in E_{k+1}$, $\beta \notin E_k$ and $q \notin E_k$. Then, for some $a_i, b_i \in E_k, c_i \notin E_k$

$$\vdash \bigwedge_i a_i \wedge \bigwedge_i b_i \wedge \bigwedge_i -Lc_i \rightarrow L\beta \vee q.$$

If for some i_0 :

$-$ $a_{i_0} = L\beta \vee q$, then because of E_k has the $P(k)$ property, $\beta \in E_{k-1}$ or $q \in E_{k-1}$, a contradiction. Hence, for all $i, a_i \neq L\beta \vee q$.

$-$ $a_{i_0} = L\beta$, or $b_{i_0} = L\beta$, then E_k has no ae-model, a contradiction. Hence, for all $i, a_i \neq L\beta$ and $b_i \neq L\beta$.

Clearly for all $i, a_i \neq q$.

Observe now, that each ae-model v for E_k such that $v(q) = \bot$ contradicts to the assumption that the above formula is a tautology. Hence $P_1(k+1)$ holds for all E_{k+1}.

Proof of $P_2(k+1)$: Let $-L\alpha \vee q \in E_{k+1}$, $\alpha \in E_k$ and $q \notin E_k$. Then

$$\vdash \bigwedge_i a_i \wedge \bigwedge_i Lb_i \wedge \bigwedge_i -Lc_i \rightarrow -L\alpha \vee q,$$

where all $a_i, b_i \in E_k$ and $c_i \notin E_k$.

If for some i_0, $a_{i_0} = -L\alpha \vee q$, then since for E_k the $P(k)$ property holds we infer that, $\alpha \notin E_{k-1}$ or $q \in E_{k-1}$. Since $q \notin E_k$ and $-L\alpha \vee q \in E_k$ we conclude that $\alpha \notin E_{k-1}$ and $\alpha \in E_k$. By the definition of E_k, a formula of the form

$$\bigwedge_i a_i \wedge \bigwedge_i b_i \wedge \bigwedge_i -Lc_i \rightarrow \alpha,$$

where all $a_i, b_i \in E_{k-1}$, $c_i \notin E_{k-1}$, is a tautology. Notice, that for all $i, a_i \neq \alpha$ and $b_i \neq \alpha$. Take as an ae-model of E_{k-1} the valuation v such that $v(\alpha) = \bot$. Then v rejects the above formula and we conclude that $\alpha \notin E_k$, a contradiction. Hence, for all $i, a_i \neq -L\alpha \vee q$.

Clearly, for all i, $a_i \neq -L\alpha$. Otherwise, E_k has no ae-model. Also every $c_i \neq \alpha$. Hence, each ae-model v of E_k such that $v(q) = \bot$ contradicts to the assumption that $-L\alpha \vee q \in E_{k+1}$. Thus E_{k+1} has the $P(k+1)$ property.

Similarly as for E_1, we can show now that

$$\text{if } L\alpha \in E_{k+1} \text{ then } \alpha \in E_{k+1},$$

$$\text{if } -L\alpha \in E_{k+1} \text{ then } \alpha \notin E_{k+1}.$$

Thus, we conclude that if E_k has an ae-model, then the set E_{k+1} also does. Hence, we have proved that if A is a good set, then for the set $T = Cn(A \cup X)$ being a better set than A there is a family $\mathcal{E}$ of consistent sets E_n, that was to be shown.

Finally, we show that the converse of Theorem 4.1 is also true. Namely,

Theorem 4.3. *If A has an ae-expansion then A is a good set.*
Proof: Suppose that A has an ae-expansion T, that is,

$$T = Cn(A \cup LT \cup -L\overline{T}).$$

We show that the set $Cn(A \cup X)$ is a better set than A, where $X = \{p \in \mathcal{L} : \alpha \rightarrow p \in SUBF(A) \cap T \text{ and } \alpha \in T\}$.

To prove that $Cn(A \cup X)$ has an ae-model we claim the conditions (1) and (2) of Lemma 2.1 to be satisfied by $Cn(A \cup X)$.

To this end, let us observe that $Cn(A \cup X) \subseteq T$. Now, let $-L\alpha \in Cn(A \cup X)$ and $\alpha \in Cn(A \cup X)$. Since T is a stable set, we conclude that $-L\alpha \in T$ and $L\alpha \in T$, a contradiction. Suppose now that $L\alpha \in Cn(A \cup X)$. Then

$$\vdash \bigwedge_i a_i \wedge \bigwedge_i p_i \rightarrow L\alpha,$$

where for all i, $a_i \in A$, $p_i \in X$. Let $\alpha \notin Cn(A \cup X)$. If $L\alpha \notin A$ then the valuation v such that $v(A) = v(X) = \top$ and $v(L\alpha) = \bot$ rejects the above formula. Hence, $L\alpha \in A$ and $\alpha \notin Cn(A)$. In virtue of Proposition 3.2, A has no ae-expansion, a contradiction which allows to infer that the set $Cn(A \cup X)$ has an ae-model.

Now, take as a stable extension of $Cn(A \cup X)$ the set T being an ae-expansion of A and let $L\beta \vee q \in SUBF(A) \cap T$ and $A \nvdash \beta$. If $\beta \in T$ then

$$\vdash \bigwedge_i a_i \wedge \bigwedge_i Lb_i \wedge \bigwedge_i -Lc_i \rightarrow \beta,$$

where all $a_i \in A$, $b_i \in T$ and $c_i \notin T$. Clearly, for all i, $a_i \neq \beta$. By the induction on the complexity of β we can show that for every i, $Lb_i \neq \beta$. Let v be a valuation such that for all i, $v(\beta) = v(Lc_i) = \bot$ and $v(A) = v(Lb_i) = \top$. It is obvious that v rejects the above formula. Hence, $\beta \notin T$ and $-L\beta \in T$, which, by Proposition 4.1 (1) proves, that $q \in X$. Observe that the condition (3) of Definition 4.1 is trivial. Hence, we have completed the proof that the set $Cn(A \cup X)$ is a better set than A. Thus A is a good set, that was to be shown.

Corollary 4.1. *For any set A, the following are equivalent:*
(1) *A is a good set.*
(2) *A has an ae-expansion.*

Acknowledgement

The author wishes to thank to the referees for their valuable remarks.

References

[C] B.F Chellas, *Modal Logic:An introduction*, Cambridge University Press, 1980.

[K] K.Konolige, *On the Relation between Default Theories and Autoepistemic Logic*, Artificial Intelligence

[M] W.Marek, *Stable Theories in Autoepistemic Logic* Fundamenta Informaticae, (1989).

[Mo] R.Moore, *Semantical Considerations on Non-monotonic Logic* Artificial Intelligence 25(1), 75-94, (1985).

[R] C.M.Rauszer, *Stable Autoepistemic Expansions* The Proceedings of the Third International Symposium on Methodologies for Intelligent Systems, Ras Z. and Saitta L. eds. North Holland, pp. 476-484 (1988), Torino, Italy,

[S] R.Stalnaker, *A Note on Non-momotonic modal Logic* unpublished manuscript, Department of Philosophy, Cornell University.

On the existence of fixpoints in Moore's autoepistemic logic and the non-monotonic logic of McDermott and Doyle

Robert F. Stärk
Institut für Informatik und angewandte Mathematik, Universität Bern
Längassstrasse 51, CH-3012 Bern, <staerk@iam.unibe.ch>

Abstract

In [6] Moore has introduced a logic to represent the beliefs of ideal rational agents, called autoepistemic logic. This logic was presented as an improvement of the non-monotonic logic of McDermott and Doyle in [4]. We give a new method to characterize the fixpoints in both logics and thus obtain decision procedures for several problems in this context. Although the two logics are conceptually very different our method is very uniform.

1 Introduction

Non-monotonic systems often use non-constructive fixpoint definitions which do not directly provide procedures for computing the derivable formulas. In the case of the autoepistemic logic of Moore in [6] it is not even clear what it means that a formula is derivable from a set of premises. Only some conditions the set of "derivable" formulas should have are stated.

He wants to model the beliefs of an ideal rational agent. The beliefs are represented in a mathematical framework, the language of propositional modal logic. The intuitive interpretation of the modal operator $\mathsf{B}\,\varphi$ or $\mathsf{L}\,\varphi$ is 'φ is believed'. However, there are authors (Marek and Truszczyński [3]) who formulate the same logic with the modal operator K and $\mathsf{K}\,\varphi$ means 'φ is known'. Therefore we will use the neutral operator $\Box$ and write $\Box\varphi$ for $\mathsf{L}\,\varphi$ or $\mathsf{K}\,\varphi$. We are not interested in the meaning of $\Box\varphi$ but in the connection between $\Box\varphi$ and φ.

Moore's autoepistemic logic and the non-monotonic logic of McDermott and Doyle are based on fixpoint equations of the form

$$T = \{\varphi \mid A \cup \Phi(T) \vdash \varphi\},$$

where A is a set of axioms and Φ is an operator defined on the class of theories. We will present a new method which allows to construct solutions of this equation, called

extensions, for a given operator Φ and a set of axioms A. The idea is, that to decide, if one should take a formula φ to T, one solves this problem first for all formulas ψ such that $\Box\psi$ is a subexpression of φ.
We present this method for Moore's autoepistemic logic and the non-monotonic logic of McDermott and Doyle. In both cases we obtain procedures which allow to compute the extensions of a given theory and to decide whether a formula belongs to some of them. Similar results were obtained at the same time by Marek, Shvarts and Truszczyński in [2].

2 Syntax and basic notions

We start with the syntax of propositional autoepistemic logic. The language of AE logic will also be used in the other logical systems which we will investigate. As basic symbols we have an infinite list of propositional variables, $p_0, p_1, \ldots$, indicated by the meta variables p, q, r, the logical symbols $\neg, \wedge, \vee, \rightarrow$ and the modal operator $\Box$. Greek letters φ, ψ, χ range over *formulas* and Latin letters A, S, T range over sets of formulas called *AE theories* or *modal theories* or simply *theories*.
In AE logic formulas are just one of the following forms:

$$p,\ \neg\varphi,\ \varphi \wedge \psi,\ \varphi \vee \psi,\ \varphi \rightarrow \psi,\ \Box\varphi.$$

Literals are formulas of the form $p, \neg p, \Box\varphi$ or $\neg\Box\varphi$. An AE formula is called *objective* if it does not contain the $\Box$ operator. It is clear that every formula is a propositional combination of propositional variables and formulas of the form $\Box\varphi$. We define $\mathrm{sub}(\varphi)$ as the set of all *subexpressions* of φ and $\mathrm{at}(\varphi)$ as the set of all *propositional atoms* of φ. Formally:

$$\begin{array}{lll@{\qquad}lll}
\mathrm{sub}(p) & := & \{p\}, & \mathrm{at}(p) & := & \{p\}, \\
\mathrm{sub}(\neg\varphi) & := & \{\neg\varphi\} \cup \mathrm{sub}(\varphi), & \mathrm{at}(\neg\varphi) & := & \mathrm{at}(\varphi), \\
\mathrm{sub}(\varphi * \psi) & := & \{\varphi * \psi\} \cup \mathrm{sub}(\varphi) \cup \mathrm{sub}(\psi), & \mathrm{at}(\varphi * \psi) & := & \mathrm{at}(\varphi) \cup \mathrm{at}(\psi), \\
\mathrm{sub}(\Box\varphi) & := & \{\Box\varphi\} \cup \mathrm{sub}(\varphi), & \mathrm{at}(\Box\varphi) & := & \{\Box\varphi\},
\end{array}$$

where $*$ is one of $\wedge, \vee$ or $\rightarrow$. We shall also apply the operators 'sub' and 'at' to theories T. This means:

$$\mathrm{sub}(T) := \bigcup_{\varphi \in T} \mathrm{sub}(\varphi), \qquad \mathrm{at}(T) := \bigcup_{\varphi \in T} \mathrm{at}(\varphi).$$

Thus $\mathrm{sub}(T)$ is the set of all subexpressions of formulas of T and $\mathrm{at}(T)$ is the set of all propositional atoms of formulas of T and obviously $\mathrm{at}(T) \subset \mathrm{sub}(T)$.
The notion $T \vdash \varphi$ means that φ is derivable from the theory T in classical propositional logic, where the formulas $\Box\psi$ are treated as ordinary propositional variables.
An *AE interpretation* is a function I which assigns a truth value **true** or **false** to every propositional variable and every formula of the form $\Box\varphi$. The function I can then be extended to arbitrary formulas in the usual way. We write $I \models \varphi$ for $I(\varphi) = \mathbf{true}$. It is clear that the truth value of a formula φ under I depends only on values that I takes on $\mathrm{at}(\varphi)$. The operator $\Box$ has no special meaning in AE interpretations but it will play an important role in *AE extensions*, which we will introduce in the next section.

3 Autoepistemic extensions

Following Moore ([6, 5]) we define AE extensions by a fixpoint equation. Let A be a set of axioms of an ideal rational agent.

Definition 1 *A theory T is an* AE extension *of a set A iff*

$$T = \left\{ \varphi \,\middle|\, A \cup \{\Box\psi \mid \psi \in T\} \cup \{\neg\Box\psi \mid \psi \notin T\} \vdash \varphi \right\}.$$

The idea is that an AE extension of a set A is a candidate for the belief set of an ideal introspective agent with premises A. In the original paper of Moore AE extensions were called *stable expansions.*
One of the reasons that AE extensions are good candidates for the belief set of an ideal introspective agent is the following. If T is an AE extension of some set A then T satisfies the conditions:

(S1) $T \vdash \varphi \implies \varphi \in T$,

(S2) $\varphi \in T \implies \Box\varphi \in T$,

(S3) $\varphi \notin T \implies \neg\Box\varphi \in T$.

Theories which satisfy these conditions are called *stable theories* and (S1)–(S3) are the *stability conditions.* It is easy to see that any stable theory T contains the following modal axiom schemata:

K. $\Box(\varphi \to \psi) \to (\Box\varphi \to \Box\psi)$,

T. $\Box\varphi \to \varphi$,

4. $\Box\varphi \to \Box\Box\varphi$,

5. $\neg\Box\varphi \to \Box\neg\Box\varphi$.

Further T contains all tautologies and is closed under the necessitation rule and modus ponens.

$$\frac{\varphi}{\Box\varphi} \qquad\qquad \frac{\varphi \to \psi \qquad \varphi}{\psi}$$

Therefore any theorem of the modal logic $S5$ is in T and further $S5$ is contained in the intersection of all stable theories. Conversely, it was shown in Moore [5] that the intersection of all stable theories is exactly $S5$.

We shall not go further into the theory of stable sets and modal logic, but we are interested in the following questions:

1. Does a given theory A have an AE extension?
2. How many AE extensions are there for a given theory A? Are they consistent? How can we compute them?
3. Is it decidable whether a given formula φ belongs to one or all AE extensions of a finite theory A?

These problems were solved by Moore [5], Marek and Truszczyński [3] and Niemelä [7], but we present here a new method which allows to characterize the AE extensions of a theory A by means of admissible labelings. In Section 6 we will apply this method to the non-monotonic logic of McDermott and Doyle and we will show how it can be generalized to abstract logics. Since the general principle is not very intuitive we will apply it first to the special case of AE logic.

4 Autoepistemic labelings

The idea is that an AE labeling represents a possible extension of a set A restricted to the class of all boxed subexpressions of A. Admissible labelings are functions which satisfy in addition some natural conditions. We shall prove that the admissible labelings of a theory correspond exactly to the AE extensions of the theory.

Definition 2 *An* AE labeling *of a theory A is a function*

$$\lambda\colon \{\varphi \mid \Box\varphi \in \mathrm{sub}(A)\} \to \{0,1\}.$$

It is called admissible for A *if for all formulas φ with $\Box\varphi \in \mathrm{sub}(A)$: $\lambda(\varphi) = 1$ iff*

$$A \cup \{\Box\psi \mid \Box\psi \in \mathrm{sub}(A), \lambda(\psi) = 1\} \cup \{\neg\Box\psi \mid \Box\psi \in \mathrm{sub}(A), \lambda(\psi) = 0\} \vdash \varphi.$$

Later we will prove that for an admissible labeling λ of A there is exactly one AE extension T of A such that a formula φ with $\Box\varphi \in \mathrm{sub}(A)$ belongs to T iff $\lambda(\varphi) = 1$.
In the definition of AE labeling we do not require that the theory A is finite but in practice only AE labelings of finite theories are interesting. Assume now that A is a finite theory. Then $\mathrm{sub}(A)$ is finite and there are 2^n possible labelings of A, if n is the number of different boxed formulas in $\mathrm{sub}(A)$. If λ is a labeling of A then the set

$$A \cup \{\Box\psi \mid \Box\psi \in \mathrm{sub}(A), \lambda(\psi) = 1\} \cup \{\neg\Box\psi \mid \Box\psi \in \mathrm{sub}(A), \lambda(\psi) = 0\}$$

is finite and it is decidable if λ is admissible.
To test whether λ is admissible one has to solve the following problems. If $\Box\varphi$ is a subexpression of A and $\lambda(\varphi) = 1$ then one has to test for provability, i. e. one has to solve a problem of **TAUT**. If $\Box\varphi$ is a subexpression of A and $\lambda(\varphi) = 0$ then one has to test for non-provability, i. e. one has to solve a problem of **SAT**.
If we want to show that a given formula φ belongs to an AE extension of a theory A then the following elementary locality principle plays a central role. The lemma says that to solve this problem we only have to know the local part of the extension which is determined by the propositional atoms of A and φ. The lemma will be used in the next sections.

Lemma 3 *(Locality principle) Let A be a theory and φ be a formula and let S be a set of non complementary literals. Then the following statements are equivalent:*

a) $A \cup S \vdash \varphi$.

b) $A \cup \{\psi \in S \mid \mathrm{at}(\psi) \subset \mathrm{at}(A) \cup \mathrm{at}(\varphi)\} \vdash \varphi$.

Proof The direction from (b) to (a) is trivial. Now we assume that $A \cup S \vdash \varphi$ and that I is an AE interpretation such that

$$I \models A \cup \{\psi \in S \mid \mathrm{at}(\psi) \subset \mathrm{at}(A) \cup \mathrm{at}(\varphi)\}.$$

Let J be the new AE interpretation with $J(\chi) := I(\chi)$ for propositional atoms $\chi \in \mathrm{at}(A) \cup \mathrm{at}(\varphi)$ and $J(\chi) := \mathbf{true}$, if $\chi \in S$, and $J(\chi) := \mathbf{false}$, if $\neg\chi \in S$, for propositional atoms $\chi \in \mathrm{at}(S) \setminus (\mathrm{at}(A) \cup \mathrm{at}(\varphi))$. Then for all formulas $\psi \in A$ the value of $J(\psi)$ is equal to $I(\psi)$. Therefore $J \models A \cup S$ and this implies $J \models \varphi$. Since $I(\varphi) = J(\varphi)$ we have $I \models \varphi$. I was arbitrary, thus

$$A \cup \{\psi \in S \mid \mathrm{at}(\psi) \subset \mathrm{at}(A) \cup \mathrm{at}(\varphi)\} \vdash \varphi$$

and the direction from (a) to (b) is proved. □

In the case of AE extensions the locality principle says that the following statements are equivalent:

$$A \cup \{\Box\psi \mid \psi \in T\} \cup \{\neg\Box\psi \mid \psi \notin T\} \vdash \varphi \qquad \text{iff}$$
$$A \cup \{\Box\psi \mid \Box\psi \in \mathrm{at}(A) \cup \mathrm{at}(\varphi), \psi \in T\} \cup \{\neg\Box\psi \mid \Box\psi \in \mathrm{at}(A) \cup \mathrm{at}(\varphi), \psi \notin T\} \vdash \varphi.$$

An immediate consequence of the locality principle is the following corollary about the consistency of AE extensions. For an agent it is important to know whether an AE extension of his axioms is consistent or not.

Corollary 4 *Let T be an AE extension of a set A. Then T is consistent iff the set $A \cup \{\Box\psi \mid \Box\psi \in \mathrm{at}(A), \psi \in T\} \cup \{\neg\Box\psi \mid \Box\psi \in \mathrm{at}(A), \psi \notin T\}$ is consistent.*

Proof Let T be an AE extension of A. Then $T \vdash \bot$ iff

$$A \cup \{\Box\psi \mid \psi \in T\} \cup \{\Box\psi \mid \neg\psi \notin T\} \vdash \bot \qquad \text{iff}$$
$$A \cup \{\Box\psi \mid \Box\psi \in \mathrm{at}(A), \psi \in T\} \cup \{\neg\Box\psi \mid \Box\psi \in \mathrm{at}(A), \psi \notin T\} \vdash \bot.$$

The corollary follows from the fact that a theory T is consistent iff $T \nvdash \bot$. □

5 Autoepistemic extensions and admissible labelings

We want to prove that the admissible labelings of a theory A correspond exactly to the AE extensions of the theory and we can identify the infinite AE extensions with admissible labelings which are finite objects. The next theorem states that an AE extension induces an admissible labeling of the theory.

Theorem 5 *Let T be an AE extension of the theory A and let*

$$\lambda\colon \{\varphi \mid \Box\varphi \in \mathrm{sub}(A)\} \to \{0,1\}$$

be the AE labeling defined by $\lambda(\varphi) = 1$ iff $\varphi \in T$. Then λ is admissible for A.

Proof Let T be an AE extension of A and let λ be as required above. By definition of AE extension

$$T = \left\{ \varphi \,\middle|\, A \cup \{\Box\psi \mid \psi \in T\} \cup \{\neg\Box\psi \mid \psi \notin T\} \vdash \varphi \right\}.$$

We have to show that λ is admissible. Let $\Box\varphi \in \mathrm{sub}(A)$ and $\lambda(\varphi) = 1$. By definition of λ we have $\varphi \in T$ and therefore $A \cup \{\Box\psi \mid \psi \in T\} \cup \{\neg\Box\psi \mid \psi \notin T\} \vdash \varphi$. By the locality principle it follows that

$$A \cup \{\Box\psi \mid \Box\psi \in \mathrm{at}(A) \cup \mathrm{at}(\varphi), \psi \in T\} \cup \{\neg\Box\psi \mid \Box\psi \in \mathrm{at}(A) \cup \mathrm{at}(\varphi), \psi \notin T\} \vdash \varphi$$

and, since $at(A) \cup \mathrm{at}(\varphi) \subset \mathrm{sub}(A)$,

$$A \cup \{\Box\psi \mid \Box\psi \in \mathrm{sub}(A), \lambda(\psi) = 1\} \cup \{\neg\Box\psi \mid \Box\psi \in \mathrm{sub}(A), \lambda(\psi) = 0\} \vdash \varphi.$$

The first condition of admissibility is satisfied. Now let $\Box\varphi \in \mathrm{sub}(A)$ and $\lambda(\varphi) = 0$. By definition of λ we have $\varphi \notin T$ and $A \cup \{\Box\psi \mid \psi \in T\} \cup \{\neg\Box\psi \mid \psi \notin T\} \nvdash \varphi$ and by the monotonicity of the propositional calculus

$$A \cup \{\Box\psi \mid \Box\psi \in \mathrm{sub}(A), \lambda(\psi) = 1\} \cup \{\neg\Box\psi \mid \Box\psi \in \mathrm{sub}(A), \lambda(\psi) = 0\} \nvdash \varphi.$$

Therefore λ is admissible. $\Box$

The next theorem says that every admissible labeling of a modal theory A determines a unique AE extension. So we obtain an upper bound for the number of AE extensions of a finite theory. It is bounded by the number of admissible labelings of the theory.

Theorem 6 *Let λ be an admissible labeling of a theory A. Then there is exactly one AE extension T of A such that for all formulas $\Box\varphi \in \mathrm{sub}(A)$: $\varphi \in T$ iff $\lambda(\varphi) = 1$.*

Proof Let λ be an admissible labeling of A.
(i) Existence: We try to extend the labeling λ to all formulas. The labeling λ is only defined on formulas φ with $\Box\varphi \in \mathrm{sub}(A)$. On the remaining formulas we define λ by recursion on the length of the formula like the operators 'sub' and 'at' at the beginning of this paper. Let φ be a formula with $\Box\varphi \notin \mathrm{sub}(A)$. Then

$$\lambda(\varphi) := \begin{cases} 1, & \text{if } A \cup \{\Box\psi \mid \Box\psi \in \mathrm{sub}(A) \cup \mathrm{at}(\varphi), \lambda(\psi) = 1\} \cup \\ & \qquad \{\neg\Box\psi \mid \Box\psi \in \mathrm{sub}(A) \cup \mathrm{at}(\varphi), \lambda(\psi) = 0\} \vdash \varphi; \\ 0, & \text{otherwise.} \end{cases}$$

This definition makes sense since if $\Box\psi \in \mathrm{at}(\varphi)$ then the length of ψ is less than the length of φ. Finally we put $T := \{\varphi \mid \lambda(\varphi) = 1\}$ and we claim that T is the desired extension of A. Let $S := \{\Box\psi \mid \psi \in T\} \cup \{\neg\Box\psi \mid \psi \notin T\}$. In a first step we show that $\varphi \in T$ implies $A \cup S \vdash \varphi$. Let $\varphi \in T$ and $\Box\varphi \in \mathrm{sub}(A)$. Then $\lambda(\varphi) = 1$ and since λ is admissible

$$A \cup \{\Box\psi \mid \Box\psi \in \mathrm{sub}(A), \lambda(\psi) = 1\} \cup \{\neg\Box\psi \mid \Box\psi \in \mathrm{sub}(A), \lambda(\psi) = 0\} \vdash \varphi$$

and therefore $A \cup S \vdash \varphi$. If $\varphi \in T$ and $\Box\varphi \notin \mathrm{sub}(A)$ then $\lambda(\varphi) = 1$ and

$$A \cup \{\Box\psi \mid \Box\psi \in \mathrm{sub}(A) \cup \mathrm{at}(\varphi), \lambda(\psi) = 1\} \cup \\ \{\neg\Box\psi \mid \Box\psi \in \mathrm{sub}(A) \cup \mathrm{at}(\varphi), \lambda(\psi) = 0\} \vdash \varphi$$

and therefore $A \cup S \vdash \varphi$. Now we show that $\varphi \notin T$ implies $A \cup S \not\vdash \varphi$. Let $\varphi \notin T$ and $\Box\varphi \in \text{sub}(A)$. Then $\lambda(\varphi) = 0$ and since λ is admissible

$$A \cup \{\Box\psi \mid \Box\psi \in \text{sub}(A), \lambda(\psi) = 1\} \cup \{\neg\Box\psi \mid \Box\psi \in \text{sub}(A), \lambda(\psi) = 0\} \not\vdash \varphi.$$

Since $\text{at}(A) \cup \text{at}(\varphi) \subset \text{sub}(A)$ it follows by the locality principle that $A \cup S \not\vdash \varphi$. If $\varphi \notin T$ and $\Box\varphi \notin \text{sub}(A)$ then $\lambda(\varphi) = 0$ and

$$A \cup \{\Box\psi \mid \Box\psi \in \text{sub}(A) \cup \text{at}(\varphi), \lambda(\psi) = 1\} \cup \\ \{\neg\Box\psi \mid \Box\psi \in \text{sub}(A) \cup \text{at}(\varphi), \lambda(\psi) = 0\} \not\vdash \varphi$$

and again by the locality principle $A \cup S \not\vdash \varphi$.
(ii) Uniqueness: Let T_1 and T_2 be two AE extensions of A such that for all formulas $\Box\varphi \in \text{sub}(A)$: $\varphi \in T_1$ iff $\lambda(\varphi) = 1$ iff $\varphi \in T_2$. For formulas φ with $\Box\varphi \notin \text{sub}(A)$ we prove by induction on the length of the formula that $\varphi \in T_1$ iff $\varphi \in T_2$. Let $\Box\varphi \notin \text{sub}(A)$. We assume by induction hypothesis that for all ψ with $\Box\psi \in \text{at}(\varphi)$: $\psi \in T_1$ iff $\psi \in T_2$. Then

$$\begin{aligned}
\varphi \in T_1 &\iff A \cup \{\Box\psi \mid \psi \in T_1\} \cup \{\neg\Box\psi \mid \psi \notin T_1\} \vdash \varphi \\
&\iff A \cup \{\Box\psi \mid \Box\psi \in \text{sub}(A) \cup \text{at}(\varphi), \psi \in T_1\} \cup \\
&\qquad \{\neg\Box\psi \mid \Box\psi \in \text{sub}(A) \cup \text{at}(\varphi), \psi \notin T_1\} \vdash \varphi \\
&\iff A \cup \{\Box\psi \mid \Box\psi \in \text{sub}(A) \cup \text{at}(\varphi), \psi \in T_2\} \cup \\
&\qquad \{\neg\Box\psi \mid \Box\psi \in \text{sub}(A) \cup \text{at}(\varphi), \psi \notin T_2\} \vdash \varphi \\
&\iff A \cup \{\Box\psi \mid \psi \in T_2\} \cup \{\neg\Box\psi \mid \psi \notin T_2\} \vdash \varphi \\
&\iff \varphi \in T_2
\end{aligned}$$

In these equivalencies we have used the locality principle. □

The unique extension of A defined by λ is called $T_\lambda(A)$. The following theorem on objective theories was proved directly by Marek [3] and Konolige [1]. We obtain it as a corollary.

Corollary 7 *Every objective theory has exactly one AE extension.*

Proof An objective theory has exactly one AE labeling which is admissible: the empty labeling. □

The first part of the proof of Theorem 6 gives a method how one can construct this unique extension. We do not repeat this construction here but we mention another immediate consequence of Theorems 5 and 6.

Corollary 8 *If S and T are two AE extensions of a theory A which agree on all formulas φ with $\Box\varphi \in \text{sub}(A)$ then S and T are equal.*

Examples

a) Let $A := \{\Box p\}$. Then there are two possible labelings of A, one with $\lambda(p) = 1$ and the other with $\lambda(p) = 0$. The first is not admissible since $A \cup \{\Box p\} \not\vdash p$ and the second is not admissible since $A \cup \{\neg\Box p\} \vdash p$. We can conclude that A has no AE extension.
b) Let $A := \{\Box p \to p\}$. Again there are two possible labelings of A. In the first case $\lambda(p) = 1$ and $A \cup \{\Box p\} \vdash p$, in the second case $\lambda(p) = 0$ and $A \cup \{\neg\Box p\} \not\vdash p$. Therefore

both labelings are admissible and we can conclude that A has two AE extensions, one that does contain p and one that does not contain p. Both extensions are consistent.
c) The next example shows that AE logic is non-monotonic. Let $A := \{\neg\Box p \to q\}$ and $B := \{\neg\Box p \to q,\ p\}$. Using the above methods it is easy to see that both theories have exactly one extension. We call them T_A and T_B. We have $A \subset B$ and $q \in T_A$, but $q \notin T_B$.
d) Let $A := \{\neg\Box p \to q,\ \neg\Box q \to p\}$. Then there are four possible labelings of A.

1. $\lambda(p) = 1, \lambda(q) = 1$: This labeling is not admissible since $A \cup \{\Box p, \Box q\} \nvdash p$.
2. $\lambda(p) = 1, \lambda(q) = 0$: This labeling is admissible since $A \cup \{\Box p, \neg\Box q\} \vdash p$ and $A \cup \{\Box p, \neg\Box q\} \nvdash q$.
3. $\lambda(p) = 0, \lambda(q) = 1$: This labeling is admissible since $A \cup \{\neg\Box p, \Box q\} \nvdash p$ and $A \cup \{\neg\Box p, \Box q\} \vdash q$.
4. $\lambda(p) = 0, \lambda(q) = 0$: This labeling is not admissible since $A \cup \{\neg\Box p, \neg\Box q\} \vdash p$.

We can conclude that A has exactly two AE extensions given by the admissible labelings (2) and (3).

We finish this section with an algorithm which allows to decide if a formula belongs to an AE extension of a theory. We assume that a finite theory A and an admissible labeling λ of A are given and we want to answer the question if a formula φ belongs to the extension $T_\lambda(A)$ of A defined by λ.

algorithm $ae(A, \lambda, \varphi)$: boolean;
input:
  A: finite modal theory;
  λ: admissible labeling of A;
  φ: modal formula;
output:
  if $\varphi \in T_\lambda(A)$ then true else false;
begin
  if $A \cup \{\Box\psi \mid \Box\psi \in \mathrm{sub}(A),\ \lambda(\psi) = 1\} \cup$
    $\{\Box\psi \mid \Box\psi \in \mathrm{at}(\varphi) \setminus \mathrm{sub}(A),\ ae(A, \lambda, \psi) = \mathrm{true}\} \cup$
    $\{\neg\Box\psi \mid \Box\psi \in \mathrm{sub}(A),\ \lambda(\psi) = 0\} \cup$
    $\{\neg\Box\psi \mid \Box\psi \in \mathrm{at}(\varphi) \setminus \mathrm{sub}(A),\ ae(A, \lambda, \psi) = \mathrm{false}\} \vdash \varphi$
  then return true **else** false
end.

This algorithm is recursive, but it terminates always since the relation

$$\varphi \prec \psi :\Longleftrightarrow \Box\varphi \in \mathrm{at}(\psi)$$

is wellfounded. The correctness of the algorithm follows by the locality principle. It is clear that the algorithm is not optimally written since for a formula $\Box\psi \in \mathrm{at}(\varphi) \setminus \mathrm{sub}(A)$ the value $ae(A, \lambda, \psi)$ is computed two times.
So the questions at the end of Section 3 are all answered: Question (1) is equivalent to the question: 'Does a theory A have an admissible labeling?' Question (2) is equivalent to: 'How many admissible labelings has a theory A?' The consistency of an extension can be decided with help of Corollary 4. Question (3) can be answered with the algorithm above.

6 The non-monotonic logic of McDermott and Doyle

In this section we shall show that the methods developed in the previous sections are applicable to the propositional non-monotonic logic of McDermott and Doyle in [4]. Some of our results were already proved by Shvarts in [8] but the point of this section are not the theorems but the uniform transformation of notions like extension and labeling and the correspondence of extensions and admissible labelings.

McDermott and Doyle use a language with a modal operator M and their non-monotonic logic is concerned with fixpoints of the form

$$T = \left\{\varphi \,\middle|\, A \cup \{\mathsf{M}\,\psi \mid \neg\psi \notin T\} \vdash \varphi\right\}.$$

The intended meaning of $\mathsf{M}\,\psi$ is 'if $\neg\psi$ is not provable then one can infer $\mathsf{M}\,\psi$' or in other words $\mathsf{M}\,\psi$ means 'ψ is consistent with the assumptions'. In our formal language $\mathsf{M}\,\psi$ is defined as $\neg\Box\neg\psi$ and we have to change the fixpoint definition a little bit. We use NM as a shortcut for 'non-monotonic logic in the sense of McDermott and Doyle'.

Definition 9 *A theory T is an* NM extension *of a set A iff*

$$T = \left\{\varphi \,\middle|\, A \cup \{\neg\Box\psi \mid \psi \notin T\} \vdash \varphi\right\}.$$

The difference in the definition of AE and NM extensions is that in NM logic the positive part $\{\Box\psi \mid \psi \in T\}$ is omitted. The NM extensions of a set A are only closed under the two stability conditions (S1) and (S3) and not under (S2) and therefore they do not have the nice properties of AE extensions. The definition of an admissible labeling for NM logic is straightforward.

Definition 10 *An* NM labeling *of a theory A is a function*

$$\lambda\colon \{\varphi \mid \Box\varphi \in \mathrm{sub}(A)\} \to \{0,1\}.$$

It is called admissible for A *if for all formulas φ with $\Box\varphi \in \mathrm{sub}(A)$: $\lambda(\varphi) = 1$ iff $A \cup \{\neg\Box\psi \mid \Box\psi \in \mathrm{sub}(A), \lambda(\psi) = 0\} \vdash \varphi$.*

One can test if a NM extension is consistent in the same way as for AE extensions. For a NM extension T of a set A only the part of the extension which is concerned with formulas φ where $\Box\varphi$ is a propositional atom of A is relevant.

Corollary 11 *Let T be an NM extension of the set A. Then T is consistent iff the set $A \cup \{\neg\Box\psi \mid \Box\psi \in \mathrm{at}(A), \psi \notin T\}$ is consistent.*

The following two theorems are the NM versions of the Theorems 5 and 6. Their proofs are analogous to the proofs of the Theorems 5 and 6. Moreover they follow from a general principle which we will explain below.

Theorem 12 *Let T be an NM extension of the theory A and let*

$$\lambda\colon \{\varphi \mid \Box\varphi \in \mathrm{sub}(A)\} \to \{0,1\}$$

be the NM labeling defined by $\lambda(\varphi) = 1$ iff $\varphi \in T$. Then λ is admissible for A.

Theorem 13 *Let λ be an admissible NM labeling of a theory A. Then there is exactly one NM extension T of A such that for all formulas $\Box\varphi \in \mathrm{sub}(A)$: $\varphi \in T$ iff $\lambda(\varphi) = 1$.*

Corollary 14 *Every objective theory has exactly one NM extension.*

The notions of AE and NM extensions of a set A are defined by the following general scheme:

$$T = \{\varphi \mid A \cup \Phi(T) \vdash \varphi\},$$

where Φ is an operator which assigns to a theory T a new theory $\Phi(T)$. In the case of AE logic the operator is

$$\Phi_{ae}(T) = \{\Box\psi \mid \psi \in T\} \cup \{\neg\Box\psi \mid \psi \notin T\},$$

in the case of NM logic the operator is

$$\Phi_{nm}(T) = \{\neg\Box\psi \mid \psi \notin T\}.$$

We will now outline the fixpoint theory of an abstract operator Φ. We need the following notions:

$$S\lceil T := \{\varphi \in S \mid \mathrm{at}(\varphi) \subset T\},$$

$$\varphi \prec \psi :\Longleftrightarrow \Box\varphi \in \mathrm{at}(\psi), \qquad \varphi_{\prec} := \{\psi \mid \psi \prec \varphi\}, \qquad T_{\prec} := \bigcup_{\varphi \in T} \varphi_{\prec}.$$

Thus $S\lceil T$ is the restriction of S to formulas with propositional atoms from the set T and $\varphi \prec \psi$ means that $\Box\varphi$ is a propositional atom of ψ and therefore the truth value of φ in an extension must be computed before the truth value of ψ can be determined. Let now Φ be an operator with the following two properties[1]

$$A \cup \Phi(T) \vdash \varphi \quad \Longrightarrow \quad A \cup \Phi(T)\lceil(\mathrm{at}(A) \cup \mathrm{at}(\varphi)) \vdash \varphi, \tag{1}$$

$$\varphi \in \Phi(T) \quad \Longleftrightarrow \quad \varphi \in \Phi(T \cap \varphi_{\prec}). \tag{2}$$

It is easy to see that Φ_{ae} and Φ_{nm} satisfy these two conditions. More generally if $\Phi(T)$ is a set of non complementary literals then condition (1) is always satisfied. Condition (2) implies that $\Phi(T)\lceil S = \Phi(T \cap S_{\prec})$ for any theories S and T.
A theory T is called a *Φ-extension* of a set of axioms A iff it satisfies the fixpoint equation

$$T = \{\varphi \mid A \cup \Phi(T) \vdash \varphi\}.$$

A set $T_0 \subset \mathrm{sub}(A)_{\prec}$ is called *admissible* for A iff for all formulas $\varphi \in \mathrm{sub}(A)_{\prec}$:

$$\varphi \in T_0 \Longleftrightarrow A \cup \Phi(T_0)\lceil \mathrm{sub}(A) \vdash \varphi.$$

It can now be shown that the Φ-extensions of a modal theory A correspond exactly to the admissible sets for A. The proofs are more or less the same as in the special case of AE logic. One direction is easy. If T is a Φ-extension of A then $T \cap \mathrm{sub}(A)_{\prec}$ is admissible for A, and if T_1 and T_2 are two Φ-extensions of A which agree on $\mathrm{sub}(A)_{\prec}$ then they are equal. The difficult part is the construction of a Φ-extension from an admissible set which

---

[1] $A \cup T\lceil S = A \cup (T\lceil S)$

we will outline now. Let T_0 be an admissible set for A. One defines by recursion on the length of a formula a function f from the set of formulas into $\{0,1\}$. For $\varphi \in \mathrm{sub}(A)_{\prec}$ let

$$f(\varphi) := \begin{cases} 1, & \text{if } \varphi \in T_0; \\ 0, & \text{if } \varphi \notin T_0, \end{cases}$$

for $\varphi \notin \mathrm{sub}(A)_{\prec}$ let

$$f(\varphi) := \begin{cases} 1, & \text{if } A \cup \Phi(T_0 \cup \{\psi \mid \psi \prec \varphi, f(\psi) = 1\}) \vdash \varphi; \\ 0, & \text{otherwise.} \end{cases}$$

Finally one puts $T := \{\psi \mid f(\psi) = 1\}$ and one can prove that $T = \{\varphi \mid A \cup \Phi(T) \vdash \varphi\}$ and therefore T is a Φ-extension of A with $T \cap \mathrm{sub}(A)_{\prec} = T_0$.

7 Conclusion

In this paper we have seen that there are uniform decision procedures for Moore's autoepistemic logic and the non-monotonic logic of McDermott and Doyle. But it seems that AE logic is much more appropriate for applications in artificial intelligence. One reason that AE logic is very popular is that it has a Kripke semantics in the following sense: A PL interpretation is a function I which assigns to every propositional variable p_i a truth value **true** or **false**. A Kripke structure is a set $\mathcal{K}$ of PL interpretations. If $\mathcal{K}$ is a Kripke structure and I is a PL interpretation (not necessarily in $\mathcal{K}$) then the relation $\mathcal{K}, I \models \varphi$ is defined by:

$$\begin{array}{lll} \mathcal{K}, I \models p_i & :\Longleftrightarrow & I(p_i) = \textbf{true}, \\ \mathcal{K}, I \models \neg\varphi & :\Longleftrightarrow & \mathcal{K}, I \not\models \varphi, \\ \mathcal{K}, I \models \varphi \wedge \psi & :\Longleftrightarrow & \mathcal{K}, I \models \varphi \text{ and } \mathcal{K}, I \models \psi, \\ \mathcal{K}, I \models \Box\varphi & :\Longleftrightarrow & \text{for all } J \in \mathcal{K}\text{: } \mathcal{K}, J \models \varphi. \end{array}$$

This is the usual truth definition for $K45$ models. Further $\mathcal{K}, I \models T$ if for all $\varphi \in T$: $\mathcal{K}, I \models \varphi$, and $\mathcal{K} \models T$ if for all $I \in \mathcal{K}$: $\mathcal{K}, I \models T$. The theory $\mathrm{Th}(\mathcal{K})$ of a structure $\mathcal{K}$ is the set of all formulas φ such that for all $I \in \mathcal{K}$: $\mathcal{K}, I \models \varphi$. Let A be a set of axioms. $\mathcal{K}$ is called an *AE Kripke model* of A if $\mathcal{K} \models A$ and for every PL interpretation I: if $\mathcal{K}, I \models A$ then $I \in \mathcal{K}$.

Now Moore has shown in [5] that a theory T is an AE extension of some set A iff there exists an AE Kripke model $\mathcal{K}$ of A such that $T = \mathrm{Th}(\mathcal{K})$.

In general if Φ is an operator satisfying (1) and (2) of Section 6 then the associated Φ-logic has not a Kripke semantics as AE logic. Consider for example the following resource bounded operator Φ_n. We denote the length of a formula φ by $|\varphi|$. Let $n \in \mathbb{N}$.

$$\Phi_n(T) := \{\Box\varphi \mid \Box \in T, |\varphi| < n\} \cup \{\neg\Box\varphi \mid \varphi \notin T, |\varphi| < n\}.$$

A Φ_n-extension of some set A can then be interpreted as the set of beliefs of an agent whose resources are bounded by n. He believes a sentence φ only if he can prove it from the set A of his axioms using classical propositional logic and positive and negative introspection for sentences of length less than n. Of course, our fixpoint theory is applicable to Φ_n-extensions too but the Φ_n-extensions cannot be characterized as easily as the AE extensions by means of Kripke models.

References

[1] K. Konolige. On the relation between default and autoepistemic logic. *Artificial Intelligence*, 35:343–382, 1988.

[2] W. Marek, G. F. Shvarts, and M. Truszczyński. Modal nonmonotonic logics: ranges, characterization, computation. Preprint, 1990.

[3] W. Marek and M. Truszczyński. Autoepistemic logic. Technical Report 115-88, Department of Computer Science, University of Kentucky, 1988.

[4] D. McDermott and J. Doyle. Non-monotonic logic I. *Artificial Intelligence*, 13:41–72, 1980.

[5] R. C. Moore. Possible-world semantics for autoepistemic logic. In *Proceedings 1984 Non-monotonic Reasoning Workshop*, pages 344–354, New Paltz, NY, 1984.

[6] R. C. Moore. Semantical considerations on nonmonotonic logic. *Artificial Intelligence*, 25:75–94, 1985.

[7] I. Niemelä. Decision procedure for autoepistemic logic. In E. Lusk and R. Overbeek, editors, *9th International Conference on Automated Deduction*, pages 675–684. Springer, 1988. Lecture Notes in Computer Science 310.

[8] G. F. Shvarts. Fixed points in the propositional nonmonotonic logic. *Artificial Intelligence*, 38:199–206, 1989.

On the Tracking of Loops in Automated Deductions *

M. E. Szabo
Department of Mathematics
Concordia University
Montreal, Canada, H3G 1M8

1 Introduction

The *Graphical Evaluation and Review Technique (GERT)* is a general method for attaching numerical complexity measures to *discrete processes*. The method is based on the idea that a discrete process consists of different possible steps and that the transition from one step to another involves both a probability that the transition will take place and a cost of carrying out the transition. The probabilities involved are assumed to be stochastic and the cost functions of different steps are considered to be stochastically independent. The processes are allowed to involve the *looping* between arbitrary steps. The principal tool of investigation of *GERT* is Mason's rule for closed stochastic networks, which provides the desire numerical measure for a given process.

The purpose of this paper is to begin the study the complexity of the *proof process* with the techniques provided by *GERT*. In this initial investigation, we demonstrate the usefulness of this technique for the analysis of the complexity of a subsystem of Gentzen's cut-free sequent calculus for propositional intuitionistic logic. We assume that the system has been automated and we therefore allow the proof process to involve *false starts* and the required *deletion* of subtrees that fail to lead to axioms of the system. The probabilities and cost functions used in this study have only a formal meaning. As the results show, they can be chosen to describe precisely the complexity of the path structure of the trees underlying the formal proofs of the system. At the end of the paper,

*This research was partially supported by Grant A-8224 of the Natural Sciences and Engineering Research Council of Canada. The results in this paper were first presented at the Computer Science Logic Conference, held in October 1990 in Heidelberg, West Germany.

we discuss possible generalizations of the probabilities and cost functions and the usefulness of such generalizations for the study of the complexity of proofs.

2 The deductive system $\Delta(V)$

As our logic we take the subsystem of Gentzen's cut-free system **LJ** that is needed to characterize symmetric monoidal closed categories (without multiplicative unit) (cf. [3]). The choice of this paradigm is motivated by our attempt to minimize the extent to which the description of the technique is obscured by the intricacies of the syntactic considerations required for a richer logic. In particular, we avoid having to account for the multiple interactions of thinnings, contractions, and operational rules. The program of systematic extensions of this logic to other systems according to the algebraic objectives pursued in [3] provides a useful guide to the methodical study of richer systems. It should be noted that in the algebraic context, the multiplicative unit is crucial. In the deductive context, it would merely play the rule of a *truth* constant and can therefore be safely disregarded for our present purposes. Thus the underlying language of $\Delta(V)$ is based on a countable set $V = \{X_1, \ldots, X_n, \ldots\}$ of variables and the propositional connectives $\wedge$ and $\Rightarrow$.

2.1 The axioms and rules of inference

As in **LJ**, the axioms and rules of $\Delta(V)$ are written in *sequent form*, i.e., in the form $\Gamma \longrightarrow \phi$, where Γ is a finite sequence of formulas, and ϕ is an individual formula.

2.1.1 The axioms

The axioms of $\Delta(V)$ are all sequents of the form $X \longrightarrow X$, where $X \in V$.

2.1.2 The rules of inference

The rules of inference of $\Delta(V)$ are:

$$\frac{\Gamma A B \Delta \longrightarrow C}{\Gamma B A \Delta \longrightarrow C}\ (L_\pi) \qquad \frac{\Gamma \longrightarrow A \quad \Delta \longrightarrow B}{\Gamma\Delta \longrightarrow A \wedge B}\ (R_\wedge)$$

$$\frac{\Gamma A B \Delta \longrightarrow C}{\Gamma A \wedge B \Delta \longrightarrow C}\ (L_\wedge) \qquad \frac{\Gamma A \Delta \longrightarrow B}{\Gamma\Delta \longrightarrow A \Rightarrow B}\ (R_\Rightarrow)$$

$$\frac{\Gamma \longrightarrow A \quad \Delta B \Theta \longrightarrow C}{\Delta \Gamma A \Rightarrow B \Theta \longrightarrow C}\ (L_\Rightarrow)$$

2.1.3 Remark

In order to simplify the exposition, we will collapse consecutive applications of L_π into a single operation requiring a single step, and think of $L_\pi(\Gamma)$ as a sequence of formulas obtained from Γ by the permutation of some of the formulas in Γ. We restrict L_π to permutations of distinct formulas.

2.2 Proofs and deductions

In this paper, we write proofs both as top-down and as bottom-up trees, depending on whether we are thinking of the proof process as an automated proof searching process (top-down) or as an automated proof generating process (bottom-up). The notations are of course mirror images of each other.

Definition 2.1 *A proof $\Pi \in \Delta(V)$ of a sequent $\Gamma \longrightarrow \phi$ is a Genzten tree whose root is $\Gamma \longrightarrow \phi$, whose leaves are axioms, and whose intervening sequents are in Π by virtue of the rules of $\Delta(V)$.*

2.2.1 Top-down proofs

$$\begin{array}{c}
A \Rightarrow B, C \longrightarrow A \Rightarrow (B \wedge C) \\ \hline
A, A \Rightarrow B, C \longrightarrow (B \wedge C) \\ \hline
\begin{array}[t]{c} A, A \Rightarrow B \longrightarrow B \\ \hline A \longrightarrow A \quad B \longrightarrow B \end{array} \qquad C \longrightarrow C
\end{array}$$

Figure 1: A top-down proof of $A \Rightarrow B, C \longrightarrow A \Rightarrow (B \wedge C)$.

and

$$\begin{array}{c}
A \Rightarrow B, C \longrightarrow A \Rightarrow (B \wedge C) \\ \hline
A, A \Rightarrow B, C \longrightarrow (B \wedge C) \\ \hline
A \longrightarrow A \qquad \begin{array}[t]{c} B, C \longrightarrow B \wedge C \\ \hline B \longrightarrow B \quad C \longrightarrow C \end{array}
\end{array}$$

Figure 2: Another top-down proof of $A \Rightarrow B, C \longrightarrow A \Rightarrow (B \wedge C)$.

are examples of two top-down proofs of the same sequent, with $A, B, C \in V$.

Definition 2.2 *A deduction $\Phi \in \Delta(V)$ is a Genzten tree with false starts which becomes a proof when its false starts are deleted.*

Definition 2.3 *A false start is a Gentzen tree whose root is a provable sequent and which contains at least one leaf which is neither an axiom nor the premiss of an instance of a rule of inference of $\Delta(V)$.*

The erroneous steps of false starts in deductions will be enclosed in frames. Immediately above a frame sits a derivable sequent of $\Delta(V)$. Immediately below the frame are either other false starts or, ultimately, the premiss(es) of a rule of inference. We *delete* a false start by erasing the framed part from the deduction and replacing it by an *inference line*. Thus

$$
\cfrac{
  \cfrac{
    \cfrac{
      \boxed{\cfrac{\longrightarrow A \qquad ABC \longrightarrow B \wedge C}{AB \longrightarrow B \quad C \longrightarrow C}}
    }{A \longrightarrow A \qquad BC \longrightarrow B \wedge C}
  }{B \longrightarrow B \quad C \longrightarrow C}
}{}
$$

$$
\cfrac{A \Rightarrow B, C \longrightarrow A \Rightarrow (B \wedge C)}{A, A \Rightarrow B, C \longrightarrow B \wedge C}
$$

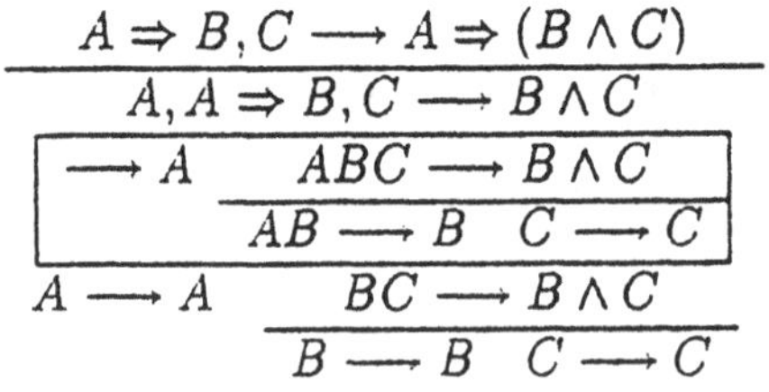

Figure 3: A false start.

is a deduction of $A \Rightarrow B, C \longrightarrow A \Rightarrow (B \wedge C)$ containing a false start. By erasing the false start, we obtain the proof in Figure 2.

2.3 The priority relation >

In order to develop stochastic strategies for efficient proof searches, we define the notion of *normal proofs*. In a sense, these proofs are the simplest proofs of provable sequents. We define the normalization process in terms of a *reduction algorithm* on proofs. By specifying a *priority relation* $>$ on the rules of $\Delta(V)$, we are providing a mechanical procedure for simplifying proofs.

$>$	$L_\wedge$	$L_\Rightarrow$	$R_\wedge$	$R_\Rightarrow$	L_π
$L_\wedge$	$R > L$	$L_\wedge$	$L_\wedge$	$R_\Rightarrow$	$L_\wedge$
$L_\Rightarrow$	$L_\wedge$	$R > L$	$R_\wedge$	$R_\Rightarrow$	L_π
$R_\wedge$	$L_\wedge$	$R_\wedge$	$\times$	$\times$	L_π
$R_\Rightarrow$	$R_\Rightarrow$	$R_\Rightarrow$	$\times$	$R > L$	$R_\Rightarrow$
L_π	$L_\wedge$	L_π	L_π	$R_\Rightarrow$	$R > L$

Table 1: The priority relation $>$.

According to this relation, $R_\wedge > L_\Rightarrow$, for example, indicates that if both the rules $R_\wedge$ and $L_\Rightarrow$ are applicable, then rules $R_\wedge$ takes precedence over $L_\Rightarrow$.

The relation $>$ is symmetric. It is undefined for certain combinations of rules. As pointed out in 2.1.3, the rule L_π is to be applied only if the syntax requires it for the derivation of a sequent. With the exception of $L_\Rightarrow$ and $R_\wedge$, we give ***operational rules*** priority over L_π. We also agree to use L_π as sparingly as possible. In addition, repetitions of $L_\wedge$, $L_\Rightarrow$, L_π, and $R_\Rightarrow$ are to be carried out from right to left whenever a choice exists. We write $R > L$ to indicate that right has priority over left.

Although the logic $\Delta(V)$ is the logic of symmetric monoidal closed categories, the priority relation $>$ differs from the reduction relation with the same name in [3]. The reason for the difference is that in the present context, normal proofs are intended to have the simplest possible path structure, whereas in [3], normal proofs are intended to reveal as transparently as possible the intended algebraic meaning of the proofs.

2.3.1 Generic Illustrations

$$\cfrac{\cfrac{ABCD \longrightarrow E}{AB, C \wedge D \longrightarrow E}}{A \wedge B, C \wedge D \longrightarrow E} \quad > \quad \cfrac{\cfrac{ABCD \longrightarrow E}{A \wedge B, CD \longrightarrow E}}{A \wedge B, C \wedge D \longrightarrow E}$$

Figure 4: $L_\pi - L_\pi (R > L)$

$$\cfrac{E \longrightarrow F \quad \cfrac{ABC \longrightarrow D}{A, B \wedge C \longrightarrow D}}{E, F \Rightarrow A, B \wedge C \longrightarrow D} \quad > \quad \cfrac{\cfrac{E \longrightarrow F \quad ABC \longrightarrow D}{E, F \Rightarrow A, BC \longrightarrow D}}{E, F \Rightarrow A, B \wedge C \longrightarrow D}$$

Figure 5: $L_\wedge > L_\Rightarrow$

$$\cfrac{\cfrac{AB \longrightarrow C}{A \wedge B \longrightarrow C} \quad \cfrac{DE \longrightarrow F}{D \wedge E \longrightarrow F}}{A \wedge B, D \wedge E \longrightarrow C \wedge F} \quad >$$

$$\cfrac{\cfrac{\cfrac{AB \longrightarrow C \quad DE \longrightarrow F}{ABDE \longrightarrow C \wedge F}}{AB, D \wedge E \longrightarrow C \wedge F}}{A \wedge B, D \wedge E \longrightarrow C \wedge F}$$

Figure 6: $L_\wedge > R_\wedge$

$$\dfrac{\dfrac{ABC \longrightarrow D}{AB \longrightarrow C \Rightarrow D}}{A \wedge B \longrightarrow C \Rightarrow D} \quad > \quad \dfrac{\dfrac{ABC \longrightarrow D}{A \wedge B, C \longrightarrow D}}{A \wedge B \longrightarrow C \Rightarrow D}$$

Figure 7: $R_{\Rightarrow} > L_{\wedge}$

$$\dfrac{\dfrac{ABCD \longrightarrow E}{AB, C \wedge D \longrightarrow E}}{BA, C \wedge D \longrightarrow E} \quad > \quad \dfrac{\dfrac{ABCD \longrightarrow E}{BA, CD \longrightarrow E}}{AB, C \wedge D \longrightarrow E}$$

Figure 8: $L_{\wedge} > L_{\pi}$

$$\dfrac{A \longrightarrow B \quad \dfrac{C \longrightarrow D \quad E \longrightarrow F}{C, D \Rightarrow E \longrightarrow F}}{A, B \Rightarrow C, D \Rightarrow E \longrightarrow F} \quad > \quad \dfrac{\dfrac{A \longrightarrow B \quad C \longrightarrow D}{A, B \Rightarrow C \longrightarrow D} \quad E \longrightarrow F}{A, B \Rightarrow C, D \Rightarrow E \longrightarrow F}$$

Figure 9: $L_{\Rightarrow} - L_{\Rightarrow}(R > L)$

$$\dfrac{E \longrightarrow F \quad \dfrac{A \longrightarrow B \quad C \longrightarrow D}{AC \longrightarrow B \wedge D}}{E, F \Rightarrow A, C \longrightarrow B \wedge D} \quad >$$

$$\dfrac{\dfrac{E \longrightarrow F \quad A \longrightarrow B}{E, F \Rightarrow A \longrightarrow B} \quad C \longrightarrow D}{E, F \Rightarrow A, C \longrightarrow B \wedge D}$$

Figure 10: $R_{\wedge} > L_{\Rightarrow}$

$$\dfrac{D \longrightarrow E \quad \dfrac{AB \longrightarrow C}{A \longrightarrow B \Rightarrow C}}{D, E \Rightarrow A \longrightarrow B \Rightarrow C} \quad > \quad \dfrac{\dfrac{D \longrightarrow E \quad AB \longrightarrow C}{D, E \Rightarrow A, B \longrightarrow C}}{D, E \Rightarrow A \longrightarrow B \Rightarrow C}$$

Figure 11: $R_{\Rightarrow} > L_{\Rightarrow}$

$$\dfrac{E \longrightarrow F \quad \dfrac{ABC \longrightarrow D}{ACB \longrightarrow D}}{E, F \Rightarrow A, CB \longrightarrow D} \quad > \quad \dfrac{\dfrac{E \longrightarrow F \quad ABC \longrightarrow D}{E, F \Rightarrow A, BC \longrightarrow D}}{E, F \Rightarrow A, CB \longrightarrow D}$$

Figure 12: $L_{\pi} > L_{\Rightarrow}$

$$\dfrac{\dfrac{AB \longrightarrow C}{BA \longrightarrow C} \quad D \longrightarrow E}{BAD \longrightarrow C \wedge E} \quad > \quad \dfrac{\dfrac{AB \longrightarrow C \quad D \longrightarrow E}{ABD \longrightarrow C \wedge E}}{BAD \longrightarrow C \wedge E}$$

Figure 13: $L_\pi > R_\wedge$

$$\dfrac{\dfrac{\Gamma A \Delta A \Lambda \longrightarrow B}{\Gamma A \Delta \Lambda \longrightarrow A \Rightarrow B}}{\Gamma \Delta \Lambda \longrightarrow A \Rightarrow (A \Rightarrow B)} \quad > \quad \dfrac{\dfrac{\Gamma A \Delta A \Lambda \longrightarrow B}{\Gamma \Delta A \Lambda \longrightarrow A \Rightarrow B}}{\Gamma \Delta \Lambda \longrightarrow A \Rightarrow (A \Rightarrow B)}$$

Figure 14: $R_\Rightarrow - R_\Rightarrow (R > L)$

$$\dfrac{\dfrac{ABC \longrightarrow D}{AB \longrightarrow C \Rightarrow D}}{BA \longrightarrow C \Rightarrow D} \quad > \quad \dfrac{\dfrac{ABC \longrightarrow D}{BAC \longrightarrow D}}{BA \longrightarrow C \Rightarrow D}$$

Figure 15: $R_\Rightarrow > L_\pi$

$$\dfrac{ABCD \longrightarrow E}{BACD \longrightarrow E} \quad > \quad \dfrac{\dfrac{ABCD \longrightarrow E}{ABDC \longrightarrow E}}{BADC \longrightarrow E}$$

Figure 16: $L_\pi - L_\pi$

2.4 Normal Proofs

Of all possible proofs $\Pi \in \Delta(V)$ of a sequent $\Gamma \longrightarrow \phi$, we intend to isolate those that have the simplest path structure.

Definition 2.4 *A proof $\Pi \in \Delta(V)$ of a sequent $\Gamma \longrightarrow \phi$ is normal if the order of application of the rules of inference in Π satisfies the conditions of the priority relation $>$.*

The following result follows by structural induction:

Proposition 2.5 *The priority relation $>$ determines an algorithm for converting any proof $\Pi \in \Delta(V)$ into a unique normal proof.*

2.4.1 Example

$$\dfrac{\dfrac{B \longrightarrow B \qquad \dfrac{\dfrac{A \longrightarrow A \quad C \longrightarrow C}{A, A \Rightarrow C \longrightarrow C} \qquad D \longrightarrow D}{A, A \Rightarrow C, D \longrightarrow C \wedge D}}{A, A \Rightarrow C, B, B \Rightarrow D \longrightarrow C \wedge D}}{A, B, A \Rightarrow C, B \Rightarrow D \longrightarrow C \wedge D}$$

$>$

$$\dfrac{\dfrac{B \longrightarrow B \qquad \dfrac{A \longrightarrow A \qquad \dfrac{C \longrightarrow C \quad D \longrightarrow D}{CD \longrightarrow C \wedge D}}{A, A \Rightarrow C, D \longrightarrow C \wedge D}}{A, A \Rightarrow C, B, B \Rightarrow D \longrightarrow C \wedge D}}{A, B, A \Rightarrow C, B \Rightarrow D \longrightarrow C \wedge D}$$

$>$

$$\dfrac{\dfrac{A \longrightarrow A \qquad \dfrac{B \longrightarrow B \qquad \dfrac{C \longrightarrow C \quad D \longrightarrow D}{CD \longrightarrow C \wedge D}}{C, B, B \Rightarrow D \longrightarrow C \wedge D}}{A, A \Rightarrow C, B, B \Rightarrow D \longrightarrow C \wedge D}}{A, B, A \Rightarrow C, B \Rightarrow D \longrightarrow C \wedge D}$$

Figure 17: Normalization steps.

Definition 2.6 *Two proofs Π_1 and Π_2 are path-equivalent if they have the same number of paths and for each length n, the number of paths of length n in Π_1 is the same as that in Π_2.*

It is easy to see that the three proofs of $A, B, A \Rightarrow C, B \Rightarrow D \longrightarrow C \wedge D$ in Figure 25, for example, are path-equivalent. The following theorem and its corollary are proved by structural induction:

Theorem 2.7 *If Π_1 and Π_2 in $\Delta(V)$ are two normal proofs of the same sequent, then Π_1 and Π_2 are path-equivalent.*

Corollary 2.8 (Uniqueness of normal proofs) *If Π_1 and Π_2 in $\Delta(V)$ are two normal proofs of the same sequent, then $\Pi_1 = \Pi_2$.*

3 Stochastic networks

A stochastic network is a directed graph in which each edge is endowed with both a probabilistic function determining the likelihood of an edge being realized in the course of an activity controlled by the network, and a cost function determining the cost of such a realization.

We shall present an algorithm for converting a network into a stochastically equivalent simple arrow. Thus we are able to compute the likelihood of the realization of the entire network in terms of the likelihood of the realization of such an arrow. By embedding this arrow in a closed network, we can use Mason's rule for closed stochastic networks to define the *mean* and *variance* of a network (cf. [2]) and hence of a proof or derivation. The main result of the present paper is the proof that there exist appropriate functions for which the Mason measures of stochastic networks yield good numerical estimates for the path complexity of proofs and deductions.

Definition 3.1 *A stochastic network is a network in which each arrow (i,j) comes equipped with a function*

$$w(i,j)(s) = p(i,j)M(i,j)(s) \tag{1}$$

consisting of a probability $p(i,j)$ and a moment-generating function

$$M(i,j) = e^{sY(i,j)}f(Y(i,j)), \tag{2}$$

in which the $Y(i,j)$ are independent random variables, the functions f are probability density functions, and in which the sum of the probabilities of all simple arrows with the same initial node equals 1.

Definition 3.2 *All canonical probabilities $p(i,j)$ in networks for proofs are either 0, $\frac{1}{2}$ or 1, and all canonical moment-generating functions $M(i,j)(s)$ are exponential functions e^{ns} for some non-negative integer n.*

Definition 3.3 *A simple arrow of a stochastic network consists of a pair of nodes i and j, connected by an arrow (i,j), equipped with a w-function $w(i,j)$, as shown in Figure 18.*

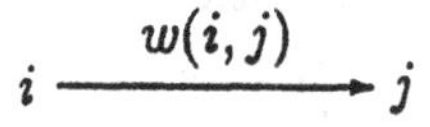

Figure 18: A simple arrow.

Definition 3.4 *Two concatenated arrows of a stochastic network consist of three nodes i, j, and k, connected by two arrows (i,j) and (j,k), equipped with w-functions $w(i,j)$ and $w(j,k)$, as shown in Figure 19.*

$$i \xrightarrow{\ w(i,j)\ } j \xrightarrow{\ w(j,k)\ } k \;\equiv\; i \xrightarrow{\ w(i,k)\ } k$$

Figure 19: Two concatenated arrows.

Definition 3.5 *Two parallel arrows of a stochastic network consist of nodes i and j, connected by two arrows (i,j), with respective probabilities p_a and p_b and w-functions w_a and w_b, as shown in Figure 20.*

$$i \overset{p_a, w_a}{\underset{p_b, w_b}{\rightrightarrows}} j \;\equiv\; i \xrightarrow{\ w(i,j)\ } j$$

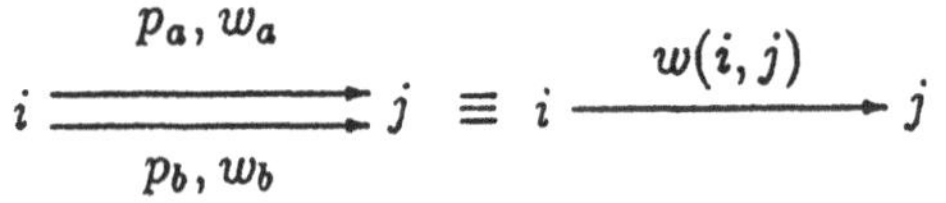

Figure 20: Two parallel arrows.

Definition 3.6 *A self-loop of a stochastic network consists of a single node i and an arrow (i,i), equipped with a w-function $w(i,i)$, as shown in Figure 21.*

$$i \overset{p_a, w_a}{\underset{p_a, w_a}{\rightleftarrows}} i \xrightarrow{\ p_b, w_b\ } j \;\equiv\; i \xrightarrow{\ w(i,j)\ } j$$

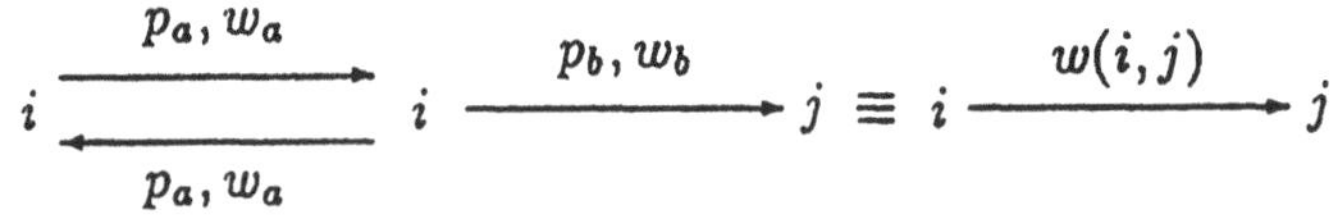

Figure 21: A self-loop.

3.1 Mason's rule for first and second order networks

For the purpose of measuring the performance of proofs and deductions in $\Delta(V)$ we use Mason's rule for closed stochastic networks. We simplify the exposition by dealing only with networks containing first and second order loops. The general rule is discussed in [1] and [2].

Definition 3.7 *A closed stochastic network E is a network of the form shown in Figure 22.*

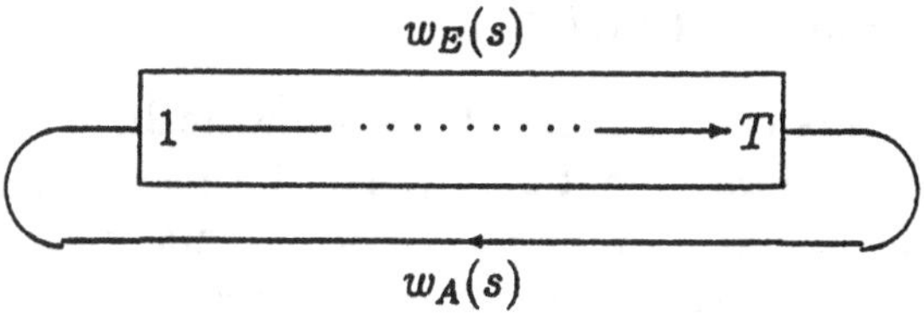

Figure 22: A closed stochastic network.

All loops of the networks required for proofs will be "first order" and all loops required for the sample deductions will be "first" and "second order". In both cases, the canonical probabilities of the prongs of a fork are $\frac{1}{n}$, where n is the number of prongs of the fork, and the cost parameters of the non-trivial arrows of the networks are assumed to have a value of 1.

Definition 3.8 *A first order loop in a stochastic network is a sequence of consecutive arrows in which the initial node of the first arrow is the same as the terminal node of the last arrow and in which all other nodes are distinct.*

Definition 3.9 *A second order loop is a set consisting of two disjoint first-order loops.*

Definition 3.10 *A first order network is a closed stochastic network all of whose loops are first order.*

Definition 3.11 *A second order network is a closed stochastic network which has only first and second order loops.*

Definition 3.12 *For any first order loop L of a stochastic network, the weight $w(L)$ is defined to be*

$$w(L) = \prod_{(i,j)\in L} w(i,j) \tag{3}$$

Definition 3.13 (Mason's Rule for First Order Networks) *If $F_1, \ldots, F_n$ are the loops of a first order stochastic network, then*

$$1 - w(F_1) - \cdots - w(F_n) = 0 \tag{4}$$

Definition 3.14 (Mason's Rule for Second Order Networks) *If the loops $F_1, \ldots, F_n$ are the first order loops and $S_1, \ldots, S_m$ the second order loops of a second order stochastic network, then*

$$1 - w(F_1) - \cdots - w(F_n) + w(S_1) + \cdots + w(S_m) = 0 \tag{5}$$

3.2 The reduction algorithm

All stochastic networks can be reduced to stochastically equivalent simple arrows. The reduction algorithm not only yields stochastically equivalent simple arrows but, equally importantly, computes the w–functions of the entire networks. The algorithm consists of the following steps:

3.2.1 Reduction of concatenated arrows

Since the random variables in the labels of different arrows are assumed to be independent, the concatenated arrows can be replaced by stochastically equivalent simple arrows connecting the nodes i and k:

$$\begin{aligned} w(i,k)(s) &= (p(i,j)M(i,j)(s))(p(j,k)M(j,k)(s)) \\ &= w(i,j)(s)w(j,k)(s). \end{aligned} \tag{6}$$

3.2.2 Reduction of parallel arrows

The stochastic equivalence of these arrows to the single arrow with w-function $w(i,j)$ is based on the assumption that

$$p(i,j) = p_a + p_b \quad \text{and that} \quad M(i,j)(s) = \frac{p_a M_a(s) + p_b M_b(s)}{p_a + p_b},$$

so that

$$\begin{aligned} w(i,j)(s) &= (p_a + p_b)\frac{p_a M_a(s) + p_b M_b(s)}{p_a + p_b} \\ &= w_a(s) + w_b(s) \end{aligned} \tag{7}$$

3.2.3 Reduction of self-loops

A self-loop is interpreted as constituting an infinite sequence of parallel chains, with each chain consisting of a finite sequence of concatenations of the looping arrow. Thus by using Equations (6) and (7), we have

$$\begin{aligned} w(i,j) &= w_b + w_a w_b + w_a^2 w_b + \cdots \\ &= w_b(1 - w_a)^{-1} \end{aligned} \tag{8}$$

Any stochastic network consisting of the different chains of concatenations of a self-loop is thus stochastically equivalent to a simple arrow with the w-function (8).

3.2.4 Reduction of networks

Using Equations (6), (7), and (8), we can reduce any stochastic network to a network consisting of a stochastically equivalent simple arrow in a finite number of steps.

3.3 Moments of stochastic networks

In the case of a stochastic network E with nested loops, we use Mason's rule to compute the function w_E. In order to close a stochastic network, we introduce an auxiliary arrow from T to R, with w-function $w_A(s)$, as shown in Figure 22.

If $Y(i,j) = a$ in Equation (2) in Section 3 is constant, then $M(i,j) = e^{sa}$, and if $a = 0$ or $s = 0$, then $M(i,j)(s) = 1$. Since $w_E(s) = p_E M_E(s)$, we have $p_E = w_E(0)$, so that

$$M_E(s) = \frac{w_E(s)}{p_E} = \frac{w_E(s)}{w_E(0)}. \tag{9}$$

The value of $w_E(0)$ can easily be computed using Mason's rule. By forming the j-th derivative of $M_E(s)$ with respect to s and evaluating the resulting function at $s = 0$, we obtain the *j-th moment* $\mu_{j,E}$ of $M_E(s)$ about the origin.

Definition 3.15 *The network complexity of a stochastic network with associated w-function $w_E(s)$ is the first moment $\mu_{1,E}$.*

4 Mason measures for proofs

As indicated earlier, our plan is to associate with each proof (deduction) $\Pi \in \Delta(V)$ a closed stochastic network $E(\Pi)$. The definition is by induction on proofs:

Definition 4.1 *Let $X = (X, X) \in \Pi$ be an axiom of $\Delta(V)$ and let T be a fixed new symbol. Then*

$$X \overset{w(X,T)}{\longrightarrow} T \in E(\Pi) \tag{10}$$

is an arrow with w-function

$$w(X,T) = e^0 = 1. \tag{11}$$

Definition 4.2 *Let P be the premiss and C the conclusion of a one-premiss rule of $\Delta(V)$ and let $(P, C) \in \Pi$. Then*

$$C \overset{w(C,P)}{\longrightarrow} P \in E(\Pi) \tag{12}$$

is an arrow with w-function

$$w(C,P)(s) = e^s. \tag{13}$$

Definition 4.3 *Let P_l and P_r be the left and right premisses and C the conclusion of a two-premiss rule of $\Delta(V)$ and let $(P_l, P_r, C) \in \Pi$. Then*

$$C \overset{w(C,P_l)}{\longrightarrow} P_l \in E(\Pi) \tag{14}$$

$$C \overset{w(C,P_r)}{\longrightarrow} P_r \in E(\Pi) \tag{15}$$

are arrows with w-functions

$$cw(C, P_l)(s) = \frac{1}{2}e^s \tag{16}$$

$$w(C, P_r)(s) = \frac{1}{2}e^s. \tag{17}$$

Definition 4.4 *Let $\Phi \in \Pi$ be the sequent proved by Π. Then*

$$T \overset{w_A}{\longrightarrow} \Phi \in E(\Pi) \tag{18}$$

is the formal arrow required to close the network $E(\Pi)$. By Mason's rule,

$$w_A = \frac{1}{w_{E(\Pi)}}, \tag{19}$$

where $w_{E(\Pi)}$ is the w-function of the arrow stochastically equivalent to the network $E(\Pi)$.

Definition 4.5 (Mason measure) *For any proof $\Pi \in \Delta(V)$, the Mason measure $\mu(\Pi)$ of the proof Π is equal to the Mason measure $\mu(E(\Pi))$ of the network $E(\Pi)$ associated with Π.*

4.1 Example 1

Consider the mirror image Π_1 the first proof in the normalization example above and its underlying path structure:

$$\dfrac{AB, A \Rightarrow C, B \Rightarrow D \longrightarrow C \wedge D}{\dfrac{A, A \Rightarrow C, B, B \Rightarrow D \longrightarrow C \wedge D}{B \longrightarrow B \qquad \dfrac{A, A \Rightarrow C, D \longrightarrow C \wedge D}{\dfrac{A, A \Rightarrow C \longrightarrow C}{A \longrightarrow A \quad C \longrightarrow C} \qquad D \longrightarrow D}}}$$

Figure 23: Top-down proof of $AB, A \Rightarrow C, B \Rightarrow D \longrightarrow C \wedge D$.

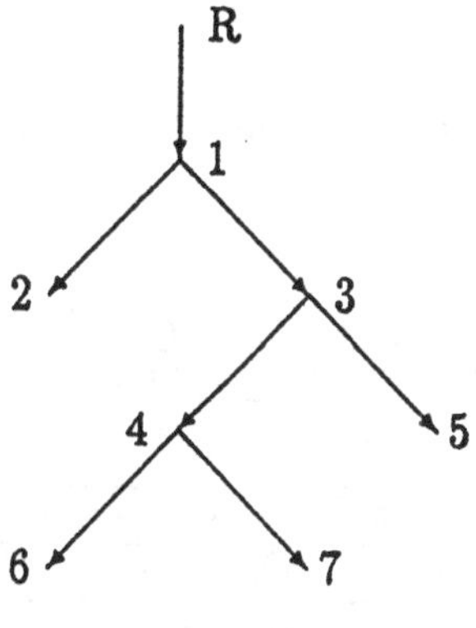

Figure 24: Path structure of Π_1.

We turn the structure in Figure 24 into a network by introducing a terminal node T and formal terminal arrows $2 \longrightarrow T$, $6 \longrightarrow T$, $7 \longrightarrow T$, and $5 \longrightarrow T$ with constant w-functions $1e^{0s} = 1$. We then close the network by introducing an additional formal arrow $T \longrightarrow R$ with w-function w_A. We denote the resulting closed network by $E(\Pi_1)$. The w-functions of the non-trivial arrows of $E(\Pi_1)$ are as specified in Definition 3.2, i.e., $w(R,1) = e^s$, and $w(i,j) = \frac{1}{2}e^s$ for all other arrows.

All four loops of $E(\Pi_1)$ are first order:

$$L_1 = R \to 1 \to 2 \to T \to R \tag{20}$$
$$L_2 = R \to 1 \to 3 \to 4 \to 6 \to T \to R \tag{21}$$
$$L_3 = R \to 1 \to 3 \to 4 \to 7 \to T \to R \tag{22}$$
$$L_4 = R \to 1 \to 3 \to 5 \to T \to R \tag{23}$$

The respective w-functions of L_i $(i = 1,2,3,4)$ are

$$w(L_1) = \frac{1}{2}e^{2s}, \quad w(L_2) = \frac{1}{8}e^{4s}, \quad w(L_3) = \frac{1}{8}e^{4s}, \quad w(L_4) = \frac{1}{4}e^{3s} \tag{24}$$

By the reduction algorithm, the w-function of the network $E(\Pi_1)$ is

$$w(E(\Pi_1)) = \frac{1}{2}e^{2s} + \frac{1}{4}e^{3s} + \frac{1}{4}e^{4s} \tag{25}$$

According to Definition 3.15, the path complexity of Π_1 is

$$\mu(\Pi_1) = \left.\frac{dw}{ds}\right|_{s=0} = 2\frac{3}{4} \tag{26}$$

4.2 Example 2

Let Π_2, Π_3, and Π_4 be the proofs occurring in the following normalization process:

$$\begin{array}{c}
AB, A \Rightarrow C, B \Rightarrow D \longrightarrow C \wedge D \\ \hline
AB, B \Rightarrow D, A \Rightarrow C \longrightarrow C \wedge D \\ \hline
B \longrightarrow B \qquad AD, A \Rightarrow C \longrightarrow C \wedge D \\ \hline
DA, A \Rightarrow C \longrightarrow C \quad D \longrightarrow D \\ \hline
A \longrightarrow A \qquad DC \longrightarrow C \wedge D \\ \hline
CD \longrightarrow C \wedge D \\ \hline
C \longrightarrow C \quad D \longrightarrow D
\end{array}$$

$>$

$$\begin{array}{c}
AB, A \Rightarrow C, B \Rightarrow D \longrightarrow C \wedge D \\ \hline
A, A \Rightarrow C, B, B \Rightarrow D \longrightarrow C \wedge D \\ \hline
B \longrightarrow B \qquad A, A \Rightarrow C, D \longrightarrow C \wedge D \\ \hline
A \longrightarrow A \qquad CD \longrightarrow C \wedge D \\ \hline
C \longrightarrow C \quad D \longrightarrow D
\end{array}$$

$>$

$$\begin{array}{c}
AB, A \Rightarrow C, B \Rightarrow D \longrightarrow C \wedge D \\ \hline
A, A \Rightarrow C, B, B \Rightarrow D \longrightarrow C \wedge D \\ \hline
A \longrightarrow A \qquad CB, B \Rightarrow D \longrightarrow C \wedge D \\ \hline
B \longrightarrow B \qquad CD \longrightarrow C \wedge D \\ \hline
C \longrightarrow C \quad D \longrightarrow D
\end{array}$$

Figure 25: Normalization.

The path structures of Π_2, Π_3, and Π_4 are identical and are given by:

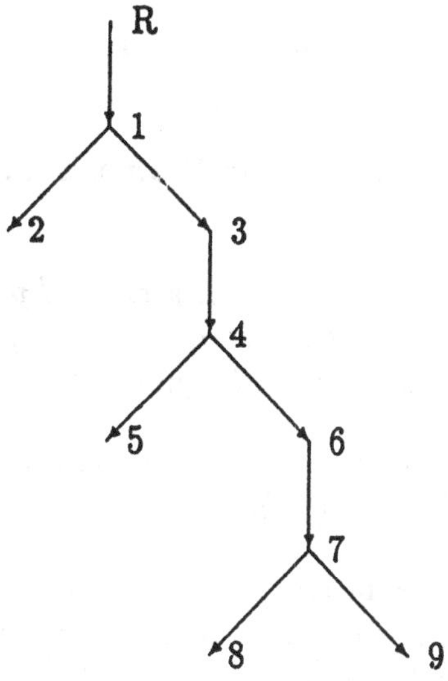

Figure 26: The path structure of Π_2, Π_3, and Π_4.

As in Example 4.1, we complete the network $E(\Pi_2)$ by introducing a terminal node T and formal terminal arrows $2 \longrightarrow T$, $5 \longrightarrow T$, $8 \longrightarrow T$, and $9 \longrightarrow T$ with constant w-functions $1e^{0s} = 1$. We then close the network by introducing an additional formal arrow $T \longrightarrow R$ with w-function w_A. The w-functions of the non-trivial arrows of $E(\Pi_1)$ are again as specified in Definition 3.2, i.e., $w(R,1) = w(3,4) = w(6,7) = e^s$, and $w(i,j) = \frac{1}{2}e^s$ for all other arrows.

Since a proof has no false starts, all four loops of $E(\Pi_2)$ are again first order:

$$L_1 = R \rightarrow 1 \rightarrow 2 \rightarrow T \rightarrow R \tag{27}$$
$$L_2 = R \rightarrow 1 \rightarrow 3 \rightarrow 4 \rightarrow 5 \rightarrow T \rightarrow R \tag{28}$$
$$L_3 = R \rightarrow 1 \rightarrow 3 \rightarrow 4 \rightarrow 6 \rightarrow 7 \rightarrow 8 \rightarrow T \rightarrow R \tag{29}$$
$$L_4 = R \rightarrow 1 \rightarrow 3 \rightarrow 4 \rightarrow 6 \rightarrow 7 \rightarrow 9 \rightarrow T \rightarrow R \tag{30}$$

with w-functions

$$w(L_1) = \frac{1}{2}e^{2s}, \;\; w(L_2) = \frac{1}{4}e^{4s}, \;\; w(L_3) = \frac{1}{8}e^{6s}, \;\; w(L_4) = \frac{1}{8}e^{6s} \tag{31}$$

By the reduction algorithm, the w-functions of the networks $E(\Pi_i)$ $(i = 2, 3, 4)$ are

$$w(E(\Pi_2)) = w(E(\Pi_3)) = w(E(\Pi_4)) = \frac{1}{2}e^{2s} + \frac{1}{4}e^{4s} + \frac{1}{4}e^{6s} \tag{32}$$

According to Definition 3.15, the path complexity of $E(\Pi_i)$ for $(i = 2, 3, 4)$, is

$$\mu(\Pi_i) = \frac{dw}{ds}\bigg|_{s=0} = 3\frac{1}{2} \tag{33}$$

The Mason measures of the proofs in Examples 4.1 and 4.2 reveal typical properties of proofs:

Proposition 4.6 *If* $\Pi \in \Delta(V)$ *is a normal proof and* p *is a path in* $E(\Pi)$, *then* $w(p) = \frac{1}{2^m}e^{ns}$, *where* m *is the number of applications of two-premiss rules and* n *is the number of applications of all rules giving rise to* p.

Corollary 4.7 *Two proofs* Π_1 *and* Π_2 *of a sequent are path equivalent proofs of a sequent* $\Gamma \longrightarrow \phi$ *iff* $\mu(\Pi_1) = \mu(\Pi_2)$.

This shows that the given definition of $\mu(\Pi)$ meets our expectations since it characterizes, as desired, the path equivalence of proofs. An induction on the defining clauses of the priority relation $>$ shows that normal proofs have been defined correctly in the sense that they achieve maximal deductive efficiency with respect to Mason measures:

Theorem 4.8 (Minimal Measure Theorem) *If* Π *is a normal proof of a sequent* $\Gamma \longrightarrow \phi$, *then the Mason measure* $\mu(\Pi)$ *is a minimum on the set of canonical measures of proofs of* $\Gamma \longrightarrow \phi$.

4.2.1 Example 3

We now illustrate the use of a second order network to measure the loop complexity of a deduction with a false start. Then we delete the false start and compare the canonical Mason measure of the resulting proof with the different realization costs α of the given derivation.

Let Δ_1 be the derivation

$$\cfrac{\cfrac{\cfrac{\cfrac{\cfrac{\cfrac{B \longrightarrow B \quad C \longrightarrow C}{BC \longrightarrow B \wedge C}}{A \longrightarrow A \quad \boxed{\cfrac{\cfrac{AB \longrightarrow B \quad C \longrightarrow C}{\longrightarrow A \quad ABC \longrightarrow B \wedge C}}{}}}}{A, A \Rightarrow B, C \longrightarrow B \wedge C}}{A \Rightarrow B, C \longrightarrow A \Rightarrow (B \wedge C) \quad D \longrightarrow D}}{A \Rightarrow B, C, D \longrightarrow (A \Rightarrow (B \wedge C)) \wedge D}}$$

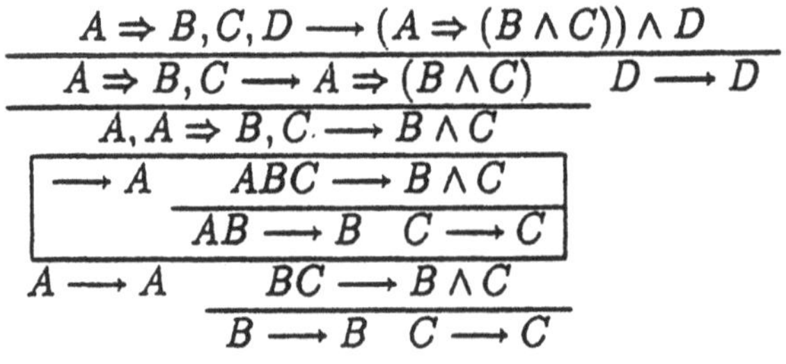

Figure 27: Sample derivation Δ_1.

The path structure of Δ_1 is

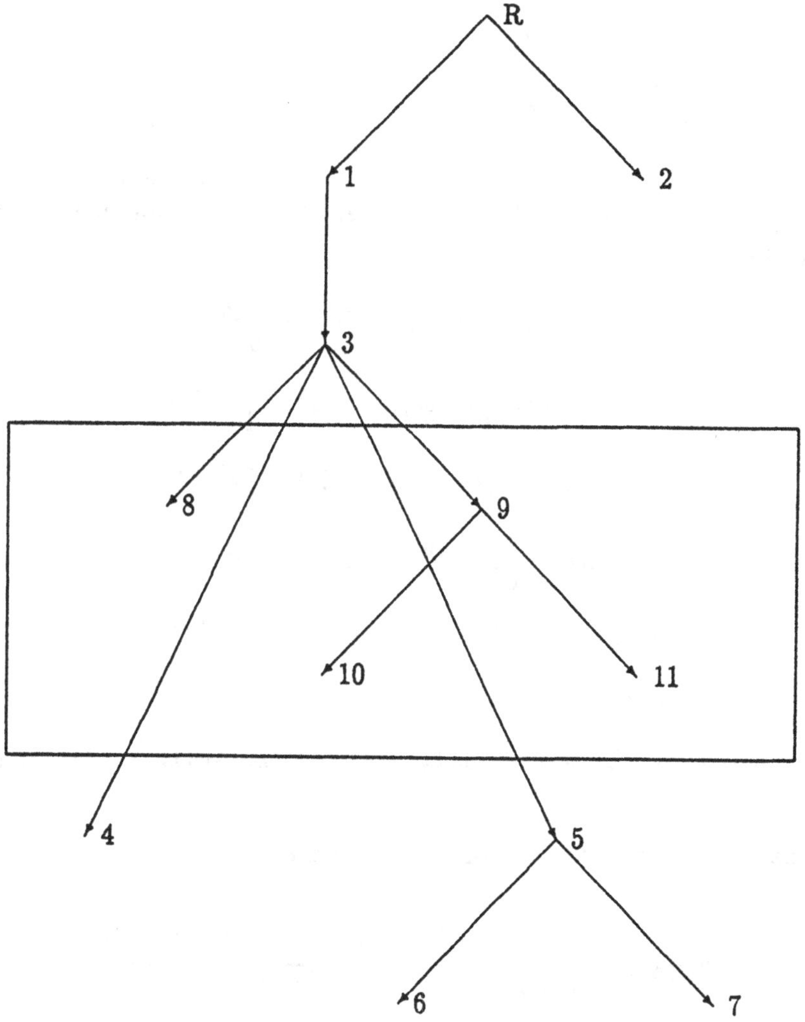

Figure 28: The path structure of Δ_1.

The network $E(\Delta_1)$ to consist of the following first and second order loops:

$$L_1 = R \longrightarrow 2 \longrightarrow T \longrightarrow R \tag{34}$$

$$L_2 = R \longrightarrow 1 \longrightarrow 3 \longrightarrow 8 \longrightarrow T \longrightarrow R \tag{35}$$

$$L_3 = R \longrightarrow 1 \longrightarrow 3 \longrightarrow 9 \longrightarrow 10 \longrightarrow T \longrightarrow R \tag{36}$$

$$L_4 = R \longrightarrow 1 \longrightarrow 3 \longrightarrow 9 \longrightarrow 11 \longrightarrow T \longrightarrow R \tag{37}$$

$$L_5 = 3 \longrightarrow 4 \longrightarrow 3 \tag{38}$$

$$\begin{aligned} L_6 &= 3 \longrightarrow 5 \longrightarrow 6 \longrightarrow 3 && (39)\\ L_7 &= 3 \longrightarrow 5 \longrightarrow 7 \longrightarrow 3 && (40)\\ L_8 &= L_1 L_5 && (41)\\ L_9 &= L_1 L_6 && (42)\\ L_{10} &= L_1 L_8 && (43) \end{aligned}$$

We let $p(3,4) = p(3,5) = p(3,8) = p(3,9) = \frac{1}{4}$ and define the cost of the arrows $4 \longrightarrow 3$, $6 \longrightarrow 3$, and $7 \longrightarrow 3$ be α. Then the w-functions of the loops of Δ_1 are

$$w(L_1) = \frac{1}{2}e^{s}w_A, \;\; w(L_2) = \frac{1}{8}e^{3s}w_A \tag{44}$$

$$w(L_3) = \frac{1}{16}e^{4s}w_A, \;\; w(L_4) = \frac{1}{16}e^{4s}w_A \tag{45}$$

$$w(L_5) = \frac{1}{4}e^{(1+\alpha)s}, \;\; w(L_6) = \frac{1}{8}e^{(2+\alpha)s} \tag{46}$$

$$w(L_7) = \frac{1}{8}e^{(2+\alpha)s}, \;\; w(L_8) = \frac{1}{8}e^{(2+\alpha)s}w_A \tag{47}$$

$$w(L_9) = \frac{1}{16}e^{(3+\alpha)s}w_A, \;\; w(L_{10}) = \frac{1}{16}e^{(3+\alpha)s}w_A \tag{48}$$

By Mason's rule,

$$1-w(L_1)-w(L_2)-w(L_3)-w(L_4)-w(L_5)-w(L_6)-w(L_7)+w(L_8)+w(L_9)+w(L_{10}) = 0.$$

The w-function of the network $E(\Delta_1)$ is therefore

$$w_{E(\Delta_1)} = \frac{1}{w_A} = \frac{4e^{s} + e^{3s} + e^{4s} - e^{(2+\alpha)s} - e^{(3+\alpha)s}}{8 - 2e^{(1+\alpha)s} - 2e^{(2+\alpha)s}} \tag{49}$$

Since $w_{E(\Delta_1)}(0) = 1$, we get the following Mason measure for Δ_1:

$$\mu(\Delta_1) = 3 + \frac{\alpha}{2} \tag{50}$$

If we let $\alpha = 1$, then $\mu(\Delta_1) = 3\frac{1}{2}$ is the canonical Mason measure of Δ_1.

4.2.2 Example 4

By deleting the false start from the deduction Δ_1 we obtain the proof Π_5:

$$\begin{array}{c} A \Rightarrow B, C, D \longrightarrow (A \Rightarrow (B \wedge C)) \wedge D \\ \hline A \Rightarrow B, C \longrightarrow A \Rightarrow (B \wedge C) \quad D \longrightarrow D \\ \hline A, A \Rightarrow B, C \longrightarrow B \wedge C \\ \hline A \longrightarrow A \quad BC \longrightarrow B \wedge C \\ \hline B \longrightarrow B \quad C \longrightarrow C \end{array}$$

Figure 29: The top-down proof Π_5.

The path structure of Π_5 is:

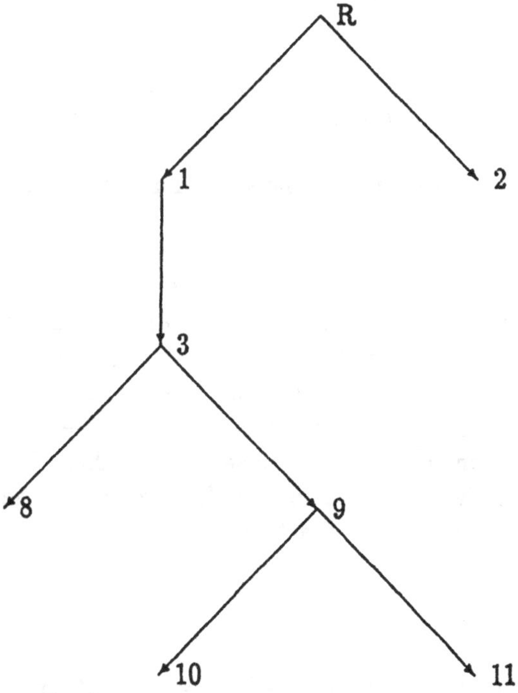

Figure 30: The path structure of Π_5.

The network $E(\Pi_5)$ consists of the following first order loops of $E(\Delta_1)$:

$$L_1 = R \longrightarrow 2 \longrightarrow T \longrightarrow R \qquad (51)$$

$$L_2 = R \longrightarrow 1 \longrightarrow 3 \longrightarrow 8 \longrightarrow T \longrightarrow R \qquad (52)$$

$$
\begin{aligned}
L_3 &= R \longrightarrow 1 \longrightarrow 3 \longrightarrow 9 \longrightarrow 10 \longrightarrow T \longrightarrow R && (53) \\
L_4 &= R \longrightarrow 1 \longrightarrow 3 \longrightarrow 9 \longrightarrow 11 \longrightarrow T \longrightarrow R && (54)
\end{aligned}
$$

The probabilities and cost functions of $E(\Pi_5)$ are the same as those in $E(\Delta_1)$, with two exceptions: $p(3,4) = p(3,5) = \frac{1}{2}$. The w-functions of the loops of $E(\Pi_1)$ therefore are

$$w(L_1) = \frac{1}{2}e^{s}w_A, \quad w(L_2) = \frac{1}{4}e^{3s}w_A \tag{55}$$

$$w(L_3) = \frac{1}{8}e^{4s}w_A, \quad w(L_4) = \frac{1}{8}e^{4s}w_A \tag{56}$$

By Mason's rule,

$$1 - w(L_1) - w(L_2) - w(L_3) - w(L_4) = 0.$$

Thus the w-function of the network $E(\Pi_5)$ is

$$w_{E(\Pi_5)} = \frac{2e^{s} + e^{3s} + e^{4s}}{4}. \tag{57}$$

Since $w_{\Delta_1}(0) = 1$, we get the following Mason measure for Π_5:

$$\mu(\Pi_5) = 2\frac{1}{4} \tag{58}$$

This shows that for all α, $\mu(\Delta_1) > \mu(\Pi_5)$, as expected. If we let $\alpha = 0$ and consider the "backtracking arrows" $4 \longrightarrow 3$, $6 \longrightarrow 3$, and $7 \longrightarrow 3$ on a par with the auxiliary terminal arrows from 8, 10, and 11 to T, we can think of $\mu(\Delta_1)$ as the pure measure that determines merely the cost of the proof Π_5 and the additional cost incurred by the false start in Δ_1. If we let $\alpha = 1$, we can interpret $\mu(\Delta_1)$ as the canonical cost of Π_5, the cost of the false start, and the cost of the deletion of the false start from Δ_1.

5 Discussion

The ideas presented in this paper are independent of any particular implementation of the Gentzen algorithm for detecting or creating proofs in $\Delta(V)$. On the other hand, the detailed recording of false starts makes sense only if the proof detection process is not prematurely terminated by the prover if a false start is suspected. We therefore think of the results presented in this paper as relevant mainly to automated deductions.

The use of Mason's measures for proofs and deductions is a general method that can be applied to more powerful deductive systems than the system $\Delta(V)$ used as a paradigm in this paper. Although the canonical Mason measures presented provide an elegant assessment of the structural complexity of proofs, the probabilities and cost functions of these measures can be generalized to take into account complexities such as the structure of the subformulas involved in proofs and costs such as the required CPU time for the execution of the rules of inference. These generalizations are dealt with in [6].

References

[1] P. GRAHAM, *Profit Probability Analysis of Research and Development Expenditures,* The Journal of Industrial Engineering **16** (1965) 186-191.

[2] [2] D. T. PHILLIPS AND A. GARROWIA-DIAZ, *Fundamentals of Network Analysis,* Prentice-Hall, New York, 1981.

[3] M. E. SZABO, *Algebra of Proofs,* North-Holland Publishing Company, Amsterdam, 1978.

[4] M. E. SZABO, *Stochastic Performance Measures for Rule-Based Expert Systems,* Information Sciences. In Press.

[5] M. E. SZABO, *On the Stochastic Complexity of Loops in Rule-Based Expert Systems,* Information Sciences. In Press.

[6] M. E. SZABO, *Mason Measures for Proofs.* In Preparation.

The Gap-Language-Technique Revisited

Heribert Vollmer
Theoretische Informatik (FB 20)
Johann-Wolfgang-Goethe-Universität
D-6000 Frankfurt am Main

Abstract

Generalizing work of Schöning and others concerning gap language constructs recognizable in polynomial time we examine structural properties of reducibilities defined for various "lower" or "parallel" complexity classes. Finally we show how the proof techniques for the above can be used to show the existence of easy complexity cores for sets which cannot be decided in logarithmic space.

1 Introduction

In 1944, Post raised the question of whether there exists a recursively enumerable, non recursive non complete set. His question was answered independently and almost simultaneously in 1956/57 by Friedberg and Muchnik (see e. g. [11, 12]). Their results required a new proof technique which became known as the finite injury priority method.

An analogous subrecursive question in the **P** vs. **NP**-context, whether there exists a non polynomial-time decidable non **NP**-complete set in **NP**, was, in a sense (i. e. under the assumption $\mathbf{P} \neq \mathbf{NP}$), solved in 1975 by Richard Ladner [7]. His result was generalized in several directions in the meantime in a number of papers [8, 5, 13], the last of them presenting what is nowadays called "The Uniform Diagonalization Theorem" [2]. This theorem allows in a fairly general context the construction of different polynomial time many-one degrees in various complexity classes.

The most extensive study of questions of the above kind was done by K. Ambos-Spies in [1]. Based on two diagonalization theorems ("diagonalization lemma" and "meet lemma") a number of very interesting and fundamental results are proved about the degree structure of various polynomial time reducibilities.

But what remains open in the above cited papers is the question whether similar results hold for classes which are contained in **P**. A step in this direction was done by Diana Schmidt [15]. She obtained some results about the structure of logspace reducibility.

In this paper we also address the just mentioned question and first show, how to obtain slightly more elegant proofs of the results of [15] and how to translate the results of [1] in the logspace-context. Our proofs heavily depend on a refinement of the "gap language technique" introduced in [13].

Our technique doesn't stop at logspace-reductions, but can be applied for so called constant-depth-reducibility, so we get results even for very low complexity classes ("parallel complexity classes", because of their strong connection to parallel computation models as the PRAM), e. g.: Suppose $TC^0 \neq NC^1$, then there exist infinitely many *cd*-degrees between TC^0 and NC^1. For classes which were separated in the past, for instance AC^0 and TC^0, we definitely obtain (without any assumption in the above form) a rich structure of different degrees between them.

But our refinement of the gap language technique does not only lead us to simple and short proofs of results about logspace reducibility and constant-depth reducibility. Perhaps the most interesting application is the following:

Take any set $A \notin L$. Then, for every Turing machine M deciding A, i.e. $L(M) = A$, there are infinitely many inputs w, such that M uses more than logarithmic space on these inputs. (Otherwise, M could be patched to work in logspace.) Can the two quantifiers in the last sentence be swapped? In other words, is there an infinite set X, such that *any* algorithm for A uses more than logarithmic space on almost all members of X? Such a set X is called a *logspace complexity-core*.

In the context of polynomial time, complexity-cores are very well examined. Lynch [9] showed the existence of a recursive (analoguously defined polynomial) complexity-core for every set $A \notin P$. Orponen and Schöning [10] improved upon this result and showed the existence of a complexity-core for every $A \notin P$, which itself is in DTIME($t(n)$) for any time-constructible superpolynomial function t.

1988 Book et al. [3] examined complexity-cores in very general contexts for various complexity classes. They proved, concerning logspace, the existence of a recursive logspace complexity-core, but couldn't bound it's complexity, since the technique of [10], based on gap language methods, didn't seem to be applicable in logspace. But our refinement used to show our first results is strong enough for this kind of theorem, too. We present for any given set $A \notin L$ a logspace complexity core X, which is "arbitrarily close" to L. I.e. the members of X are hard with respect to membership in A, but membership in X itself can be decided easily.

In the next section, we will present the relevant definitions and our refinement of the gap language technique. Then we will proceed demonstrating how to obtain structural results about logspace reducibility. In section 4 we will address the question of complexity cores. In section 5 we will generalize our results from section 3 to constant-depth reducibility, and finally, in the last section, we will give a short conclusion.

2 Definitions

We assume the reader is familiar with standard complexity classes and complexity theory notions, as time and space bounded classes (**L**, **NL**, **P**, **NP**), reduciblity concepts (especially *logspace*-reducibility) and their properties. Our notation here is standard (see for example [2]). Further we presuppose some basic knowledge about parallel complexity classes like $\mathbf{AC}^0$, $\mathbf{TC}^0$, $\mathbf{NC}^1$ and **NC**, and the reducibility concepts involved here [4, 6].

2.1 Constructibility

Every one familiar with complexity theory knows about the concept of space- or time-constructibility. Here we need a slightly refined definition.

(2.1) Definition: *A function* $f\colon \mathbb{N} \to \mathbb{N}$ *is said to be* strongly space-constructible *if and only if* f *is computable in space* $\log \circ f$.

This means that there is a Turing machine M computing f which on input n uses no more than $\log f(n)$ tape cells, i.e. the space needed by M is bounded above by the logarithm of *the value of the output of* M; or in other words, M uses no more tape cells than are needed to write down $f(n)$ (which can be done in space $\log f(n)$).

That this is not a too severe restriction is the claim of the following lemma.

(2.2) Lemma: *For each function* $f\colon \mathbb{N} \to \mathbb{N}$ *there is a function* $f' \geq f$*, which is strongly space-constructible.*

Proof. Choose $f_1 \geq f$ to be space-constructible in the traditional way. Let M be the Turing machine witnessing this fact.

Define a Turing machine M' to operate as follows:

input x:
  compute $f_1(x)$ by simulating $M(x)$;
  print $2^{f_1(x)}$.

Let $f' = f_{M'}$.

Clearly $f'(x) \geq f(x)$ for all inputs x; and M' operates in space $f_1(x) = \log f'(x)$, since the simulation of M on input x requires exactly $f_1(x)$ tape cells. □

The following notion was implicit in [7, 8] and first defined explicitly by Schöning in [13].

Let $f^{(n)}$ denote the function obtained by n-fold application of f.

(2.3) Definition: *For a given recursive function f, such that $f(n) > n$ for all $n \in \mathbb{N}$, define the language $G[f]$ to be*

$$G[f] = \{x \in \Sigma^* \mid f^{(n)}(0) \leq |x| < f^{(n+1)}(0) \text{ for some } n \equiv 0 \pmod 2\}$$

If we require the parameter function f in our language $G[f]$ to be constructible in the above sense, we get the following result, which we will often use in the future.

(2.4) Lemma: *If $f: \mathbb{N} \to \mathbb{N}$ is strongly space-constructible and $f(n) > n$ for all $n \in \mathbb{N}$, then $G[f] \in$ **L**.*

Proof. Let M be the machine showing the constructibility of f. Consider the following Turing machine M':

On input x, $|x| = n$, first M' marks exactly $\log n$ tape cells. (This can be done reading x from left to right, counting the length of x, and upon reading the last symbol of x marking the space used by the counter.) Now consecutively compute 0, $f(0)$, $f(f(0))$, ... by simulating M, until some k is found which satisfies $f^{(k)}(0) \leq n < f^{(k+1)}(0)$. Accept if and only if $k \equiv 0 \pmod 2$. To compute 0, $f(0)$, $f(f(0))$, ..., $f^{(k)}(0)$, which are all less than n, we surely need no more space than $\log n$. In computing $f^{(k+1)}(0)$, which can be considerably greater than n, we just have to wait, until machine M tries to leave the marked region. Then we know for sure, that $f^{(k+1)}(0)$ will be greater than $2^{\log n}$.

Since in all these computations the same space can be used, M' doesn't exceed the resource bounds of the class **L**. □

2.2 Presentability

(2.5) Definition: *A language class C is* recursively presentable, *if and only if there is an effective enumeration M_1, M_2, M_3, ... of deterministic and total Turing machines, such that $C = \{L(M_i) \mid i \geq 1\}$.*

Nearly all "reasonable" complexity classes are recursively presentable [13]. We are especially interested in the following cases.

(2.6) Lemma: *The following classes are recursively presentable:*

1. **L**, **P**
2. **NC**
3. *the set of* **P**-*complete problems under logspace-reductions.*

Proof. Similar to the results in [13]. □

3 The degree structure of logspace reducibility

We only state now that the results from the preceding section allow us to prove all of the well-known diagonalization theorems mentioned in the introduction for logspace (instead of polynomial time); and these theorems lead us to a number of very interesting structural properties of logspace reducibility. All these results are obtained in a completely analogous way to the proofs of [1].

For concreteness we only state (without proofs) the most simple and direct results, some of which are already present in [15].

(3.1) Corollary: *Suppose* $\mathrm{P} \neq \mathrm{L}$. *Then there exists a language* $L \in \mathrm{P}$ *which is neither logspace-complete for* P *nor in* L.

An analogous result holds for the class NC:

(3.2) Corollary: *If* $\mathrm{P} \neq \mathrm{NC}$, *then there exists a language* $L \in \mathrm{P}$, *which is neither logspace-complete for* P *nor in* NC.

So we get the existence of problems which are not efficiently parallelizable but also not P-complete.

Let $\aleph_0$ denote the cardinality of the set of natural numbers.

(3.3) Corollary: *There is a rich structure of* $\aleph_0$ *different* $\leq_m^{\mathrm{L}}$*-degrees between* P *and* L *(assuming* $\mathrm{P} \neq \mathrm{L}$*).*

(3.4) Definition: *Two sets* A_1, A_2 *form a* minimal pair for L *if and only if* A_1 *and* A_2 *are not in* L *and every set logspace-reducible to both* A_1 *and* A_2 *is in* L.

(3.5) Corollary: *There is a minimal pair for* L, *which can be decided in* P *(assuming* $\mathrm{P} \neq \mathrm{L}$*).*

(3.6) Corollary: *There is a minimal pair for* NC *under logspace reductions, which can be decided in* P *(assuming* $\mathrm{P} \neq \mathrm{NC}$*).*

For more subtle and far-reaching questions, concerning among others a generalization of the last corollaries ("meet lemma") and a general result about lattice embeddings into the partial ordering of the logspace degrees, the interested reader is refered to [1] and asked to translate the results given there from the polynomial time context into the logspace context.

4 Complexity Cores

In this section we apply our generalized gap language technique to another problem:

(4.1) Definition: *Let A be recursive and X be of infinite cardinality. We call X a* logspace complexity-core *for A, if and only if for every Turing machine M deciding A and every $c \in \mathbb{N}$,*

$$|\{x \in X \mid \text{space}_M(x) \leq c \log |x|\}| < \infty.$$

Here $\text{space}_M(x)$ denotes the number of tape cells, which M uses on input x.

(4.2) Definition: $H(M, s) = \{x \mid \text{space}_M(x) > s(|x|)\}$.

I.e. X is a logspace complexity-core for A, if and only if for all Turing machines M deciding A and $c \in \mathbb{N}$,

$$\{x \in X \mid x \notin H(M, \lambda n.c \log n)\} = X - H(M, \lambda n.c \log n)$$

is finite. (We use Church's lambda-notation to define functions from terms. If $t(x)$ is a term with free variable x, then $\lambda x.t(x)$ denotes the associated partial function.)

In the proof of our result, we need a very special kind of enumerating Turing machines for some language.

(4.3) Lemma: *Let A be recursive.*

There is an enumeration $\langle M_i \rangle$ of deterministic and total Turing machines, such that:

1. *For all $i \geq 1$, $L(M_i) = A$ or $|L(M_i)| < \infty$.*
2. *For all $i \geq 1$, if $|L(M_i)| < \infty$, then $|\overline{H(M_i, \lambda n.c \log n)}| < \infty$.*
3. *For all Turing machines M, such that $L(M) = A$, and all $c \in \mathbb{N}$, there are $i, c' \in \mathbb{N}$, such that*

$$H(M_i, \lambda n.c' \log n) \subseteq H(M, \lambda n.c \log n).$$

Proof. Let $A = L(M_0)$ for some Turing machine M_0; and let $s(n) \geq n$ be a space-constructible space-bound of M_0. Let $\langle N_i \rangle$ be an enumeration of all Turing machines operating in space s.

Now define:

$M_i(x)$, $|x| = n$:
  if $\forall y, s(|y|) \leq 2^n : y \in L(M_0) \Leftrightarrow y \in L(N_i)$
  then simulate $N_i(x)$
  else mark n tape cells and reject.

Propositions (1) and (2) are immediate.

Now take any M, $L(M) = A$, and any $c \in \mathbb{N}$ and define:

$M'(x)$:
simulate in parallel $M_0(x)$ and $M(x)$;
if any of the two stops, **then**
stop and reject or accept according to this machine.

Since M_0 operates in space s, M' also operates in space s. So $M' = N_i$ for some i. Now, machine M_i uses (for some c') no more than $c' \log n$ tape cells on those inputs of length n, which are accepted by N_i. All in all we have:

$$H(M_i, \lambda n.c' \log n) \subseteq H(N_i, \lambda n.c \log n) = H(M', \lambda n.c \log n) \subseteq H(M, \lambda n.c \log n);$$

so our enumeration fulfills property (3). □

Our result now is:

(4.4) Theorem: *Let A be a recursive set not in* **L**. *Then A has a logspace complexity core $X \in$* DSPACE$(s(n))$ *for any superlogarithmic space-constructible function s.*

Proof. Let $\langle M_i \rangle$ be the enumeration of the previous lemma.

Define

$$f(j) = (\mu k > j)(\exists x, |x| = k : j \log k \leq s(k) \wedge \forall i \leq j : x \in H(M_i, \lambda n.j \log n).$$

f is total, since for all j,

$$\bigcap_{i=1}^{j} H(M_i, \lambda n.j \log n)$$

is infinite (otherwise combining the M_is, $i \leq j$, would show $A \in$ **L**, in contradiction to our assumption).

Let $g \geq f$ be strongly space-constructible.

Define a Turing machine M to operate as follows:

input x, $|x| = n$:
let $k = g^{(m)}(0)$, where $g^{(m)}(0) < |x| \leq g^{(m+1)}(0)$;
accept iff $k \log n \leq s(n) \wedge \forall i \leq k : x \in H(M_i, \lambda n.k \log n)$.

Now let $X = L(M)$.

The computation of k needs only logspace, so surely $X \in$ DSPACE$(s(n))$.

X is infinite, since for all m, there is one x of length between $g^{(m)}(0)$ and $g^{(m+1)}(0)$, which is accepted by M. This follows from the definition of f.

Let M be any machine deciding A, and let $c \in \mathbb{N}$. Since for every i and every $c' \in \mathbb{N}$, all but finitely many $x \in X$ are in $H(M_i, \lambda n.c' \log n)$, $X - H(M, \lambda n.c \log n)$ is finite by the lemma from the last section; and hence X is a complexity core for A. □

5 Inside NC

The reducibility notion used in the preceding sections was that of *logspace reducibility.* In this section we show how our results apply even for some very strong notion of reducibility, i.e. that of *constant-depth reduciblity.* Thus our techniques apply within NC and even within NC^1.

We first define a very strong constructibility notion.

(5.1) Definition: *Call a function* $f: \mathbb{N} \to \mathbb{N}$ **strongly time-constructible,** *if and only if* f *can be computed in time* $\log \circ f$ *(i.e. in time logarithmic in the value of the output).*

In other words: f is strongly time-constructible if the time needed to compute f is dominated by the time needed to write down the output.

Still, these functions can majorize every other recursive function.

(5.2) Lemma: *For every recursive function* $f: \mathbb{N} \to \mathbb{N}$, *there is a strongly time-constructible function* $f' \geq f$.

Proof. Analogously to the corresponding lemma in section 2. □

(5.3) Lemma: *Suppose* f *is strongly time-constructible and* $f(n) \geq n^{1+\varepsilon}$ *for all* $n \in \mathbb{N}$ *and for some fixed* $\varepsilon > 0$. *Then* $G[f] \in \mathrm{LT}$, *i.e.* $G[f]$ *can be computed in logarithmic time by a deterministic Turing machine with a random access input tape.*

Proof. On input x, first compute $n = |x|$. An indexing Turing machine can do so in time $\log |x|$ by one-sided binary search [17]. Then again compute $0, f(0), f(f(0)), \ldots$, until some k is found s.t. $f^{(k)}(0) \leq |x| < f^{(k+1)}(0)$. What is the time needed for this procedure? The time to calculate $f^{(k)}(n)$ is surely $\leq \log n$. Since $f(n) \geq n^{1+\varepsilon}$, we have $f^{(k-1)}(0) \leq n^{1/(1+\varepsilon)}$ and therefore the time needed to compute this value is at most $1/(1+\varepsilon) \log n$. Repeating this argument we see that the whole time needed is at most

$$\log n + \sum_{i=0}^{k} \frac{1}{(1+\varepsilon)^i} \log n = O(\log n).$$

□

In this section we want to obtain results for parallel complexity classes, for instance for TC^0 and NC^1; so we are interested in *cd*-reducibility. But it will be convenient not to use these reductions directly but to introduce yet another, even stronger, reducibility:

(5.4) Definition [17]: $A \leq_m^{\mathrm{LT}} B$ *if and only if there is a function* $\pi: \Sigma^* \to \Sigma^*$ *with the following properties:*

1. $x \in A \iff \pi(x) \in B$.

2. $|\pi(x)| \leq p(|x|)$ *for some polynomial* p.

3. The function $\varphi(x, i) =$ *the i-th bit of* $\pi(x)$ *is computable in logarithmic time by a deterministic indexing Turing machine.*

(5.5) Proposition: $A \leq_m^{LT} B \quad \Longrightarrow \quad A \leq_{cd} B$

Proof. A logtime Turing machine can be simulated by a poly-size constant depth circuit by brute force. Combine the machines for different bits and feed them as input to an oracle gate. □

So the results mentioned in section 3 even translate to very low complexity classes (since the gap constructs now can be recognized in logarithmic time). A rich structure of different degrees exists between classes like $\mathbf{AC^0}$, $\mathbf{TC^0}$, $\mathbf{NC^1}$, **L**, **NL**, $\mathbf{NC^2}$, $\mathbf{NC^3}$ etc.

6 Conclusion

We presented a refinement of Schöning's gap language technique [13] and applied it in the contexts of degree structure of subpolynomial reducibilities and complexity cores. Thus we extended results of [7, 13, 14, 1, 15, 10, 3]. It should be mentioned that parts of our results of the first sections were obtained independently and almost simultaneously by M. Serna in [16]. In this paper, and also in the paper of D. Schmidt [15], on which [16] is based, a complicated condition called "recursive gap closed" is required to be shown before applying Schöning's technique. We drop this condition and require instead "strong constructibility" of the functions used. We think that this condition is more intuitive and easier to prove. We thus obtain stronger results.

Another advantage of our approach is that besides examining structural properties of subpolynomial reducibilities it also permits some other application of the gap language technique, concerning complexity cores, as presented in section 4. We obtain all these results in a uniform manner, relying in all cases on strong constructibility.

Our paper shows that some very familiar techniques and problems in the **P** vs. **NP** context, e.g. constructibility, are not dependent on the question of the power of nondeterminism versus determinism, but also appear in very different other settings; so they seem to be instances of a more general principle. This we believe opens the door to serious complexity theory research about the so called parallel complexity classes, which evolved in the recent past and are of immense importance in the theory of algorithms.

References

[1] K. AMBOS-SPIES, Polynomial time degrees of NP-sets; in: E. BÖRGER, *Trends in Theoretical Computer Science*, Computer Science Press, Rockville 1988, 95–142.

[2] J. L. BALCÁZAR, J. DÍAZ, J. GABARRÓ, *Structural Complexity I;* Springer, Berlin, 1988.

[3] R. BOOK, D.-Z. DU, D. RUSSO, On polynomial and generalized complexity cores; *Proc. 3rd Structure* (1988), 236–250.

[4] A. K. CHANDRA, L. STOCKMEYER, U. VISHKIN, Constant depth reducibility; *SIAM J. Comput.* (13) (1984), 423–439.

[5] P. CHEW, M. MACHTEY, A note on structure and looking back applied to the relative complexity of computable functions; *J. Comput. System Sci.* 22 (1981), 53–59.

[6] S. A. Cook, A taxonomy of problems with fast parallel Algorithms; *Inf. & Contr.* 64 (1985), 2–22.

[7] R. E. LADNER, On the structure of polynomial time reducibility, *J. ACM* 16 (1975), 155–171.

[8] L. H. LANDWEBER, R. J. LIPTON, E. L. ROBERTSON, On the structure of sets in NP and other complexity classes; *Theoret. Comput. Sci.* 1 (1975), 103–123.

[9] N. LYNCH, On reducibility to complex or sparse sets; *J.ACM* 22 (1975), 341–345.

[10] P. ORPONEN, U. SCHÖNING, The density and complexity of polynomial cores for intractable sets; *Inf. & Contr.* 70 (1986), 54–68.

[11] H. ROGERS JR., *Theory of Recursive Functions and Effective Computability;* McGraw-Hill, New York, 1967.

[12] R. I. SOARE, *Recursively Enumerable Sets and Degrees;* Springer, Berlin, 1986.

[13] U. SCHÖNING, A uniform approach to obtain diagonal sets in complexity classes; *Theoret. Comput. Sci.* 18 (1982), 95–103.

[14] U. SCHÖNING, Minimal pairs for P, *Theoret. Comput. Sci.* 31 (1984), 41–48.

[15] D. SCHMIDT, The recursion-theoretic structure of complexity classes; *Theoret. Comput. Sci.* 38 (1985), 143–156.

[16] M. SERNA, *The parallel approximability of* P-*complete problems;* Tesis doctoral, Facultat d'Informàtica de Barcelona (1990).

[17] J. TORÁN, *Structural properties of the counting hierarchies;* Tesis doctoral, Facultat d'Informàtica de Barcelona (1988).

Lecture Notes in Computer Science

For information about Vols. 1–454
please contact your bookseller or Springer-Verlag

Vol. 455: C.A. Floudas, P.M. Pardalos, A Collection of Test Problems for Constrained Global Optimization Algorithms. XIV, 180 pages. 1990.

Vol. 456: P. Deransart, J. Maluszyński (Eds.), Programming Language Implementation and Logic Programming. Proceedings, 1990. VIII, 401 pages. 1990.

Vol. 457: H. Burkhart (Ed.), CONPAR '90 – VAPP IV. Proceedings, 1990. XIV, 900 pages. 1990.

Vol. 458: J.C.M. Baeten, J.W. Klop (Eds.), CONCUR '90. Proceedings, 1990. VII, 537 pages. 1990.

Vol. 459: R. Studer (Ed.), Natural Language and Logic. Proceedings, 1989. VII, 252 pages. 1990. (Subseries LNAI).

Vol. 460: J. Uhl, H.A. Schmid, A Systematic Catalogue of Reusable Abstract Data Types. XII, 344 pages. 1990.

Vol. 461: P. Deransart, M. Jourdan (Eds.), Attribute Grammars and their Applications. Proceedings, 1990. VIII, 358 pages. 1990.

Vol. 462: G. Gottlob, W. Nejdl (Eds.), Expert Systems in Engineering. Proceedings, 1990. IX, 260 pages. 1990. (Subseries LNAI).

Vol. 463: H. Kirchner, W. Wechler (Eds.), Algebraic and Logic Programming. Proceedings, 1990. VII, 386 pages. 1990.

Vol. 464: J. Dassow, J. Kelemen (Eds.), Aspects and Prospects of Theoretical Computer Science. Proceedings, 1990. VI, 298 pages. 1990.

Vol. 465: A. Fuhrmann, M. Morreau (Eds.), The Logic of Theory Change. Proceedings, 1989. X, 334 pages. 1991. (Subseries LNAI).

Vol. 466: A. Blaser (Ed.), Database Systems of the 90s. Proceedings, 1990. VIII, 334 pages. 1990.

Vol. 467: F. Long (Ed.), Software Engineering Environments. Proceedings, 1969. VI, 313 pages. 1990.

Vol. 468: S.G. Akl, F. Fiala, W.W. Koczkodaj (Eds.), Advances in Computing and Information – ICCI '90. Proceedings, 1990. VII, 529 pages. 1990.

Vol. 469: I. Guessarian (Ed.), Semantics of Systeme of Concurrent Processes. Proceedings, 1990. V, 456 pages. 1990.

Vol. 470: S. Abiteboul, P.C. Kanellakis (Eds.), ICDT '90. Proceedings, 1990. VII, 528 pages. 1990.

Vol. 471: B.C. Ooi, Efficient Query Processing in Geographic Information Systems. VIII, 208 pages. 1990.

Vol. 472: K.V. Nori, C.E. Veni Madhavan (Eds.), Foundations of Software Technology and Theoretical Computer Science. Proceedings, 1990. X, 420 pages. 1990.

Vol. 473: I.B. Damgård (Ed.), Advances in Cryptology – EUROCRYPT '90. Proceedings, 1990. VIII, 500 pages. 1991.

Vol. 474: D. Karagiannis (Ed.), Information Syetems and Artificial Intelligence: Integration Aspects. Proceedings, 1990. X, 293 pages. 1991. (Subseries LNAI).

Vol. 475: P. Schroeder-Heister (Ed.), Extensions of Logic Programming. Proceedings, 1989. VIII, 364 pages. 1991. (Subseries LNAI).

Vol. 476: M. Filgueiras, L. Damas, N. Moreira, A.P. Tomás (Eds.), Natural Language Processing. Proceedings, 1990. VII, 253 pages. 1991. (Subseries LNAI).

Vol. 477: D. Hammer (Ed.), Compiler Compilers. Proceedings, 1990. VI, 227 pages. 1991.

Vol. 478: J. van Eijck (Ed.), Logics in Al. Proceedings, 1990. IX, 562 pages. 1991. (Subseries in LNAI).

Vol. 480: C. Choffrut, M. Jantzen (Eds.), STACS 91. Proceedings, 1991. X, 549 pages. 1991.

Vol. 481: E. Lang, K.-U. Carstensen, G. Simmons, Modelling Spatial Knowledge on a Linguistic Basis. IX, 138 pages. 1991. (Subseries LNAI).

Vol. 482: Y. Kodratoff (Ed.), Machine Learning – EWSL-91. Proceedings, 1991. XI, 537 pages. 1991. (Subseries LNAI).

Vol. 483: G. Rozenberg (Ed.), Advances In Petri Nets 1990. VI, 515 pages. 1991.

Vol. 484: R. H. Möhring (Ed.), Graph-Theoretic Concepts In Computer Science. Proceedings, 1990. IX, 360 pages. 1991.

Vol. 485: K. Furukawa, H. Tanaka, T. Fullsaki (Eds.), Logic Programming '89. Proceedings, 1989. IX, 183 pages. 1991. (Subseries LNAI).

Vol. 486: J. van Leeuwen, N. Santoro (Eds.), Distributed Algorithms. Proceedings, 1990. VI, 433 pages. 1991.

Vol. 487: A. Bode (Ed.), Distributed Memory Computing. Proceedings, 1991. XI, 506 pages. 1991.

Vol. 488: R. V. Book (Ed.), Rewriting Techniques and Applications. Proceedings, 1991. VII, 458 pages. 1991.

Vol. 489: J. W. de Bakker, W. P. de Roever, G. Rozenberg (Eds.), Foundations of Object-Oriented Languages. Proceedings, 1990. VIII, 442 pages. 1991.

Vol. 490: J. A. Bergstra, L. M. G. Feljs (Eds.), Algebraic Methods 11: Theory, Tools and Applicatlons. VI, 434 pages. 1991.

Vol. 491: A. Yonezawa, T. Ito (Eds.), Concurrency: Theory, Language, and Architecture. Proceedings, 1989. VIII, 339 pages. 1991.

Vol. 492: D. Sriram, R. Logcher, S. Fukuda (Eds.), Computer-Aided Cooperative Product Development. Proceedings, 1989 VII, 630 pages. 1991.

Vol. 493: S. Abramsky, T. S. E. Maibaum (Eds.), TAPSOFT '91. Volume 1. Proceedings, 1991. VIII, 455 pages. 1991.

Vol. 494: S. Abramsky, T. S. E. Maibaum (Eds.), TAPSOFT '91. Volume 2. Proceedings, 1991. VIII, 482 pages. 1991.

Vol. 495: 9. Thalheim, J. Demetrovics, H.-D. Gerhardt (Eds.), MFDBS '91. Proceedings, 1991. VI, 395 pages. 1991.

Vol. 496: H.-P. Schwefel, R. Männer (Eds.), Parallel Problem Solving from Nature. Proceedings, 1991. XI, 485 pages. 1991.

Vol. 497: F. Dehne, F. Fiala. W.W. Koczkodaj (Eds.), Advances in Computing and Intormation - ICCI '91 Proceedings, 1991. VIII, 745 pages. 1991.

Vol. 498: R. Andersen, J. A. Bubenko jr., A. Sølvberg (Eds.), Advanced Information Systems Engineering. Proceedings, 1991. VI, 579 pages. 1991.

Vol. 499: D. Christodoulakis (Ed.), Ada: The Choice for '92. Proceedings, 1991. VI, 411 pages. 1991.

Vol. 500: M. Held, On the Computational Geometry of Pocket Machining. XII, 179 pages. 1991.

Vol. 501: M. Bidoit, H.-J. Kreowski, P. Lescanne, F. Orejas, D. Sannella (Eds.), Algebraic System Specification and Development. VIII, 98 pages. 1991.

Vol. 502: J. Bārzdiņš, D. Bjørner (Eds.), Baltic Computer Science. X, 619 pages. 1991.

Vol. 503: P. America (Ed.), Parallel Database Systems. Proceedings, 1990. VIII, 433 pages. 1991.

Vol. 504: J. W. Schmidt, A. A. Stogny (Eds.), Next Generation Information System Technology. Proceedings, 1990. IX, 450 pages. 1991.

Vol. 505: E. H. L. Aarts, J. van Leeuwen, M. Rem (Eds.), PARLE '91. Parallel Architectures and Languages Europe, Volume I. Proceedings, 1991. XV, 423 pages. 1991.

Vol. 506: E. H. L. Aarts, J. van Leeuwen, M. Rem (Eds.), PARLE '91. Parallel Architectures and Languages Europe, Volume II. Proceedings, 1991. XV, 489 pages. 1991.

Vol. 507: N. A. Sherwani, E. de Doncker, J. A. Kapenga (Eds.), Computing in the 90's. Proceedings, 1989. XIII, 441 pages. 1991.

Vol. 508: S. Sakata (Ed.), Applied Algebra, Algebraic Algorithms and Error-Correcting Codes. Proceedings, 1990. IX, 390 pages. 1991.

Vol. 509: A. Endres, H. Weber (Eds.), Software Development Environments and CASE Technology. Proceedings, 1991. VIII, 286 pages. 1991.

Vol. 510: J. Leach Albert, B. Monien, M. Rodríguez (Eds.), Automata, Languages and Programming. Proceedings, 1991. XII, 763 pages. 1991.

Vol. 511: A. C. F. Colchester, D.J. Hawkes (Eds.), Information Processing in Medical Imaging. Proceedings, 1991. XI, 512 pages. 1991.

Vol. 512: P. America (Ed.), ECOOP '91. European Conference on Object-Oriented Programming. Proceedings, 1991. X, 396 pages. 1991.

Vol. 513: N. M. Mattos, An Approach to Knowledge Base Management. IX, 247 pages. 1991. (Subseries LNAI).

Vol. 514: G. Cohen, P. Charpin (Eds.), EUROCODE '90. Proceedings, 1990. XI, 392 pages. 1991.

Vol. 515: J. P. Martins, M. Reinfrank (Eds.), Truth Maintenance Systems. Proceedings, 1990. VII, 177 pages. 1991. (Subseries LNAI).

Vol. 516: S. Kaplan, M. Okada (Eds.), Conditional and Typed Rewriting Systems. Proceedings, 1990. IX, 461 pages. 1991.

Vol. 517: K. Nökel, Temporally Distributed Symptoms in Technical Diagnosis. IX, 164 pages. 1991. (Subseries LNAI).

Vol. 518: J. G. Williams, Instantiation Theory. VIII, 133 pages. 1991. (Subseries LNAI).

Vol. 519: F. Dehne, J.-R. Sack, N. Santoro (Eds.), Algorithms and Data Structures. Proceedings, 1991. X, 496 pages. 1991.

Vol. 520: A. Tarlecki (Ed.), Mathematical Foundations of Computer Science 1991. Proceedings, 1991. XI, 435 pages. 1991.

Vol. 521: B. Bouchon-Meunier, R. R. Yager, L. A. Zadek (Eds.), Uncertainty in Knowledge-Bases. Proceedings, 1990. X, 609 pages. 1991.

Vol. 522: J. Hertzberg (Ed.), European Workshop on Planning. Proceedings, 1991. VII, 121 pages. 1991. (Subseries LNAI).

Vol. 523: J. Hughes (Ed.), Functional Programming Languages and Computer Architecture. Proceedings, 1991. VIII, 666 pages. 1991.

Vol. 524: G. Rozenberg (Ed.), Advances in Petri Nets 1991. VIII, 572 pages. 1991.

Vol. 525: O. Günther, H.-J. Schek (Eds.), Large Spatial Databases. Proceedings, 1991. XI, 471 pages. 1991.

Vol. 526: T. Ito, A. R. Meyer (Eds.), Theoretical Aspects of Computer Software. Proceedings, 1991. X, 772 pages. 1991.

Vol. 527: J.C.M. Baeten, J. F. Groote (Eds.), CONCUR '91. Proceedings, 1991. VIII, 541 pages. 1991.

Vol. 528: J. Maluszynski, M. Wirsing (Eds.) Programming Language Implementation and Logic Programming. Proceedings, 1991. XI, 433 pages. 1991.

Vol. 529: L. Budach (Ed.) Fundamentals of Computation Theory. Proceedings, 1991. XII, 426 pages. 1991.

Vol. 530: D. H. Pitt, P.-L. Curien, S. Abramsky, A. M. Pitts, A. Poigné, D. E. Rydeheard (Eds.), Category Theory and Computer Science. Proceedings, 1991. VII, 301 pages. 1991.

Vol. 531: E. M. Clarke, R. P. Kurshan (Eds.), Computer-Aided Verification. Proceedings, 1990. XIII, 372 pages. 1991.

Vol. 532: H. Ehrig, H.-J. Kreowski, G. Rozenberg (Eds.), Graph Grammars and Their Application to Computer Science. Proceedings, 1990. X, 703 pages. 1991.

Vol. 533: E. Börger, H. Kleine Büning, M. M. Richter, W. Schönfeld (Eds.), Computer Science Logic. Proceedings, 1990. VIII, 399 pages. 1991.